Membrane Biophysics

Hongda Wang · Guohui Li
Editors

Membrane Biophysics

New Insights and Methods

Springer

Editors
Hongda Wang
Changchun Institute of Applied Chemistry
Chinese Academy of Sciences
Changchun
China

Guohui Li
Dalian Institute of Chemical Physics
Chinese Academy of Sciences
Dalian, Liaoning
China

ISBN 978-981-10-6822-5 ISBN 978-981-10-6823-2 (eBook)
https://doi.org/10.1007/978-981-10-6823-2

Library of Congress Control Number: 2017955268

Printed on acid-free paper

This Springer imprint is published by Springer Nature
The registered company is Springer Nature Singapore Pte Ltd.
The registered company address is: 152 Beach Road, #21-01/04 Gateway East, Singapore 189721, Singapore

Preface

Cell membranes are frontiers between cells' surroundings and their interiors. They not only serve a structural and containment role for the cell but they also regulate many essential cellular activities, such as material transport, signal transduction, cell–cell interaction, and enzymatic reactions. From the first lipid bilayer membrane structure proposed in 1925, to the widely accepted fluid mosaic model put forward by Singer and Nicolson in 1972, our understanding of the structures and functions of cell membranes is continuing to improve as well. The membrane is a very complicated supramolecular liquid crystalline structure that is composed of phospholipids, proteins, and sugars. Each component has a variety of types and functions, for example, membrane proteins. About a third of all cellular proteins are membrane proteins which are involved in transport, catalysis, and recognition. Thus, when different components organize together in the relative ratio and composition, the structural complicacy and functional distinctiveness it brings to the cell membrane is conceivable. Moreover, the highly dynamic property of the membrane also makes it more difficult to study. Therefore, the exploration of the structure and function of cell membranes has never ceased over the years.

Consistent with other scientific studies, every major breakthrough in the field of cell membranes is due to the application of new techniques and approaches in biophysics, cell biology, biochemistry, and molecular biology. Particularly in recent years, many traditional techniques have been continuously improved and perfected, such as light microscopy, electron microscopy (EM), x-ray crystallography, nuclear magnetic resonance (NMR), mass spectroscopy (MS), infrared (IR), and Raman spectrum. These techniques have allowed researchers to gain spectacular insights into the biological and physical properties of cell membranes. Moreover, emerging technologies, including high-resolution atomic force microscopy (AFM) and super-resolution fluorescence microscopy (SRFM), are being used as appropriate tools by which the characteristic of the membrane can be better understood. Besides these experimental tools, computer simulation of detailed atomic models based on realistic microscopic interactions represents a powerful approach to explore the structure and dynamics of cell membranes.

The idea for this book was to introduce a broad collection of methods to study the biophysics and function of cell membranes including materials, protocols, and notes, and present the recent advances and new insights into the field of membrane science. The book is divided into thirteen parts. General membrane composition, functions, and other properties are summarized in Chap. 1. Chapter 2 reviews the major historical discoveries and theories that tackled the existence and structure of cell membranes, and discusses the strengths and weaknesses of the related techniques by which different membrane models have been proposed. From Chap. 3 to Chap. 12, different classic and cutting-edge techniques of biophysics and chemistry are presented in detail, including the basic principle of the technique, the overall situation and development trends of the technique, the typical and successful applications of the technique in studying membrane science, and how to utilize the technique to solve related problems. At last, Chap. 13 describes some other modern methods for studying biomembranes which are more targeted.

Our aim was to provide readers with a comprehensive but still concise collection of methods to study membrane biophysics including both basic protocols of general applications and more specialized methods for novel techniques that were not presented systematically before. We sincerely hope that readers will gain further understanding of membrane organization and the relationship between membrane structure and functions. Meanwhile, the detailed procedures of each approach will serve as a laboratory guide to both students of biophysics and expert researchers.

We would like to thank all the authors for their concise and clear contributions which make working with them a great pleasure. We also want to express our deepest gratitude to the editors, for their patience and encouragement throughout the editing process.

This book is dedicated to Harry and Memary for the lifelong investigation of cell membranes.

Changchun, Jilin, China Hongda Wang
June 2017

Contents

Contributors

Matthias Amrein Department of Cell Biology and Anatomy, Snyder Institute of Chronic Diseases, University of Calgary, Calgary, AB, Canada

Mingjun Cai State Key Laboratory of Electroanalytical Chemistry, Changchun Institute of Applied Chemistry, Chinese Academy of Sciences, Changchun, Jilin, China

Longxing Cao State Key Laboratory of Molecular Biology, National Center for Protein Science Shanghai, Shanghai Science Research Center, CAS Center for Excellence in Molecular Cell Science, Shanghai Institute of Biochemistry and Cell Biology, Chinese Academy of Sciences, Shanghai, China

Haishuang Chang State Key Laboratory of Molecular Biology, National Center for Protein Science Shanghai, Shanghai Science Research Center, CAS Center for Excellence in Molecular Cell Science, Shanghai Institute of Biochemistry and Cell Biology, Chinese Academy of Sciences, Shanghai, China

Junling Chen State Key Laboratory of Electroanalytical Chemistry, Changchun Institute of Applied Chemistry, Chinese Academy of Sciences, Changchun, Jilin, China

Huiying Chu Laboratory of Molecular Modeling and Design, State Key Laboratory of Molecular Reaction Dynamics, Dalian Institute of Chemical Physics, Chinese Academy of Science, Dalian, Liaoning, China

Jing Gao State Key Laboratory of Electroanalytical Chemistry, Changchun Institute of Applied Chemistry, Chinese Academy of Sciences, Changchun, Jilin, China

Hua He State Key Laboratory of Heavy Oil Processing and Center for Bioengineering and Biotechnology, China University of Petroleum (East China), Qingdao, China

Yongning He State Key Laboratory of Molecular Biology, National Center for Protein Science Shanghai, Shanghai Science Research Center, CAS Center for Excellence in Molecular Cell Science, Shanghai Institute of Biochemistry and Cell Biology, Chinese Academy of Sciences, Shanghai, China

Fang Huang State Key Laboratory of Heavy Oil Processing and Center for Bioengineering and Biotechnology, China University of Petroleum (East China), Qingdao, China

Xiue Jiang State Key Laboratory of Electroanalytical Chemistry, Changchun Institute of Applied Chemistry, Chinese Academy of Sciences, Changchun, China

Dianfan Li National Center for Protein Science Shanghai, Shanghai Science Research Center, CAS Center for Excellence in Molecular Cell Science, Shanghai Institute of Biochemistry and Cell Biology, Chinese Academy of Sciences, Shanghai, China; University of Chinese Academy of Sciences, Shanghai, China

Guohui Li Laboratory of Molecular Modeling and Design, State Key Laboratory of Molecular Reaction Dynamics, Dalian Institute of Chemical Physics, Chinese Academy of Science, Dalian, Liaoning, China

Yan Li Laboratory of Molecular Modeling and Design, State Key Laboratory of Molecular Reaction Dynamics, Dalian Institute of Chemical Physics, Chinese Academy of Science, Dalian, Liaoning, China

Chao-Zi Liu Key Laboratory of Analytical Chemistry for Living Biosystems, Institute of Chemistry Chinese Academy of Sciences, Beijing, China; University of Chinese Academy of Sciences, Beijing, China

Zong-Xiu Nie Key Laboratory of Analytical Chemistry for Living Biosystems, Institute of Chemistry Chinese Academy of Sciences, Beijing, China; University of Chinese Academy of Sciences, Beijing, China

Yuping Shan School of Chemistry and Life Science, Advanced Institute of Materials Science, Changchun University of Technology, Changchun, China

Yan Shi School of Medicine, Tsinghua-Peking Center for Life Sciences, Institute for Immunology and Department of Basic Medical Sciences, Tsinghua University, Beijing, China; Department of Microbiology, Immunology & Infectious Diseases and Snyder Institute, University of Calgary, Calgary, AB, Canada

Hongda Wang State Key Laboratory of Electroanalytical Chemistry, Changchun Institute of Applied Chemistry, Chinese Academy of Sciences, Changchun, Jilin, China

Shenlin Wang College of Chemistry and Molecular Engineering, Peking University, Beijing, China; Beijing NMR Centre, Peking University, Beijing, China

Xiaojuan Wang State Key Laboratory of Heavy Oil Processing and Center for Bioengineering and Biotechnology, China University of Petroleum (East China), Qingdao, China

Lie Wu State Key Laboratory of Electroanalytical Chemistry, Changchun Institute of Applied Chemistry, Chinese Academy of Sciences, Changchun, China

Tie Xia School of Medicine, Tsinghua-Peking Center for Life Sciences, Institute for Immunology and Department of Basic Medical Sciences, Tsinghua University, Beijing, China

Xiaojun Xu College of Chemistry and Molecular Engineering, Peking University, Beijing, China; Beijing NMR Centre, Peking University, Beijing, China

Yufei Yang College of Chemistry and Molecular Engineering, Peking University, Beijing, China; Beijing NMR Centre, Peking University, Beijing, China

Jialu Zha National Center for Protein Science Shanghai, Shanghai Science Research Center, CAS Center for Excellence in Molecular Cell Science, Shanghai Institute of Biochemistry and Cell Biology, Chinese Academy of Sciences, Shanghai, China; University of Chinese Academy of Sciences, Shanghai, China

Ling-Peng Zhan Key Laboratory of Analytical Chemistry for Living Biosystems, Institute of Chemistry Chinese Academy of Sciences, Beijing, China; University of Chinese Academy of Sciences, Beijing, China

Yuebin Zhang Laboratory of Molecular Modeling and Design, State Key Laboratory of Molecular Reaction Dynamics, Dalian Institute of Chemical Physics, Chinese Academy of Science, Dalian, Liaoning, China

Chapter 1
Composition and Function of Cell Membranes

Mingjun Cai, Jing Gao and Hongda Wang

1.1 Introduction

The cell membrane encloses the cell, defines its boundaries, and maintains the essential differences between the cytosol and the extracellular environment. Inside eukaryotic cells, the membranes of the nucleus, endoplasmic reticulum, Golgi apparatus, mitochondria, and other membrane-enclosed organelles maintain the characteristic differences between the contents of each organelle and the cytosol. All these membranes form enclosed structure and a related net called cytoplasmic membrane system. Ion gradients across membranes, established by the activities of specialized membrane proteins, can be used to synthesize adenosine triphosphate (ATP), to drive the transport of selected solutes across the membrane, or as in nerve and muscle cells, to produce and transmit electrical signals. In all cells, the cell membrane also contains proteins those act as sensors of external signals, allowing the cell to change its behavior in response to environmental cues, including signals from other cells; these protein sensors, or receptors, transfer information rather than molecules across the membrane. Cell membranes are dynamic, fluid structures, and most of their molecules move about in the plane of the membrane.

M. Cai · J. Gao · H. Wang (✉)
State Key Laboratory of Electroanalytical Chemistry, Changchun Institute of Applied Chemistry, Chinese Academy of Sciences, No. 5625 Renmin Street, Changchun 130022, Jilin, People's Republic of China
e-mail: hdwang@ciac.ac.cn

H. Wang and G. Li (eds.), *Membrane Biophysics*,
https://doi.org/10.1007/978-981-10-6823-2_1

1.2 Composition of Cell Membranes

Despite their different functions, all biological membranes have a general structure: a very thin film of lipid and protein molecules, linking together mainly by non-covalent interactions [1]. The lipid molecules are arranged as a continuous double layer (Fig. 1.1). This lipid bilayer provides the basic fluid structure of the membrane and serves as a relatively impermeable barrier to the passage of most water-soluble molecules. Most membrane proteins span the lipid bilayer and mediate nearly all of the other functions of the membrane, including the transport of specific molecules, and the catalysis of membrane-associated reactions such as ATP synthesis. In the cell membrane, some transmembrane proteins serve as structural links that connect the cytoskeleton through the lipid bilayer to either the extracellular matrix or an adjacent cell, while others serve as receptors to detect and transduce chemical signals in the cell's environment [2]. It takes many kinds of membrane proteins to enable a cell to function and interact with its environment, and it is estimated that about 30% of the proteins encoded in an animal's genome are membrane proteins [3–5].

1.2.1 Lipids

Lipid bilayer constitutes the fundamental structure of all the cell membranes. The lipids in cell membranes combine two very different properties in a single molecule: Each lipid has a hydrophilic ("water-loving") head and a hydrophobic ("water-fearing") tail. The most abundant lipids in cell membranes are the

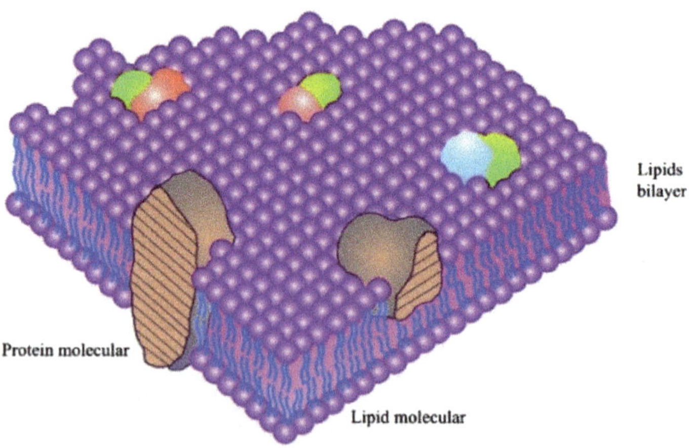

Fig. 1.1 Fluid-mosaic membrane model of cell membrane structure. Reprinted from Ref. [4], Copyright 2014, with permission from Elsevier

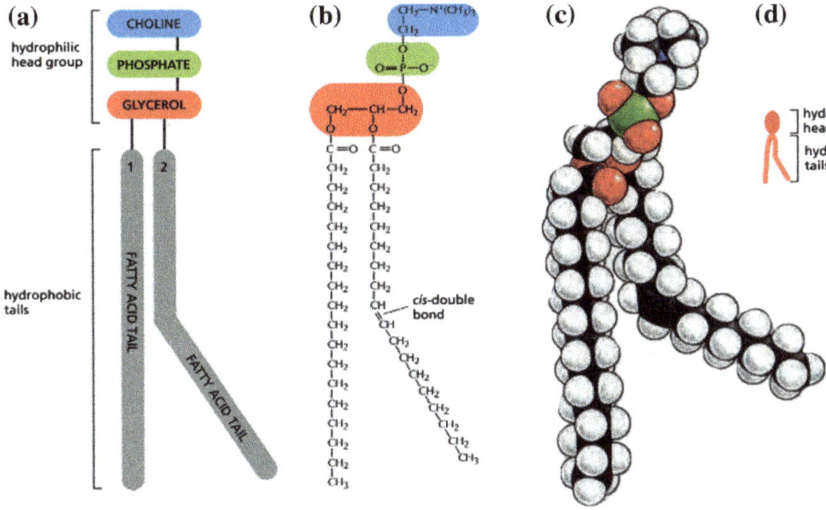

Fig. 1.2 Structure of phosphatidylcholine, a typical phospholipid molecule. **a** Scheme of molecular structure, **b** Formula, **c** Space-filling model, and **d** Symbol. Reprinted from Ref. [6] by permission of Garland Science Taylor & Francis Group

phospholipids, which have a phosphate-containing hydrophilic head linked to a pair of hydrophobic tails (Fig. 1.2). The bilayer structure is maintained by hydrophobic and Van der Waals force.

Lipid molecules constitute about 50% of the mass of most animal cell membranes, and nearly all of the remainder are proteins. There are approximately 5×10^6 lipid molecules in a 1×1 μm area of lipid bilayer or about 10^9 lipid molecules in the cell membrane of a small animal cell. Lipids contained in typical cell membrane are phospholipids (include phosphoglycerides, sphingolipids) and steroids.

1.2.1.1 Phosphoglycerides

Phosphoglycerides derivatives of glycerol 3-phosphate are the most abundant class of lipids in most membrane. A typical phosphoglyceride molecule consists of a polar head attached to the phosphate group, and a hydrophobic tail composed of two fatty acyl chain esterified to the two hydroxyl groups in glycerol phosphate. The fatty acyl chains may differ in the number of carbons (they normally contain between 14 and 24 carbon atoms) and their degree of saturation (0, 1, or 2 double bonds). The length and the saturation degree of the fatty acid tails are important in regulating the fluidity of the membrane (Fig. 1.2).

Phosphoglycerides are classified according to the nature of their head groups. In phosphatidylcholines, for example, the head group consists of choline; a positively charged alcohol is esterified to the negatively charged phosphate. In other

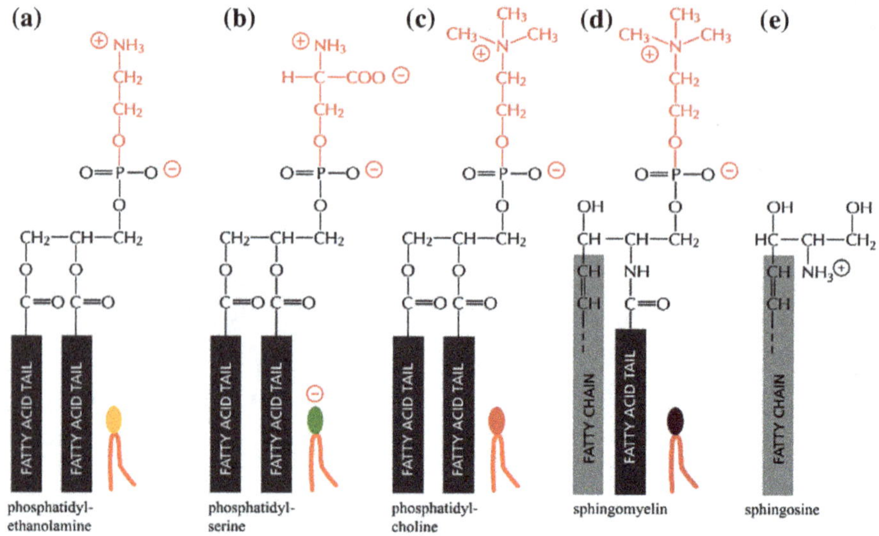

Fig. 1.3 Four major phospholipids in mammalian cell membranes. Different head groups are represented by different colors in the symbols. The lipid molecules shown in (**a–c**) are phosphoglycerides, which are derived from glycerol. The molecule in (**d**) is sphingomyelin, which is derived from sphingosine (**e**) and is therefore a sphingolipid. Note that only phosphatidylserine carries a net negative charge, and the other three are electrically neutral at physiological pH, carrying one positive and one negative charge. Reprinted from Ref. [6] by permission of Garland Science Taylor & Francis Group

phosphoglycerides, an OH-containing molecule such as ethanolamine, serine, or the sugar derivative inositol is linked to the phosphate group. The negatively charged phosphate group and the positively charged or the hydroxyl groups on the head group interact with water (Fig. 1.3).

1.2.1.2 Sphingolipids

The second class of membrane lipids is sphingolipid. Sphingolipid is derived from sphingosine rather than glycerol (Fig. 1.3d, e). Sphingosine is a long acyl chain with an amino group (NH_2) and two hydroxyl groups (OH) at one end. In sphingomyelin, the most common sphingolipid, a fatty acid tail is attached to the amino group, and a phosphocholine group is attached to the terminal hydroxyl group. Thus, the overall structure of sphingomyelin is quite similar to that of phosphatidylcholine, and therefore sphingomyelin is a kind of phospholipid. Other sphingolipids are amphipathic glycolipids whose polar head groups are sugars (Fig. 1.4). Together, the phospholipids phosphatidylcholine, phosphatidylethanolamine, phosphatidylserine, and sphingomyelin constitute more than half the mass of lipid in most mammalian cell membranes (Table 1.1).

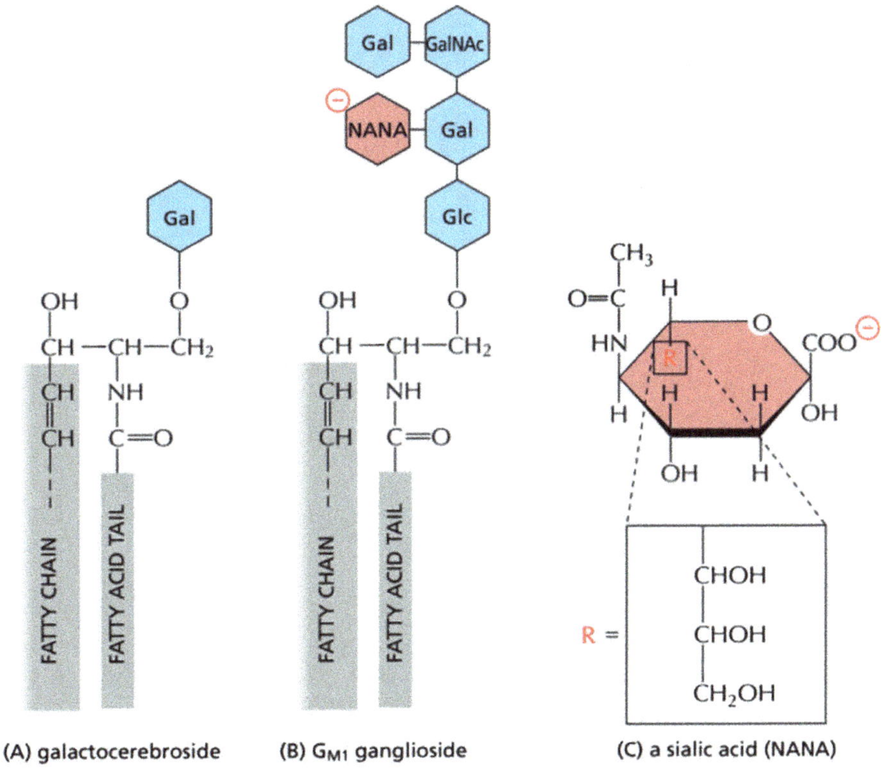

Fig. 1.4 Structure of glycolipid molecules in cell membrane. **a** Galactocerebroside is called a neutral glycolipid because the sugar that forms its head group is uncharged. **b** A ganglioside always contains one or more negatively charged sialic acid moiety. There are various types of sialic acid; in human cells, it is mostly N-acetylneuraminic acid, or NANA), whose structure is shown in (**c**). Whereas in bacteria and plants almost all glycolipids are derived from glycerol, as are most phospholipids, in animal cells almost all glycolipids are based on sphingosine, as is the case for sphingomyelin (see Fig. 1.3). Gal = galactose; Glc = glucose, GalNAc = N-acetylgalactosamine; these three sugars are uncharged. Reprinted from Ref. [6] by permission of Garland Science Taylor & Francis Group

Glucosylcerebroside,the simple glycosphingolipid, contains a single glucose unit attached to sphingosine. In the complex glycosphingolipids called ganglioside, one or two branched sugar chains containing sialic acid groups are attached to sphingosine. Glycolipids and glyceroglycolipid constitute 2–10% of the total lipid in cell membrane. The outer layer of the glycolipids plays an important role in protecting cell.

1.2.1.3 Cholesterol

Cholesterol and its analogues constitute the third important class of membrane lipids. The structure of cholesterol is a four-ring hydrocarbon and a hydroxyl

Table 1.1 Compares the lipid compositions of several biological membranes

Lipid	Percentage of total lipid by weight					
	Liver cell membrane	Red blood cell membrane	Myelin	Mitochondrion (inner and outer membranes)	Endoplasmic reticulum	*E. colibacterium*
Cholesterol	17	23	22	3	6	0
Phosphatidylethanolamine	7	18	15	28	17	70
Phosphatidylserine	4	7	9	2	5	Trace
Phosphatidylcholine	24	17	10	44	40	0
Sphingomyelin	19	18	8	0	5	0
Glycolipids	7	3	28	Trace	Trace	0
Others	22	14	8	23	27	30

Note that bacterial cell membranes are often composed of one main type of phospholipid and contain no cholesterol. Reprinted from Ref. [6] by permission of Garland Science Taylor & Francis Group

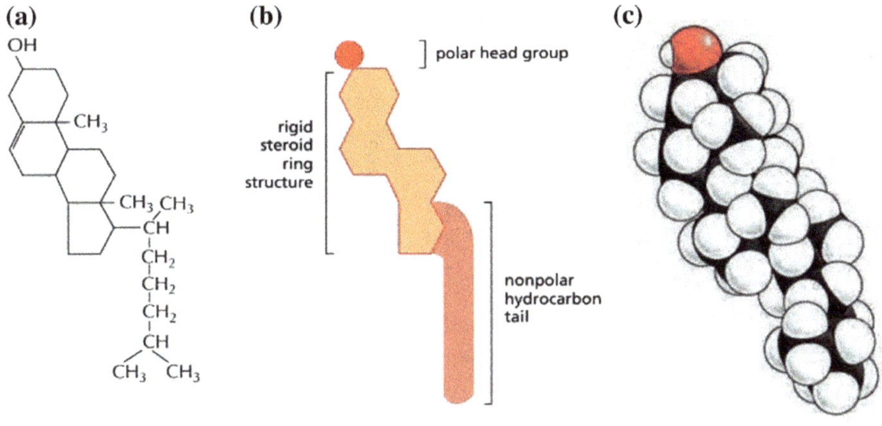

Fig. 1.5 Structure of cholesterol. Cholesterol is represented **a** Formula, **b** Symbol, **c** Space-filling model. Reprinted from Ref. [6] by permission of Garland Science Taylor & Francis Group

Fig. 1.6 Cholesterol in a lipid bilayer. Schematic drawing (to scale) of a cholesterol molecule interacting with two phospholipid molecules in one monolayer of a lipid bilayer. Reprinted from Ref. [6] by permission of Garland Science Taylor & Francis Group

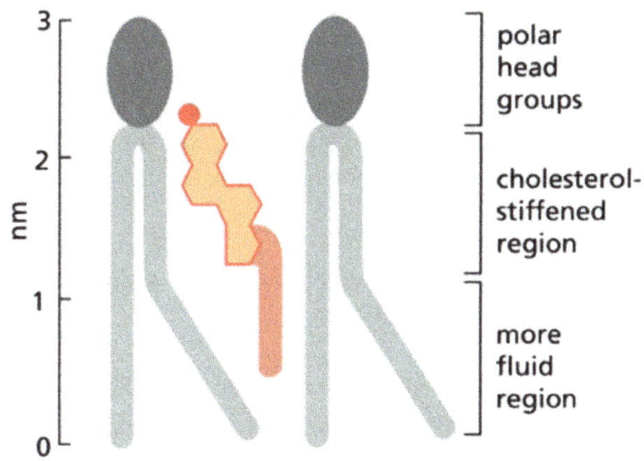

substituent on one ring (Fig. 1.5). Although the composition of cholesterol is almost entire hydrocarbon, it is amphipathic since its hydroxyl group can interact with water. The cholesterol molecules orient themselves in the bilayer with their hydroxyl group close to the polar head groups of adjacent phospholipid molecules (Fig. 1.6). Cholesterol is abundant in the plasma membrane of mammalian cell but is absent in most prokaryotic cells. At the high concentrations found in eukaryotic cell membranes, cholesterol prevents the hydrocarbon chains from aggregating and crystallizing, therefore keeps the bilayer fluidity.

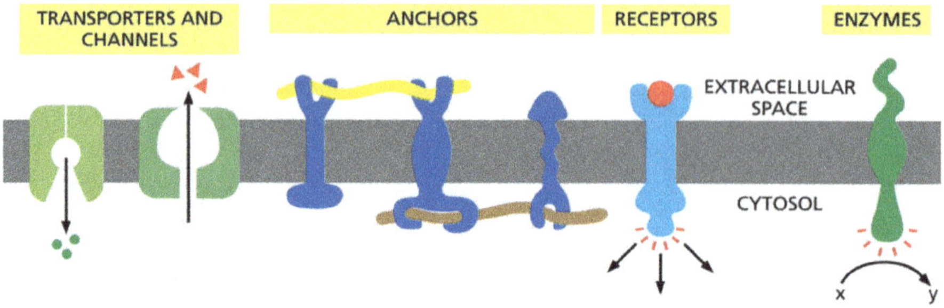

Fig. 1.7 Cell membrane proteins have a variety of functions. Reprinted from Ref. [6] by permission of Garland Science Taylor & Francis Group

1.2.2 Membrane Proteins

Although the lipid bilayer provides the basic structure of biological membranes, the membrane proteins perform most of the membrane's specific tasks and therefore give each type of cell membrane its characteristic functional properties. Membrane proteins serve many functions. Some transport particular nutrients, metabolites, and ions across the lipid bilayer. Others anchor the membrane to macromolecules on either side. Still others function as receptors those detect chemical signals in the cell's environment and relay them into the cell interior or work as enzymes to catalyze specific reactions at the membrane (Fig. 1.7).

1.2.2.1 Membrane Proteins Can Be Associated with the Lipid Bilayer in Various Ways

Proteins can be associated with the lipid bilayer of a cell membrane in any one of the ways illustrated in Fig. 1.8. Like their lipid neighbors, membrane proteins are amphiphilic, having hydrophobic and hydrophilic regions. Many membrane proteins extend through the lipid bilayer, and hence are called transmembrane proteins, with part of their mass on either side (Fig. 1.8, examples 1, 2, and 3). Their hydrophobic regions pass through the membrane and interact with the hydrophobic tails of the lipid molecules in the interior of the bilayer, where they are sequestered away from water. Their hydrophilic regions are exposed to water on either side of the membrane. The covalent attachment of a fatty acid chain that inserts into the cytosolic monolayer of the lipid bilayer increases the hydrophobicity of some of these transmembrane proteins (Fig. 1.8, example 1).

Other membrane proteins are located entirely in the cytosol and are attached to the cytosolic monolayer of the lipid bilayer, either by an amphiphilic α-helix exposed on the surface of the protein (Fig. 1.8, example 4) or by one or more covalently attached lipid chains (Fig. 1.8, example 5). Yet, other membrane proteins are entirely exposed at the external cell surface being attached to the lipid

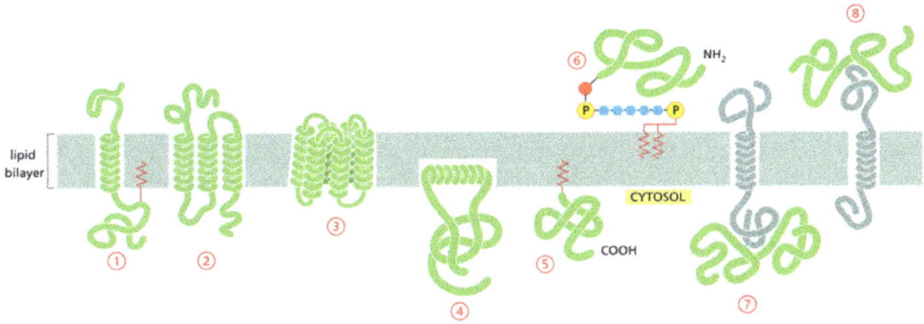

Fig. 1.8 Various ways in which proteins associate with the lipid bilayer. Most membrane proteins are thought to extend across the bilayer as ① a single α-helix, ② as multiple α-helices, or ③ as a rolled-up β-sheet (a β-barrel). Some of these "single-pass" and "multi-pass" proteins have a covalently attached fatty acid chain inserted in the cytosolic lipid monolayer ①. Other membrane proteins are exposed at only one side of the membrane. ④ Some of these are anchored to the cytosolic surface by an amphiphilic α-helix that partitions into the cytosolic monolayer of the lipid bilayer through the hydrophobic face of the helix. ⑤ Others are attached to the bilayer solely by a covalently bound lipid chain, either a fatty acid chain or a prenyl group in the cytosolic monolayer or, ⑥ via an oligosaccharide linker, to phosphatidylinositol in the non-cytosolic monolayer called a GPI anchor. ⑦, ⑧ Finally, membrane-associated proteins are attached to the membrane only by non-covalent interactions with other membrane proteins. Reprinted from Ref. [6] by permission of Garland Science Taylor & Francis Group

bilayer only by a covalent linkage (via a specific oligosaccharide) to a lipid anchor in the outer monolayer of the cell membrane (Fig. 1.8, example 6).

The lipid-linked proteins in example 5 in Fig. 1.8 are made as soluble proteins in the cytosol and are subsequently anchored to the membrane by the covalent attachment of the lipid group. The proteins in example 6, however, are made as single-pass membrane proteins in the endoplasmic reticulum (ER). While still in the ER, the transmembrane segment of the protein is cleaved off and a glycosylphosphatidylinositol (GPI) anchor is added, leaving the protein bound to the non-cytosolic surface of the ER membrane solely by this anchor; transport vesicles eventually deliver the protein to the cell membrane.

By contrast to these examples, membrane-associated proteins do not extend into the hydrophobic interior of the lipid bilayer at all; they are instead bound to either face of the membrane by non-covalent interactions with other membrane proteins (Fig. 1.8, examples 7 and 8). Many of the proteins of this type can be released from the membrane by relatively gentle extraction procedures, such as exposure to solutions of very high or low ionic strength or of extreme pH, which interfere with protein–protein interactions but leave the lipid bilayer intact; these proteins are often referred to as peripheral membrane proteins. Transmembrane proteins and many proteins held in the bilayer by lipid groups or hydrophobic polypeptide regions that insert into the hydrophobic core of the lipid bilayer cannot be released in these ways.

1.2.2.2 A Polypeptide Chain Usually Crosses the Lipid Bilayer as an α-Helix

All membrane proteins have a unique orientation in the lipid bilayer, which is essential for their function. For a transmembrane receptor protein, for example, the part of the protein that receives a signal from the environment must be on the outside of the cell, whereas the part that passes along the signal must be in the cytosol (Fig. 1.7). This orientation is a consequence of the way in which membrane proteins are synthesized. Specialized membrane-spanning segments of the polypeptide chain (Fig. 1.8) connect the portions of a transmembrane protein located on either side of the lipid bilayer. These segments, which run through the hydrophobic environment of the interior of the lipid bilayer, are composed largely of amino acids with hydrophobic side chains. Because these side chains cannot form favorable interactions with water molecules, they prefer to interact with the hydrophobic tails of the lipid molecules, where no water is present.

In contrast to the hydrophobic side chains, however, the peptide bonds that join the successive amino acids in a protein are normally polarized, making the polypeptide backbone hydrophilic. Because water is absent from the interior of the bilayer, atoms forming the backbone are driven to form hydrogen bonds with one another. Hydrogen bonding is maximized if the polypeptide chain forms a regular α-helix, and so the great majority of the membrane-spanning segments of polypeptide chains traverse the bilayer as α-helices. In these membrane-spanning α-helices, the hydrophobic side chains are exposed on the outside of the helix, where they contact the hydrophobic lipid tails and the polypeptide backbone hydrogen bonds generated within the helix (Fig. 1.9).

In many transmembrane proteins, the polypeptide chain crosses the membrane only once (Fig. 1.8 ①). Many of these single-pass transmembrane proteins are receptors for extracellular signals. Other transmembrane proteins function as channels, forming aqueous pores across the lipid bilayer to allow small, water-soluble molecules to cross the membrane. Proteins with a single transmembrane α-helix cannot form such channels. Instead, they usually consist of a series of α-helices that cross the bilayer a number of times (Fig. 1.8 ②). In many of these multi-pass transmembrane proteins, one or more of the membrane-spanning regions are amphipathic-formed α-helices those contain both hydrophobic and hydrophilic amino acid side chains. These amino acids tend to be arranged so that the hydrophobic side chains are located on one side of the helix, while the hydrophilic side chains are concentrated on the other side. In the hydrophobic environment of the lipid bilayer, α-helices of this sort packed side by side in rings with the hydrophobic side chains exposed to the lipids of the membrane and the hydrophilic side chains forming the lining of a hydrophilic pore through the lipid bilayer.

Although the α-helix is by far the most common form in which a polypeptide chain crosses a lipid bilayer, β-sheet is another form, while the polypeptide chain is rolled into a cylinder, this keglike structure called a β-barrel (Fig. 1.8 ③). As expected, the amino acid side chains face the inside of the barrel and therefore line as the hydrophilic aqueous channel, while those on the outside of the barrel, which

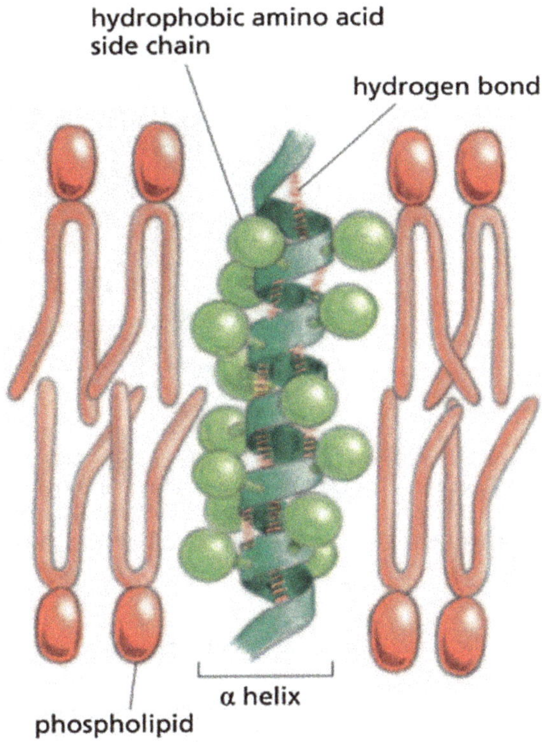

Fig. 1.9 A transmembrane polypeptide chain usually crosses the lipid bilayer as an α-helix. Reprinted from Ref. [6] by permission of Garland Science Taylor & Francis Group

contact the hydrophobic core of the lipid bilayer, are exclusively hydrophobic. The most striking example of a β-barrel structure is found in the porin proteins, which form large, water-filled pores in mitochondrial and bacterial outer membranes (Fig. 1.10).

1.2.2.3 Membrane Proteins Can Be Solubilized in Detergents

To understand a protein comprehensively, one needs to investigate its structure in detail. For membrane proteins, special problems are presented. Most biochemical procedures are designed for studying molecules in aqueous solution. Membrane proteins, however, are built to operate in an environment that is partly aqueous and partly fatty, and taking them out of this environment and purifying them while preserving their essential structure is not an easy task.

Before an individual protein can be studied in detail, it must be separated from all the other cell proteins. For most membrane proteins, the first step in this separation process involves solubilizing the membrane with agents that destroy the lipid bilayer by disrupting hydrophobic associations. The most widely used disruptive agents are detergents. These small, amphipathic, lipid-like molecules differ from membrane phospholipids, and in that they have only a single hydrophobic tail.

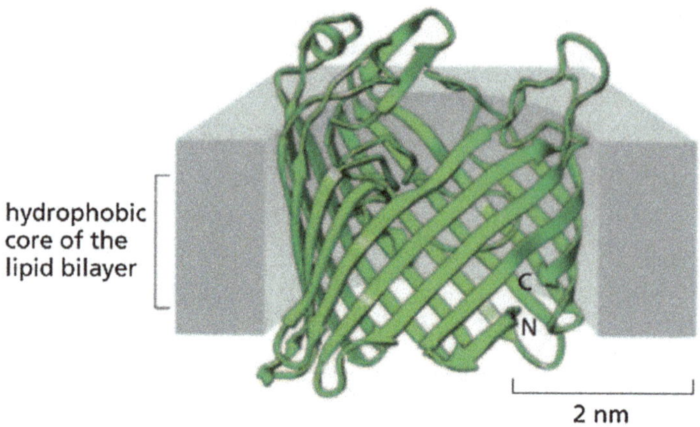

Fig. 1.10 Prion proteins form water-filled channels in the outer membrane of a bacterium. The protein illustrated is from *E. coli*, and it consists of a 16-stranded β-sheet curved around on itself to form a transmembrane water-filled channel. The three-dimensional structure was determined by X-ray crystallography. Reprinted from Ref. [6] by permission of Garland Science Taylor & Francis Group

Their polar (hydrophilic) ends can be either charged (ionic), as in sodium dodecyl sulfate (SDS), or uncharged (nonionic), as in octylglucoside and Triton (Fig. 1.11). Because they have one tail, detergent molecules are shaped like cones; in water, they thus tend to aggregate into small clusters called micelles, rather than forming a bilayer as do the phospholipids, which with their two tails are more cylindrical in shape.

When mixed in great excess with membranes, the hydrophobic ends of detergent molecules interact with the membrane-spanning hydrophobic regions of the transmembrane proteins, as well as with the hydrophobic tails of the phospholipid molecules, thereby disrupting the lipid bilayer and separating the proteins from most of the phospholipids. Because the other end of the detergent molecule is hydrophilic, these interactions bring the membrane proteins into solution as protein–detergent complexes; at the same time, the detergent solubilizes the phospholipids (Fig. 1.12). The protein–detergent complexes can then be separated from one another and from the lipid–detergent complexes for further analysis.

1.2.2.4 Membrane Proteins Often Function as Large Complexes

Many membrane proteins function as part of multi-component complexes, several of which have been studied by X-ray crystallography. One is a bacterial photosynthetic reaction center, which was the first membrane protein complex to be crystallized and analyzed by X-ray diffraction. Many of the membrane protein complexes involved in photosynthesis, proton pumping, and electron transport are even larger than the photosynthetic reaction center. The enormous photosystem II

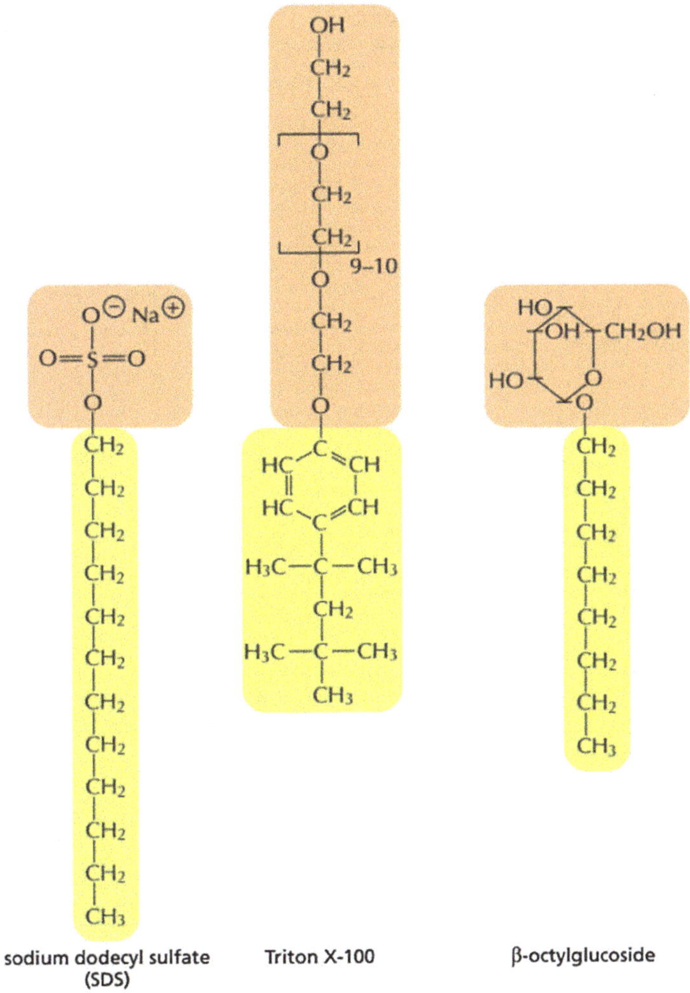

Fig. 1.11 Structure and function of detergents. Three commonly used detergents are sodium dodecyl sulfate (SDS), an anionic detergent, and triton X-100 and β-octylglucoside, two nonionic detergents. Triton X-100 is a mixture of compounds in which the region in brackets is repeated between 9 and 10 times. The hydrophobic portion of each detergent is shown in yellow, and the hydrophilic portion is shown in orange. Reprinted from Ref. [6] by permission of Garland Science Taylor & Francis Group

complex from cyanobacteria, for example, contains 19 protein subunits and well over 60 transmembrane helices. Membrane proteins are often arranged in large complexes, not only for harvesting various forms of energy, but also for transducing extracellular signals into intracellular ones.

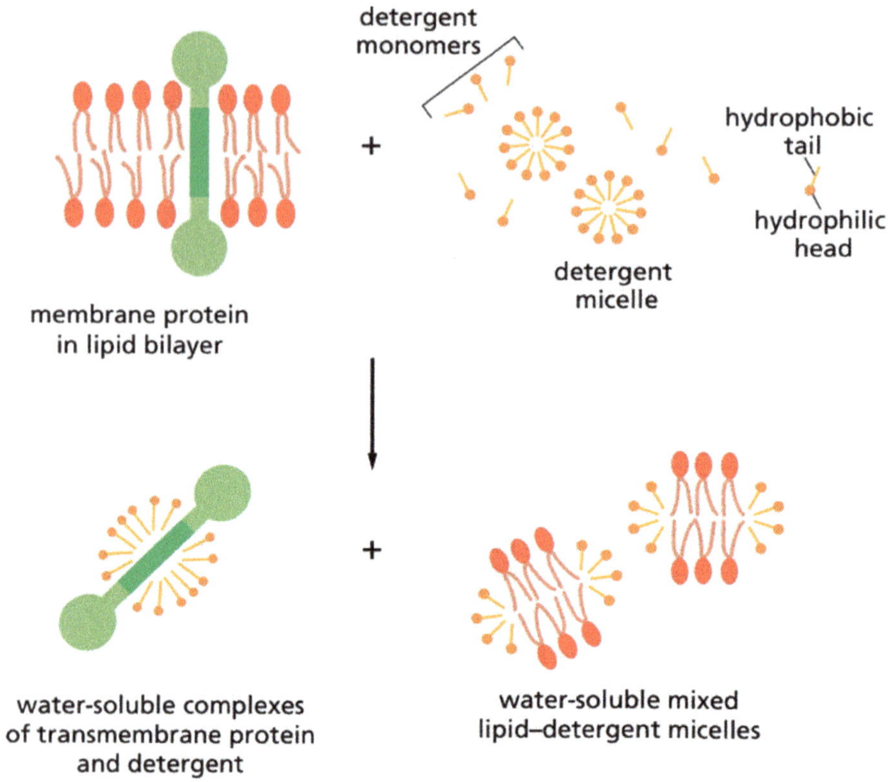

Fig. 1.12 A mild detergent such as Triton X-100 can solubilize membrane proteins. The detergent molecules (gold) are shown as monomers and micelles, the form in which detergent molecules tend to aggregate in water. The detergent disrupts the lipid bilayer and brings the proteins into solution as protein–detergent complexes. As illustrated, the phospholipids in the membrane are also solubilized by the detergents, forming lipid detergent micelles. Reprinted from Ref. [6] by permission of Garland Science Taylor & Francis Group

1.2.2.5 We Know the Complete Structure of Relatively Few Membrane Proteins

For many years, much of what we knew about the structure of membrane proteins was learned by indirect means. The standard method for determining a protein's three-dimensional structure directly is X-ray crystallography but this requires ordered crystalline arrays of the molecule. Because membrane proteins have to be purified in detergent micelles that are often heterogeneous in size, it is harder to crystallize membrane proteins than the soluble proteins those inhabit the cell cytosol or extracellular fluids. Nevertheless, with recent advances in protein preparation and X-ray crystallography, the structures of an increasing number of membrane proteins have now been determined with high resolution.

One example is the structure of bacteriorhodopsin, which first revealed exactly how α-helices cross the lipid bilayer. Bacteriorhodopsin is a small protein (about 250 amino acids) found in large amounts in the cell membrane of an archaea, called Halobacteriumhalobium that lives in salt marshes. Bacteriorhodopsin is a pump protein, a class of transmembrane protein that pumps H^+ (protons) out of the cell, and the pumping required energy is directly from sunlight (see Fig. 1.7). Each bacteriorhodopsin molecule contains a single light-absorbing nonprotein molecule, called retinal that gives the protein and the bacterium a deep purple color. This small hydrophobic molecule is covalently attached to one of bacteriorhodopsin's seven transmembrane α-helices. When retinal absorbs a photon of light, it changes shape, and in doing so, it causes the protein embedded in the lipid bilayer to undergo a series of small conformational changes. These changes result in the transfer of one H^+ from the retinal to the outside of the bacterium. The retinal is then regenerated by taking up an H^+ from the cytosol, returning the protein to its original conformation so that it can repeat the cycle. The overall outcome is the movement of one H^+ from inside to outside the cell.

In the presence of sunlight, thousands of bacteriorhodopsin molecules pump H^+ out of the cell, generating a concentration gradient of H^+ across the cell membrane. The cell uses this proton gradient to store energy and convert it into ATP.

1.2.3 Membrane Carbohydrate

Some of the lipids in the outer layer of the cell membrane have sugars covalently attached to them. The same is true for most of the proteins in the cell membrane, named glycoproteins. The short chains of sugars linked onto glycoproteins are called oligosaccharides. It also occurs as the polysaccharide chains of integral membrane proteoglycan molecules found mainly outside the cell, as part of the extracellular matrix. The carbohydrate layer protects cells against mechanical and chemical damage and also helps to keep cells at a distance, preventing unwanted protein–protein interactions.

The membrane carbohydrates help membrane proteins to form and maintain the correct packing, three-dimensional configures, and to transfer to a correct position. Both the diversity and the exposed position of the oligosaccharides on the cell surface make them especially well-suited to function in special cell recognition processes. For example, some glycolipids are important blood group determinants, and cell membrane bound-lectins recognize specific oligosaccharides on cell surface and mediate a variety of transient cell–cell adhesion processes in sperm–egg interaction, blood clotting, lymphocyte recirculation, and inflammatory response.

1.3 Structure Characters of Cell Membranes

The phospholipid molecules diffuse freely in cell membrane and form a bilayer framework structure with their hydrophobic nonpolar tails buried in the interior and their hydrophilic heads exposed to water. The main character of cell membrane is asymmetry and fluidity.

1.3.1 Membrane Asymmetry

All cellular membranes have an internal face designated as the cytosolic face (the surface oriented toward the interior of the compartment) and an external face designated as the exoplasmic face (the surface presented to the environments). Therefore, cell membranes are asymmetrical [7]: the two halves of the bilayer often contain different kinds of phospholipids and glycolipids. In addition, the proteins embedded in the bilayer have a specific orientation.

The asymmetry of the lipid bilayer is functionally important. The lipid compositions of the two monolayers of the lipid bilayer in many membranes are dramatically different. In the human red blood cell (erythrocyte) membrane, for example, almost all of the phospholipid molecules those have choline $(CH_3)_3N^+CH_2CH_2OH$ in their head group (phosphatidylcholine and sphingomyelin) are in the outer monolayer, whereas almost all of those contain a terminal primary amino group (phosphatidylethanolamine and phosphatidylserine) are in the inner monolayer (Fig. 1.13). Because the negatively charged phosphatidylserine is located in the inner monolayer, there is a significant difference in charge between the two halves of the bilayer.

Among lipids, those show the most dramatically lopsided distribution in cell membranes are the glycolipids, which are located mainly in the cell membrane and only in the exoplasmic half of the bilayer (Fig. 1.13). Their glucose groups face the

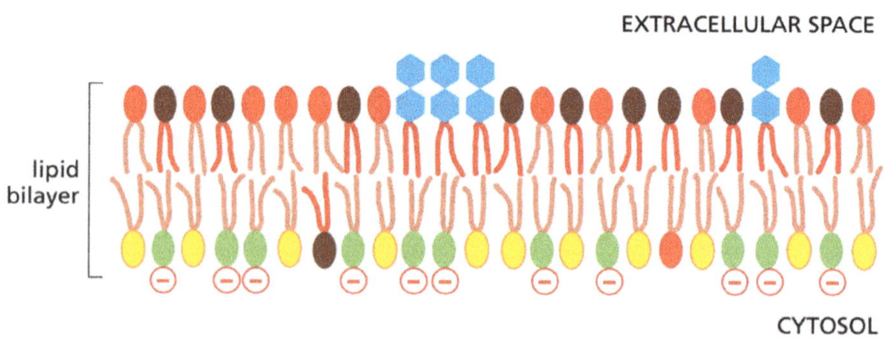

Fig. 1.13 Asymmetrical distribution of phospholipids and glycolipids in the lipid bilayer of human red blood cells. Glycolipids are drawn with hexagonal polar head groups (blue). Reprinted from Ref. [6] by permission of Garland Science Taylor & Francis Group

extracellular, where they form part of a continuous coat of carbohydrate that surrounds and protects animal cells. Glycolipid molecules acquire their glucose groups in the Golgi apparatus, where the enzymes engineer this confined chemical modification. These enzymes are oriented such that glucose is added only to lipid molecules in the exoplasmic half of the bilayer. Once a glycolipid molecule has been created in this way, it remains trapped in this monolayer, as there are no flippases that transfer glycolipids to the cytosolic side. Thus, when a glycolipid molecule is finally delivered to the cell membrane, it displays its glucose to the exterior of the cell.

Other lipid molecules show different types of asymmetric distributions, which relate to their specific functions. For example, the inositol phospholipids, a minor component of the cell membrane, have a special role in relaying signals from the cell surface to the cell interior; thus, they are concentrated in the cytosolic half of the lipid bilayer.

Lipid asymmetry is functionally important, especially in converting extracellular signals into intracellular portion [8]. Many cytosolic proteins bind to specific lipid head groups found in the cytosolic monolayer of the lipid bilayer. In other cases, lipid head groups must first be modified to create protein-binding sites at a particular time and place. Various lipid kinases can add phosphate groups at distinct positions on the inositol ring, creating binding sites that recruit specific proteins from the cytosol to the membrane. The cell membrane contains various phospholipases that are activated by extracellular signals to cleave specific phospholipid molecules, generating fragments of these molecules that act as short-lived intracellular mediators [9, 10].

The distributions of the membrane proteins include the extrinsic protein, and intrinsic protein is asymmetric. The asymmetry of the membrane protein referring to the distribution of each type of membrane proteins on the cytoplasmic membrane has obvious directivity.

The receptor protein and carrier protein on the cell surface process the signal transduction, and material transportation must be according to certain direction. The enzymatic reaction, which relates to the cytoplasmic membrane, is to react in the certain side of the cell membrane. As glycolipids, glycoprotein oligosaccharide chain is located in the exoplasmic face of the membrane, and they interact with extracellular matrix, cell growth factor, lectin and antibody.

1.3.2 Membrane Fluidity

Membrane fluidity referring the viscosity of the lipid bilayer is the physical state of the lipid molecules within the cell membrane, which is essential for the membrane function, and has to be maintained within certain limits. It allows molecules to diffuse rapidly in the plane of the bilayer and interact with each other; thus, many basic cellular processes such as cell movement, cell growth, cell division, secretion can take place.

Temperature and membrane composition are the main determinants for membrane fluidity. At the relatively warm temperature (37 °C), the lipid bilayer is in a relatively fluid state. When lowering the temperature solely there comes a point (phase transition temperature) when the lipid is converted to a frozen crystalline gel in which the movement of the phospholipid fatty acid chains is greatly limited.

How the fluidity a lipid bilayer is at a given temperature depends on its phospholipid composition and, in particular, on the nature of the hydrocarbon tails: the closer and more regular the packing of the tails, the more viscous and less fluid the bilayer will be. Two major properties of hydrocarbon tails affect how tightly they pack in the bilayer: their length and the number of double bonds they contain. A shorter chain length reduces the tendency of the hydrocarbon tails to interact with one another and therefore increases the fluidity of the bilayer. The hydrocarbon tails of membrane phospholipids vary in length between 14 and 24 carbon atoms, with 18–20 atoms being most usual. Most phospholipids contain one hydrocarbon tail that has one or more double bonds between adjacent carbon atoms and a second tail with single bonds only. The chain that harbors a double bond does not contain the maximum number of hydrogen atoms that could, in principle could be attached to its carbon backbone; it is thus said to be unsaturated with respect to hydrogen. The hydrocarbon tail with no double bonds has a full complement of hydrogen atoms and is said to be saturated. Each double bond in an unsaturated tail creates a small kink in the tail, which makes it more difficult for the tails to pack against one another. For this reason, lipid bilayers that contain a large proportion of unsaturated hydrocarbon tails are more fluid than those with lower proportions.

In animal cells, membrane fluidity is modulated by the inclusion of the sterol cholesterol. This molecule is present in especially large amounts in the cell membrane, where it constitutes approximately 20% of the lipids in the membrane by weight. Because cholesterol molecules are short and rigid, they fill the spaces between neighboring phospholipid molecules left by the kinks in their unsaturated hydrocarbon tails (Fig. 1.6). By this way, cholesterol tends to stiffen the bilayer, making it less flexible, as well as less permeable.

Lateral diffusion and rotation are the main movement way of the membrane protein. As the lipids, the flip-flop mobility is rarely occurring between the lipid bilayer. Membrane protein mobility was first demonstrated using cell fusing technique. Accompanied by the time of fusing progresses, the membrane proteins from two species were seen to gradually move within the membrane to the opposite hemisphere and finally distribute uniformly in the entire hybrid cell membrane [11].

The early studies seem to show that integral membrane proteins were capable of moving freely within the membrane. However, other techniques such as fluorescence recovery after photobleaching (FRAP) show their restrictions on the protein mobility. The rate of the fluorescence recovery indicates the diffusion rate of the mobile molecules. The results of FRAP studies indicated that a significant portion of membrane proteins was not free to diffuse back into the irradiated circle, meaning that the movement of certain proteins on cell membrane is restricted. Some membrane proteins assemble into large complex in which individual protein molecules are relatively fixed and diffuse very slowly.

1.4 The Basic Functions of Cell Membranes

Cell membrane separates and protects the cell from the outside environment. Lipid bilayer determines the basic structure of cell membranes; proteins are responsible for most membrane functions, serving as specific receptors, enzymes, transporters, and so on. Proteins are responsible for the transport of ions and other small water-soluble molecules. The dynamic association of membrane-bending proteins confers on membranes their characteristic three-dimensional shapes. The mains functions of cell membrane are summarized as follows [12–14]:

(1) Barrier membrane. The cell membrane, on the one hand, provides a stable internal environment for ensuring the normal operation of cell. On the other hand, it can provide binding sites for various enzymes, making the highly efficient and orderly response.
(2) Permselective membrane. The cell membrane provides a barrier for the exchange of substances on both side, including the metabolic and metabolites. Material transmembrane transport is selective.
(3) Signal transmission. The receptor protein on the cell membrane can identify and bind the specific ligand, for complex signaling and coordinated responses.
(4) Mediated the interaction between cell–cell and cell–extracellular matrix. The cell membrane provides cellular recognition sites, mediated or mucoid between cells and cells, cells and extracellular matrix. The receptors of pathogens such as viruses and the specific host cells of infection are also present in the cell membrane.

Limited to current techniques, there are still many unanswered questions about cell membranes. It is not clear how the cell membranes efficiently and precisely accomplished the intricate functions. There are many structural details of the cell membrane waiting for exploration. Future investigations on cell membranes will involve membrane modifications to improve health or correct membrane dynamic functions in transport, signaling, biosynthesis, remodeling, and other membrane functions. The resolution of these problems will depend on the establishment of new techniques. We are looking forward to more advanced technologies those will facilitate the study of the structure and function of cell membrane and membrane proteins.

Acknowledgements This work was supported by the National Key R&D Program of China (No. 2017YFA0505300), the National Natural Science Foundation of China (No. 21525314, 21503213, 21703231, 21721003).

References

1. Singer SJ, Nicolson GL (1972) The fluid mosaic model of the structure of cell membranes. Science 175(4023):720–731
2. Kusumi A, Fujiwara TK, Chadda R et al. (2012) Dynamic organizing principles of the plasma membrane that regulate signal transduction: commemorating the fortieth anniversary of Singer and Nicolson's fluid-mosaic model. In: Schekman R (ed) Annual review of cell and developmental biology, vol 28. pp 215–250. doi:10.1146/annurev-cellbio-100809-151736
3. Vereb G, Szollosi J, Matko J et al (2003) Dynamic, yet structured: the cell membrane three decades after the Singer-Nicolson model. Proc Natl Acad Sci USA 100(14):8053–8058
4. Nicolson GL (2014) The fluid-mosaic model of membrane structure: still relevant to understanding the structure, function and dynamics of biological membranes after more than 40 years. Biochim et Biophys Acta-Biomembr 1838(6):1451–1466
5. Goni FM (2014) The basic structure and dynamics of cell membranes: an update of the Singer-Nicolson model. Biochim et Biophys Acta-Biomembr 1838(6):1467–1476
6. Bruce A, Alexander J, Julian L et al (2014) Molecular biology of the cell, 6th edn. Garland Science, New York
7. Rothman JE, Lenard J (1977) Membrane asymmetry. Science 195(4280):743–753
8. Carquin M, D'Auria L, Pollet H et al (2016) Recent progress on lipid lateral heterogeneity in plasma membranes: from rafts to submicrometric domains. Prog Lipid Res 62:1–24
9. Simons K, Ikonen E (1997) Functional rafts in cell membranes. Nature 387(6633):569–572
10. Lingwood D, Simons K (2010) Lipid rafts as a membrane-organizing principle. Science 327 (5961):46–50
11. Simons K, Vaz WLC (2004) Model systems, lipid rafts, and cell membranes. Annu Rev Biophys Biomolec Struct 33:269–295
12. Bruce A, Dennis B, Karen H et al (2013) Essential cell biology, 4th edn. Garland Science, New York
13. Zhai Z, Wang X, Ding M (2011) Cell biology, 4th edn. Higher Education Press, Beijing
14. Li Y, Wu C, Shen D (2014) Cell biology, 2nd edn. Fudan University Press, Shanghai

Chapter 2
History and Traditional Techniques of Studying the Structure of Cell Membranes

Jing Gao and Hongda Wang

2.1 Introduction

All cells are surrounded by a membrane layer called cell membrane or plasma membrane. Cell membranes consist of membrane lipid and proteins and are crucial for the life of the cell by defining the cell boundaries, therefore, maintaining the differences between the intracellular and extracellular environment. The cell membrane is not only the boundary of the unit of life, but also a specific compartment that harbors many essential cell functions including the provision of a stable internal environment, transportation of substances between cells and the environment, energy conversion, and signal transduction [1–3]. Due to its key role in a variety of activities, the cell membrane is one of the most attractive topics for multidisciplinary studies, including studies that combine chemistry, biology, and physics. For chemists, cell membranes are supermolecular structures that contain thousands of types of lipids, proteins, and saccharides [4]. Biologists consider cell membranes to be the first barrier to defend cells from external harms, and cell membranes are related to multiple functions and diseases [5]. From the physicists' point of view, how the basic atoms and molecules can form a multi-functional and high-efficiency system is related to the origin of life [6].

After the basic composition of the cell membrane was discovered, scientists have paid more attention to the structure and function of the whole supermolecular complexes. The studying history of cell membrane structure can be traced back to the early twentieth century. Many scientists coming from different fields contributed to our understanding of membranes along the centuries. We can find that every breakthrough of membrane structure is always accompanied by the progress of

J. Gao · H. Wang (✉)
State Key Laboratory of Electroanalytical Chemistry, Changchun Institute of Applied Chemistry, Chinese Academy of Sciences, No. 5625 Renmin Street, Changchun 130022, Jilin, People's Republic of China
e-mail: hdwang@ciac.ac.cn

H. Wang and G. Li (eds.), *Membrane Biophysics*,
https://doi.org/10.1007/978-981-10-6823-2_2

science and technology. In this part, we will review the main discoveries that led to our current knowledge of biological membranes and the related techniques which were used to explore different membrane models. We will also discuss the strengths and weaknesses of the techniques and methods, and the suitability of different membrane models in various biological systems. This, we hope, will provide a more complete account of the discovery of membrane structure and open new perspectives for further uncovering the more refined and precise membrane structure at the sub-cellular and molecular level.

2.2 Membrane Models and Related Techniques

The properties and functions of the cell membrane are dependent on the structure of its own component molecules. The details of how these various kinds of lipids, proteins, and saccharides are organized to form a complex and ordered cell membrane are highly important. The investigation of this question requires a number of technological developments, which have been the basis of new understanding of cell membrane structure. Since last century, researchers have been working on the structure and function of cell membranes and proposed distinct membrane models to explain why cell membranes can play so many functions. Among them, the fluid mosaic model is one of the most classical models, which is widely accepted by biologists. Even so, the long history of studying cell membranes tells us that the membrane structure is far from completely understood. There are still open questions that remain to be resolved, such as the location and interaction of proteins in the membrane. Therefore, summarizing the development of cell membrane studies will help us more appropriately apply different membrane models to diverse disciplinary fields and discover novel models to explain what are happening or will happen in the membrane. Readers will find a timeline displaying the most dramatic contributions to membrane structure models up to the early twenty-first century in Fig. 2.1.

2.2.1 Lipid Bilayer Model

From a historical perspective, monomolecular lipid films at the air–water interface (so-called Langmuir films) are the oldest membrane systems ever reported. The earliest record about this particular system (reported as oil films floating in water) came from Hammurabi period's (eighteenth century BC) cuneiform-written Babylonian clay tablets [7]. Babylonians used it as one of their forms of divination, based on pouring oil on water and observing the shapes originated from the subsequent oil spreading. This application was also used in Greek, Chinese, and ancient Japanese cultures. The most often mentioned ancient application of lipid films is the calming effect of ripples on the water surface caused by spreading oil,

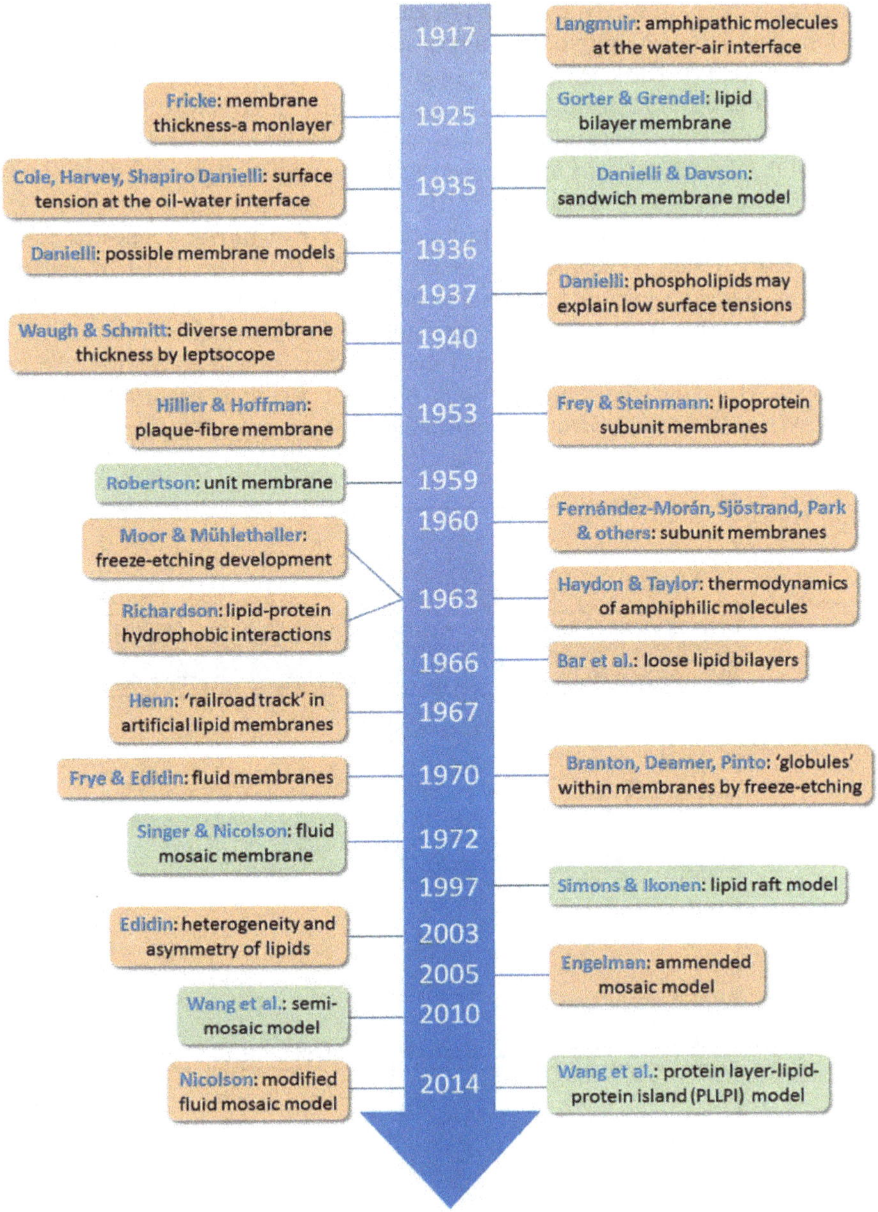

Fig. 2.1 Summary of the main contributions related to cell membrane discovery between the first studies on cell membrane structures and the development of membrane models. The events are approximately stacked in chronological order from top to bottom. The orange boxes represent some major research influent to the history, and the green boxes reflect representative membrane models

which was already observed by sailors in ancient times and described by Plutarch, Aristotle, and Pliny. Pouring oil on stormy seas is believed to stem from the 429 to 651 AD period [7]. The scientific investigation of this phenomenon was initiated in 1744 by Benjamin Franklin, followed by contributions of Lord Rayleigh and Agnes Pockels. Until 1917, Langmuir presented the first evidence of the monomolecular nature of the lipid films by measuring the surface covered by lipid monolayers at the interface between water and air (Fig. 2.2) [7–11].

Langmuir films were used in a historically key experiment connected to our contemporary vision of a biological membrane. In 1925, two Dutch physicians, Gorter and Grendel explored the molecular structure of membranes [12]. They extracted the lipids from an erythrocyte sample; since these cells were known to lack internal membranes, they assumed that all lipids should come from the cell envelopes. Measuring the spread of lipids on water was done using a Langmuir's trough. When Gorter and Grendel compared the surface covered by the lipids to the

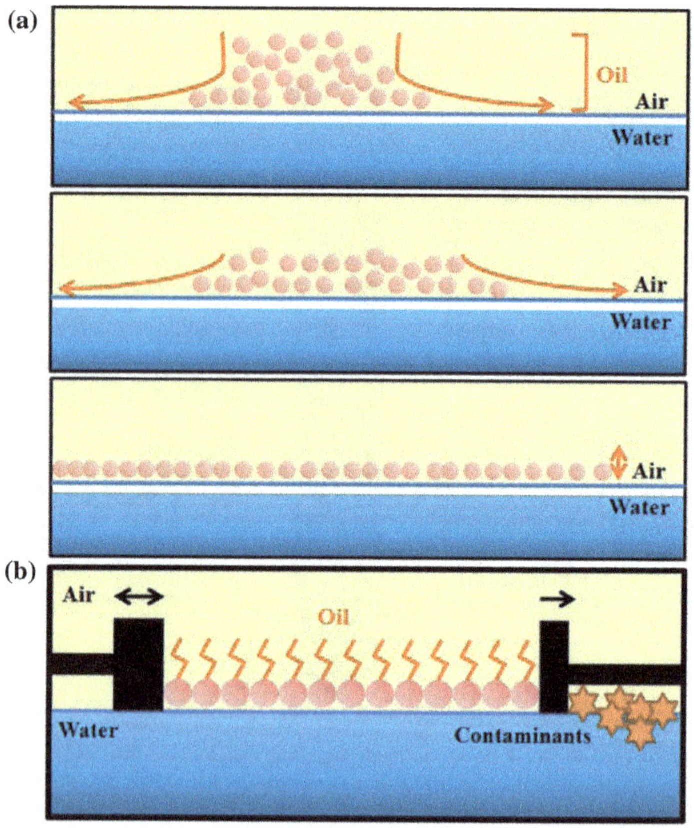

Fig. 2.2 Oil spreads at air–water interfaces. **a** Oil molecules spontaneously spread on the air–water interface until they form a layer one molecule thick. **b** Langmuir trough allows to precisely measure the surface that these monolayers can spread depending on the applied pressure

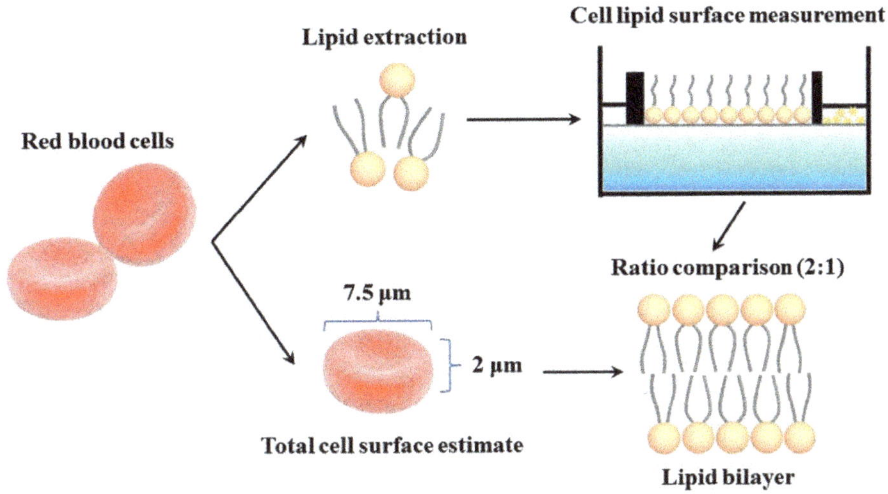

Fig. 2.3 Surface measurement of membrane lipid monolayers as a way to determine membrane structure. Summary of the method, consisting in the comparison between the area of lipids extracted from erythrocyte membranes and the estimated surface area of cells. The ratio of 2:1 indicated the lipid bilayer structure

estimated sum of cell surfaces, they found a 2:1 ratio (Fig. 2.3) [11]. As a result, they concluded that cells were surrounded by a lipid membrane with two molecules thickness with hydrophobic components in the internal part of the membrane and the hydrophilic components in the external part [12].

This study has been commonly cited as the most conclusive argument in favor of the lipid bilayer nature of cell membranes. In spite of the elegance of this work, it is important to balance its contribution to the field because of its errors from both technical and theoretical grounds. First, the extraction technique employed could not isolate the totality of erythrocyte lipids from the samples [13]. Second, the equation that they used to estimate the surface of the erythrocytes also underestimated the cell area [13]. Some historical reviews on membrane discovery have argued that it was fortunate that the two errors neutralized each other in order to give credit to the lipid bilayer hypothesis that we now recognize in current membrane models. When calculating the surface covered by lipids, Gorter and Grendel measured the first detected pressure (i.e., the maximal continuous surface covered by an amount of lipid). A more accurate ratio for the lipid surface was measured at the collapse pressure in 1960s, which was 1.3:1 [13]. This ratio conformed to loose lipid bilayers, whereas the modern fluid mosaic model supports the idea that membrane lipids are tightly packed and the excess space is actually occupied by membrane proteins.

Disregarding their experimental mistakes, Gorter and Grendel's work is the first documented evidence that cell lipids are arranged in a bilayer configuration. The contribution of lipid bilayer model can be helpful in the light of later progress in membrane studies. However, it should be noticed that at that time the presence of proteins as components of biological membranes was unknown.

2.2.2 Davson–Danielli Model

The formulation of the lipid bilayer model had opened the door to the molecular description of cell membrane structure. In an attempt to confirm or refute the lipid bilayer postulate, one of the first lines of research to be explored was the measurement of membrane thickness. In 1925, Fricke measured the static capacitance per surface unit with an estimation of the cell membrane dielectric constant. He extrapolated that the thickness of erythrocyte and yeast cell membranes was about 4 nm [14, 15]. This thickness supported the hypothesis of a lipid monolayer but not a bilayer. Although using the dielectric constant in these studies brought some questions, there were no enough tests to refute the estimation by Fricke at that time [16]. Until 1940, the invention of leptoscope provided a relatively accurate testing of measuring membrane thickness. It could measure very thin membranes by comparing the intensity of light reflected from a sample to the intensity of a membrane standard of known thickness [17]. The thickness it measured ranged from 8.6 to 23.2 nm. As a result, the lower value was compatible with the lipid bilayer, whereas the higher one supported the presence of supplementary superimposed layers [18]. Yet, the thickness of the membrane was only revealed after two decades when electron microscopy developed. The observation of membrane sections using the new instrument established the now accepted ~8 nm value for standard cell membranes [19].

Although the actual thickness of the membrane remained unclear during that time, studies in this field still took a huge step because the most influential membrane structure model introduced, namely the sandwich membrane model (Fig. 2.4). The genesis of this model followed previous studies on surface tension of various cells

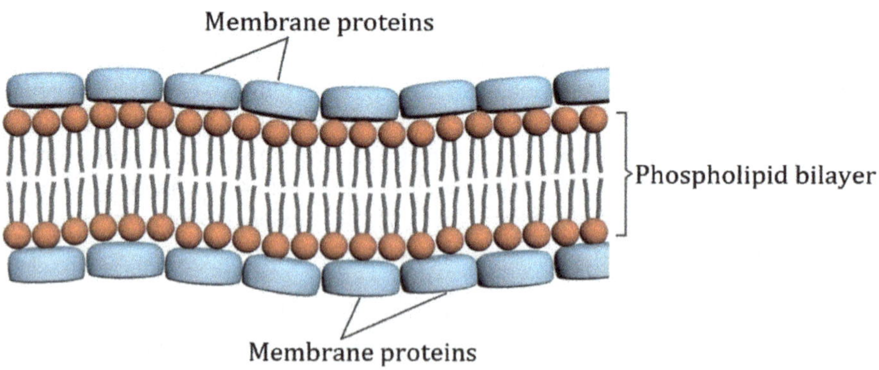

Fig. 2.4 Davson–Danielli "sandwich membrane model"

[20, 21]. As the surface tension values appeared to be much lower than would be expected for an oil–water interface, it was assumed that some substance was responsible for lowering the interfacial tensions in the surface of cells [21]. At this time, membranes were known to contain substantial quantities of proteins [22, 23] but little had been said about their position in cell membranes. Therefore, in 1935 Danielli and Davson proposed a sandwich membrane model (the Davson–Danielli model or the protein–lipid–protein model) [24]; in this model, the lipid layer was covered by a thin protein layer on both sides. This observation was significant progress in understanding better the compositional nature of biological membranes.

The Davson-Danielli model immediately became popular, and it dominated cell membrane studies for the following 30 years. There were some light polarization and X-ray diffraction studies to support the model. In order to provide insights into membrane structure, these methods required repeated structures, so the samples used were myelinized axons. These analyzes showed an alternation of protein and lipid layers in support of the sandwich model [25]. Unfortunately, we know now that this biological material is physiologically very specific and can hardly be compared to a regular cell membrane.

Despite the fact that the superposition of protein and lipid layers was characterized in this model, the authors were aware of its existent problems. In the surface tension experiments, they used triglyceride oils and other non-miscible lipids. These substances are very different from most natural amphipathic cell membrane components, and thus they are not suitable for a realistic comparison to cell membranes. Soon after the authors postulated the sandwich model, they added amphipathic cell membrane components such as fatty acids, cholesterol, or phospholipids to non-miscible mixtures, which significantly decreased the interfacial tension between water and highly hydrophobic substances [26]. Other contemporary debates were about membrane permeability. As the ion permeability developed and active transporters discovered, the authors acknowledged the three popular possibilities of that time: pores, simple diffusion, and the existence of some proteins that can span the membrane.

In summary, the Davson-Danielli model took into account the difference between the physical properties measured for natural membranes and membranes composed of pure lipids, for example, the low values of surface tension measured in the 1920s and 1930s in plasma membranes compared to lipid bilayers. However, although the researchers were conscious of the permeability of ions, sugars, and other hydrophilic solutes, the model they postulated implied that the protein layer did not interact with the hydrophobic parts of the lipid bilayer and that the protein layer formed by simple physical adsorption.

2.2.3 Unit Membrane Model

The direct characterization of membrane structure did not progress much until new techniques allowed dramatic discoveries. Electron microscopy emerged in the

1930s, however, sharper resolutions that allowed the direct observation of cell membranes was obtained till 1950s. Sjöstrand et al. demonstrated the “railroad track” structure of membrane cross sections after staining with heavy metals, which stained the phosphates of phospholipid headgroups and washed the proteins out [27]. The structure was known as two dense lines separated by a middle lighter space. Yet, which one among the whole structure or one of the dense parts should be considered as the quintessential cell membrane? In 1959, Robertson compared a collection of cross-section pictures and observed that the whole railroad track was consistent in a variety of cells [28] (Fig. 2.5). He thought that the railroad track fit the Davson–Danielli model and extended it as the unit membrane or the Davson–Danielli–Robertson (DDR) model. The model supported the concept of a “unit membrane” with a thickness of 6–8 nm where the proteins were coating the surface of the lipid bilayer in an extended β-sheet configuration, corresponding to the width of myelin sheaths observed in X-ray diffraction experiments. Robertson stressed two points in the model: first, that the three layers observed in the electron microscope cross-section shots were part of the same structure, the cell membrane, regardless of the other cell envelopes that might exist; and second, that this structure was universally shared among all biological membranes [29].

The unit membrane model assigned the bilayer structure to all membrane systems, including cell membranes and cellular organelle membranes. However, many researchers called into questioning the idea that one unique membrane model could account for all biological membranes because it seemed contradictory to the large membrane diversity that was being discovered at that time [31, 32]. For example, it had been observed that the protein-lipid ratio from different membranes could vary from 1:4 to 5:1, suggesting that the amount of membrane proteins was not always enough to entirely cover the cell surface [31]. In addition, the lack of resolution prevented the observation of the railroad track in some bacteria, thus implying that the three-layer structure was either not always visible or not universal [33].

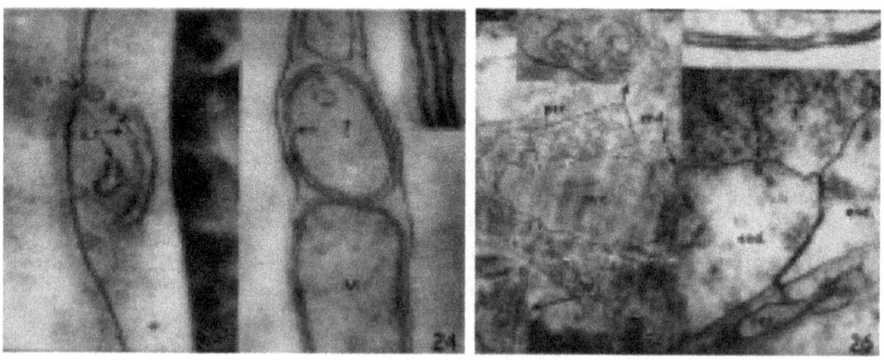

Fig. 2.5 Unit membrane concept. Railroad track images from thin section electron microscopy of osmium-fixed cells: two dark lines separated by a clear region. Robertson summarized the available data and used many new examples to make the point that all cell membranes have a common structure. Reprinted by permission from Macmillan Publishers Ltd: Ref. [30], copyright 2003

In 1966, observations made by Green and Benson challenged the unit membrane model [34, 35]. Working independently with inner mitochondria and chloroplast membranes, these authors demonstrated: first, that subunits not railroad tracks containing lipid and proteins can be separated from the whole membrane; and second, that these subunits can be reconstituted to regain activity. Based on these observations, a different model was postulated where lipids represent a solvent for embedded globular proteins, i.e., opposite to that postulated in the unit membrane model [35, 36]. Electron microscopy images from viral capsides also lent credit to the view that membranes could be made up of subunits [37]. From a historical perspective, it is noteworthy that these models emphasized the dominant role of proteins as the major structural components of membranes [38]. The so-called subunits that were being observed most likely corresponded to the functional proteins that dominated the mitochondrial and chloroplastic membranes, for example, ATPase or photosystems.

The unit membrane model was confronted and there was no consensus as to agreeing on a general model for biological membranes [36], i.e., a model that includes common relevant features existing in different biological membranes. In spite of the debates, the change in perspective that allowed the prevalence of the fluid mosaic model was highly impacted by the gathering of conclusive evidence from new techniques.

2.2.4 *Fluid Mosaic Model*

With the development of the freeze-fracture technique and immuno-electron microscopy, researchers identified antibody-recognizing isolated membrane proteins and proteins that spanned the lipid bilayer. Thus, Singer and Nicolson proposed their celebrated fluid mosaic model (FMM) of biological membranes in 1972 [39], and this model has since been a central paradigm in membrane science. The basics of the model have remained the same ever since the membrane is a lipid bilayer with hydrophilic parts on the sides and hydrophobic parts in the interior; proteins can interact with surface through transient polar contacts, but a lot of proteins are partially or totally embedded in the lipid bilayer where their hydrophobic parts also interact with the hydrophobic parts of lipids (Fig. 2.6).

Considering the big controversy going on at the beginning of the 1970s, the FMM incorporated for the first time many relevant experimental facts and the development of techniques: (1) the permeability and transport studies that predicted enzyme-like transmembrane proteins [41]; (2) the apparent lack and limited transverse mobility of lipids to make up complete bilayers [13], thus pointing out to the participation of proteins in the membrane plane; (3) electron microscopy pictures, including freeze-etching technique and immuno-EM studies that identified antibody-recognizing isolated membrane proteins [42]; (4) the stability of artificial lipid bilayers that supported them as suitable and sufficient components to make up structures similar in the biological membranes [43]; (5) the favorable conformations

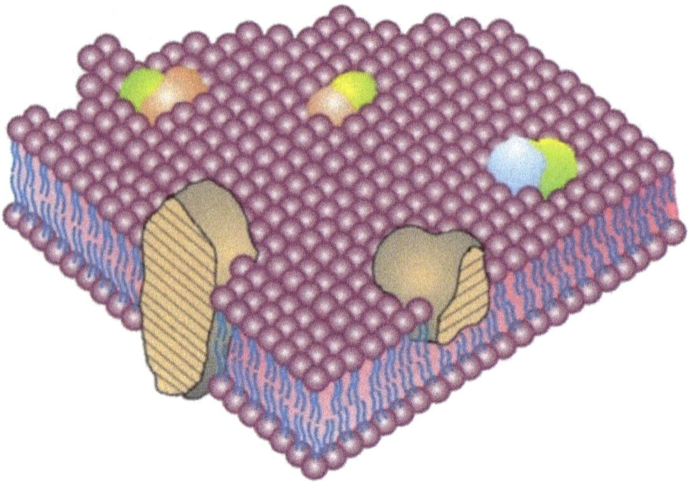

Fig. 2.6 Fluid mosaic membrane model as originally proposed in 1972. In this cross-sectional view, the solid bodies with stippled cut surfaces represent globular integral membrane proteins, which at long range are randomly distributed in the plane of the membrane. At short range, some may form specific aggregates. Reprinted from Ref. [40], Copyright 2014, with permission from Elsevier

predicted for the membrane proteins, such as α-helical, globular, or membrane spanning [38].

As a key property, the Singer–Nicolson model on one hand assigned to the lipid bilayer component of membranes a certain degree of fluidity. The fluidity concept was meant to characterize the lipid bilayer as a kind of pseudo-two-dimensional liquid in which both lipids and membrane-associated proteins display sufficient lateral mobility to allow for function. On the other hand, it emphasized the mosaic nature of proteins, which could span the lipid bilayer. The overall random appearance of this lipid–protein fluid composite made the membrane look like a mosaic. The specific features and important contributions of the model are demonstrated in detail as follows.

First, the model recovered the lipid bilayer concept from Danielli and Davson as mentioned above. Membrane lipids are amphipathic. They possess both a hydrophobic and a hydrophilic moiety. Because of this amphipathic character, in an aqueous medium they can organize themselves on both sides of an imaginary plane, with the hydrophobic portions facing each other, and the polar moieties oriented to the outer, aqueous space.

Second, membrane proteins can be associated either to the lipid bilayer polar headgroups (peripheral proteins) or to the hydrophobic matrix (integral proteins). This idea of proteins embedded in a hydrophobic milieu had never been proposed in a clear and explicit way before. Moreover, the acknowledgment of the thermodynamic hydrophobic constraints improved our understanding of membrane proteins, for example, the first tridimensional structure of a transmembrane protein—the archaeal bacteriorhodopsin [44], the development of the patch-clamp technique which allowed the understanding of single ion channels [45, 46], and the late discovery of the aquaporins [47, 48].

Third, both lipids and proteins are in constant motion. In principle, three main modes of motion could be considered, rotational, translational, and transbilayer. Rotational motion occurs essentially around an axis perpendicular to the plane of the membrane. Both lipids and proteins rotate around their long axis. Protein rotation had been considered but not given much attention. It was experimentally demonstrated by Chapman and co-workers [49]. Translational diffusion occurs along the plane of the membrane, unhindered by diffusion barriers [50]. Transbilayer (or flip-flop) diffusion, though in theory possible, would not occur because of the energy barrier presented by the bilayer hydrophobic core to the polar groups of lipids and proteins.

Finally, the asymmetry of membranes has also proven to be a fruitful characteristic [51]. This is a direct consequence of the just mentioned lack of transbilayer motion. Asymmetry means that the two sides of a membrane are not identical. Lipids exhibit a relative asymmetry [52], i.e., the fraction of a given lipid in one of the monolayers is different from that in the other. A well-known example is phosphatidylserine, that is found almost entirely in the inner side of human red blood cell membranes. Conversely most, but not all, phosphatidylcholine occur in the outer monolayer [53]. Protein asymmetry is absolute, every single protein molecule in a membrane occurs with exactly the same orientation. Integral proteins are anchored to one or both side of the membrane immediately after synthesis and insertion into the bilayer.

Although FMM has been remarkably consistent with data collected on biological membranes since 1925, it was inevitable that the original model could not explain aspects of membrane dynamics. It had become obvious that the original model needed to be modified or augmented to reflect the emergence of new finds on extracellular and intracellular mechanisms that can affect the lateral distributions and movements of plasma membrane components [54, 55]. Thus, Nicolson elaborated the basic model in 1976 (Fig. 2.7) [55]. It added the interactions of extracellular matrix and membrane-associated cytoskeletal components with cell membranes and their potential influence on the mobility and distribution of transmembrane glycoproteins. Moreover, Nicolson pointed out the possibility that less mobile lipid–protein or lipid–lipid domains might exist in membranes as frozen or semi-frozen islands of less mobile lipids in a sea of fluid phospholipids. The revision of the FMM produced a few years after the original model was less homogeneous and contained additional information not included in the original model. However, this revised model retained all of the basic elements at the nanoscale level, and strictly speaking, it did not propose a new concept of membrane structure.

The FMM successfully explains some characteristics of cell membranes, such as fluidity and protein mosaicism. However, techniques at that time had their limits, for example, immuno-EM only detected the separated proteins that labeled with antibodies in the membrane. Therefore, membrane domains and cell

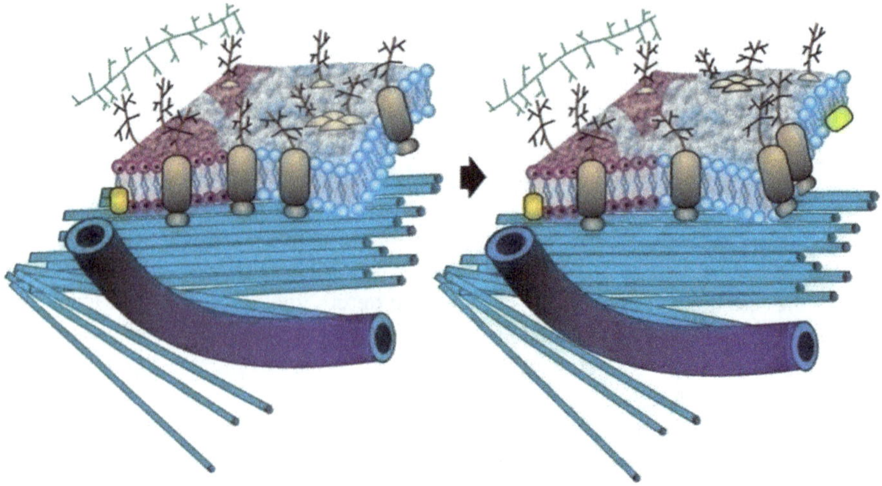

Fig. 2.7 Modified schematic version of the fluid mosaic membrane model of biological membrane structure, as proposed in 1976. In this version, different snapshots in time are represented by the two panels. Some integral membrane glycoproteins are relatively free to diffuse laterally in the membrane plane within a fluid lipid region or domain, whereas others are "anchored" or relatively impeded by a cytoskeletal assemblage or an ordered or solid lipid phase. Reprinted from Ref. [40], Copyright 2014, with permission from Elsevier

membrane-associated structures were not yet discovered, and some membrane functions, such as multiple protein-associated signal transduction and membrane endocytosis, were not explained clearly as well. Nonetheless, the FMM has proven its usefulness in describing basic cell membrane structure for over 40 years, and it remains the most explanatory hypothesis to understand biological membranes.

2.2.5 *Lipid Raft Model*

Approximately at the same time when the FMM was proposed, some lipid biophysicists reported experimental evidence that lipids could laterally segregate in membranes under certain temperature and compositional conditions and could form distinct lipid domains with particular structural characteristics [56–58]. Marcelja in 1976 [59] and Sackmann in 1984 [60] anticipated the possible role of different membrane regions induced by lipid–protein interactions as a physical basis for membrane-mediated processes. This discussion was also addressed by various researchers focused in the membrane biophysics research area [61–63]. Although novel experimental data and discussions were presented about the structure and dynamics of biological membranes, this view was largely ignored by most of life scientists. Until substantial evidence regarding protein clusters from biophysical studies emerged, such as centrifugation, detergent extraction, and cholesterol depletion by methyl-beta-cyclodextrin (MβCD) [64], Simons and Ikonen in 1997 postulated particular functional aspects of specialized domains called lipid rafts in biological membranes [65]. The lipid raft hypothesis proposes that the lipid bilayer is not a structurally passive solvent, but that the preferential association between sphingolipids, sterols, and specific proteins bestows cell membranes with lateral segregation potential [66] (Fig. 2.8).

The formation of lipid rafts is a dynamic and reversible process that confers functional signaling properties to cell membranes. As mentioned, lipid rafts are characterized by enrichments of cholesterol and sphingolipids, which are held together by hydrogen bonds, charge pairing, and hydrophobic and van der Waals forces [67–69]. Their constituents can quickly exchange with bulk membrane lipids as well as with lipids in other rafts [70]. Lipid rafts are generally nanosized domains (<300 nm diameter, most ~10–200 nm) that can contain integral and peripheral membrane proteins. These segregated domains are presumed to be behind a variety of membrane processes, such as formation of proteins clusters, signal transduction, endocytosis, cell polarization, and motility [2, 71–73].

The lipid raft model has long suffered controversy since there is little direct quantitative experimental evidence that such domains exist. The traditional method to study lipid rafts and their association with certain membrane proteins is based on the observation of the detergent-resistant membranes (DRMs), composed of mainly sphingomyelin, other saturated phospholipids, cholesterol, and GPI-anchored proteins [64, 71, 74]. However, this technique was severely criticized, because it utilized detergents that might induce phase separation and/or affect the partitioning

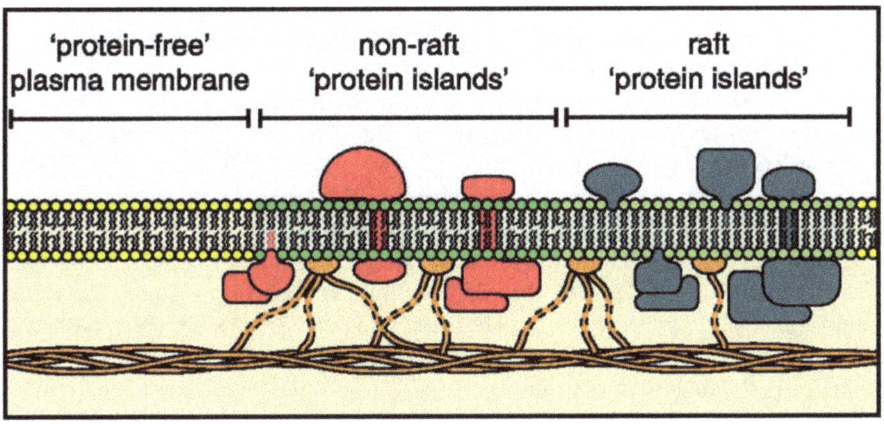

Fig. 2.8 Lipid raft domains in cell membranes. In this model, all membrane-associated proteins are clustered in protein islands (green lipids) that are surrounded by a sea of protein-free membrane (yellow lipids). The islands can be subdivided into raft and non-raft islands, which is also illustrated by their lipid composition (bright-green and dark-green lipids, respectively) and protein contents (gray and red proteins, respectively). The lipid raft domains are connected to the cytoskeleton (orange), most likely by actin. Reprinted with permission from Ref. [66]. Copyright (2006) National Academy of Sciences, U.S.A

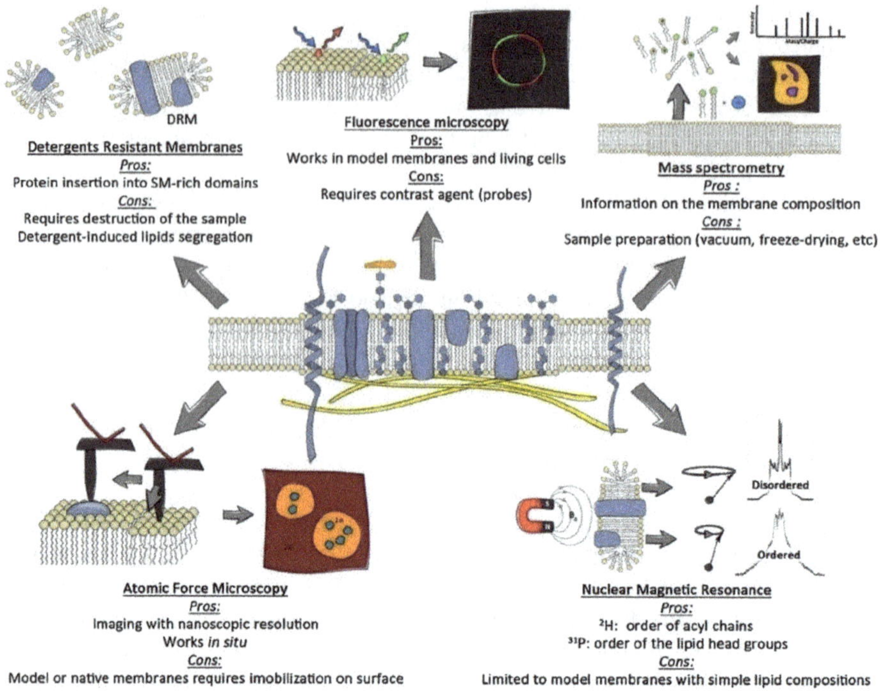

Fig. 2.9 Strengths and weaknesses of the most common methods for characterization of lipid rafts. Reprinted from Ref. [88], Copyright 2014, with permission from Elsevier

of membrane proteins to a given phase [75–77]. Therefore, this disruptive measure tells us little about native membrane organization.

Since then, a variety of other techniques to study the phase behavior of model or native membranes in situ have been developed (Fig. 2.9). Conventional fluorescence microscopy has consistently failed to reveal large-scale, laterally segregated structures enriched in a major raft component, GPI-anchored proteins [72, 73]. One reason is its limited resolution, which is ~300 nm in lateral direction. Hence, the much smaller lipid rafts and/or extremely dynamic or simply compositional fluctuations could not be detected. In addition, light microscopy requires the use of contrast agents, fluorescent probes, because biological systems are poor in intrinsic fluorescent species. Probably, the most direct method to study lipid rafts is based on monitoring chemical composition of the cell membranes with mass spectrometry [78–80]. However, this method works under ultra-high vacuum and with freeze-dried samples. Atomic force microscopy (AFM) provides a nanoscopic resolution for imaging of membranes in aqueous phase [72, 81–84] and can clearly distinguish domains of liquid ordered phase due to their characteristic membrane width, being thicker by ~0.5 nm as compared to the liquid disordered phase. Nevertheless, AFM technique works well only in model or isolated native membranes immobilized on surfaces. Nuclear magnetic resonance provides direct information on the order of the lipid head groups and fatty acid chains, but it is limited to model membranes with simple lipid compositions [85–87].

The recent development of super-resolution imaging further confirmed the existence of lipid rafts [89, 90]. One of the studies utilized stimulated emission depletion microscopy (STED)-fluorescence correlation spectroscopy as a main experimental technique [89]. It showed the presence of nanoscopic domains of sizes around 20 nm where plasma membrane proteins dwelled in periods of 10–20 ms. The study also demonstrated that these special domains owe in part their existence to sphingolipids in the cholesterol-enriched rafts.

The proposal of the lipid raft model has ascertained the structural and functional roles of cell membranes in signal transduction, protein-mediated endocytosis, and viral infection. However, membrane rafts remain one of the most controversial issues in biophysics, which is in part because the methods for their detection are still not perfect enough. In recent years, emerging super-resolution fluorescence microscopy, correlative AFM, and fluorescence microscopy have provided an unprecedented opportunity to study the cell membranes at the single-molecule level. We believe that the composition of lipid rafts and the interactions between multiple rafts will be revealed in the near future.

2.2.6 Modified Fluid Mosaic Model

More than forty years have passed since the fluid mosaic model proposed. During this period, myriads of experimental data have been produced that relate to the structure and dynamics of biomembranes. Although basic membrane structure of the FMM has been severely tested and found worthy, the FMM requires to be revised and modified, which would add the data published in the last forty years and some of the misconceptions inherent in the models of 1970s [39, 91].

In 2005, Engelman proposed an update of the Singer–Nicolson model [92]. He put forward the idea that "membranes are more mosaic than fluid." Transmembrane proteins are so frequent in membranes that hardly any lipid molecule in the bilayer is left unperturbed (Fig. 2.10). Apart from the high density of transbilayer proteins in the membrane, he also pointed out three other features that are now considered to occur in most cell membranes. First, integral proteins exist voluminous extramembrane domains. This conclusion has been proved by the increasing data of the three-dimensional structures of many membrane proteins [93]. Second, contacts between proteins are more frequent. This phenomenon is observed in multiple events, for example, G-protein-mediated signaling, in which a receptor protein will physically interact with a G-protein that, in turn, will transiently bind and activate a cyclic ATPase, thus triggering a cellular response to the signal [94]. Third, the thickness of the lipid bilayer is uneven. Because the transmembrane portions of intrinsic membrane proteins exhibit a rough surface, and the whole domains have a noncritical size with respect to the lipids, membrane lipids are perturbed by the

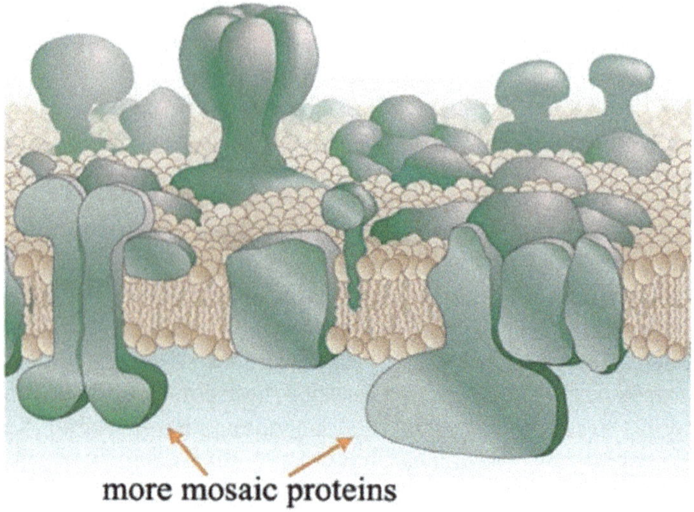

Fig. 2.10 An amended and updated version of the fluid mosaic model with more mosaic proteins. Reprinted by permission from Macmillan Publishers Ltd: Ref. [92], copyright 2005

proteins. The lateral diffusion of lipids is hindered, lipid alkyl chains are disordered, and the bilayer thickness is irregular [95–97].

Engelman's updated version of the FMM focuses more on the mosaic nature of transmembrane proteins. Without doubt, lipids have also been the key elements in the membrane. In 2008, Escribá et al. [98] published a review that shows how membrane lipids, and the structures they form, participate, and regulate numerous important cellular activities. In the paper, they discussed the influence of membrane asymmetry on the membrane lipid organization. The asymmetry of biological membranes includes lateral and cross-sectional asymmetry. The cross-sectional asymmetry reflects the lipid composition of each leaflet (Fig. 2.11). Membrane heterogeneity is further achieved by lateral asymmetry in which membrane regions (basal, lateral, apical), and different specialized membrane regions or microdomains (lipid rafts, caveolae, coated pits, synaptosomes, etc.) [99] extend the mosaic nature of membranes. The formation of these domains in part results from the nonideal mixing of lipids in membranes and in some cases, it is enhanced by the participation of certain cytoskeletal structures that underpin the lipid bilayer and that restrict the traffic of proteins and lipids.

With the development of various techniques, our understanding of the structure and function of the cell membranes is more thorough and comprehensive. In 2014, Nicolson, one of the founders of the FFM, has also presented a modified FMM that takes into account many of the numerous contributions that have been made since the 1970s (Fig. 2.12) [40]. This model incorporates recent information on both membrane proteins and lipids, containing membrane domains, lipid rafts, and

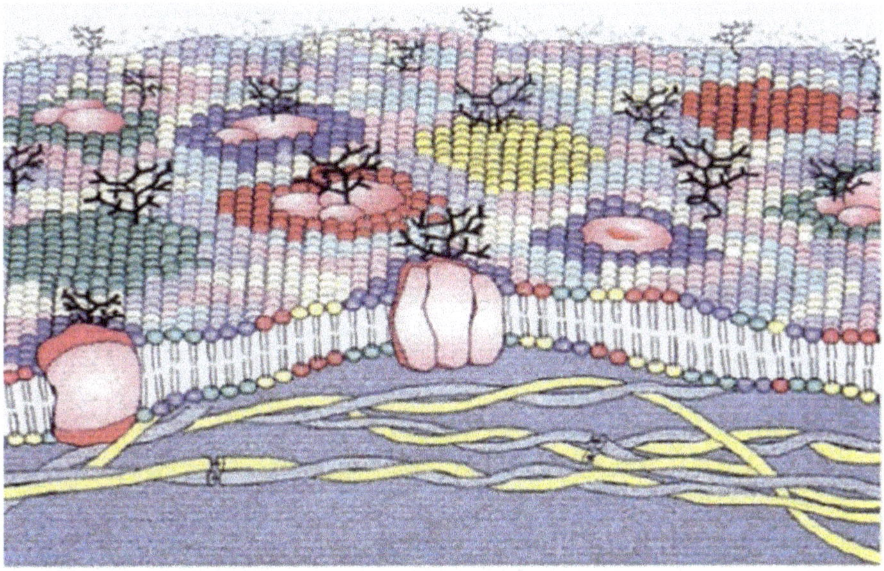

Fig. 2.11 Schematic illustration of a modification of the fluid mosaic membrane model. Different lipids are indicated in various colors forming specialized domains around integral membrane proteins and glycoproteins as well as being asymmetrically distributed across the membrane. Reproduced from Ref. [98] by permission of John Wiley & Sons Ltd

cytoskeletal fencing that were unknown in the 1970s. In this figure, the membrane has very crowded structure, with integral membrane proteins separated by only a few layers of phospholipid molecules. Although the dynamic properties of the membrane cannot be presented in a static medium, the author incorporated most of the important information into the model. In the figure, different types of interactions occur with integral membrane proteins and glycoproteins, membrane lipids, and membrane-associated cytoskeletal systems, and extracellular matrix components. These nonrandom interactions are important in controlling the mobilities, distributions, and aggregation states of membrane components.

In a word, the straightforward application of the Singer–Nicolson model as a frame of events is impossible without introducing new concepts. New concepts have the following attributes emphasizing the mosaic and dynamic properties of the cell membrane: (1) lipid domain structure, (2) cytoskeletal or other cytosolic interactions, or (3) homo- and hetero-associations with other integral proteins. Diffusion, intermolecular forces, and ever-changing membrane potential, and extracellular influences can dynamically generate and destroy supermolecular structures. Hence, the cell membrane model will undoubtedly become more and more complex, just as the original concepts of membrane structure have evolved into more and more complex structures.

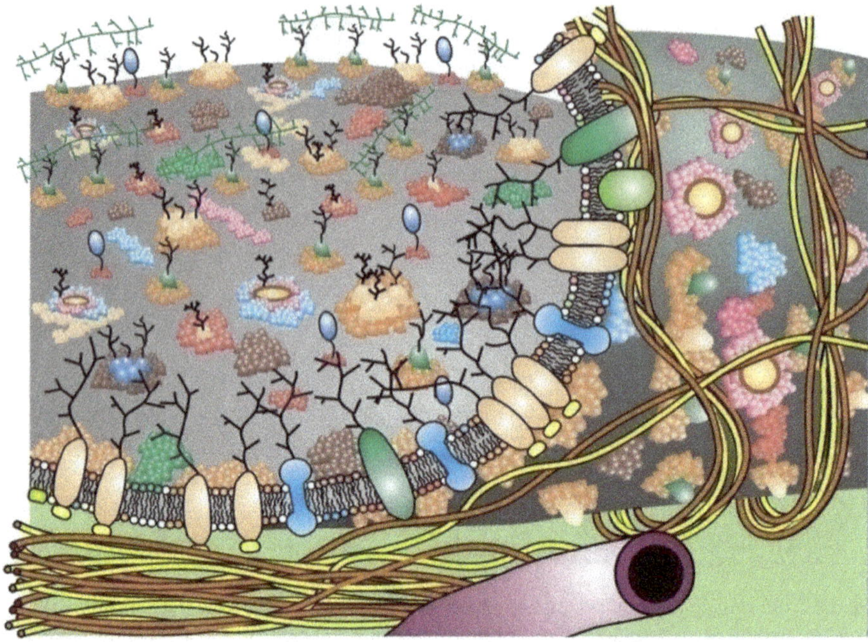

Fig. 2.12 An updated fluid mosaic membrane model that contains information on membrane domain structures and membrane-associated cytoskeletal and extracellular structures. Different integral proteins, glycoproteins, lipids, and oligosaccharides are represented by different colors, and where the membrane has been peeled-up to view the inner membrane surface cytoskeletal fencing. Reprinted from Ref. [40], Copyright 2014, with permission from Elsevier

2.3 Conclusions

In light of the long history of cell membrane studies, we must realize that the membrane structure is far from completely understood. Biological details are more complicated than the resolving power of a simple model, which describes generalized, uniform behavior of molecules in the membrane. Moreover, we should know that one model cannot fit every cellular membrane under all conditions. For example, the fluid mosaic model has been cited as the most successful general model of biological membranes; however, it does not provide adequate explanations for every cell membrane structure, such as lipid rafts. Therefore, the cell membrane structure needs to be refined and modified continually and considered on each specific case, such as different cell types and external environments.

Another fact of note is that every achievement made in the field of the cell membrane depends on the development of new techniques and experimental approaches. From traditional methods such as electron microscopy, X-ray crystallograph, fluorescence microscopy, nuclear magnetic resonance spectroscopy, and mass spectroscopy, to modern methods like atomic force microscopy (AFM), single molecular tracking, and super-resolution fluorescence microscopy (SRFM), all the techniques have provided and will provide the significant information of the membrane structure. Not only that, recent developed combined technologies, have also integrated the strengths of two kinds of instruments, which will offer complemented information and facilitate solving the problem from multiple angles. For example, the combined AFM and SRFM can directly visualize the whole cell membrane in situ (AFM) and have the potent ability to identify the local architecture of cell membranes (SRFM). Based on these new techniques, Wang et al. have proposed a semi-mosaic model for the red blood membrane and a protein layer–lipid–protein island (PLLPI) model for the nucleated cell membrane, which will be introduced in Chap. 3.

To thoroughly study the cell membrane, not only the techniques require faster development, questions regarding the membrane structure that were not involved before should also be focused on, including the accurate location of membrane proteins in the membrane, the membrane protein relationships, and the mechanism that underlies the formation of the protein pattern. These open questions will continue to intrigue investigators for some time.

Acknowledgements This work was supported by the National Key R&D Program of China (No. 2017YFA0505300), the National Natural Science Foundation of China (No. 21525314, 21703231, 21721003).

References

1. Raposo G, Stoorvogel W (2013) Extracellular vesicles: exosomes, microvesicles, and friends. J Cell Biol 200(4):373–383
2. Simons K, Toomre D (2000) Lipid rafts and signal transduction. Nat Rev Mol Cell Biol 1 (1):31–39
3. Glancy B, Balaban RS (2012) Role of Mitochondrial Ca^{2+} in the regulation of cellular energetics. Biochemistry 51(14):2959–2973
4. Sacchettini JC, Baum LG, Brewer CF (2001) Multivalent protein-carbohydrate interactions. A new paradigm for supermolecular assembly and signal transduction. Biochemistry 40 (10):3009–3015
5. Vandenabeele P, Galluzzi L, Vanden Berghe T et al (2010) Molecular mechanisms of necroptosis: an ordered cellular explosion. Nat Rev Mol Cell Biol 11(10):700–714
6. Szostak JW, Bartel DP, Luisi PL (2001) Synthesizing life. Nature 409(6818):387–390
7. Dynarowicz-Latka P, Dhanabalan A, Oliveira ON (2001) Modern physicochemical research on Langmuir monolayers. Adv Coll Interface Sci 91(2):221–293
8. Langmuir I (1917) The constitution and fundamental properties of solids and liquids. II. liquids. J Am Chem Soc 39:1848–1906
9. Pockels A (1894) On the spreading of oil upon water. Nature 50:223–224
10. Rayleigh L (1899) The theory of anomalous dispersion. Phil Mag 48:151
11. Lombard J (2014) Once upon a time the cell membranes: 175 years of cell boundary research. Biol Direct 9:32
12. Gorter E, Grendel F (1925) On bimolecular layers of lipoids on the chromocytes of the blood. J Exp Med 41(4):439–443

13. Bar RS, Deamer DW, Cornwell DG (1966) Surface area of human erythrocyte lipids—reinvestigation of experiments on plasma membrane. Science 153(3739):1010–1012
14. Fricke H (1925) The electric capacity of suspensions with special reference to blood. J Gen Physiol 9(2):137–152
15. Fricke H, Curtis HJ (1934) Electric impedance of suspensions of yeast cells. Nature 134: 102–103
16. Danielli JF (1935) The thickness of the wall of the red blood corpuscle. J Gen Physiol 19 (1):19–22
17. Waugh DF, Schmitt FO (1940) Investigations of the thickness and ultrastructure of cellular membranes by the analytical leptoscope. Cold Spring Harb Symp Quant Biol 8:233–241
18. Davson H (1962) Growth of concept of paucimolecular membrane. Circulation 26(5): 1022–1037
19. Robertson JD (1957) New observations on the ultrastructure of the membranes of frog peripheral nerve fibers. J Biophys Biochem Cytol 3(6):1043–1048
20. Cole KS (1932) Surface forces of the Arbacia egg. J Cell Comp Physiol 1(1):1–9
21. Danielli JF, Harvey EN (1935) The tension at the surface of mackerel egg oil, with remarks on the nature of the cell surface. J Cell Comp Physiol 5(4):483–494
22. Halliburton WD, Friend WM (1889) The stromata of the red corpuscles. J Physiol 10(6): 532–549
23. Jorpes E (1932) The protein component of the erythrocyte membrane or stroma. Biochem J 26:1488–1503
24. Danielli JF, Davson H (1935) A contribution to the theory of permeability of thin films. J Cell Comp Physiol 5(4):495–508
25. Schmitt FO, Clark GL, Mrgudich JN (1934) X-ray diffraction studies on nerve. Science 80:567–568
26. Danielli JF (1937) The relations between surface pH, ion concentrations and interfacial tension. Proc R Soc Ser B-Biol Sci 122(827):155–174
27. Sjostrand FS, Andersson-Cedergren E, Dewey MM (1958) The ultrastructure of the intercalated discs of frog, mouse and guinea pig cardiac muscle. J Ultrastruct Res 1(3): 271–287
28. Robertson JD (1959) The ultrastructure of cell membranes and their derivatives. Biochem Soc Symp 16:3–43
29. Robertson JD (1960) The molecular structure and contact relationships of cell membranes. Prog Biophys Mol Biol 10:343–418
30. Edidin M (2003) Timeline—Lipids on the frontier: a century of cell-membrane bilayers. Nat Rev Mol Cell Biol 4(5):414–418
31. Branton D, Park RB (1968) Papers on biological membrane structure. Bioscience. Little, Brown and Company, Boston
32. Korn ED (1966) Structure of biological membranes. Science 153(3743):1491–1498
33. Robinow CF (1962) On plasma membrane of some bacteria and fungi. Circulation 26 (5):1092–1104
34. Benson AA (1966) On orientation of lipids in chloroplast and cell membranes. J Am Oil Chem Soc 43(5):265–270
35. Green DE, Perdue JF (1966) Membranes as expressions of repeating units. Proc Natl Acad Sci USA 55(5):1295–1302
36. Stoecken W, Engelman DM (1969) Current models for structure of biological membranes. J Cell Biol 42(3):613–646
37. Hechter O (1965) Role of water structure in molecular organization of cell membranes. Fed Proc 24(2S15):S91–102
38. Morange M (2013) What history tells us XXX. The emergence of the fluid mosaic model of membranes. J Biosci 38(1):3–7
39. Singer SJ, Nicolson GL (1972) The fluid mosaic model of the structure of cell membranes. Science 175(4023):720–731

40. Nicolson GL (2014) The fluid-mosaic model of membrane structure: still relevant to understanding the structure, function and dynamics of biological membranes after more than 40 years. Biochim et Biophys Acta-Biomembr 1838(6):1451–1466
41. Mitchell P (1957) General theory of membrane transport from studies of bacteria. Nature 180 (4577):134–136
42. Branton D (1966) Fracture faces of frozen membranes. Proc Natl Acad Sci USA 55(5): 1048–1056
43. Bangham AD, Standish MM, Watkins JC (1965) Diffusion of univalent ions across lamellae of swollen phospholipids. J Mol Biol 13(1):238–252
44. Henderson R, Unwin PNT (1975) Three-dimensional model of purple membrane obtained by electron-microscopy. Nature 257(5521):28–32
45. Neher E, Sakmann B (1976) Single-channel currents recorded from membrane of denervated frog muscle-fibers. Nature 260(5554):799–802
46. Neher E (1992) Ion channels for communication between and within cells. Science 256 (5056):498–502
47. Benga G, Popescu O, Pop VI et al (1986) Para-(chloromercuri) benzenesulfonate binding by membrane-proteins and the inhibition of water transport in human-erythrocytes. Biochemistry 25(7):1535–1538
48. Preston GM, Carroll TP, Guggino WB et al (1992) Appearance of water channels in Xenopus oocytes expressing red-cell CHIP28 protein. Science 256(5055):385–387
49. Razi Naqvi K, Gonzalez-Rodriguez J, Cherry RJ et al (1973) Spectroscopic technique for studying protein rotation in membranes. Nat: New Biol 245(147):249–251
50. Edidin M (1974) Rotational and translational diffusion in membranes. Ann Rev Biophys Bioeng 3:179–201
51. Rothman JE, Lenard J (1977) Membrane asymmetry. Science 195(4280):743–753
52. Daleke DL (2003) Regulation of transbilayer plasma membrane phospholipid asymmetry. J Lipid Res 44(2):233–242
53. Zwaal RFA, Comfurius P, Vandeenen LLM (1977) Membrane asymmetry and blood-coagulation. Nature 268(5618):358–360
54. Bretscher MS (1973) Membrane structure: some general principles. Science 181(4100): 622–829
55. Nicolson GL (1976) Transmembrane control of receptors on normal and tumor-cells.1. Cytoplasmic influence over cell-surface components. Biochem Biophys Acta 457(1):57–108
56. Gebhardt C, Gruler H, Sackmann E (1977) Domain-structure and local curvature in lipid bilayers and biological-membranes. Zeitschrift Fur Naturforschung C-a J Biosci 32(7–8): 581–596
57. Shimshick EJ, McConnell HM (1973) Lateral phase separation in phospholipid membranes. Biochemistry 12(12):2351–2360
58. Simons K, Van Meer G (1988) Lipid sorting in epithelial-cells. Biochemistry 27(17): 6197–6202
59. Marcelja S (1976) Lipid-mediated protein interaction in membranes. Biochem Biophys Acta 455(1):1–7
60. Sackmann E (1984) Biological membranes. Physical basis for trigger processes and membrane structures. Academic Press, London
61. Opdenkamp JAF (1979) Lipid asymmetry in membranes. Annu Rev Biochem 48:47–71
62. Thompson TE, Tillack TW (1985) Organization of glycosphingolipids in bilayers and plasma-membranes of mammalian-cells. Ann Rev Biophys Biophys Chem 14:361–386
63. Vaz WLC, Almeida PFF (1993) Phase topology and percolation in multiphase lipid bilayers—is the biological membrane a domain mosaic. Curr Opin Struct Biol 3(4):482–488
64. Brown DA, Rose JK (1992) Sorting of GPI-anchored proteins to glycolipid-enriched membrane subdomains during transport to the apical cell-surface. Cell 68(3):533–544
65. Simons K, Ikonen E (1997) Functional rafts in cell membranes. Nature 387(6633):569–572

66. Lillemeier BF, Pfeiffer JR, Surviladze Z et al (2006) Plasma membrane-associated proteins are clustered into islands attached to the cytoskeleton. Proc Natl Acad Sci USA 103 (50):18992–18997
67. Ramstedt B, Slotte JP (2006) Sphingolipids and the formation of sterol-enriched ordered membrane domains. Biochim Et Biophys Acta-Biomembr 1758(12):1945–1956
68. van Meer G, Voelker DR, Feigenson GW (2008) Membrane lipids: where they are and how they behave. Nat Rev Mol Cell Biol 9(2):112–124
69. Simons K, Gerl MJ (2010) Revitalizing membrane rafts: new tools and insights. Nat Rev Mol Cell Biol 11(10):688–699
70. Quinn PJ, Wolf C (2009) The liquid-ordered phase in membranes. Biochim Biophys Acta-Biomembr 1788(1):33–46
71. Brown DA, London E (2000) Structure and function of sphingolipid- and cholesterol-rich membrane rafts. J Biol Chem 275(23):17221–17224
72. Jacobson K, Mouritsen OG, Anderson RGW (2007) Lipid rafts: at a crossroad between cell biology and physics. Nat Cell Biol 9(1):7–14
73. Lingwood D, Simons K (2010) Lipid rafts as a membrane-organizing principle. Science 327 (5961):46–50
74. Lingwood D, Simons K (2007) Detergent resistance as a tool in membrane research. Nat Protoc 2(9):2159–2165
75. Heerklotz H (2002) Triton promotes domain formation in lipid raft mixtures. Biophys J 83 (5):2693–2701
76. Lichtenberg D, Goni FM, Heerklotz H (2005) Detergent-resistant membranes should not be identified with membrane rafts. Trends Biochem Sci 30(8):430–436
77. Sot J, Bagatolli LA, Goni FM et al (2006) Detergent-resistant, ceramide-enriched domains in sphingomyelin/ceramide bilayers. Biophys J 90(3):903–914
78. Boxer SG, Kraft ML, Weber PK (2009) Advances in imaging secondary ion mass spectrometry for biological samples. Ann Rev Biophys 38:53–74. doi:10.1146/annurev.biophys.050708.133634
79. Kraft ML, Weber PK, Longo ML et al (2006) Phase separation of lipid membranes analyzed with high-resolution secondary ion mass spectrometry. Science 313(5795):1948–1951
80. Lozano MM, Liu Z, Sunnick E et al (2013) Colocalization of the ganglioside G(M1) and cholesterol detected by secondary ion mass spectrometry. J Am Chem Soc 135(15): 5620–5630
81. Chiantia S, Ries J, Kahya N et al (2006) Combined AFM and two-focus SFCS study of raft-exhibiting model membranes. ChemPhysChem 7(11):2409–2418
82. Giocondi MC, Vie V, Lesniewska E et al (2000) In situ imaging of detergent-resistant membranes by atomic force microscopy. J Struct Biol 131(1):38–43
83. Goksu EI, Vanegas JM, Blanchette CD et al (2009) AFM for structure and dynamics of biomembranes. Biochim Biophys Acta-Biomembr 1788(1):254–266
84. Shaw JE, Epand RF, Epand RM et al (2006) Correlated fluorescence-atomic force microscopy of membrane domains: structure of fluorescence probes determines lipid localization. Biophys J 90(6):2170–2178
85. Filippov A, Oradd G, Lindblom G (2003) The effect of cholesterol on the lateral diffusion of phospholipids in oriented bilayers. Biophys J 84(5):3079–3086
86. Guo W, Kurze V, Huber T et al (2002) A solid-state NMR study of phospholipid-cholesterol interactions: sphingomyelin-cholesterol binary systems. Biophys J 83(3):1465–1478
87. Soni SP, LoCascio DS, Liu Y et al (2008) Docosahexaenoic acid enhances segregation of lipids between raft and nonraft domains: H-2-NMR study. Biophys J 95(1):203–214
88. Klymchenko AS, Kreder R (2014) Fluorescent probes for lipid rafts: from model membranes to living cells. Chem Biol 21(1):97–113
89. Eggeling C, Ringemann C, Medda R et al (2009) Direct observation of the nanoscale dynamics of membrane lipids in a living cell. Nature 457(7233):1159–U1121
90. Wang Y, Gao J, Guo X et al (2014) Regulation of EGFR nanocluster formation by ionic protein-lipid interaction. Cell Res 24(8):959–976

91. Nicolson GL (1976) Transmembrane control of the receptors on normal and tumor cells: I. cytoplasmic influence over cell surface components. Biochem Biophys Acta 457(1):57–108
92. Engelman DM (2005) Membranes are more mosaic than fluid. Nature 438(7068):578–580
93. Abrahams JP, Leslie AGW, Lutter R et al (1994) Structure at 2.8 A resolution of F1-ATPase from bovine heart-mitochondria. Nature 370(6491):621–628
94. Golebiewska U, Scarlata S (2010) The effect of membrane domains on the G protein-phospholipase C beta signaling pathway. Crit Rev Biochem Mol Biol 45(2):97–105
95. Cortijo M, Alonso A, Gomezfernandez JC et al (1982) Intrinsic protein-lipid interactions—infrared spectroscopic studies of gramicidin-A, bacteriorhodopsin and Ca^{2+}-ATPase in biomembranes and reconstituted systems. J Mol Biol 157(4):597–618
96. Arrondo JLR, Goni FM (1998) Infrared studies of protein-induced perturbation of lipids in lipoproteins and membranes. Chem Phys Lipid 96(1–2):53–68
97. Holt A, Killian JA (2010) Orientation and dynamics of transmembrane peptides: the power of simple models. Eur Biophys J Biophys Lett 39(4):609–621
98. Escriba PV, Gonzalez-Ros JM, Goni FM et al (2008) Membranes: a meeting point for lipids, proteins and therapies. J Cell Mol Med 12(3):829–875
99. Escriba PV (2006) Membrane-lipid therapy: a new approach in molecular medicine. Trends in Mol Med 12(1):34–43

Chapter 3
Imaging Membranes by High-Resolution Atomic Force Microscopy

Mingjun Cai, Jing Gao and Hongda Wang

3.1 Introduction

Cell membranes possess crucial functions in a living cell, such as compartmentalizing the cell from the environment, cell signaling, and solute transporting. The structure of cell membranes at the molecular level is a fundamental question in cell biology. The study of their structure is vital for various applications, including drug screening, cancer treatment, signal transduction control. Hypotheses that include the liquid mosaic model, lipid raft model, and protein domain model have been constructed in the last four decades (see Chap. 2). However, the structure of cell membranes is still a controversial topic because these models are based on indirect evidences or non-native conditions. Atomic force microscope takes advantage of a microfabricated cantilever with a sharp tip to scan the surface of the samples, and the deflection of the cantilever is utilized to record the information of the surface properties. AFM allows imaging of cellular membranes at a spatial resolution of a few nanometers, giving real three-dimensional imaging. One of the greatest advantages of AFM is that it can image the samples in solutions under the physiological conditions, which makes it very important in biological applications. Furthermore, recent advances in AFM technique extend its application to recognize specific molecules in heterogeneous samples at the single-molecule level, which is called topography and recognition imaging (TREC). In this chapter, we summarizes the new progress regarding membrane structure using in situ AFM, which may shed light on the study of the structure and functions of cell membranes.

M. Cai · J. Gao · H. Wang (✉)
State Key Laboratory of Electroanalytical Chemistry, Changchun Institute of Applied Chemistry, Chinese Academy of Sciences, No. 5625, Renmin Street, Changchun, Jilin 130022, People's Republic of China
e-mail: hdwang@ciac.ac.cn

H. Wang and G. Li (eds.), *Membrane Biophysics*,
https://doi.org/10.1007/978-981-10-6823-2_3

3.2 Atomic Force Microscopy and Related Techniques

Atomic force microscopy, which was invented in 1986 by Binnig et al., is an important member of the family of scanning probe microscopy techniques and provides molecular-scale imaging of non-conducting materials [1]. In contrast to optical microscopy and electron microscope (EM), AFM overcomes the wavelength limit to provide images of samples in various environments with three-dimensional topography. The lateral and vertical resolutions of AFM reach 1 and 0.1 nm, respectively. AFM has been used in various areas of science and technology with excellent results [2–4].

3.2.1 Principle of Atomic Force Microscopy

The setup and working principle of AFM is shown in Fig. 3.1a. The basic components include a piezoelectric scanner, flexible cantilever containing a sharp probe, laser, photodiode detector, and feedback electronics [5]. The principle of AFM is simple whereby the movements of a flexible cantilever with a tiny tip detects the weak interactions between atoms on the tip and the sample surface. As the probe approaches to the sample surface, the attraction force between the sample and the tip induces either a deflection of the cantilever probe (contact mode) or an amplitude change (tapping mode). A photodetector detects the shift or amplitude change and converts this signal to an electrical signal read by a computer; as a result, the surface information is recorded. The imaging modes of AFM are divided into the following major categories according to whether the tip contacts the sample: the contact mode AFM and the tapping mode (acoustic AC (AAC) mode) AFM [6]. There is another type of tapping mode AFM that the cantilever is attached with magnetic particle and then the cantilever is driven directly by a magnetic field named Magnetic AC mode (MAC mode) AFM [7]. The better signal-to-noise means that cantilevers with much lower force-constant and much smaller vibrating amplitudes can be used. This decreases damage to the sample and preserves asperities on the probe, contributing to greatly improved resolution in liquid environments [8].

The contact mode was developed firstly, with the highest resolution, and is commonly used. In this mode, the AFM tip contacts the sample surface and is scanned using a piezoelectric scanner (Fig. 3.1b). The feedback electronics adjust the vertical position of the piezoelectric scanner to maintain a constant laser deflection of the cantilever at a predetermined set point value, thereby maintaining constant force on the sample. The adjustment in the vertical position required to maintain a constant deflection is recorded and used to generate a topographical

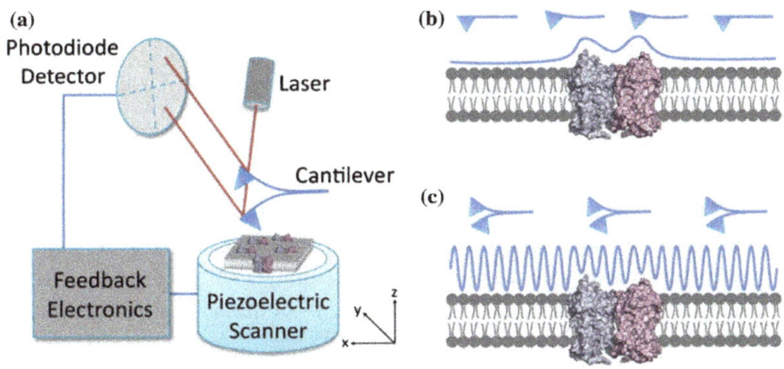

Fig. 3.1 Schematic diagram of atomic force microscopy. **a** Components of an atomic force microscope. **b** Contact mode imaging. **c** Tapping mode or MAC mode imaging. Reprinted from Ref. [5], Copyright 2014, with permission from Elsevier

image of the sample. The contact mode is able to obtain high-resolution images in both air and liquid. However, because of the existence of surface shear stress, this mode may deform soft samples; thus, it is not ideal for imaging unstable and flexible samples.

The tapping mode pioneered by Hansma's group uses a constant driving force (acoustic, magnetic) to vibrate the cantilever at a certain frequency [9]. The oscillation amplitude is monitored, and the feedback electronics adjust the vertical position of the piezoelectric scanner to maintain constant amplitude. The amplitude of the cantilever changes as the height of the sample changes (Fig. 3.1c). The tapping mode (AAC mode) is more suitable for softer or loosely adsorbed samples. The resolution of the tapping mode AFM is similar to the contact mode AFM. The advantage of the tapping mode AFM is the short contact time and the minimum contact force, which effectively prevents damage to the samples during tip scanning. The tapping mode AFM (or MAC mode) in liquid is often used to image live biological samples (such as cells and bacteria) and to monitor the progress of reactions in solution [10].

The comparisons of the two main imaging modes clearly indicate that the tapping mode AFM prevents the shear force-induced destruction of samples that occurs in contact mode. Therefore, with the advantages of high resolution and minimum destruction to the sample, the tapping mode (or MAC mode) is the most widely used AFM mode in biological fields [11–14].

3.2.2 *Molecular Recognition Imaging Based on AFM*

AFM provides incomparable advantages for obtaining topography information regarding the surface of a sample. However, for a complex sample with multiple components, the height measurements of AFM are unable to differentiate between the species (e.g., for a mixture of two proteins, we cannot distinguish them from their shape and height). Molecular recognition imaging (TREC) is a technique for the identification of a particular molecule in a complex [15, 16]. In principle, the AFM tip is functionalized with specific molecules (e.g., antibodies) to recognize the counterpart molecules on the surface (e.g., antigens). High-resolution topography and recognition images are simultaneously obtained. As shown in Fig. 3.2a, an AFM tip tethered to an antibody scans the surface; when there is a specific interaction between the sample and the antibody on the AFM tip, the amplitude of the AFM cantilever will be reduced, and feedback systems adjust the cantilever position (ΔA) to maintain a constant amplitude (Fig. 3.2b, c). This peak change corresponds to the molecular recognition signal from the specific interactions. Molecular recognition imaging has excellent specificity and reproducibility, and it further expands the biological applications of AFM. AFM and TREC have excellent properties for the study of cell membrane structure [17, 18].

3.2.3 *In Situ Imaging Using a Flow Through Liquid Cells*

In situ imaging is the greatest advantage of AFM compared with EM. To maintain biological samples in their native states, in situ imaging of cell membranes is

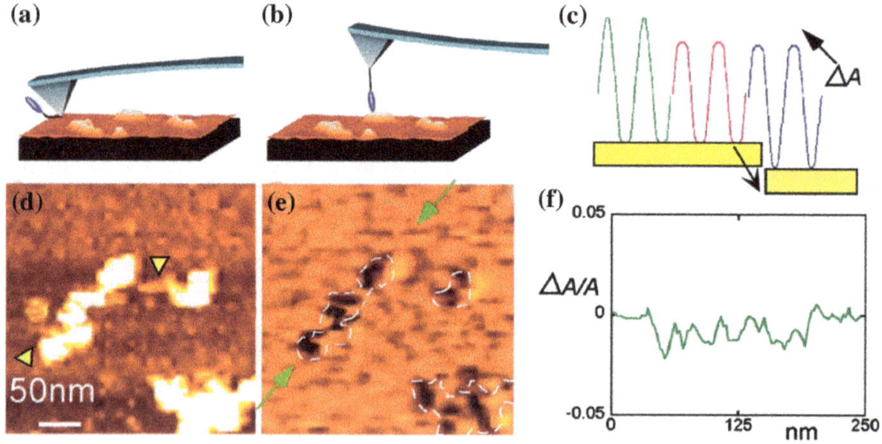

Fig. 3.2 Principle of topography and recognition imaging. An AFM tip tethered to an antibody scans the sample **a**, **b**. Antigen-antibody binding causes a transient reduction in the oscillation amplitude of the tip **c**. The AFM feedback system converts this change to a recognition signal. **d**, **e** A topographic image of MMTV chromatin arrays and the corresponding recognition image. **f** A plot of the peak signal, ΔA, for the portion of the recognition image between the green arrows in **e**. Reprinted with permission from Ref. [15] with permission Copyright (2004) National Academy of Sciences

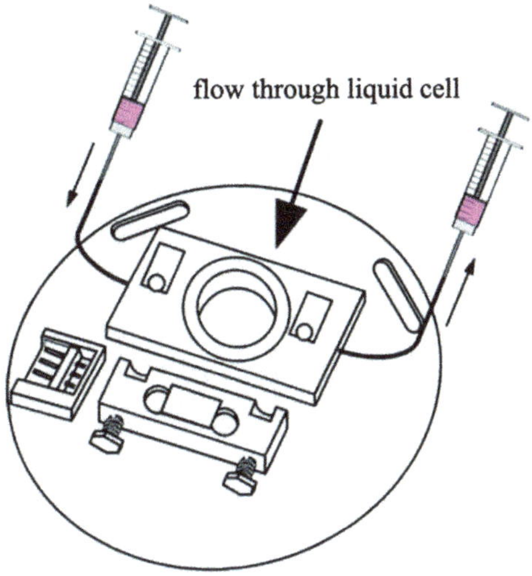

Fig. 3.3 Schematic of flow through liquid cell. Reprinted from Ref. [20], Copyright 2007, with permission from Elsevier

desirable because a dry sample will be deformed and will not represent the physiological features. Environmental control can be realized using a flow through liquid cell [19, 20]. As shown in Fig. 3.3, using a syringe to inject the appropriate solution from one end and removing the same volume solution from the other end can change the buffer conditions (e.g., the salt concentration and pH) during imaging, which is rather important to observe biomolecules at work. It is also important to maintain a constant solution volume and temperature during imaging because a small temperature change may cause drift between the AFM probe and the sample, which thus affects repeated images in the same location. In liquid environments, high-resolution imaging of biological molecules and dynamic changes between individual molecules are important features for biologists. The flow through liquid cell technique allows the real-time study of dynamic behaviors of single-molecule interactions [21–23].

3.3 Imaging the Native Membrane Proteins and Cell Membrane by AFM

Membrane proteins that exist in a lipid bilayer serve many critical roles in the cell including transport solutes, convert energies, and transmit information between the external environment and the inside of the cell. Membrane proteins are highly hydrophobic and are natively embedded in a lipid bilayer. Plasma membranes play important roles in protein structure and function. The study of membrane proteins often involves procedures and manipulations that move the system away from its native condition. Furthermore, membrane proteins must be labeled with fluorescent dyes or staining reagents to facilitate their detection; these processing steps also disturb the native conformation of membrane proteins. AFM is a powerful tool for imaging the membrane protein under physiological conditions. Native properties of membrane proteins can be studied since the conditions, which these proteins are normally exposed to, can be maintained during AFM [24–26].

3.3.1 Imaging the Supramolecular Organization of Membrane Proteins in Physiological Conditions

AFM can directly image the membrane protein at nanometer resolution and at single-molecule level. AFM imagings reveal the packing arrangement of the light receptor rhodopsin (a kind of G protein-coupled receptor) [27]. Rhodopsin was found to be dimeric and to form higher oligomers as shown in Fig. 3.4. The higher-order oligomerization of rhodopsin and other G protein-coupled receptors (GPCRs) has important implications for G protein recognition, binding kinetics, signal amplification, and signal termination. A white ellipse marks a rhodopsin dimer, the building block of the paracrystal. Three different surface types are evident: ① mica, ② lipid, and ③ the cytoplasmic side of the disk membrane.

The oocytes and embryos of the amphibian anuran *Xenopus laevis* (*X. laevis*) are an interesting model for the study of many developmental mechanisms because of

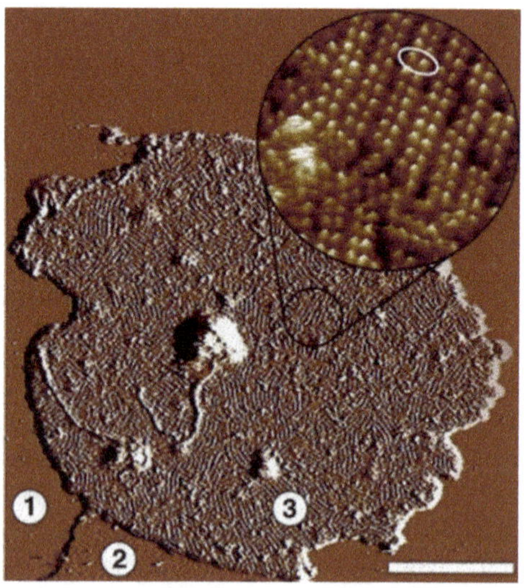

Fig. 3.4 Atomic force microscopy (AFM) of native membranes. **a** Contact mode AFM of native disk membranes isolated from mouse retinal rods. The deflection image reveals paracrystalline arrays of the G protein-coupled receptor rhodopsin. At higher magnification (round inset), rows of rhodopsin dimers are discerned. Scale bar, 200 nm. Reprinted by permission from Macmillan Publishers Ltd: Ref. [27], Copyright 2003

their dimensions and the ease with which they can be manipulated. Therefore, *X. laevis* oocytes represent a widely used biological tool in different fields of life sciences from physiology to embryology. In addition, *X. laevis* oocytes are a widely employed system for the expression and functional study of heterologous membrane proteins [28]. An AFM topography image of South African frog oocyte membranes indicated protein particles with sizes of hundreds of nanometers (Fig. 3.5), which is larger than a single protein, suggesting the membrane proteins form clusters that contain multiple proteins [29]. Functional structures (e.g., clathrin pits) perform the dynamic functions required for membrane transportation. Usukura et al. provided direct evidence of the existence of clathrin pits on the cytoplasmic side of cell membranes using high-resolution AFM imaging (Fig. 3.6) [30]. The same type of membrane protein is randomly distributed in the cytoplasmic side of membrane; this distribution was verified via the use of labeled quantum dots to locate prestin (a membrane protein) in the cytoplasmic side membrane of CHO cells [31]. The pattern of membrane proteins in the cytoplasmic face of membranes is closely associated with membrane activity. Vijayan et al. studied the changes in membrane proteins in response to an acute stress that induced the stress signaling

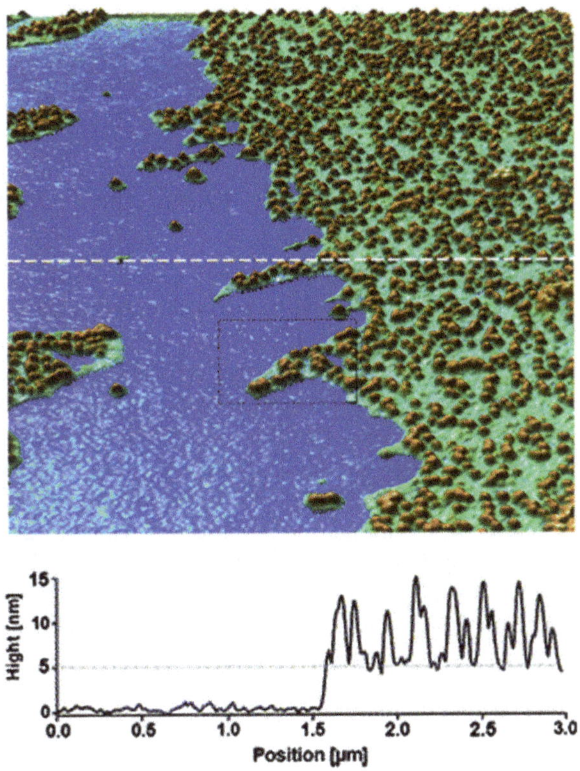

Fig. 3.5 AFM image of South African frog oocyte membranes (cytosolic side). Poly-L-lysine-coated glass (blue), the lipid bilayer membrane (turquoise), and the membrane proteins (brown). The height profile along the broken line is presented at the bottom. Reprinted from Ref. [29], with kind permission from Springer Science Business Media

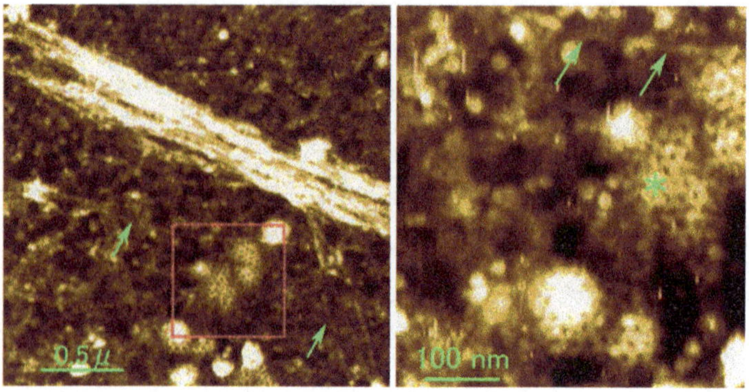

Fig. 3.6 AFM image of clathrin coats and actin filaments at the cytoplasmic side of the plasma membrane. The right shows an enlarged view of the boxed area in the left. Clathrin-coated pits are clearly observed in the boxed area. Arrows indicate actin filaments. Reprinted from Ref. [30] by permission of Oxford University Press

pathways, and the researchers demonstrated that the protein domains became larger after stressor exposure [32]. Using fluorescence imaging, Burns et al. located multiple membrane components (including immunoglobulin E receptors, cholera toxin-aggregated GM1, and clathrin) and suggested that cholesterol-enriched domains (lipid rafts) are responsible for signaling and endocytosis [33].

3.3.2 Observing Single Membrane Proteins at Work

A unique feature of the AFM is its ability to observe single membrane proteins at work. This has been used to study the conformational changes that certain membrane proteins undergo in response to external stimuli, for example, light, pH change, and the presence of ligands. Connexin 26 (Cx 26) gap junctions allow cells to communicate and respond to external signals. Upon cell death, relatively high amounts of Ca^{2+} ions are released into the extracellular space. To impede influx of this second messenger into healthy cells and false initiation of signaling cascades, gap junctions close their pores. Using in situ AFM, Muller et al. studied the fine structure of the human communication channel protein connexin 26 in a dense protein array [34]. As shown in Fig. 3.7, AFM clearly provides images of a hexamer structure of connexin 26. The Ca^{2+}-induced conformational change of the connexin 26 can be visualized at a single-molecule resolution [26].

AFM has become an important tool for structural biologists. It is the only instrument that allows surfaces of cells, supramolecular assemblies, and single molecules to be imaged in the physiological environment at nanometer-scale

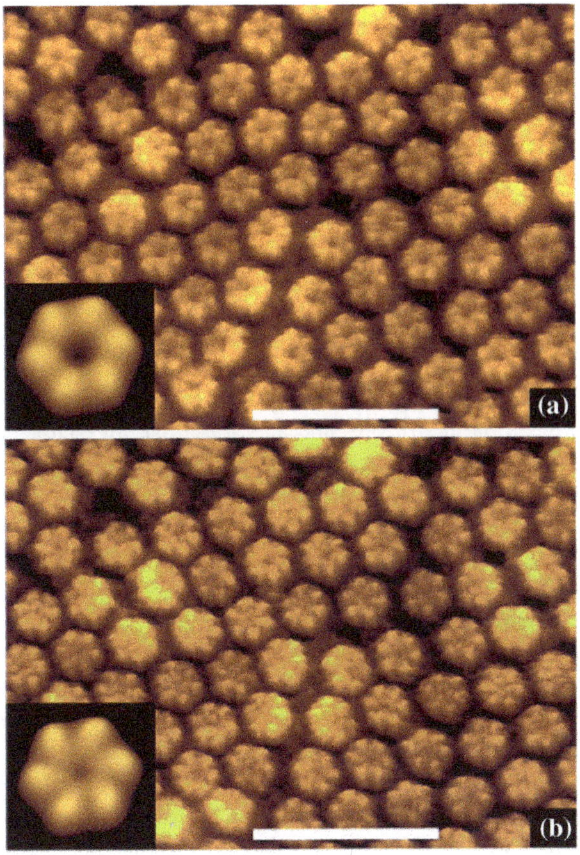

Fig. 3.7 Visualization of calcium-induced conformational changes of the extracellular connexin surface by contact mode atomic force microscopy. Topography of the extracellular surfaces of a native connexin 26 gap junction plaque imaged in buffer solution in the absence (**a**) and presence of 0.5 mM Ca^{2+} (**b**). The diameter of the extracellular connexin pore is reduced significantly in the presence of calcium. Insets in **a** and **b** are correlation averages calculated from the corresponding topographs with a lateral resolution of 12 Å. Scale bars represent 25 nm in **a** and **b**. The frame sizes of the insets are 68 Å **a** and **b**. Reprinted from Ref. [26], Copyright 2012, with permission from Elsevier

resolution. In addition, it makes manipulation of such structures at this scale possible. AFM images of native membranes and membrane proteins at molecular resolution have provided the potential to achieve high-resolution imaging of native cell membrane.

3.3.3 Examination of Functional Microdomains in Red Blood Cell Membranes

Since the proposal of the fluid mosaic model (FMM) of cell membranes in 1972 [35], some phenomena regarding cell membranes have been interpreted appropriately, such as the mosaic proteins and fluidic cell membranes (see Chap. 2). However, this model cannot explain complicated membrane functions [36].

The lipid raft model focuses on functional microdomains enriched with cholesterol, sphingomyelin, and membrane proteins [37, 38]. Many cellular functions are associated with lipid rafts, such as signal transduction, protein-mediated endocytosis, and viral infection [39, 40]. Lipid rafts are dynamic and polymorphous; thus, conventional methods are not suitable for their studies. In general, the size of lipid rafts is considered to be in the range of several hundred nanometers. Traditional methods, such as sucrose density gradient centrifugation, electrophoresis, fluorescence staining, and EM, are likely to damage the structure of lipid rafts. Fluorescence microscopy studies provided solid contributions to the understanding of lipid rafts [41]; in particular, super-resolution imaging demonstrated the dynamic characteristic of lipid rafts [42]. However, Shaw et al. determined that only visualizing membrane domains using a fluorescent probe might induce artifacts. AFM provides features of the absolute topography of lipid rafts. Because of the high resolution, the in situ, and alterable imaging conditions, AFM has high potential to resolve the unanswered questions regarding lipid rafts.

Phase separation according to the type of lipid was confirmed using a model lipid bilayer (supported DOPC/DPPC bilayer), in which an AFM image indicated that lipids enriched with cholesterol tend to form domains resistant to Triton X-100 [43]. Orsini et al. used sucrose density gradient centrifugation to isolate lipid raft domains from Triton X-100 treated cells and AFM to directly confirm that there are lipid rafts associated with specific proteins in membranes [44]. However, the whole cell lysis may induce contamination from the endomembrane system (e.g., Golgi body and endoplasmic reticulum). In situ treatments and observations of cell membranes are highly desirable for these studies.

Lipid rafts exist in most membrane systems, such as erythrocytes, nucleated cells, and plant cells [40, 45]. In red blood cell membranes, detergent-resistant membranes' (DRMs) domains were first confirmed using in situ imaging, which exhibited an irregular shape with a size of several hundred nanometers (Fig. 3.8a–c) [46]. The in situ observation of the specific extraction of cholesterol (a key component in lipid rafts) by methyl-β-cyclodextrin (M-β-CD) has provided convincing results. As shown in Fig. 3.8d–g, in situ AFM demonstrated that M-β-CD gradually eroded the membrane with time. The size of the eroded irregular patches is mainly in the range of 100–300 nm. Unlike the living cell, the cell membrane was electrostatically absorbed on the APTES-mica for AFM imaging, thus the lipid rafts can be stably imaged by AFM for long time. Additional evidence has indicated that lipid rafts exist in nucleated mammalian cell membranes. Chen et al. detected the depletion of lipid rafts from endothelial cells by M-β-CD using AFM and fluorescence microscopy [47]. The lipid raft is the functional unit and is associated with the underlying cytoskeletal network for the control of dynamic functions [48], as confirmed by AFM force measurements [49].

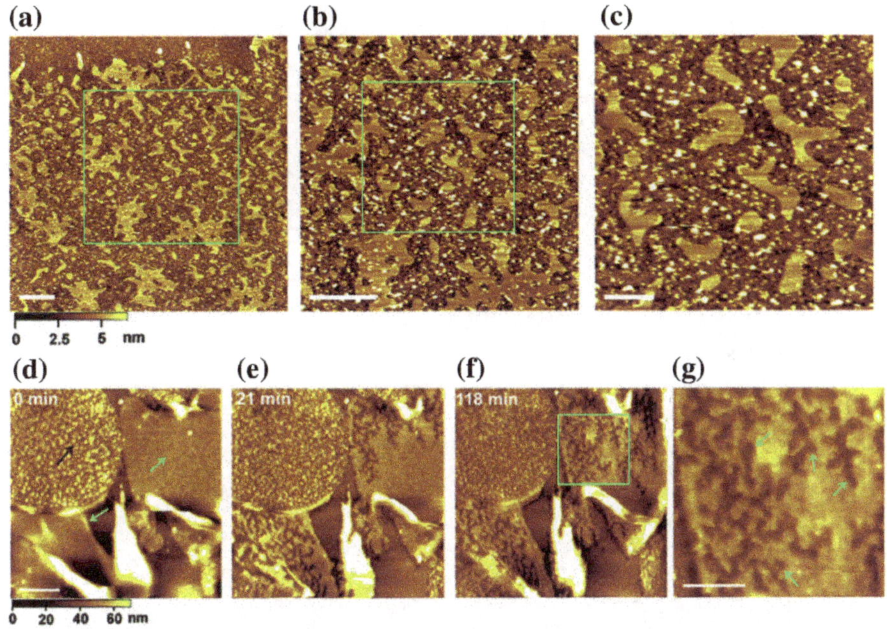

Fig. 3.8 **a–c** DRMs from the ectoplasmic side membrane. **a** AFM image of DRMs. **b**, **c** Serial magnifications of the DRMs in **a**. Scale bars are 500 nm in **a** and **b** and 200 nm in **c**. **d–g** In situ images of cell membranes treated with MβCD. **d–f** A series of images after the addition of MβCD for 0 min (**d**), 21 min (**e**), and 118 min (**f**). The green and blue arrows in **d** point to the ectoplasmic and cytoplasmic side of the cell membranes, respectively. **g** Magnified image of the green square area in **f**. The green arrows indicate the regions eroded by M-*β*-CD. Scale bars are 2 nm in **d** and 1 nm in **g**. Reproduced from Ref. [45] by permission of John Wiley & Sons Ltd

3.3.4 Imaging the Structure of HRBC Membranes by AFM

Human red blood cells (hRBCs) are small and have a distinctive flattened shape. These cells have a relatively simple and regular structure. Unlike the other cells in our body, the red blood cells lack nucleus and endometrium, so the plasma membrane of the red blood cell can serve as a model to study the structure of the cell membrane.

3.3.4.1 Imaging the Cytoplasmic Side of HRBC Membranes

The advantage of AFM is the capability to image the whole cell membrane at high resolution under nearly physiological conditions. The completely cytoplasmic side of membranes of hRBCs is simply prepared using a shearing open method. As shown in Fig. 3.9a, many membrane proteins are located in real cell membranes, and the determination of protein distribution is the major goal in the study of cell

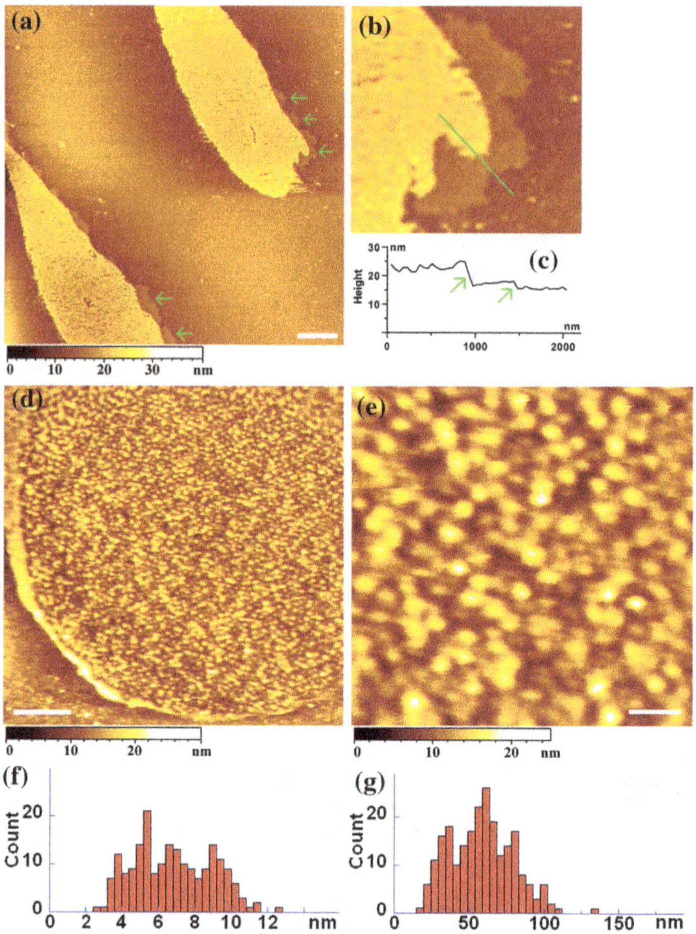

Fig. 3.9 Cytoplasmic side of hRBCs membrane. Arrows point to the free lipid bilayers under the membrane proteins. Scale bar is 3 μm. **b** Magnification of the membrane in **a**. **c** Cross-sectional analysis along the line in **b**. Arrows point to the protein layer and free lipid bilayer, respectively. **d** Image cytoplasmic side of membrane at liquid phase resolution. Scale bar is 1 μm. **e** High-resolution image of cytoplasmic side of membrane. Scale bar is 200 nm. **f** Height distribution of proteins in the cytoplasmic side of membrane. **g** Width distribution of proteins in the cytoplasmic side of membrane. Reprinted from Ref. [52], Copyright 2010, with permission from Elsevier

membrane structure. The height of the cytoplasmic side membrane between the substrate and the proteins is about 10 nm. The average roughness of the cytoplasmic side membrane is 1.9 nm. Arrows point to free lipid bilayers on the edge of the membranes with the height of 2.94 ± 0.36 nm. The magnifying image (Fig. 3.9b) shows the detail how the proteins and lipid bilayer are arranged in the membranes. The protein layer is just above the lipid bilayer. There are two clear steps shown by the cross-sectional analysis (Fig. 3.9c). The arrows point to the protein layer (left) and the free lipid bilayer (right), respectively. The area percentages of free lipid bilayers in cell membranes were from 0 to 10%, and not all the patches showed free lipid bilayers. This result indicates that fewer lipids are free in cell membranes than expected, which is consistent with the reference's statement "Membranes are more mosaic than fluid" [50].

Figure 3.9d shows the median resolution image of the cytoplasmic side cell membranes, wherein the dense particles represent the proteins in the membrane. The proteins display a broad height distribution between 2.5 and 13 nm with the multiple peaks (Fig. 3.9f). These proteins were mostly identified as transmembrane proteins (e.g., $Na^{+}-K^{+}$ ATPase, band 3) by TREC technique [51, 52]. Using the same technique, Schillers et al. identified the cystic fibrosis transmembrane conductance regulator (CFTR, a protein of the adenosine triphosphate-binding cassette transporter superfamily) in the red blood cells' cell membranes [29]. Figure 3.9e shows a higher resolution image, in which the proteins are very close to each other without the free lipid bilayer observed. The proteins show a diameter distribution between 20 and 130 nm (Fig. 3.9g). It is can be assumed that there are protein aggregations in the membrane due to the large particle size [53].

In order to study the relation between proteins and the lipid bilayer, the membrane protein was digested by trypsin. After digestion, the proteins are separated, leaving some high particles of undigested proteins in the membrane (Fig. 3.10a). Figure 3.10b shows the magnification of the image from Fig. 3.10a. Arrows point to small particles in the membrane, which could be transmembrane parts of membrane proteins. Figure 3.10c shows a cross section along the green line on the three small particles in Fig. 3.10b. The height distribution of particles above the membrane is in the range of 2–10 nm with the peak at 3–4 nm as shown in Fig. 3.10d. The average width of small particles is 13.6 ± 2.7 nm. These results

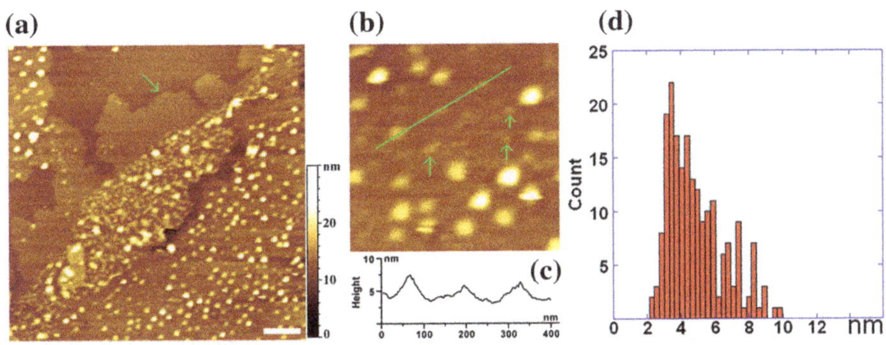

Fig. 3.10 Cytoplasmic side membrane treated with trypsin. Arrow points to the free lipid bilayer. Scale is 300 nm. **b** Magnification of the membrane in Fig. 3.2a. Arrows point to the possible peptides after trypsin digestion. **c** Cross-sectional analysis along the line in Fig. 3.2b. **d** Height distribution of particles in the membrane after trypsin digestion. Reprinted from Ref. [52], Copyright 2010, with permission from Elsevier

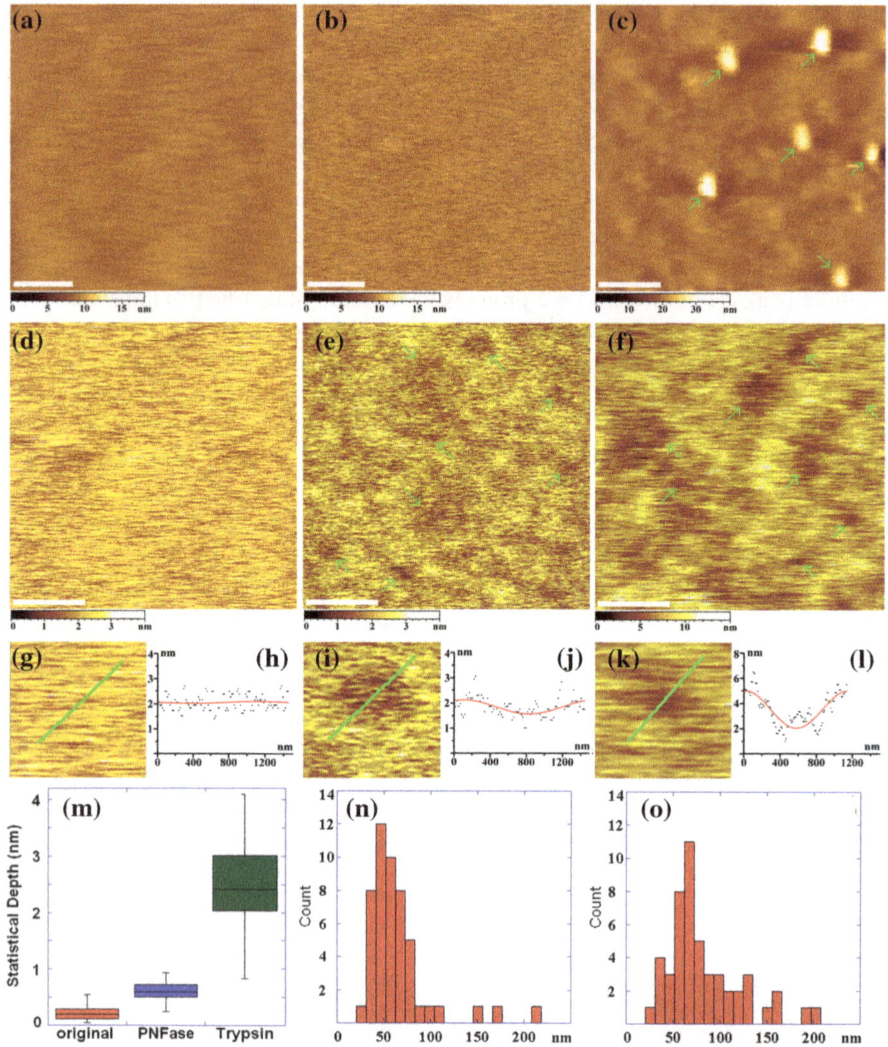

Fig. 3.11 **a** The ectoplasmic side membrane of hRBCs prepared by the approach of hypotonic lysis. Scale bar is 300 nm. **b** The ectoplasmic side membrane of hRBCs prepared by sheared open approach. Scale is 200 nm. **c** Control experiment for the smooth ectoplasmic side membrane. Green arrows indicate blood group B antibodies on the ectoplasmic side membrane. Scale is 300 nm. **d** High-resolution image of the ectoplasmic side membrane. *Z*-axis scale is 4 nm. **e** High-resolution image of the ectoplasmic side membrane treated by PNGase F. Arrows point to the presumed regions of oligosaccharide aggregations in the membrane. *Z*-axis scale is 4 nm. **f** The ectoplasmic side membrane treated by PNGase F and trypsin. Arrows indicate the regions of dug transmembrane proteins. *Z*-axis scale is 10 nm. Scale bar is 150 nm for Fig. 3.3d–f. **g** Magnification of the membrane in Fig. 3.3d. **h** Cross-sectional analysis along the line in Fig. 3.3g. **i** Magnification of the membrane from Fig. 3.3e. **j** Cross-sectional analysis along the line in Fig. 3.3i. **k** Magnification of the membrane in Fig. 3.3f. **l** Cross-sectional analysis along the line in Fig. 3.3k. **m** Statistical depths of these indents in the ectoplasmic side membrane. Red bar means the untreated ectoplasmic side membrane. Blue bar means the ectoplasmic side membrane treated by PNGase. Green bar means the ectoplasmic side membrane treated by PNGase and trypsin (marked as "Trypsin" in the fig.). **n** Distribution of the indent widths after PNGase F treatment. **o** Distribution of the indent widths after PNGase F and trypsin treatment. Reprinted from Ref. [52], Copyright 2010, with permission from Elsevier

further indicate that AFM provides an ability of high resolution to reveal the fine structure of cell membranes.

3.3.4.2 Imaging the Ectoplasmic Side of HRBC Membranes

Although the ectoplasmic side membranes of living cells are crucial for cell functions, its structure at the molecular level and native state is still controversial due to limited approaches. The ectoplasmic side of red blood cell membranes has been considered the same as the cytoplasmic side. In current cell membrane models, the transmembrane proteins are presumed to protrude out of the ectoplasmic side membrane just like they are on the cytoplasmic side membrane, and oligosaccharides cover the proteins and lipids above the lipid bilayer [36]. Recently, AFM provided a complete picture of the ectoplasmic side membranes in situ [53]. The ectoplasmic side is rather smooth; the average roughness of the ectoplasmic side membrane is approximately 0.20 nm, which is much less than the cytoplasmic side membrane (Fig. 3.11a, b). These findings suggest that the membrane protein distribution is much more asymmetrical than previously proposed. To demonstrate the ability that AFM can observe any obvious protrusion on the membrane, the blood group B antibody was adsorbed onto the ectoplasmic side membrane of B-type hRBCs. There is single antibody particle on the cell membranes, which supports that there is no protein protrusion out the membrane (Fig. 3.11c).

The fine structure of the ectoplasmic side membrane is imaged at high resolution with a small *Z*-axis scale (Fig. 3.11d). From the image, there is no any obvious protrusion or particle (Fig. 3.11g). Cross-sectional data are smoothened by a fast Fourier transform (FFT) filter as the red line in Fig. 3.11h. Measuring 50 small areas with cross-sectional analysis, the depths were calculated from the difference between the high and low points of the smoothened curves. The depth statistics of 50 curves is shown as the red bar in Fig. 3.11m, and the average depth is 0.22 ± 0.13 nm (the red bar). This result indicates that the ectoplasmic side membrane is rather smooth.

In order to confirm where the oligosaccharides and peptides on the cell membrane proved by other techniques, are the ectoplasmic side membrane was digested by PNGase F that cleaves an entire saccharide from a glycoprotein [54]. Figure 3.11e shows the image of the ectoplasmic side membrane treated by PNGase F. Interestingly, there are many indents in the membrane compared with Fig. 3.11d. From the smoothened cross-sectional line, the deep area of the curve is obvious. The statistics of 50 curves is shown as a blue bar in Fig. 3.11m with the average depth of 0.61 ± 0.14 nm. It is noteworthy that the depth value of indents is just within the range of the height of the hydrophilic lipid head (0.80 ± 0.15 nm) in the lipid bilayer despite considering the tip pressure on the sample, which indicates that oligosaccharides could actually occupy the position of hydrophilic lipid heads [55]. Furthermore, to reveal the location of transmembrane peptides, the ectoplasmic side membrane was digested by both PNGase F and trypsin that was used to cleave the transmembrane protein of intact hRBCs [56]. As shown in Fig. 3.11f, there are deeper indents in the membrane than those in Fig. 3.11e after PNGase F and trypsin treatment. The section analysis in Fig. 3.11k and l shows a typical indent from Fig. 3.11f. The average depth of indents is 2.50 ± 0.82 nm (green bar in Fig. 3.11m), which is in the range of the thickness of the lipid bilayer. The depth sequence for the three conditions in Fig. 3.11m is D_b (depths of bare membrane), $<D_p$ (depths of membrane treated by PNGase), and $<D_t$ (depths of membrane treated by PNGase F and trypsin), which obviously indicates that the cell membranes were dug after PNGase F and trypsin treatment. Although the result from PNGase F digestion is not perfect due to the low indents for the interpretation of oligosaccharide arrangement in the ectoplasmic side membrane, the trypsin digestion result shows the low noise and deep indents that clearly indicates the location of the transmembrane protein in the ectoplasmic side membrane.

The width of the indents were both in the range of 20–220 nm with the peak at 50–60 nm (Fig. 3.11 n, o), which indicates that these indents created by trypsin could be just the same sites made by PNGase F. These results could support that oligosaccharides may be spread in the middle of lipid heads rather than protruding from the membrane surface and peptides called transmembrane parts of proteins may be buried in the lipid membrane under oligosaccharides.

3.3.4.3 A New Model of Red Blood Cells

Based on the above observation, Wang et al. drew a proposed model of red blood cell membranes—semi-mosaic model. Figure 3.12a is a top view of semi-mosaic model of cell membranes, in which oligosaccharides are evenly spread in the middle of lipid heads without protein protrusion. Figure 3.12b is a bottom view of cell membranes that shows dense proteins with less free lipids. The bottoms of transmembrane proteins are inserted in the lipid bilayer. Figure 3.12c shows the lipid bilayer scheme based on the molecular structure and Ref. [55]. Lipid heads are emphasized here because they have been neglected when interpreting membrane structure. Semi-mosaic model suggests that proteins may be partly set within the lipid bilayer rather than protruding out of the ectoplasmic side cell surface [53]. The carbohydrates are located in the middle of lipid hydrophilic heads, and the membrane proteins are mainly located on the cytoplasmic side of membranes. Super-resolution imaging technique suggests that membrane proteins tend to form clusters in cholesterol-enriched domains such as lipid rafts (see Chap. 5) [57].

3.3.4.4 Imaging the Blood Cell Membranes of Other Species

In the process of biological evolution, the basic structure of red blood cells is conservative in the whole species, although there may be some changes between different species. The current cell membrane structure models are put forward according to the results of the mammalian cell membrane; studies on the membrane structures of other lower animals are not paid much attention. Compared with

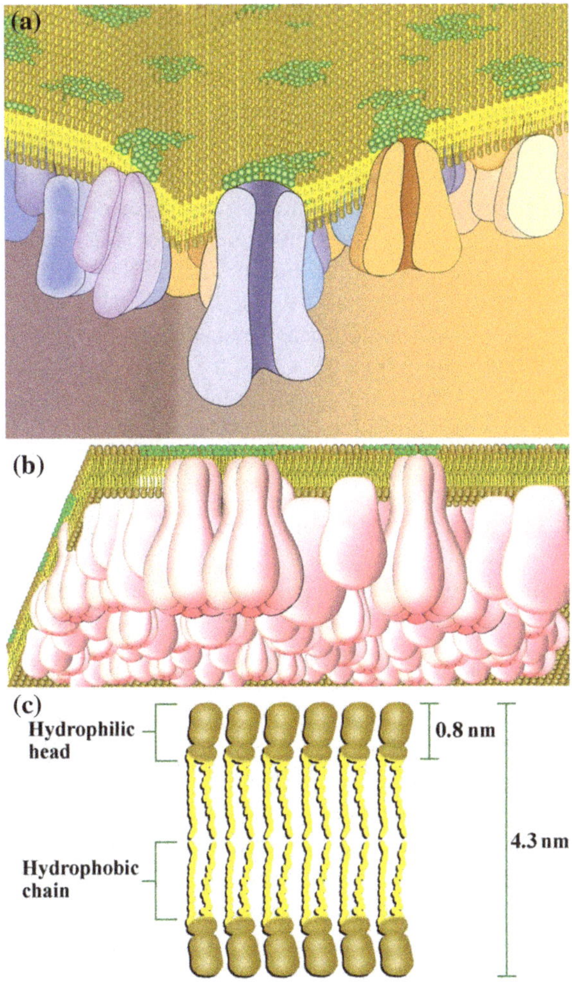

Fig. 3.12 Proposed semi-mosaic model of cell membranes. **a** Top view of cell membranes. Lipids (golden) and glycans (green) are shown on the top of the layer. Proteins are under the lipid layer. **b** Bottom view of cell membranes. **c** Scheme of the lipid bilayer. Reprinted from Ref. [52], Copyright 2010, with permission from Elsevier

hRBCs, questions on what the structure of red blood cell membrane is, what the distribution of membrane proteins is, and what the red cell membrane structure model of those lower animals is, need to be solved. Wang et al. studied red blood cell membranes in an evolutionary order from lower vertebrates (fish) [58], reptiles (turtle) [59] to higher vertebrate birds (chicken) [60]. The researchers determined that the features of these membranes are very similar to human red blood cell membranes, and clarified the semi-mosaic model of hRBCs is a common structure.

Imaging the Fish Red Blood Cell Membrane. From an evolutionary perspective, the fish erythrocytes can be considered as the prototype of the circularing nucleated, hemoglobin-bearing cell in sub-mammalian vertebrates. Compared with biconcave disk-like, non-nucleated human erythrocytes, fish erythrocytes are predominantly oval, biconvex disks with elliptical nucleus and much larger in size than human erythrocytes. As reported, fish and human erythrocyte membrane possess many features in common, including major membrane protein and phospholipid composition. How the fish membranes are structured at molecule level and where the membrane proteins are distributed?

Crucian carp erythrocytes membrane is scanned by AFM. Figure 3.13a shows the topographical image of the intact unfixed crucian carp erythrocytes, in which the erythrocytes are spread on the substrate as elegant ellipsoids with the nucleus protruding in the central region. The intact crucian carp erythrocytes are 23.7 ± 3.1 μm in long axis, 15.2 ± 3.3 μm in short axis, and 1.3 ± 0.4 μm in height, while the nucleus has the length of 9.2 ± 1.7 μm and width of 5.8 ± 1.4 μm. Without fixation, the ectoplasmic side of crucian carp erythrocyte appears quite smooth at the edge. As shown in Fig. 3.13b, the intact fixed erythrocyte display spindle shape 23.1 ± 2.4 μm in long axis, 13.4 ± 2.1 μm in short axis, and 1.8 ± 0.3 μm in height with elliptical nucleus 10.3 ± 1.8 μm long and 5.7 ± 0.8 μm wide. The ectoplasmic side surface of fixed erythrocyte appears slightly wrinkled around the nucleus caused by the fixation procedure, and a small knob (pointed by arrow) arises at the edge indicating some organelles or vesicular structure. Figure 3.13c is the high-resolution image at the edge of unfixed erythrocyte, which reveals that the native ectoplasmic side of erythrocyte membrane is very flat and soft with the height of 35.4 ± 2.9 nm. To observe detailed membrane structure, the close-up view of the square area on the membrane edge in Fig. 3.13c was obtained. As shown in Fig. 3.13d, no indents or particles are apparently visible on the membrane. The ectoplasmic side of erythrocyte membranes is determined to be rather smooth with the average roughness of 0.56 ± 0.06 nm in its native state.

To obtain clean membranes for high-resolution images, the cytoplasmic side of crucian carp membranes was further treated with high salt buffer to remove membrane skeletons, but the transmembrane proteins were kept well due to their insertion in the lipid bilayers. Figure 3.14a shows the large-scale topographical image of the cytoplasmic side of erythrocyte membranes covered by dense proteins. The height of the cytoplasmic side membrane between the substrate and the proteins is about 18.2 ± 3.0 nm. The average roughness of the cytoplasmic side membrane is measured to be 3.1 ± 0.7 nm, which is remarkably larger than that of the ectoplasmic side membrane. Figure 3.14b is the middle resolution image of the square area in Fig. 3.14a, which clearly displays the dense protein particles in the cytoplasmic side membrane. Higher resolution image of the square area in Fig. 3.14b has been achieved as shown in Fig. 3.14c, in which the proteins are standing closely to each other and exhibit variability in size and shape. These protein particles display a broad diameter distribution from 30 to 200 nm with 60% in the range of 70–115 nm as shown in Fig. 3.14d, which is a little larger than real protein particles due to the geometry of the cantilever tip [61]. As depicted in Fig. 3.14e, the height of the protein particles above the membrane varies from 1.0 to 35.0 nm with the peak at 10–13 nm, corresponding to varieties of proteins and protein aggregations in the cytoplasmic side membrane.

As shown above, crucian carp erythrocytes are oval, biconvex disks with elliptical nucleus protruding in the central region, which is very different from

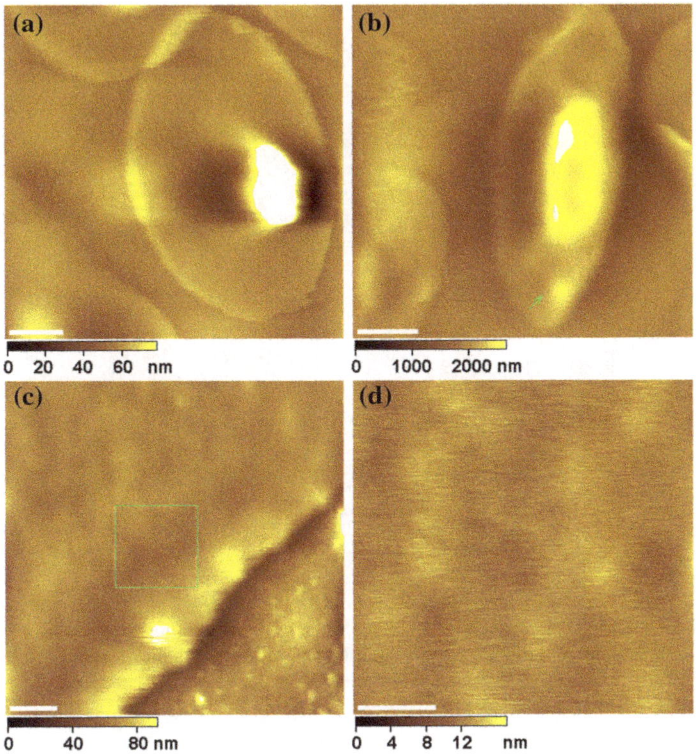

Fig. 3.13 **a** representative AFM topographic images of the intact unfixed crucian carp erythrocyte. **b** AFM topographic image of the fixed crucian carp erythrocyte. **c** Median resolution image on the edge of unfixed erythrocyte. **d** High-resolution image of the smooth ectoplasmic side membrane. The arrow points to the small knob at the edge. Scale bars: 5 μm in **a** and **b**, 1 μm in **c**, and 200 nm in **d**. Reprinted from Ref. [58], with kind permission from Springer Science + Business Media

biconcave disk-like, non-nucleated human erythrocytes. The asymmetric distribution of proteins on both sides of crucian carp membranes fits well with the semi-mosaic model of human erythrocyte. This model highlights the smooth ectoplasmic side membrane and the proteins embedded in cytoplasmic side membrane. The study of crucian carp erythrocyte membranes provides consistency of membrane structure in mammalian and lower vertebrate erythrocytes and extends the semi-mosaic model of erythrocytes membrane structure to a wider range of species [58].

Imaging the Reptile Red Blood Cell Membrane. Turtles are the most ancient of reptiles in the evolutionary context and have been phylogenetically classified both as basal to all other reptiles and as nested within them [62]. Studies on turtle erythrocytes, as an example of reptiles, are introduced here to investigate the detail information of membrane structure under near-native conditions. Differences

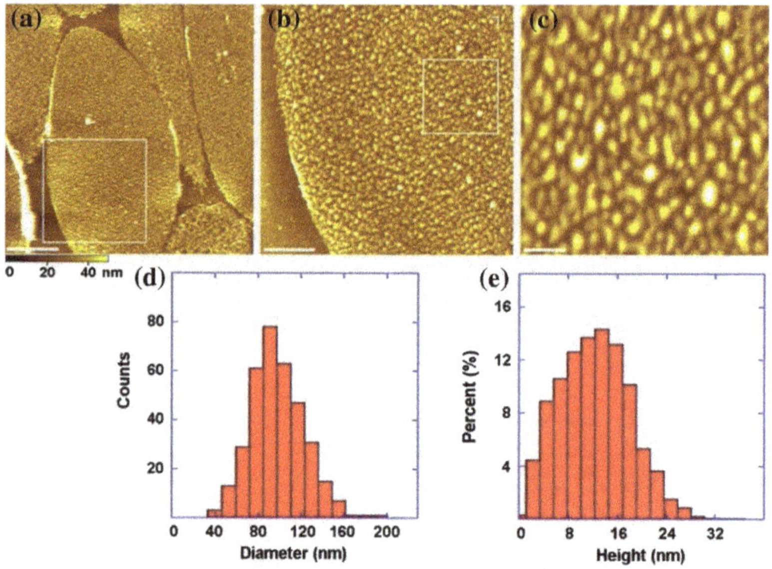

Fig. 3.14 **a** AFM topographic image of the cytoplasmic side of crucian carp erythrocyte membranes. **b**, **c** Serial magnifications of the cytoplasmic side membrane in **a**. **d**, **e** the diameter and height distribution of proteins in the cytoplasmic side membrane, respectively. Scale bars: 5 μm in **a**, 2 μm in **b**, and 500 nm in **c**. Reprinted from Ref. [58], with kind permission from Springer Science + Business Media

between erythrocytes of human and turtles are observed clearly and proved the semi-mosaic model is applicable to reptiles.

As shown in Fig. 3.15a, the intact turtle erythrocytes are biconvex oval disks with elliptical nuclei bulging in the central regions (black dashed circle) and a marginal band of microtubules along the periphery. The turtle erythrocytes have an average length of 27.5 ± 4.5 μm and an average width of 15.2 ± 1.5 μm. The average nucleus has a length of 8.0 ± 3.1 μm and a width of 4.8 ± 1.2 μm. The height between the substrate and the nucleus is 1.4 ± 0.3 μm (Fig. 3.15a, bottom).

The intact turtle erythrocytes contain few organelles in the cytoplasm, and the membrane surface appears quite smooth. Firmly adhered to the substrate, the erythrocytes flatten out with a thickness of 47.2 ± 4.1 nm at the edges (Fig. 3.15b, bottom), which enables us to obtain high-resolution images of the ectoplasmic side surface. We have acquired many stable images on the smooth edge. As shown in Fig. 3.15c, there are no obvious proteins or protrusions on the ectoplasmic side, which has an average roughness of 0.52 ± 0.11 nm in its native state. To test whether the interaction between the AFM tips and the ectoplasmic surface of living cells produces blurry images, rendering the ectoplasmic side proteins undetectable, we performed high-resolution imaging on ectoplasmic side membrane patches produced by the shearing method discussed in Sect. 3.4.4 (Fig. 3.28). It was found that the ectoplasmic side membrane patches were just as smooth as that of the intact

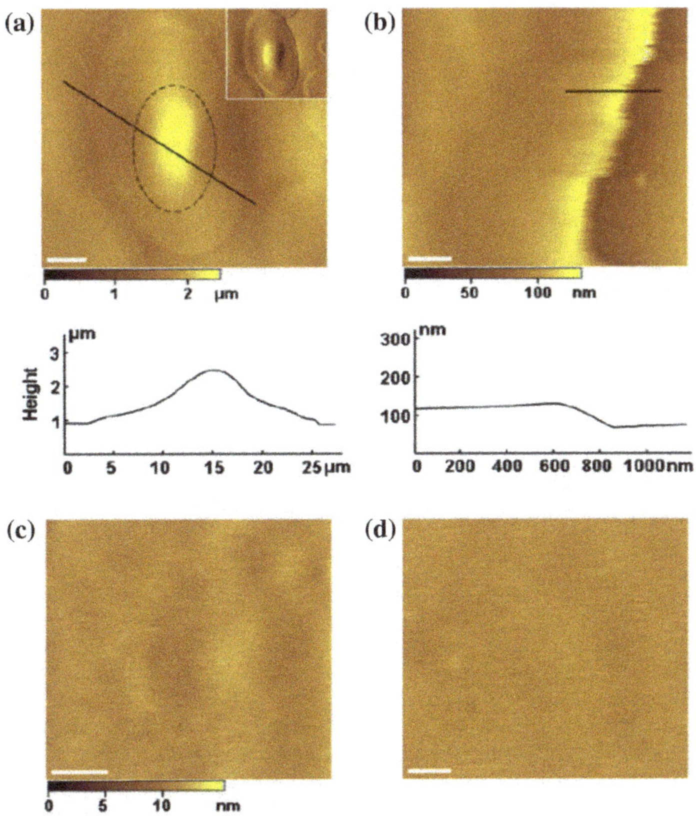

Fig. 3.15 AFM topographic images of the smooth ectoplasmic surface of the turtle erythrocytes. **a** The intact turtle erythrocyte is 27.5 ± 4.5 μm long, 15.2 ± 1.5 μm wide, and 1.4 ± 0.3 μm high. A black dashed circle encloses the nucleus (8.0 ± 3.1 μm long; 4.8 ± 1.2 μm wide). The inset is the corresponding amplitude image. The bottom is the cross-sectional analysis along the black line. **b** The smooth edge of the intact turtle erythrocyte is 47.2 ± 4.1 nm thickness. The bottom is the cross-sectional analysis along the black line. **c** High-resolution image of the smooth edge. **d** High-resolution image of the prepared ectoplasmic side membrane. The average roughness of the ectoplasmic side membrane is 0.52 ± 0.11 nm. Scale bars: 5 μm in **a**, 500 nm in **b**, 200 nm in **c**, and 100 nm in **d**. Reprinted from Ref. [59] by permission of the Korean Society for Molecular and Cellular Biology

erythrocyte (Fig. 3.15d). Thus, there appear to be no protein protrusions out of the ectoplasmic side of the turtle erythrocyte membranes.

The cytoplasmic side of turtle erythrocyte membranes was revealed to be rather rough and covered with dense proteins (Fig. 3.16a), which can be seen more clearly at higher resolution (Fig. 3.16b).There are also many particles on the substrate (arrows), which are most likely adsorbed proteins from the shearing procedure. The average roughness of the cytoplasmic side membrane (4.0 ± 0.8 nm) is considerably larger than that of the ectoplasmic side membrane. The average height of the

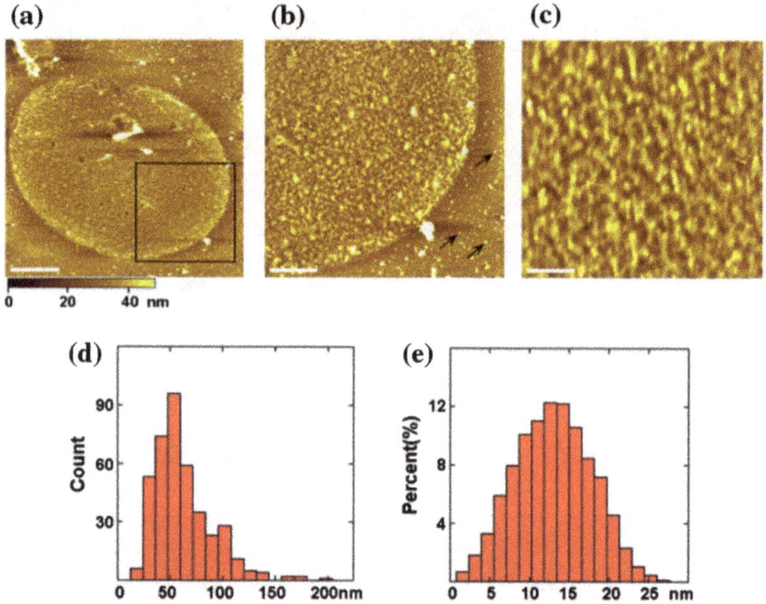

Fig. 3.16 Characterization of the protein-covered cytoplasmic side of the turtle erythrocyte membranes. **a** AFM topographic image of the cytoplasmic side membrane. **b** Magnification of the square area in **a**. Arrows point to proteins on the substrate. **c** High-resolution image of the cytoplasmic side membrane. The average height and roughness of the membranes are 18.5 ± 2.4 nm and 4.0 ± 0.8 nm, respectively. **d** The diameter distribution of proteins in the cytoplasmic side membrane varies from 15 to 200 nm with 70% falling between 40 and 80 nm. **e** The height of proteins above the membrane is from 1.0 to 27.0 nm with the peak at 10–16 nm. Scale bars; 5 μm in **a**; 2 μm in **b**; and 500 nm in **c**. Reprinted from Ref. [59] by permission of the Korean Society for Molecular and Cellular Biology

membranes between the substrate and the proteins is 18.5 ± 2.4 nm. High-resolution images of the cytoplasmic side membrane reveal crowded protein particles standing closely to each other with no visible lipid bilayer (Fig. 3.16c). The proteins in the cytoplasmic side membrane have a broad range in size and shape, with diameters of 15–200 nm and 70% fall between 40 and 80 nm (Fig. 3.16d). Figure 3.16e shows the height distribution of the protein particles above the membrane, which varies from 1.0 to 27.0 nm and has a peak at 10–16 nm. This corresponds to a diversity of proteins and protein aggregations in the cytoplasmic side membrane.

Morphologically, turtle erythrocytes share many features in common with fish erythrocytes, including the existence of the nucleus in central region with an ellipsoidal, flattened, and biconvex shape. The most prominent similarity among these erythrocyte membranes is the asymmetric distribution of proteins; there are more proteins on the cytoplasmic side membrane than on the ectoplasmic side. The turtle erythrocyte membranes were revealed by high-resolution AFM to be quite smooth without obvious proteins on the ectoplasmic side membrane, whereas rough with dense proteins on the cytoplasmic side. The asymmetric membrane structure of turtle erythrocytes corresponds well with the semi-mosaic model of human and the structure of fish erythrocyte membrane [59].

Imaging the Bird Red Blood Cell Membrane. Chicken is one of the most common and widespread domestic fowl, whose phylogenetic position within Galliformes is near the base of Phasianidae. There are more chickens in the world than any other species of birds. The study of the chicken erythrocyte membrane structure, as a representative of birds, will bridge the gap between fish and mammals.

When imaged by AFM in physiological buffer solution, the isolated chicken erythrocytes firmly attached onto the substrate represent oval disks with an elliptical nucleus highly protruding in the central region (Fig. 3.17a), which is similar to the nucleated fish erythrocyte except for the smaller size. The intact chicken erythrocyte is measured to be 14.2 ± 3.0 μm in the long axis, 8.9 ± 1.6 μm in the short axis,

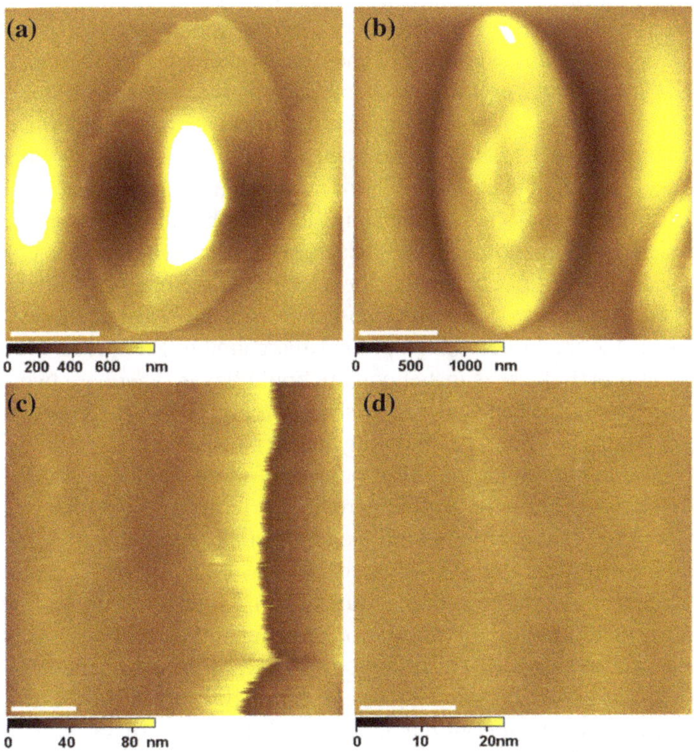

Fig. 3.17 Representative AFM topographic images of the intact and the ectoplasmic side of chicken erythrocyte membranes. **a** Unfixed chicken erythrocyte. **b** Chicken erythrocyte fixed by 4% paraformaldehyde. **c** The unfixed chicken erythrocyte appears quite smooth at the edge. **d** High-resolution image shows no detectable protrusions or particles on the smooth ectoplasmic surface. Scale bars: 5 μm in **a** and **b**, 500 nm in **c**, and 200 nm in **d**. Reproduced from Ref. [60] by permission of The Royal Society of Chemistry

and 1.2 ± 0.4 μm in height, while the nucleus has the length of 6.8 ± 1.9 μm and the width of 4.3 ± 0.7 μm. Without fixation, the ectoplasmic side of the erythrocyte appears quite smooth at the edge. Figure 3.17b shows the topographical image of the intact fixation chicken erythrocyte, in which the oval erythrocyte appears fleshy at the edge because of the fixation procedure. The intact fixed chicken erythrocyte is 16.2 ± 2.6 μm in the long axis, 8.9 ± 1.0 μm in the short axis, and 1.7 ± 0.2 μm in height with the elliptical nucleus 8.7 ± 1.2 μm long and 5.0 ± 0.6 μm wide. In both cases, the ectoplasmic side of the chicken erythrocyte is quite smooth at the edge. The intact unfixed chicken erythrocyte is measured to be 45.6 ± 1.4 nm thick at the edge.

As expected, the ectoplasmic side of chicken erythrocytes is rather smooth with no obvious proteins or wrinkles (Fig. 3.17c). A higher resolution image was achieved, as shown in Fig. 3.17d shows no detectable protrusions or particles on the smooth ectoplasmic side membrane surface. The average roughness of the ectoplasmic side of chicken erythrocyte membranes was 0.47 ± 0.05 nm, which demonstrates that the membrane surface is extremely smooth.

The cytoplasmic side of chicken erythrocyte membranes displays protein-decorated topography after removal of membrane skeletons by high salt treatment (Fig. 3.18a). The prepared erythrocyte membranes show elliptical profiles resembling those of intact erythrocytes, indicating that the upper surface membrane and cytoplasm components were totally sheared away. The height of the cytoplasmic side membrane was measured to be 16.5 ± 3.6 nm between the substrate and the proteins. The average roughness of the cytoplasmic side membrane is about 3.4 ± 0.8 nm, which is remarkably larger than that of the ectoplasmic side membrane. The arrows in Fig. 3.18a point to free lipid bilayers at the edge of the membranes with the height of 3.1 ± 0.7 nm, which is in complete accordance with the value measured by AFM. Figure 3.18b is the magnified image of the square area in Fig. 3.18a, which clearly shows that the protein layer is just above the lipid bilayer. There are two clear steps shown by the cross-sectional analysis, as shown in Fig. 3.18c the arrows point to the protein layer (right) and the free lipid bilayer (left), respectively.

Figure 3.18d is the middle resolution image of the cytoplasmic side erythrocyte membranes, showing crowded protein particles with no free lipid bilayer visible. As shown in Fig. 3.18e, the proteins are standing closely to each other and exhibit great variability in size and shape. As depicted in Fig. 3.18f, the height of the protein particles above the membrane varies from 1.0 to 35.0 nm with the peak at 10 to 13 nm, corresponding to varieties of proteins and protein aggregations in the cytoplasmic side membrane. These proteins display a broad diameter distribution from 45 to 250 nm with 75% in the range of 55 to 100 nm as shown in Fig. 3.18g.

In morphology, chicken erythrocytes are oval, biconvex disks with an elliptical nucleus protruding in the central region, which are similar to crucian carp erythrocytes and the turtle except for the smaller size. A high-resolution image by AFM on both sides of chicken erythrocyte membranes revealed that the ectoplasmic side of erythrocyte membranes is quite smooth without proteins protruding out of the cell surface, whereas the cytoplasmic side of erythrocyte membranes is very rough with dense proteins standing closely to each other above the lipid bilayer.

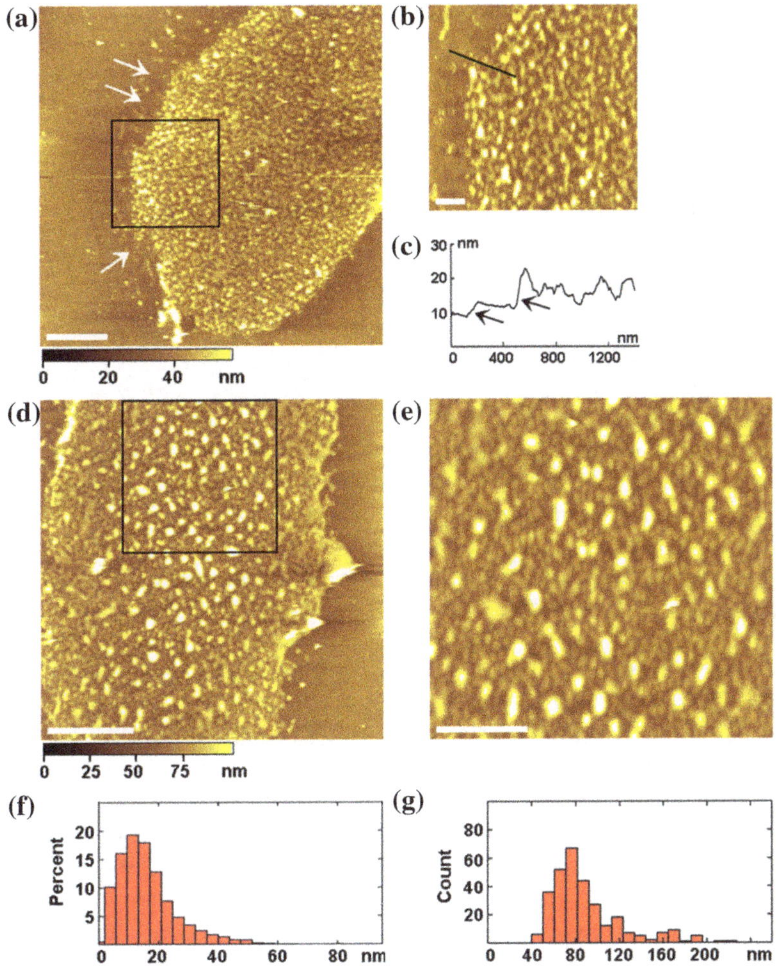

Fig. 3.18 AFM topographic images of the cytoplasmic side of chicken erythrocyte membranes. In **a**, arrows point to the free lipid bilayers under the membrane proteins. **b** Magnification of the square area in **a**. **c** Cross-sectional analysis along the line in **b**. Arrows point to the free lipid bilayer and the protein layer, respectively. **d** and **e** Median and high-resolution image of the cytoplasmic side membrane. **f** and **g** The height and diameter distribution of proteins in the cytoplasmic side membrane, respectively. Scale bars: 2 μm in **a** and **d**, 500 nm in **b** and 1 mm in **e**. Reproduced from Ref. [60] by permission of The Royal Society of Chemistry

The asymmetric distribution of proteins on both sides of chicken erythrocyte membranes fits well with the semi-mosaic model of the human and the structure of fish and turtle erythrocyte membrane, which highlights the smooth ectoplasmic side membrane and the protein-covered cytoplasmic side membrane. This study of chicken erythrocyte membranes provides consistency of the membrane structure in mammalian, bird, and lower vertebrate erythrocytes and extends the semi-mosaic model of erythrocyte membrane structures from mammals to lower vertebrates to birds. This similarity of membrane structures between mammalian, bird, and lower vertebrate erythrocytes further leads us to believe that asymmetry in the membrane structure exists universally in erythrocyte cells [18, 60].

3.3.5 Examination of Nucleated Mammalian Cell Membranes by AFM

3.3.5.1 The Ectoplasmic Side of Nucleated Mammalian Cell Membranes

To realize the complicated functions of cell membrane (e.g., signaling transduction and membrane transporting), the ectoplasmic side of nucleated mammalian cell membranes exhibit more proteins (e.g., receptors) on the surface. Using the TREC technique, Van Vliet et al. mapped vascular endothelial growth factor receptor-2, which indicated an irregular distribution of the receptors in the membrane (Fig. 3.19) [63]. Tang et al. located the human gonadotropin-releasing hormone receptor (GnRH-R) using TREC and concluded that the receptor tends to form domains or clusters those are irregularly distributed on the T24 cell surface [64].

In order to verify the feature of the native cell membrane, ectoplasmic side of living cells were directly imaged. Surprisingly, the ectoplasmic side of Madin–Darby canine kidney (MDCK) cells seemed to be very smooth without obvious protrusion (Fig. 3.20b). The roughness of membrane surface was only 1.1 ± 0.2 nm. Figure 3.20c showed the high-resolution image of the cell membrane's ectoplasmic side prepared by shearing open the cells. The ectoplasmic side was found to be rather smooth with a roughness root mean square (RMS) of

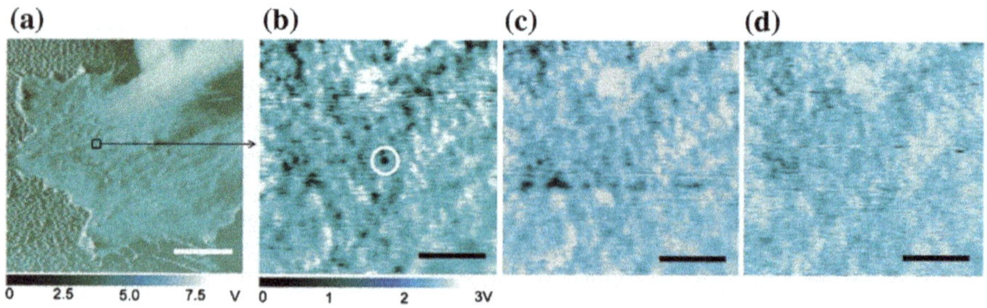

Fig. 3.19 Imaging of individual VEGFR2 receptors on fixed HUVEC surfaces. **a** Phase image of cell body and periphery. **b** Recognition image over indicated area in **a** shows strong binding events between the anti-VEGFR2 functionalized probe and the cell surface as discrete, dark spots (e.g., circled) that are ostensibly VEGFR2. **c** and **d** The capacity to block these binding events by addition of 5 μg/ml soluble anti-VEGFR2 supports this binding specificity between the probe and VEGFR2, as the number of recognition events decreases with time post blocking of 12 min (**c**) and 60 min (**d**) min. Scale bars: 10 μm, for **a**; 500 nm for **b–d**. Reprinted with permission from Ref. [63] with permission Copyright (2007) National Academy of Sciences

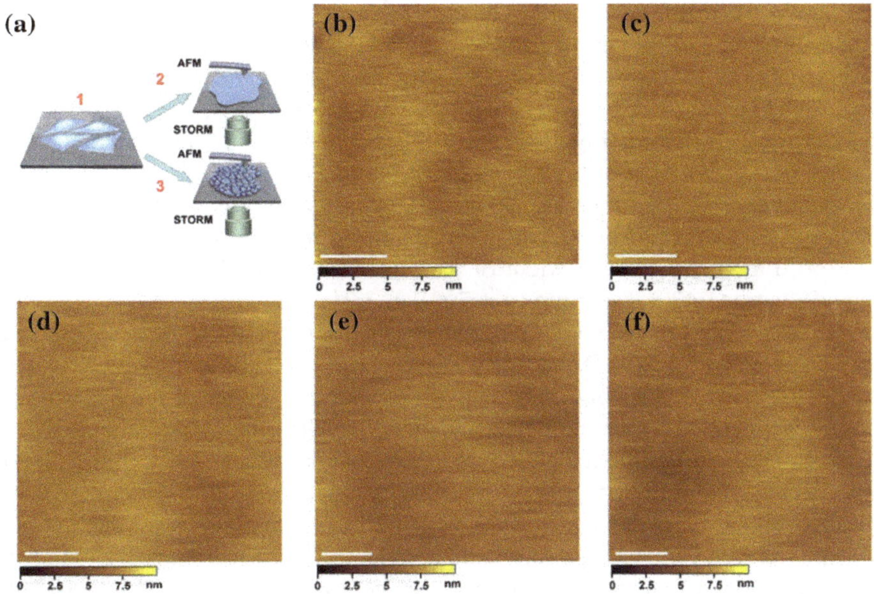

Fig. 3.20 Cells were cultured on cover slips (**a1**), (**a2**), and (**a3**). The ectoplasmic and cytoplasmic sides of membranes were prepared separately and then investigated with AFM imaging, single-molecule force spectroscopy (Chap. 4), and Super-resolution imaging (Chap. 5), respectively. Imaging the ectoplasmic side of the cell membranes from various types of mammalian cells. **b** The ectoplasmic side of MDCK cell membrane was directly imaged on a living cell. **c** The image of the ectoplasmic side of the MDCK cell membrane prepared by shearing open the cells on a cover slip. **d** The image of the ectoplasmic side of the MDCK cell membrane prepared by centrifugation. **e** and **f** The ectoplasmic side of A549 **e** and HeLa **f** cell membranes prepared by the shearing open approach, respectively. Scale bars: 100 nm in (**b–f**). Reprinted from Ref. [68], with kind permission from the Public Library of Science

0.9 ± 0.2 nm. No indents or particles were visible. As shown in Fig. 3.20d, the ectoplasmic side of the cell membrane obtained by the hypotonic-centrifugation procedure presented a roughness RMS of 1.0 ± 0.2 nm, thus showing a feature very similar to that in Fig. 3.20b, c. The result proved additional evidence of the smooth ectoplasmic side of MDCK cell membranes.

The ectoplasmic side of two other human cancer cell lines derived from different organs, A549 cells from lung and HeLa cells from cervix, by the shearing open approach to confirm whether this phenomenon was common for other types of nucleated mammalian cells. As shown in Fig. 3.20e, f, the ectoplasmic sides of cell membrane were as smooth as MDCK cells.

It is well established that proteins, such as receptors and glycosyl phosphatidyl inositol-anchored proteins (GPI-APs), are present on the ectoplasmic side of the cell membrane. Proteinase K can digest most proteins above the lipid bilayers. The ectoplasmic side of the cell membrane is treated with proteinase K and monitored the real-time changes with time-lapse AFM. The surface of undigested ectoplasmic side of the cell membrane was consistently smooth without any pits or protrusions, as indicated in the magnified image and the corresponding section analysis (Fig. 3.21a, d, e). After digestion with proteinase K, most proteins were removed, except some undigested or half-digested proteins, as indicated by the blue arrows in Fig. 3.21b, c, at the ectoplasmic side of the cell membrane. The heights between the undigested proteins and the local pits, as indicated by the green arrows in Fig. 3.21b, ranged from 1.3 to 5 nm, with the majority around 2.8 ± 0.9 nm (n = 60, Fig. 3.21b, j). The width of the pits varied from 30 to 80 nm, with the majority around 43.8 ± 7.7 nm (n = 70, Fig. 3.21b, k).

In order to know where the lipid bilayer is located, the digested membrane was further treated with MβCD that can deplete cholesterol in the lipid bilayer. After treated by MβCD, the total height of the membrane remained the same as the untreated membrane, but the depth and width of the indents were increased, as could be observed in the magnified image and the section analysis (Fig. 3.21f–i). The depths of the indents eroded by MβCD range from 2 to 10 nm, with the majority about 4.2 ± 1.1 nm (n = 90, Fig. 3.21l), which agrees with the height of the lipid bilayer. The indent widths exhibit variability between 40 and 110 nm, with the majority of indentations having an average width of around 63.1 ± 10.1 nm (n = 70, Fig. 3.21m), which was larger than that digested by proteinase K. These results demonstrate that the ectoplasmic side of the cell membrane comprises a layer of dense proteins, e.g., GPI-APs and the extracellular segment of the transmembrane proteins, associated with the lipid bilayer.

3.3.5.2 The Cytoplasmic Side Membrane of Nucleated Mammalian Cells

The distribution of proteins at the cytoplasmic side of the cell membrane is another key aspect of cell membrane structure and function. Various types of proteins can be found at the cytoplasmic side of cell membrane, such as the intracellular domains of receptors and transporters. Figure 3.22b displays the fluorescent image of the cytoplasmic side of the cell membrane, in which abundant actin filaments (green) are visible on the membrane surface (red). The cytoskeletons were disrupted by high salt treatment (Fig. 3.22c). The AFM topographical images of the cytoplasmic side of membranes before and after treatment with high salt buffer are shown in Fig. 3.22d, e, respectively. Dense actin filaments are shown as strips (Fig. 3.22d, g), while no obvious cytoskeleton can be observed in Fig. 3.22e. The average height of the membranes was 19.5 ± 2.8 nm (Fig. 3.22h). The cytoplasmic side of cell membranes was rather rough and covered with proteins, which can be seen more clearly in the magnified image (Fig. 3.22f). The roughness RMS of the cytoplasmic side of membranes was 3.7 ± 0.2 nm (Fig. 3.22f), which was much more significant than that of the ectoplasmic side (Fig. 3.20). The height of the proteins measured from top to bottom was 11.2 ± 1.9 nm (Fig. 3.22i). Based on the similarity of heights of the ectoplasmic protein layer and lipid bilayer both at

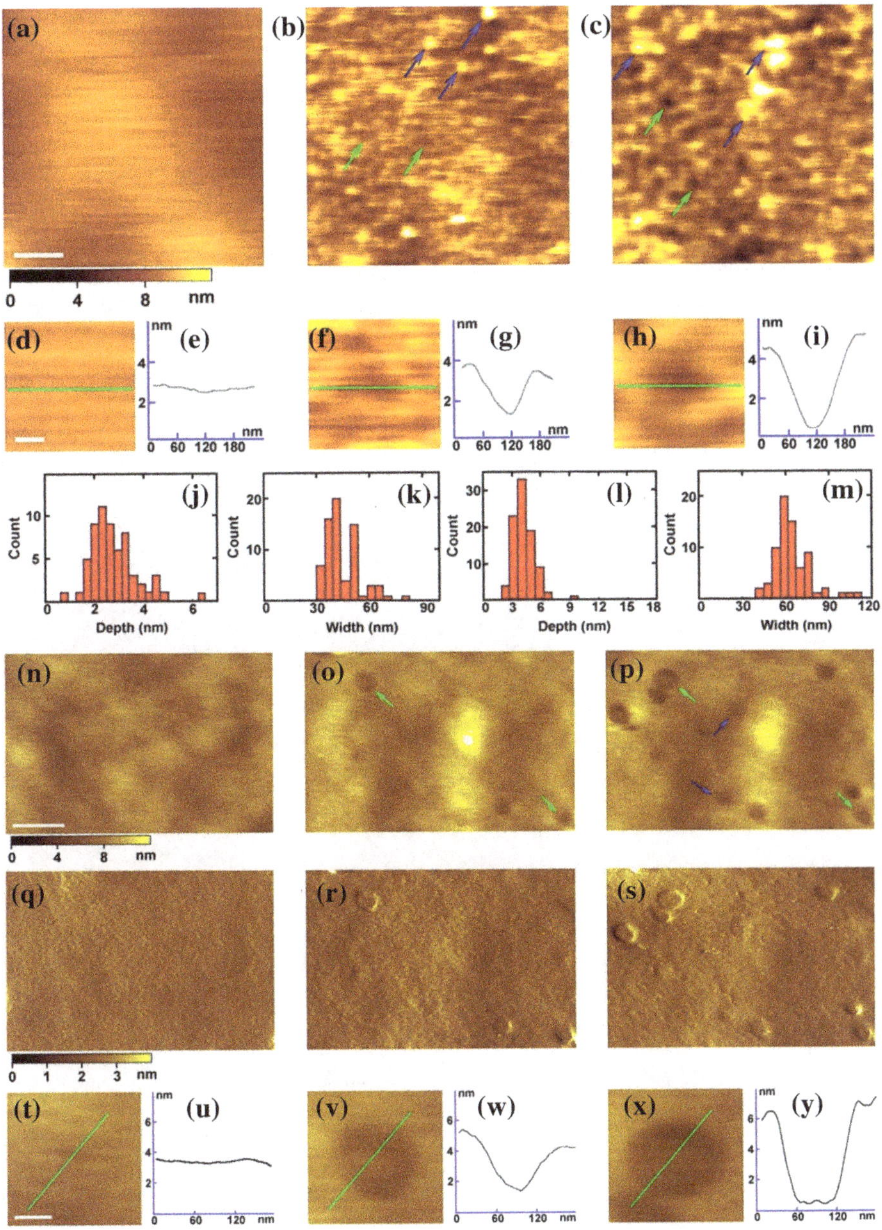

about 4 nm (Fig. 3.21), the total height of the cell membrane was calculated to be about 20 nm, consistent with the real size measured from the whole cell membrane (Fig. 3.22e). The width of the protein microdomains was 98.5 ± 8.6 nm, much larger than that of a single protein, about 20 nm measured by AFM, indicating the

◂Fig. 3.21 Digestion of the ectoplasmic side of the cell membrane with proteinase K or collagenase 3. **a** The AFM topographic image of the ectoplasmic side of the cell membrane. **b** and **c** The AFM topographic image of the ectoplasmic side of cell membranes treated with proteinase K **b** and M-β-CD **c** in sequence. **d**, **f** and **h** The magnified images from **a**, **b** and **c**, respectively, showing the gradual deepening of the pits. **e**, **g** and **i** The cross-sectional analysis along the green lines in **d**, **f** and **h**, respectively. **j** and **k** The depth and width distributions of the pits after proteinase K treatment, respectively. **l** and **m** The depth and width distributions of the pits after M-β-CD treatment, respectively. **n** The AFM topographic image of the ectoplasmic side of the cell membrane without treatment. **o** and **p** The AFM topographic image of the ectoplasmic side of the cell membrane treated in situ with collagenase 3 and M-β-CD in sequence, respectively. **q**, **r** and **s** The AFM amplitude images corresponding to **n**, **o** and **p**, respectively. **t**, **v** and **x** The magnified images of **n**, **o** and **p**, respectively. **u**, **w** and **y** The cross-sectional analysis along the green lines in **t**, **v** and **x**, respectively. Scale bars: 300 nm in **a–c**, 80 nm in **d**, **f** and **h**, 300 nm in **n–s**, and 50 nm in **t**, **v** and **x**. Reprinted from Ref. [68], with kind permission from the Public Library of Science

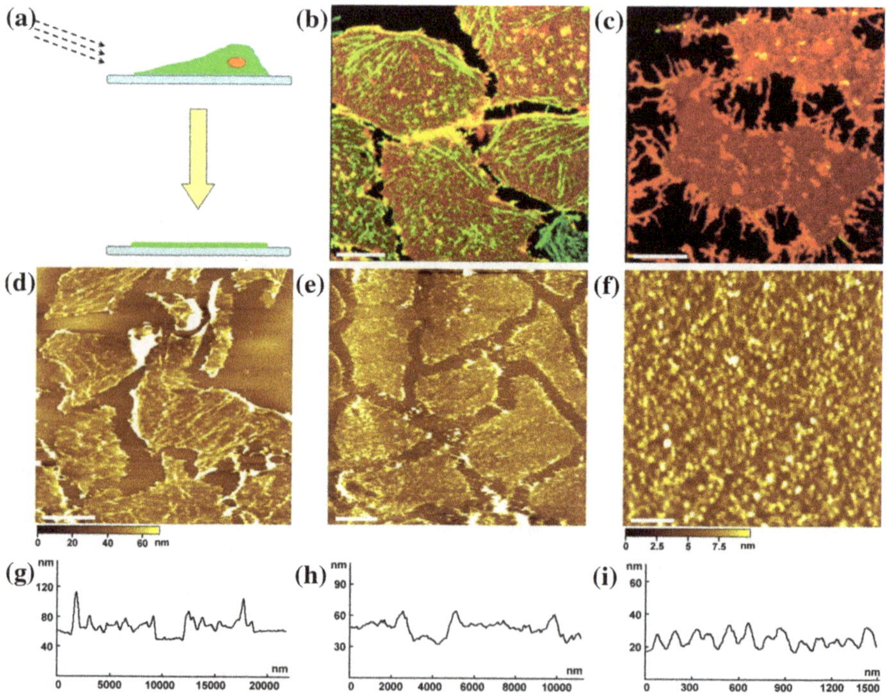

Fig. 3.22 **a** The scheme for preparing the cytoplasmic side of the cell membrane. **b** and **c** The fluorescent images of the cytoplasmic side of the cell membrane before and after incubation with high salt buffer, respectively. The red membrane patches represent the lipid bilayer labeled with DiI, and the green fibers represent the actin filaments labeled with phalloidin-FITC. **d** and **e** The AFM topographic images of the cytoplasmic side of the cell membrane before and after incubation with high salt buffer, respectively. **f** The high-magnification image of the cytoplasmic side of the cell membrane. **g–i** Cross-sectional analysis along the green line in **d–f**, respectively. Scale bars: 7 mm in **b–e**; 500 nm in **f**. Reprinted from Ref. [68], with kind permission from the Public Library of Science

presence of multiple proteins in the microdomains. The distribution of distances of the adjacent protein domains from border to border was about 53.2 ± 12.0 nm. These results demonstrate that the cytoplasmic side membrane consists of protein microdomains scattered in the lipid bilayer.

To investigate the relationship between the protein microdomains and lipid bilayer, the cytoplasmic side of membranes was treated with trypsin that could digest most membrane protein domains at the cytoplasmic side. The topographical image of the digested cytoplasmic side of the cell membrane showed that most of the proteins had been removed, thereby revealing the relative smoothness of local membrane patches (Fig. 3.23a, c). Some undigested proteins were right above the lipid bilayer, as shown by the bright dots. The height of single-layered, digested membrane patches was 8.0 ± 0.5 nm (average time $n = 30$, Fig. 3.23d), as depicted by the green arrows in Fig. 3.23a, indicating that the membrane patches are composed of the lipid bilayer (4 nm) and a dense protein layer (4 nm) at the ectoplasmic side, as mentioned previously. Double layers of digested membranes with an average height of 15.7 ± 1.9 nm ($n = 18$), as indicated by the pink arrow, can still be seen.

Triton X-100 has been widely used to destroy the lipid bilayer by interacting gently with the lipids. Then, the trypsin-digested cytoplasmic side of the cell membrane treated used 0.1% Triton X-100 to disrupt the lipid bilayer (Fig. 3.23b). As a result, the average height of the remaining membrane decreased to about 4.7 ± 0.4 nm (Fig. 3.23b, f), which was consistent with the height of the protein layer on the ectoplasmic side membrane. The proteins (Fig. 3.23b, e) on the remaining membrane surface may consist of membrane-anchoring proteins, such as GPI proteins, while the pits in the left membrane implicated sites of the transmembrane proteins, such as receptors. These results further confirm that the whole cell membrane consists of cytoplasmic side dispersed protein domains (12 nm), a lipid bilayer (4 nm), and an ectoplasmic layer of dense proteins (4 nm).

To directly clarify whether cholesterol-enriched domains (lipid rafts) exist on the cytoplasmic side of the cell membrane, the membrane treated by proteinase K. Figure 3.23g shows the cytoplasmic side cell membrane after digestion by proteinase K. The left membrane was smooth with a height of 8.0 ± 1.4 nm, which agreed perfectly with the result of trypsin treatment (Fig. 3.23a). After MβCD was injected into the AFM sample cell by the flow through liquid cell (Fig. 3.3), the lipid bilayer was quickly eroded, as shown in Fig. 3.23h. The height of the left membrane patches remained the same (about 8 nm), except for many pits. The magnified image of the green square area is shown in Fig. 3.23i. The pits eroded by MβCD were 40–200 nm in size, with the majority around 98.5 ± 25.5 nm (Fig. 3.23j), which was in good agreement with the sizes of the protein domains on the cytoplasmic side of the cell membrane, as shown in Fig. 3.22f. The depth distribution of the areas eroded by MβCD is shown in Fig. 3.23k. Two major depth distributions of the pits are evident: one is at 4.3 ± 0.3 nm, corresponding with the height of the lipid bilayer, and the other one is at 8.3 ± 0.4 nm, consistent with the total height of the lipid bilayer and ectoplasmic protein layer. Taken together, this data indicate that the cholesterol-enriched domains may be the protein microdomains on the cytoplasmic side of cell membranes (Fig. 3.22f).

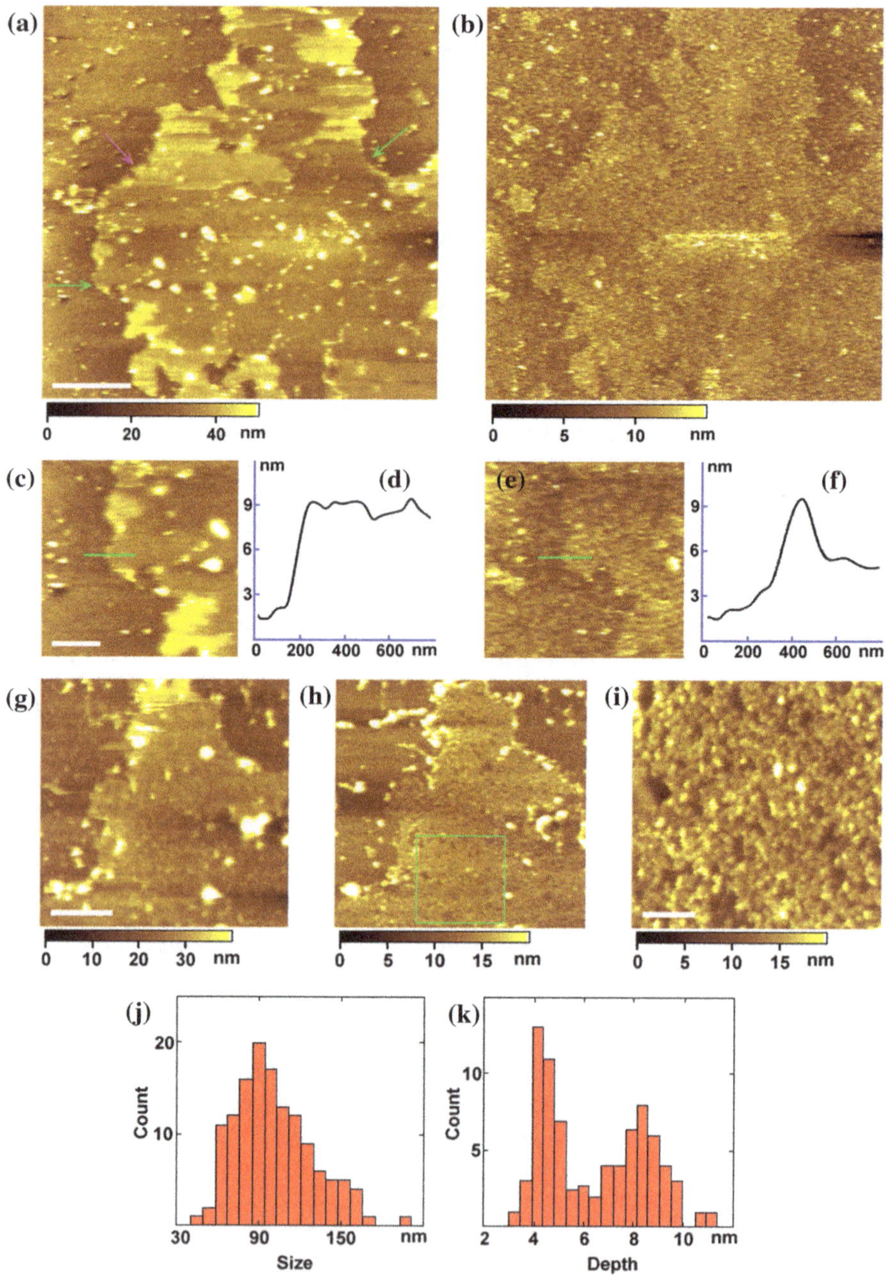
(a)
(b)
0 20 40 nm
0 5 10 nm
(c)
(d)
nm
9
6
3
0 200 400 600 nm
(e)
(f)
nm
9
6
3
0 200 400 600 nm
(g)
(h)
(i)
0 10 20 30 nm
0 5 10 15 nm
0 5 10 15 nm
(j)
Count
20
10
30 90 150 nm
Size
(k)
Count
10
5
2 4 6 8 10 nm
Depth

◀**Fig. 3.23** Cytoplasmic side of the cell membrane treated with trypsin or proteinase K. **a** AFM image of the cytoplasmic side of the cell membrane after digestion with trypsin for 1 h. Green and pink arrows indicate the single and double layers of membranes, respectively. **b** The membranes were treated with 0.1% Triton X-100 in situ. **c** and **e** The magnified images from **a** and **b**, respectively. **d** and **f** The cross-sectional analysis along the green lines in **c** and **e**, respectively. **g** and **h** The images of the cytoplasmic side of membranes after treatment with proteinase K and MβCD in sequence. **i** The magnified image of the green square area in **h**. **j**, and **k** The size and depth distributions of the pits eroded with MβCD in (**h** and **i**), respectively. Reprinted from Ref. [68], with kind permission from the Public Library of Science

3.3.5.3 A New Model of Nucleated Mammalian Cell Membranes-Protein Layer-Lipid-Protein Island (PLLPI)

Although current cell membrane models (e.g., the liquid mosaic model) tend to assign the same structure to all membranes from different cells, there is definitely a substantial difference between red blood cells and nucleated body cells. For example, the functions of nucleated mammalian cell membranes are substantially more complicated compared with human red blood cell membranes (almost no interaction with other cells and only carrying oxygen); thus, there are more receptors in their membranes compared with red blood cells. Combining in situ AFM imaging with enzyme digestion in fluid through the liquid cell, Wang et al. studied various mammalian cell membranes and demonstrated the following: (1) the proteins on the ectoplasmic side of the cell membrane form a dense protein layer (~4 nm) on top of a lipid bilayer; (2) proteins aggregate to form islands that are evenly dispersed on the cytoplasmic side of the cell membrane with a height of approximately 10–12 nm (a structural pattern similar to the cytoplasmic face cell membrane can be observed in large oocytes [65]); (3) cholesterol-rich domains exist within the cell membrane; (4) carbohydrates remain in microdomains on the ectoplasmic side of membrane; (5) exposed amino groups are asymmetrically distributed on both sides.

Based on these observations, Wang et al. [66] proposed a PLLPI model for mammalian nucleated cell membranes, as shown in Fig. 3.24. The PLLPI model emphasizes that the dense protein layer is the main functional component in the membrane in terms of mechanical properties, signaling transduction, and material transport. A protein domain could be observed on the cytoplasmic face of membranes, which may function as functional clusters (i.e., lipid rafts). It is worth noting that the PLLPI model is only for a static and basic cell membrane that lacks cell–cell interactions. In tissues, membrane proteins could tend to protrude out of the membrane surface to interact with the cell matrix [36].

The thickness of various membranes is shown in Table 3.1. The proposed "semi-mosaic" model for red blood cell membranes and the PLLPI model for nucleated mammalian membranes clearly indicate the universality of membrane asymmetry and the diversity of various membranes.

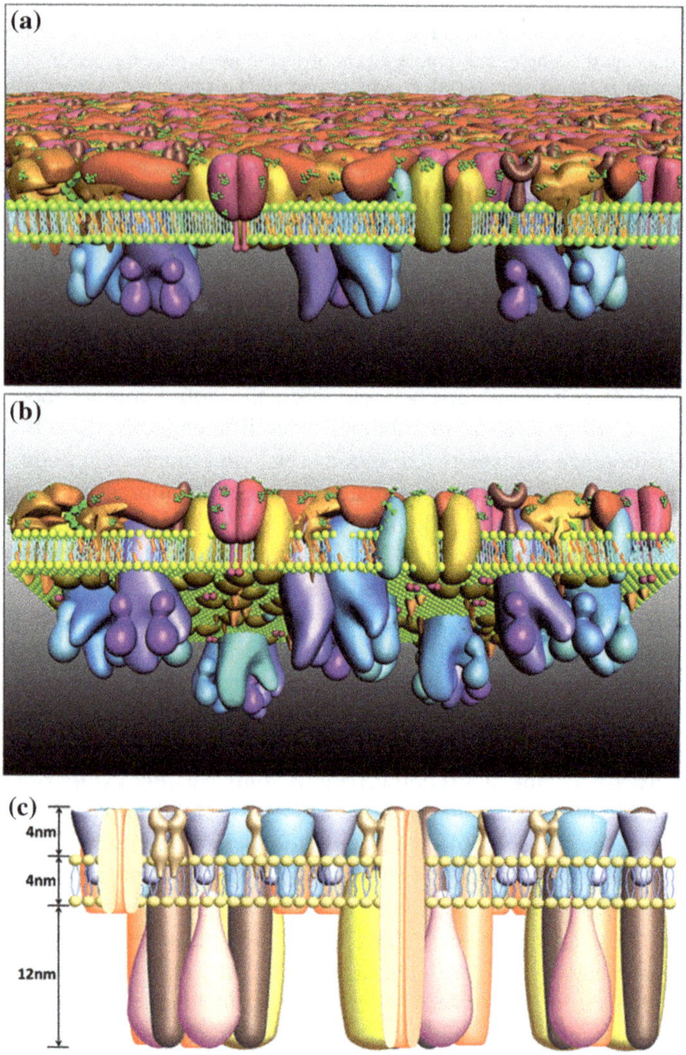

Fig. 3.24 Proposed protein layer-lipid-protein island (PLLPI) model of the cell membrane. **a** and **b** The top and bottom views of the cell membrane, respectively. The proteins on the ectoplasmic side of the cell membrane form a dense protein layer, which provides a smooth feature **a**; the proteins on the cytoplasmic side tend to form dispersed microdomains **b**. **c** The size of the cell membrane. The total height of the cell membrane is 20 nm, and the membrane is composed of the ectoplasmic protein layer (4 nm), the lipid bilayer (4 nm), and the cytoplasmic protein layer (12 nm). Reprinted from Ref. [66], with kind permission from the Public Library of Science

3.3.6 Structure of Organelle Membranes

Because of the long history of the hypothesis that all membranes in cells have a similar structure, organelle membranes were speculated to be the same as cell

Table 3.1 Comparison of membrane thicknesses and roughness (nm) [18]

	Red blood cells (nm) [53, 58–60]				Nucleated cells [66]	Golgi [67]	Mitochondrial [68]
	Human	Chicken	Turtle	Fish			
Thickness of lipid bilayer with the ectoplasmic side	2.9 ± 0.4	3.1 ± 0.7	2.5 ± 0.5	3.5 ± 0.6	8.0 ± 0.5	3.7 ± 1.6	7.1 ± 0.9
Total thickness of membranes	~10	16.5 ± 3.6	18.5 ± 2.4	18.2 ± 3.0	19.5 ± 2.8	~7.0	~10
Roughness of the ectoplasmic side membrane	0.18	0.47 ± 0.05	0.52 ± 0.11	0.56 ± 0.06	1.1 ± 0.2	0.43 ± 0.09	0.6 ± 0.2
Roughness of the cytoplasmic side membrane	1.9	3.4 ± 0.8	4.0 ± 0.8	3.1 ± 0.7	3.7 ± 0.2	1.95 ± 0.62	1.5 ± 0.4

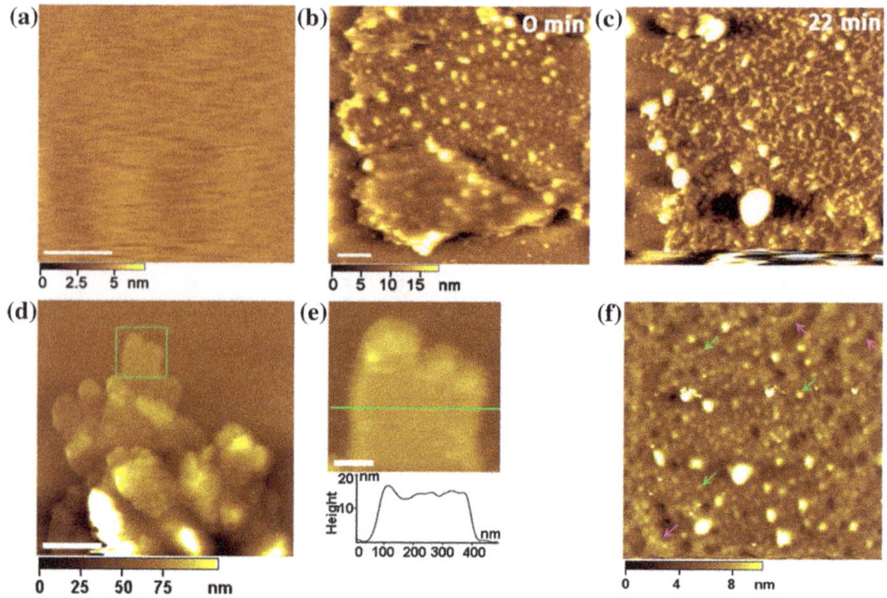

Fig. 3.25 AFM image of individual Golgi cisternae and mitochondrial membranes. **a** Image of the ectoplasmic side of Golgi apparatus membranes. The membrane is flat without protruding proteins. The scale bar is 100 nm. **b** Image of an opened Golgi cisternae. Protein domains are located in the cytoplasmic side of membranes. The scale bar is 500 nm. **c** Real-time images of a Golgi membrane treated with MβCD for 22 min. Cholesterol-enriched domains were extracted. The scale bar is 200 nm. **d** AFM topographic images of flattened mitoplasts from mitochondria subjected to a freeze–thaw treatment. The scale bar is 500 nm. **e** Higher magnification image (top) of the green square area in **d** with the cross-sectional analysis (bottom) along the green line. The membrane is flat without protruding proteins. The scale bar is 100 nm. **f** AFM topographic images of unilamellar inner mitochondrial membranes covered by dense proteins. The scale bar is 300 nm. Reproduced from Ref. [18] by permission of The Royal Society of Chemistry

membranes. The structures of membranes differ according to their functions. Two main organelle membranes (the Golgi apparatus and mitochondrial membranes) were imaged using in situ AFM (Fig. 3.25) [67, 68]. The mitochondrial and Golgi membranes exhibit features similar to red blood cell membranes, that is, the ectoplasmic surface is flat, with protruding proteins located in the cytoplasmic side (Table 3.1). In addition, whether there is a lipid raft domain in endocytotic membranes was unclear before Wang et al. directly visualized a Golgi vesicle that was eroded using Triton X-100 and MβCD [67]. The size of the lipid rafts in Golgi membranes is very similar to that in red blood cell membranes. The results from endomembrane structure studies confirm that the membrane protein distribution is asymmetrical, which is similar to the structure of red blood cell membranes.

In Table 3.1, there is the thickness of membranes (lipid bilayer + ectoplasmic side layer) without the inclusion of cytoplasmic protein particle measurements. The red blood cell and Golgi membrane exhibit very similar features. The nucleated mammalian cell membranes and mitochondrial membranes are substantially thicker than the red blood cell membranes. Notably, the thickness of a lipid bilayer is approximately 3.0 nm, which is close to the thickness of the red blood cell and Golgi membranes and indicates that there is no outer protein layer for these membranes [18].

3.4 Sample Preparation of Cell Membranes

3.4.1 Substrate for Membrane Samples

Mica is an appropriate substrate for high-resolution AFM imaging because its surface is smooth at the atomic level. However, fresh mica surfaces are negatively charged, which is not suitable for the immobilization of negatively charged cell membranes. A modification of mica by aminopropyltriethoxysilane (APTES) easily creates the positively charged surface; thus, membranes can be tightly attached on the surface for imaging. The setup for producing APTES modified mica (AP-mica) is shown in Fig. 3.26. Mica sheets with a thickness of about 0.5 mm and 20×20 mm size are cleaved with scotch tape for several times to make sure the smooth and clean surface, followed by being placed in the desiccator. After a desiccator is purged with argon for 5 min, 30 μl of APTES and 10 μl of N, N-diisopropylethylamine (DIPEA) are each placed into small containers at the bottom of the desiccator. Purged with argon for additional 5 min, the desiccator is sealed off. The mica is exposed to APTES vapor for 4 h. After this exposure, the APTES is removed from the desiccator carefully; the desiccator is purged with argon and sealed. The AP-mica is stored in the sealed desiccator under argon for used within a week [20, 52].

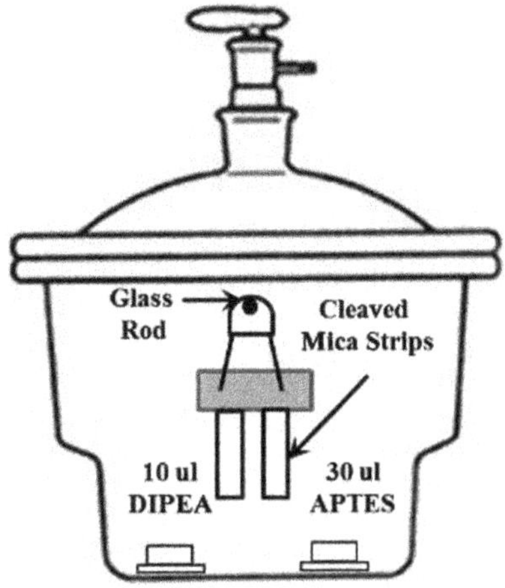

Fig. 3.26 Making AP-mica. Mica strips are clamped in metal clips suspended on a glass rod of appropriate length (to nest snugly across the upper part of the desiccator). The reagents DIPEA and APTES are placed in small containers on the bottom of the desiccator. Reprinted from Ref. [20], Copyright 2007, with permission from Elsevier

Glass cover slips also can be used for various cultured nucleated mammalian cells. Prior to cell culture, strict cleaning steps must be performed to obtain a clean surface on the cover slips. In general, glass cover slips are cleaned using a detergent (Micro-90) and sonicated in 1 M potassium hydroxide for 20 min at room temperature. The cover slips are rinsed with sterile distilled water (18 MΩ) and subsequently stored in absolute ethylalcohol. Prior to use, the cover slips are washed 3 times with sterile distilled water and dried with pure argon. The clean cover slips are placed in a culture dish as the substrate for cells culture [57, 69].

3.4.2 AFM Tip Modification

In order to identify specific proteins in a compositionally complex sample, an antibody against the protein of interest can be attached to the AFM tip. During the scanning process, the usual AFM topographic image is generated simultaneously and in exact spatial registration with a "recognition image", which identifies the locations of antibody–antigen binding events and thus the locations of the protein of interest in the field [15]. The two images can be electronically superimposed to obtain an accurate map of specific protein locations relative to the topography.

First, AFM tips are cleaned in a UV-cleaner for 15 min to get rid of any organic contamination on the tips. The tips are placed in a dish at the bottom of a desiccator and modified with APTES just as in preparing AP-mica. After the treatment process, the APTES is removed, and the treated tips (AP-tip) are stored in the sealed desiccator until used. About 1 mg of bifunctional PEG cross-linker (SensoPath Technologies, Inc.) and 5 μl of triethylamine are mixed in 1 ml $CHCl_3$. AP-modified tips are placed into the solution and treated for 2–3 h. The tips are washed with $CHCl_3$ and dried with argon. The succinimide group attaches to the amino group on the AP-tip (Fig. 3.27). For coupling of antibody, 40 μl of biotin IgG (2.2 mg/ml) was mixed with 380 μl of buffer A (100 mM NaCl, 50 mM NaH_2PO_4, 1 mM EDTA, pH 7.5, adjusted with NaOH) and 8 μl of 1 M $NaCNBH_3$ (freshly prepared by dissolving 32 mg of $NaCNBH_3$ in 500 μl of 10 mM NaOH). Tips with aldehyde-PEG linkers were immersed in this solution for 1 h; then, the tips are washed twice with buffer A and once with PBS buffer (100 mM NaCl, 50 mM Na_2HPO_4, pH 7.5) and stored in PBS at 4 °C until used (up to 3 days) [70, 71]. The functionalized tip is available for recognition imaging and force curves (see Chap. 4).

3.4.3 AFM Recognition Imaging with Modified Tips

Silicon-nitride cantilever tips for recognition imaging were modified as described above. Recognition imaging was performed in Magnetic AC mode (MAC mode) AFM 5500 with a PicoTREC recognition imaging attachment (Agilent

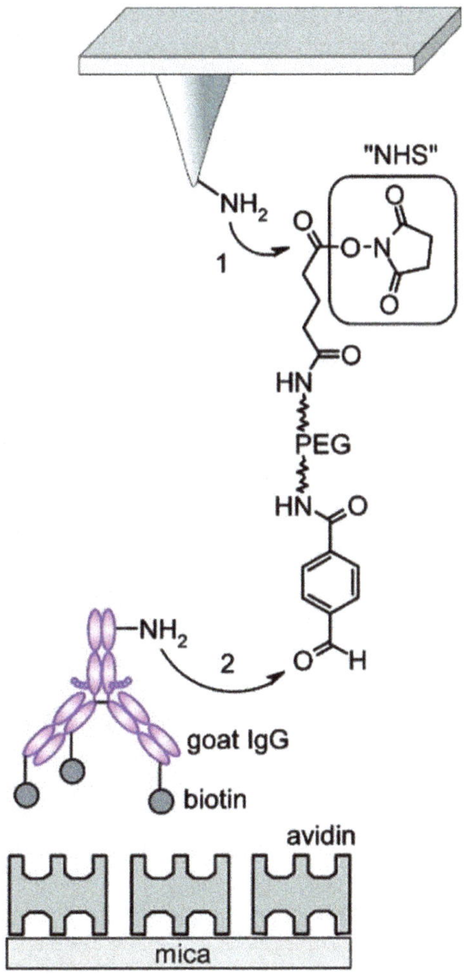

Fig. 3.27 Schematic representation of the testing system for tip antibody linking via lysine–aldehyde coupling. In the first step (in organic solvent), the NHS-ester of aldehyde–PEG–NHS reacts with the amino groups on the tip, while in the second step (in aqueous buffer), one of the antibody's lysines forms a Schiff base with the aldehyde function on the free end of the cross-linker. The Schiff base is reduced by $NaCNBH_3$ to form an irreversible bond (not shown). When brought into contact with avidin-covered mica, the biotin residues on the model antibody show reversible binding to, and unbinding from, immobilized avidin. Reprinted with the permission from Ref. [70], Copyright 2007 American Chemical Society

Technologies, Chandler, AZ). Topographic and recognition images were acquired with 6–8 nm amplitude oscillation at 9 kHz, imaging at 70% set point and scan speed at <1 Hz. For testing specificity of the recognition reaction, 50 μl of a 30 μg/ml solution of a peptide antigenic to the antibody (usually the one that was used to elicit the antibody response) is flowed into the cell by the flow through cells (Fig. 3.3) and the sample rescanned. If recognition is specific, the presence of the peptide will block the recognition to a very high degree (Fig. 3.19c, d).

Note: soft cantilevers with low frequency (6–9 kHz) should be used for recognition imaging. Stiffer cantilevers (higher frequency) do not produce good recognition images. Also, scan speeds >1 Hz should be avoided as they increase the leakage of topography into the recognition image.

It is very important to check the specificity of antibodies used for recognition imaging studies. We have noted a number of non-specific reactions, i.e., antibodies that recognize non-antigens, sometimes quite strongly, in both AFM recognition imaging (including force curves) and standard ELISA assays.

3.4.4 Preparation of Red Blood Cell Membranes

The shearing open method is appropriate for the preparation of a clean membrane with minimum damage. The following simple steps are used: two drops of blood are taken from a fingertip and centrifuged in 1 ml of PBSA buffer (136.9 mM NaCl, 2.7 mM KCl, 1.5 mM KH_2PO_4, and 8.1 mM Na_2HPO_4, pH 7.4) 5 times (1000 rpm for 2 min). A drop of red blood cells in PBSA buffer (100 μl) is subsequently deposited on the AP-mica surface for approximately 20 min of absorption. Next, PBSA is used to wash out the non-adsorbed cells. As shown in Fig. 3.28a, a syringe is adjusted to obtain a 20° to the sample surface, and 10 ml of hypotonic buffer (6.85 mM NaCl, 0.135 mM KCl, 0.075 mM KH_2PO_4, 0.405 mM Na_2HPO_4, pH 7.4) is injected to flush the mica surface and obtain a flat membrane patch [29].

Using the hypotonic lysis-centrifugation method, the whole erythrocyte ghost membranes can be prepared as follows. Briefly, erythrocytes are collected and cleaned by centrifugation in a physiological buffer solution. The cells are lysed using a low salt buffer (5 mM Na_2HPO_4, 0.2 mM EGTA) and washed 3 times by centrifugation at 20,000 g for 20 min at 0 °C until the ghost pellet becomes white. The membrane pellet is diluted in buffer for deposition on AP-mica for AFM imaging [53].

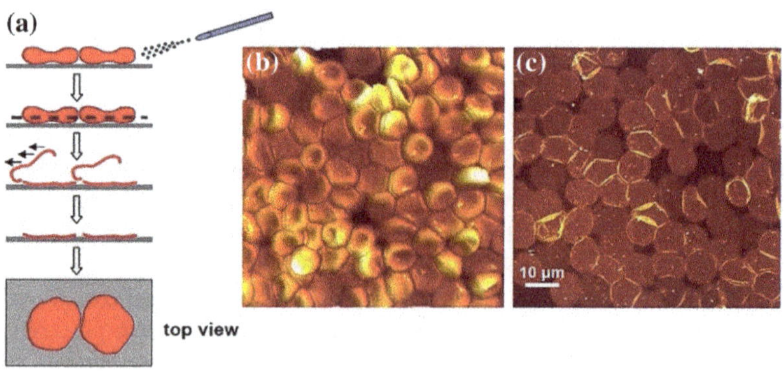

Fig. 3.28 Shearing open method of membrane preparation scheme. **a** RBCs are exposed to fluid flow-imposed shear stress to open the cells. **b** AFM images of RBCs attached to AP-mica. **c** Cytoplasmic side RBC membranes spread on surface after shear stress. Reprinted from Ref. [29] by permission of oxford university press

3.4.5 Preparation of the Nucleated Mammalian Cell Membranes

Cells were cultivated overnight on glass coverslips. The cell membranes are prepared by the shearing open method [72]. Briefly, the cells are washed twice with ice-cold buffer (20 mM PIPES, 150 mM KCl, pH 6.2), incubated with ice-cold hypotonic buffer (4 mM PIPES and 30 mM KCl, pH 6.2) for 3 min, and then sheared open by a stream of 10 ml of hypotonic buffer through a needle at an angle of 20°. The membranes are subsequently treated with high salt buffer (2 M NaCl, 1.5 mM KH_2PO_4, 2.7 mM KCl, and 1 mM Na_2HPO_4, pH 7.2) for 30 min at room temperature to remove the cytoskeletons. The prepared membranes are immediately imaged in PBSA buffer using AFM (Fig. 3.22a).

Ultrasonic stimulation is an alternative method for the preparation of the cytoplasmic side of membranes, as shown in Fig. 3.29. Cells on cover slips were washed once with 4-(2-hydroxyethyl)-1-piperazineethanesulfonic acid (HEPES)-based Ringer's solution (155 mM NaCl, 3 mM KCl, 2 mM $CaCl_2$, 1 mM $MgCl_2$, 3 mM NaH_2PO_4, 10 mM glucose in 5 mM HEPES, pH 7.4), and then with Ca^{2+}-free Ringer's solution. Subsequently, the cover slips were soaked for about 10 s in poly-lysine solution (0.5 mg/ml poly-lysine dissolved in Ca^{2+}-free Ringer's solution), and then washed three times for a few seconds each in hypotonic Ringer's solution prepared by mixing one part of Ringer's solution with two parts of distilled water. This induced cell swelling, which enabled the cells to burst easily in the following ultrasonic stimulation. Immediately after immersing in hypotonic solution, the cells were exposed to a small bubble jet by weak ultrasonic vibration in isotonic KHMgE buffer (30 mM HEPES, pH 7.4, 70 mM KCl, 3 mM $MgCl_2$, 1 mM ethylene glycol tetraacetic acid, 1 mM dithiothreitol, 0.1 mM AEBSF (4-(2-aminoethyl) benzenesulfonyl fluoride hydrochloride)). Cells unroofed by the bubble jet were washed briefly in fresh KHMgE buffer used for AFM imaging [30].

Two strategies are used to prepare the ectoplasmic side of membranes. First, the ectoplasmic side of membranes can be imaged at the flat edge of a living cell under

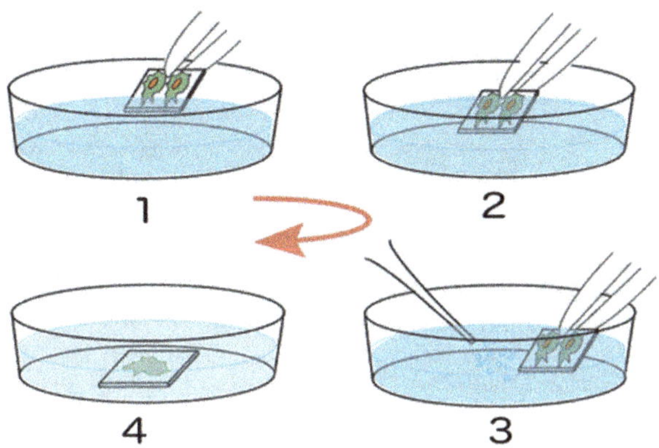

Fig. 3.29 Scheme for unroofing cells. The cells are washed three times in hypotonic, and subsequently unroofed by ultrasonic stimulation, which removes the apical cell membrane and the cytoplasm for AFM imaging. Reprinted from Ref. [30] by permission of Oxford University Press

native conditions (Fig. 3.20a). Second, the ectoplasmic side of membranes can be obtained using the hypotonic lysis-centrifugation method [66]. Briefly, cells are first incubated with 60 μM nocodazole (Sigma–Aldrich) and 20 μM cytochalasin B (Sigma–Aldrich) for 50 min at 37 °C to destroy the actin filaments and microtubules. Then, the cells are subsequently digested using trypsin (1 mg/ml) and washed with 1 ml of PBSA buffer three times. The cells are treated with 1 mg/ml DNase to digest the nuclei/DNA and then centrifuged at 3000 rpm for 10 min. The cell membrane precipitate is dissolved in PBSA buffer and deposited on AP-mica for AFM imaging.

Caution is required when imaging the ectoplasmic side of membrane because the sharp AFM cantilever can destroy the relatively soft ectoplasmic side lipid bilayer membrane and induce imaging artifacts. Tapping mode (or MAC mode) in solution and a regular tip radius (approximately 20 nm) are better for imaging a soft lipid bilayer. In addition, the use of glutaraldehyde to fix a cell may cause cross-linking of membrane proteins and produce artifacts in the membrane appearance.

It is worth noting that to obtain the native structure of cell membranes without damage, it is not recommended to fix the membrane using chemicals (e.g., glutaraldehyde, paraformaldehyde) because they may crosslink the membrane proteins and result in local destruction of the membranes (e.g., blurred images because of protein cross-linking).

Acknowledgements This work was supported by the National Key R&D Program of China (No. 2017YFA0505300), the National Natural Science Foundation of China (No. 21525314 2, 1503213, 21703231, 21721003).

References

1. Binnig G, Quate C, Gerber C (1986) Atomic force microscope. Phys Rev Lett 56:930–933
2. Alessandrini A, Facci P (2005) AFM: a versatile tool in biophysics. Meas Sci Technol 16(6): R65–R92
3. Muller DJ, Dufrene YF (2008) Atomic force microscopy as a multifunctional molecular toolbox in nanobiotechnology. Nat Nanotechnol 3(5):261–269
4. Goksu EI, Vanegas JM, Blanchette CD et al (2009) AFM for structure and dynamics of biomembranes. Biochim Biophys Acta 1788(1):254–266
5. Whited AM, Park PSH (2014) Atomic force microscopy: a multifaceted tool to study membrane proteins and their interactions with ligands. Biochim Biophys Acta 1838(1):56–68
6. Garcia R, Perez R (2002) Dynamic atomic force microscopy methods. Surf Sci Rep 47(6–8):197–301
7. Han WH, Lindsay SM, Jing TW (1996) A magnetically driven oscillating probe microscope for operation in liquids. Appl Phys Lett 69(26):4111–4113
8. Ge GL, Han D, Lin DY et al (2007) MAC mode atomic force microscopy studies of living samples, ranging from cells to fresh tissue. Ultramicroscopy 107(4–5):299–307
9. Hansma PK, Cleveland JP, Radmacher M et al (1994) Tapping mode atomic force microscopy in liquids. Appl Phys Lett 64(13):1738–1740
10. Basak S, Raman A (2007) Dynamics of tapping mode atomic force microscopy in liquids: theory and experiments. Appl Phys Lett 91(6):064107

11. Allison DP, Mortensen NP, Sullivan CJ et al (2010) Atomic force microscopy of biological samples. Wiley Interdisc Rev Nanomed Nanobiotechnology 2(6):618–634
12. Muller DJ, Dufrene YF (2011) Atomic force microscopy: a nanoscopic window on the cell surface. Trends Cell Biol 21(8):461–469
13. Heinisch JJ, Lipke PN, Beaussart A et al (2012) Atomic force microscopy—looking at mechanosensors on the cell surface. J Cell Sci 125(18):4189–4195
14. Dufrene YF, Ando T, Garcia R et al (2017) Imaging modes of atomic force microscopy for application in molecular and cell biology. Nat Nanotechnol 12(4):295–307
15. Stroh C, Wang H, Bash R et al (2004) Single-molecule recognition imaging-microscopy. Proc Natl Acad Sci USA 101(34):12503–12507
16. Hinterdorfer P, Dufrene YF (2006) Detection and localization of single molecular recognition events using atomic force microscopy. Nat Methods 3(5):347–355
17. Dufrene YF, Martinez-Martin D, Medalsy I et al (2013) Multiparametric imaging of biological systems by force-distance curve-based AFM. Nat Methods 10(9):847–854
18. Shan YP, Wang HD (2015) The structure and function of cell membranes examined by atomic force microscopy and single-molecule force spectroscopy. Chem Soc Rev 44(11): 3617–3638
19. Wang HD, Bash R, Yodh JG et al (2002) Glutaraldehyde modified mica: a new surface for atomic force microscopy of chromatin. Biophys J 83(6):3619–3625
20. Lohr D, Bash R, Wang H et al (2007) Using atomic force microscopy to study chromatin structure and nucleosome remodeling. Methods 41(3):333–341
21. Wang H, Bash R, Yodh JG et al (2004) Using atomic force microscopy to study nucleosome remodeling on individual nucleosomal arrays in situ. Biophys J 87(3):1964–1971
22. Wang HD, Obenauer-Kutner L, Lin M et al (2008) Imaging glycosylation. J Am Chem Soc 130(26):8154–8155
23. Wang HD, Dalal Y, Henikoff S et al (2008) Single-epitope recognition imaging of native chromatin. Epigenetics Chromatin 1:10
24. Muller DJ, Engel A (2007) Atomic force microscopy and spectroscopy of native membrane proteins. Nat Protoc 2(9):2191–2197
25. Braga PC, Ricci D (2011) Atomic force microscopy in biomedical research: methods and protocols. Methods Mol Biol Methods Protoc 736 (Springer Science + Business Media, LLC, Totowa, NJ):223–241
26. Fotiadis D (2012) Atomic force microscopy for the study of membrane proteins. Curr Opin Biotechnol 23(4):510–515
27. Fotiadis D, Liang Y, Filipek S et al (2003) Atomic-force microscopy: rhodopsin dimers in native disc membranes. Nature 421(6919):127–128
28. Orsini F, Santacroce M, Arosio P et al (2010) Observing *Xenopus laevis* oocyte plasma membrane by atomic force microscopy. Methods 51(1):106–113
29. Schillers H (2008) Imaging CFTR in its native environment. Pflugers Arch Euro J Physiol 456(1):163–177
30. Usukura J, Yoshimura A, Minakata S et al (2012) Use of the unroofing technique for atomic force microscopic imaging of the intra-cellular cytoskeleton under aqueous conditions. J Electron Microsc 61(5):321–326
31. Murakoshi M, Iida K, Kumano S et al (2009) Immune atomic force microscopy of prestin-transfected CHO cells using quantum dots. Pflugers Arch Eur J Physiol 457(4): 885–898
32. Dindia L, Faught E, Leonenko Z et al (2013) Rapid cortisol signaling in response to acute stress involves changes in plasma membrane order in rainbow trout liver. Am J Physiol Endocrinol Metab 304(11):E1157–E1166
33. Frankel DJ, Pfeiffer JR, Surviladze Z et al (2006) Revealing the topography of cellular membrane domains by combined atomic force microscopy/fluorescence imaging. Biophys J 90(7):2404–2413
34. Muller DJ, Hand GM, Engel A et al (2002) Conformational changes in surface structures of isolated connexin 26 gap junctions. EMBO J 21(14):3598–3607

35. Singer SJ, Nicolson GL (1972) The fluid mosaic model of the structure of cell membranes. Science 175(4023):720–731
36. Nicolson GL (2014) The fluid-mosaic model of membrane structure: still relevant to understanding the structure, function and dynamics of biological membranes after more than 40 years. Biochim Biophys Acta 1838(6):1451–1466
37. Simons K, Ikonen E (1997) Functional rafts in cell membranes. Nature 387(6633):569–572
38. Simons K, Gerl MJ (2010) Revitalizing membrane rafts: new tools and insights. Nat Rev Mol Cell Biol 11(10):688–699
39. Simons K, Toomre D (2000) Lipid rafts and signal transduction. Nat Rev Mol Cell Biol 1(1): 31–39
40. Lingwood D, Simons K (2010) Lipid rafts as a membrane-organizing principle. Science 327 (5961):46–50
41. Neumann-Giesen C, Falkenbach B, Beicht P et al (2004) Membrane and raft association of reggie-1/flotillin-2: role of myristoylation, palmitoylation and oligomerization and induction of filopodia by overexpression. Biochem J 378:509–518
42. Eggeling C, Ringemann C, Medda R et al (2009) Direct observation of the nanoscale dynamics of membrane lipids in a living cell. Nature 457(7233):1159
43. El Kirat K, Morandat S (2007) Cholesterol modulation of membrane resistance to Triton X-100 explored by atomic force microscopy. Biochimica Et Biophysica Acta-Biomembranes 1768(9):2300–2309
44. Orsini F, Cremona A, Arosio P et al (2012) Atomic force microscopy imaging of lipid rafts of human breast cancer cells. Biochim Biophys Acta 12:2943–2949
45. Mongrand S, Stanislas T, Bayer EMF et al (2010) Membrane rafts in plant cells. Trends Plant Sci 15(12):656–663
46. Cai MJ, Zhao WD, Shang X et al (2012) Direct evidence of lipid rafts by in situ atomic force microscopy. Small 8(8):1243–1250
47. Wu L, Huang J, Yu X et al (2014) AFM of the ultrastructural and mechanical properties of lipid-raft-disrupted and/or cold-treated endothelial cells. J Membr Biol 247(2):189–200
48. Chichili GR, Rodgers W (2007) Clustering of membrane raft proteins by the actin cytoskeleton. J Biol Chem 282(50):36682–36691
49. Sun M, Northup N, Marga F et al (2007) The effect of cellular cholesterol on membrane-cytoskeleton adhesion. J Cell Sci 120(13):2223–2231
50. Engelman DM (2005) Membranes are more mosaic than fluid. Nature 438(7068):578–580
51. Jiang JG, Hao X, Cai MJ et al (2009) Localization of $Na^{+}-K^{+}$ ATPases in quasi-native cell membranes. Nano Lett 9(12):4489–4493
52. Shan YP, Wang ZY, Hao XA et al (2010) Locating the Band III protein in quasi-native cell membranes. Anal Methods 2(7):805–808
53. Wang HD, Hao X, Shan YP et al (2010) Preparation of cell membranes for high resolution imaging by AFM. Ultramicroscopy 110(4):305–312
54. Plummer TH, Elder JH, Alexander S et al (1984) Demonstration of peptide-N-glycosidase-F activity in endo-beta-N-acetylglucosaminidase F preparations. J Biol Chem 259(17):700–704
55. Johnson SJ, Bayerl TM, McDermott DC et al (1991) Structure of an adsorbed dimyristoylphosphatidylcholine bilayer measured with specular reflection of neutrons. Biophys J 59(2):289–294
56. Triplett RB, Carraway KL (1972) Proteolytic digestion of erythrocytes, resealed ghosts and isolated membranes. Biochemistry 11(15):2897–2903
57. Wu JZ, Gao J, Qi M et al (2013) High-efficiency localization of Na+−K+ ATPases on the cytoplasmic side by direct stochastic optical reconstruction microscopy. Nanoscale 5(23): 11582–11586
58. Tian YM, Cai MJ, Zhao WD et al (2014) The asymmetric membrane structure of erythrocytes from Crucian carp studied by atomic force microscopy. Chin Sci Bull 59(21):2582–2587
59. Tian YM, Cai MJ, Xu HJ et al (2014) Atomic force microscopy of asymmetric membranes from turtle erythrocytes. Mol Cells 37(8):592–597

60. Tian YM, Cai MJ, Xua HJ et al (2014) Studying the membrane structure of chicken erythrocytes by in situ atomic force microscopy. Anal Methods 6(20):8115–8119
61. Allen MJ, Hud NV, Balooch M et al (1992) Tip-radius-induced artifacts in AFM images of protamine-complexed DNA fibers. Ultramicroscopy 42:1095–1100
62. Chiari Y, Cahais V, Galtier N et al (2012) Phylogenomic analyses support the position of turtles as the sister group of birds and crocodiles (Archosauria). BMC Biol 10(1):65
63. Lee S, Mandic J, Van Vliet KJ (2007) Chemomechanical mapping of ligand–receptor binding kinetics on cells. Proc Natl Acad Sci 104(23):9609–9614
64. Zhang J, Chtcheglova LA, Zhu R et al (2014) Nanoscale organization of human GnRH-R on human bladder cancer cells. Anal Chem 86(5):2458–2464
65. Lau JM, You HX, Yu L (2002) Lattice-like array particles on Xenopus oocyte plasma membrane. Scanning 24(5):224–231
66. Zhao WD, Tian YM, Cai MJ et al (2014) Studying the nucleated mammalian cell membrane by single molecule approaches. PLoS One 9(5):13
67. Xu HJ, Su WH, Cai MJ et al (2013) The asymmetrical structure of golgi apparatus membranes revealed by in situ atomic force microscope. PLoS One 8(4):e61596
68. Tian YM, Li JH, Cai MJ et al (2013) High resolution imaging of mitochondrial membranes by in situ atomic force microscopy. RSC Adv 3(3):708–712
69. Wang Y, Gao J, Guo XD et al (2014) Regulation of EGFR nanocluster formation by ionic protein-lipid interaction. Cell Res 24(8):959–976
70. Ebner A, Wildling L, Kamruzzahan ASM et al (2007) A new, simple method for linking of antibodies to atomic force microscopy tips. Bioconjug Chem 18(4):1176–1184
71. Wildling L, Unterauer B, Zhu R et al (2011) Linking of sensor molecules with amino groups to amino-functionalized AFM tips. Bioconjug Chem 22(6):1239–1248
72. Ziegler U, Vinckier A, Kernen P et al (1998) Preparation of basal cell membranes for scanning probe microscopy. FEBS Lett 436(2):179–184

Chapter 4 Detection of Membrane Mechanical Properties and Endocytosis by Single Molecule Force Spectroscopy

Yuping Shan

4.1 Introduction

AFM is an ideal method to study the surface topography of biological membranes. It allows membranes that are adsorbed to flat solid supports to be raster-scanned in physiological solutions with an atomically sharp tip [1]. Therefore, AFM is capable of observing biological molecular machines at work [2]. AFM offers great potential for characterizing single molecule and thus presents many advantages in scientific discoveries [3]. Paul Hansma's group detected adhesive interaction forces that they attributed to either the rupture of individual hydrogen bonds or hydrogen interacting with ordered water layers near the surface [4]. After this early report, a significant amount of research effort was devoted to using the AFM cantilever as a force sensing device in biological research.

Gaub. H et al. reported the first biological application of SMFS on detection of ligand-receptor pairs interaction in 1994 [5]. And since then, SMFS measurements have particularly attracted scientists' attention. We do not think it is exaggerated to state that many biologists dream to "play" with a single molecule. Along with optical tweezer experiments, AFM-based single molecule force measurements represent the dawn of what is now termed "single molecule experiments". The conceptual advance is that single molecule experiments provided different and novel information that could not be assessed by bulk experiments [6]. Over the past two decades, the SMFS technique has been focused on the manipulation of proteins [7, 8], DNA [9, 10], sugars [11], and components of bacterial surfaces [12].

Taken advantages of AFM highly sensitive tip cantilever, force–distance curve could detect forces in a large range from 10 pN to 100 nN, which is wider than the

Y. Shan (✉)
School of Chemistry and Life Science, Advanced Institute of Materials Science, Changchun University of Technology, Changchun 130012, China
e-mail: shanyp@ciac.ac.cn

H. Wang and G. Li (eds.), *Membrane Biophysics*,
https://doi.org/10.1007/978-981-10-6823-2_4

range of other competing techniques [13, 14]. SMFS measurements have been performed on a large variety of biological systems. For example, force spectroscopy measurements were performed on membrane proteins providing information about folding and membrane insertion of the hydrophobic protein domains. Measurements of the interactions between biotin and avidin [5, 15], complementary strands of DNA [16], and an antibody with its antigen [17] were conducted, respectively. On living cell, detecting the mechanical properties and nanoscale forces arising in the cellular dynamic process including endocytosis and direct penetration plays a remarkable role on understanding the physiology and function of living cells and organisms [18, 19]. In this chapter, we will describe several force spectroscopy techniques including force–distance curve, nanoindentation, and force tracing as well as their applications on biological molecules and living cell membrane. We finally provide the practical protocols in SMFS techniques on preparing AFM tips and samples, conducting force–distance/tracing curves, and data analysis.

4.2 Principle

In force spectroscopy, the AFM cantilever is moved perpendicular to the sample (*Z*-direction), and its position is recorded. Molecules adsorbed on a surface are picked up by a microscopic tip (nanometers wide) that is located on the end of an elastic cantilever. In a more sophisticated version of this experiment (chemical force microscopy) the tips are covalently functionalized with the molecules of interest [20]. A piezoelectric controller then pulls up the cantilever. If some force is acting on the elastic cantilever, for example, because some molecule is being stretched between the surface and the tip, the cantilever will deflect upward (repulsive force) or downward (attractive force). Deflection is measured by the position of a laser beam reflected by the cantilever. The cantilever can be treated as a spring and described with Hooke's law after its spring constant has been determined [21]. Therefore, force acting on the cantilever and force applied by the cantilever to the surface can be recorded with pN (10^{-12}N) sensitivity [18], and cannot achieve much better resolution only because of thermal noise.

4.2.1 Force–Distance Curve

Force–distance curve is a plot of the deflection of the cantilever, measured by a position-sensitive photodetector, versus the extension of the piezoelectric scanner. In general, force–distance (F versus d) curves are used to measure the vertical force that the tip applies to the surface while a contact-AFM image is being taken. The force F between the tip and the sample is related to the cantilever's deflection through Hooke's law:

$$F = k.\alpha.V \tag{4.1}$$

where k is the cantilever's spring constant, V is the measured cantilever's deflection in volt, and α is called the deflection sensitivity that converts cantilever's deflection from volt to nanometers. In order to quantitatively measure interaction forces, it is necessary to calibrate the bending stiffness (or spring constant) of the AFM cantilever as accurately as possible. In a force–distance curve measurement, the sample is mounted on the sample stage that is drove by applying a voltage to the piezoelectric translator, while measuring the cantilever deflection [22]. The principle of force-distance detection of the typical ligand and receptor interactions on living cells by AFM is illustrated in Fig. 4.1. In this force-distance mode, the deflection angle of the cantilever is measured as a function of the vertical position of the cantilever [23]. The tip is moved toward the cell surface (solid line, 1–2) and then retracted (dotted line, 3–4) at a constant lateral position. During tip approaching, the ligand specifically binds with a receptor, which will lead to a force signal with a distinct shape. Upon tip retraction (3), the force increases until bond rupture occurs (4) at an unbinding force (*fu*). However, for some type of forces the force signal is detected during the approaching process. Usually, a specific type of force will shows a distinct force–distance curve, and the typical force–distance curves for the corresponding interaction pathway are shown in Fig. 4.2.

4.2.2 Force Extension

Because of the high efficiency and piconewton force sensitivity, force measuring mode of the AFM has been used to study the elastic properties of polymers and proteins with mechanical functions. In the force measuring mode [24], deflections

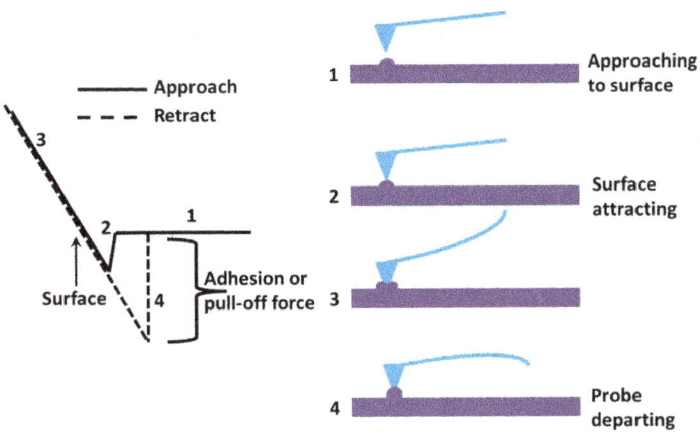

Fig. 4.1 Schematic of single molecule force–distance curve measuring

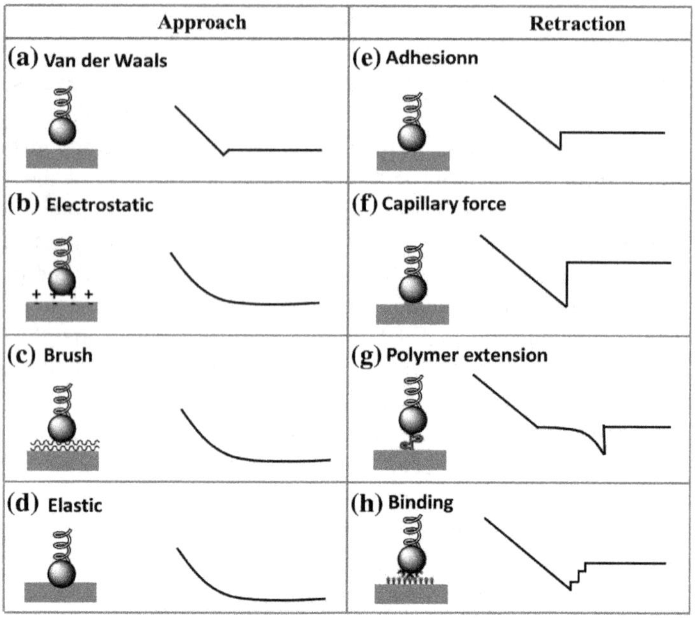

Fig. 4.2 The various detectable force types and corresponding typical force–distance curve

result from the forced extension of a sample attached to the cantilever tip. The extension data is obtained by measuring the change in displacement between the two ends of the protein or other molecule of interest. The accuracy of this data is determined by the quality of the piezo-stage. In this configuration, the cantilever is mounted above an absorptive substrate whose height is controlled by a single axis piezoelectric positioner with angstrom resolution. When the both ends of a single polymer/protein are adsorbed between the cantilever and a substrate respectively, recession of the sample substrate causes elongation of the polymer or multimodular protein, resulting in a force signal that causes deflection of the cantilever. The schematic of the force-extension mode is shown in Fig. 4.3 [25]. The exact extension of the stretched protein can be obtained only if the attachment points can be precisely located, e.g., by using microfabricated cantilevers. The exact position of the tip on the soft cantilever, relative to the fixed surface, such as a cover slip, can also be determined directly with the use of a calibrated evanescent field [26]. To describe force-extension curves, the analytical approximation is often used to fit the worm-like chain (WLC) model to experimental force-extension curves [27]:

$$F(x) = \frac{k_B T}{l_p}\left(\frac{1}{4\left(1-\frac{x}{L}\right)^2} - \frac{1}{4} + \frac{x}{L}\right) \tag{4.2}$$

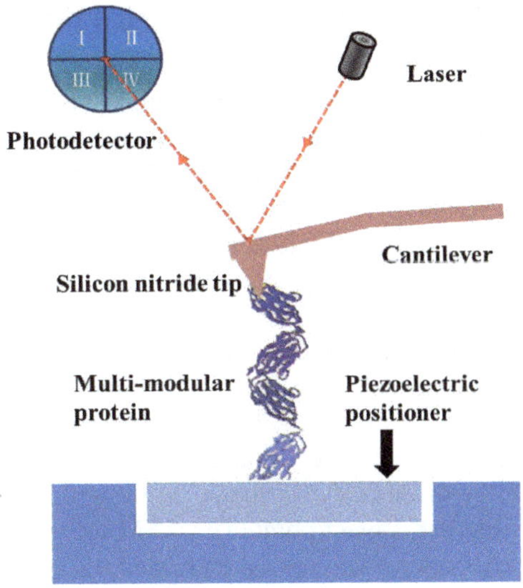

Fig. 4.3 Setup schematic of force-extension mode force spectroscopy

The parameters T is the absolute temperature, k_B presents the Boltzmann constant, L is contour length, and $\imath_p$ is persistence length.

4.2.3 Dynamic Process of Endocytosis on Living Cells

It is well known that fishing is an act of catching the fish via a bait attached on one end of the fish line. Inspired with the concept of "fishing," nanoparticles and molecules could also be attached onto a tiny AFM tip to detect the force of single nanoparticle/molecule interaction with living cells under native conditions. Therefore, AFM-based SMFS was used to detect the dynamic processes on living cell membrane, such as the endocytosis of nanoparticle and transporting of biological molecules [28, 29]. The force signal is based on the detection of small shifts of the cantilever that occurs when a tip-tethered molecule/particle was taken by the living cells. The schematic of details about detecting the endocytosis by force spectroscopy on living cell is illustrated in Fig. 4.4. (1) The AFM tip tethered with nanoparticle/molecule initially moves toward the living cell which allows the nanoparticle/molecule to be contacted with the cell membranes. (2) As the tip lands on the cell surface, the nanoparticle/molecule on the pinpoint of AFM tip contact with the cell membranes and the cellular uptake occurs, which produces a force signal during approaching process. During the subsequent tip retraction process (3), a transient force signal is observed, which points to occurrence of the retraction of the nanoparticle/molecule from cell membranes (4). In most case, the force signals of both cellular uptake and retraction of nanoparticle could be simultaneously detected, which is clearly different from the sole unbinding event in the force–distance curve.

4.2.4 Force Tracing Based on AFM

According to the principle of force–distance curve, the AFM tip approaching process may disturb the dynamic process of the cellular uptake, a static AFM tip cantilever to hold the molecule/nanoparticle is necessary. A novel AFM-based technique, force tracing, was developed, which provides an unprecedented temporal-spatial resolution for the in situ observation of dynamic process on living cell membrane surface. The setup for force tracing with an AFM instrument is shown in Fig. 4.5 [30]. A laser beam is reflected by the AFM cantilever. The photodetector detects the laser position and records the change in the cantilever deflection. The AFM tip is approached to the cell surface with contact mode of AFM; subsequently, force–distance curves measuring are performed on the living cell to locate the contact point between the AFM tip and the cell surface. In a force-distance cycle, the AFM tip will stop at the position of Z scan start that is far from the contact point. The AFM tip is slowly moved to the contact point with the feedback system, and then turned off the feedback; thus, the AFM tip is suspended

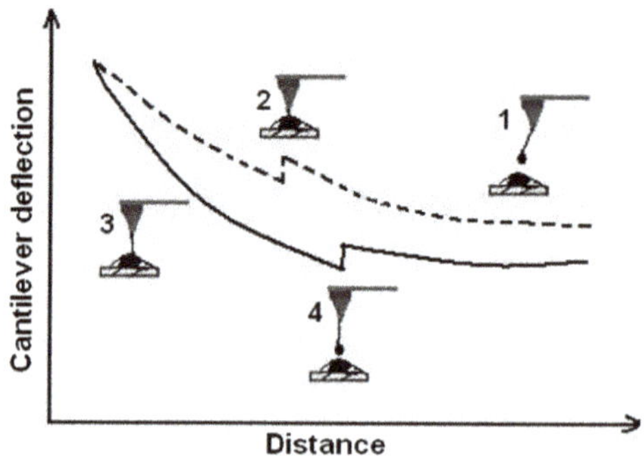

Fig. 4.4 Schematic of detecting dynamic process on living cell membrane by force spectroscopy

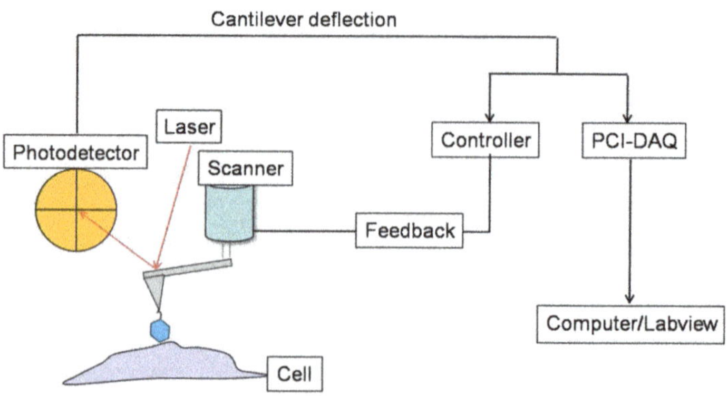

Fig. 4.5 The schematic setup of force tracing using AFM and a PCI-DAQ

on the cell surface. Upon the application of an external force (the cells take in the molecule/particle tethered on AFM tip and stretch the bifunctional crosslinker), the cantilever is deflected, and the change of vertical position of the laser on the photodetector is acquired using a PCI card that can readily monitor rapid processes down to 1 μs; this card is suitable for recording the process of viral entry into living cells. The software is written in LabVIEW, and the sampling rate is set at suitable value for recording the cellular uptake dynamic process of different nanoparticles and molecules. The displacement of the molecule/nanoparticle during the trans-membrane process includes two parts (Bending distance of the cantilever d and the extension length of the PEG linker h):

$$D = d + h \tag{4.3}$$

4.2.5 Nanoindentation on Cell Membrane

Nanoindentation is one of the tools to analyze samples at a microstructural level, which allows to measure hardness and elastic properties of materials at sub-micrometer depths [31]. It has already been applied to the major classes of materials [32] and is increasingly being used to characterize biological and biomedical materials [33–35]. In a traditional indentation test, a hard tip whose mechanical properties are known (frequently made of a very hard material like diamond) is pressed into a sample whose properties are unknown. The load placed on the indenter tip is increased as the tip penetrates further into the specimen and soon reaches a user-defined value. At this point, the load may be held constant for a period or removed. The area of the residual indentation in the sample is measured and the hardness, H, is defined as the maximum load, $P_{\max}$, divided by the residual indentation area, A_r:

$$H = \frac{P_{\max}}{A_r} \tag{4.4}$$

A complete cycle of loaded and unloaded indentation will create a load–displacement curve (such as the one shown in Fig. 4.6) [36]. This curve can be used to extract mechanical properties of the material. It has already been applied to the major classes of materials [32] and is increasingly being used to characterize biological and biomedical materials [33–35]. The ability to conduct nanoindentation studies with nanometer depth and sub-nanonewton force resolution is possible using a standard AFM setup. The AFM allows nanomechanical studies to be conducted alongside topographic analyses, without the use of other dedicated instruments. Load–displacement curves can be collected similarly for a variety of materials, and mechanical properties can be directly calculated from these curves [37]. An AFM nanoindentation experiments is shown in Fig. 4.7 [38]. During its

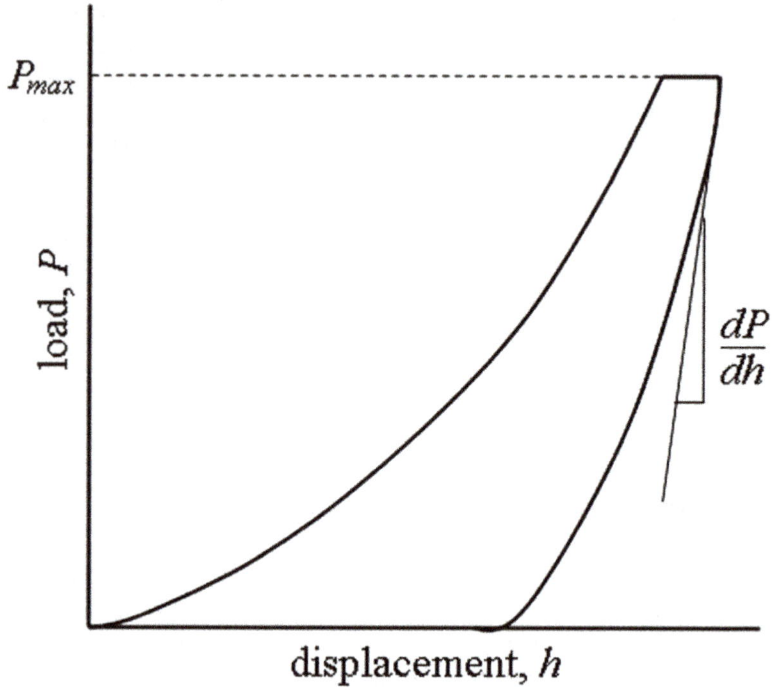

Fig. 4.6 Schematic of load–displacement curve for an instrumented nanoindentation test. Reproduced from Ref. [36] by permission of Cambridge University

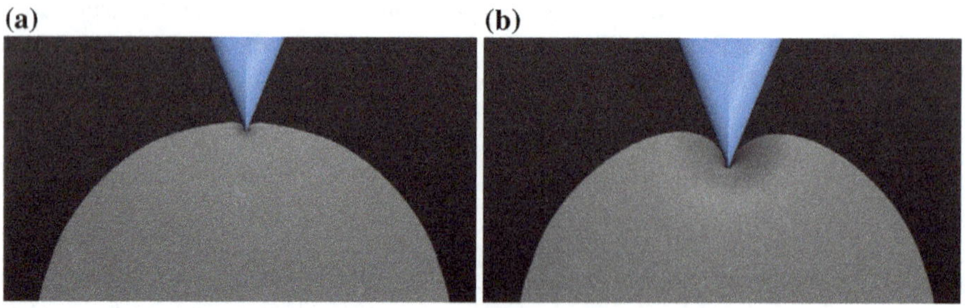

Fig. 4.7 Sketch of the indentation of the cell wall by a pyramidal shape tip. **a** The tip penetrates the wall without noticeably change in the curvature of the cell wall. **b** For a deeper indentation, the wall curvature is modified by the pyramidal tip. Reprinted from Ref. [38], Copyright 2015, with permission from Elsevier

progression into the cell, the sharp pyramidal tip of the AFM cantilever penetrates first into the cell wall with a minor modification of its curvature (Regime A, Fig. 4.7a) and then bends the cell wall, acting on a thin viscoelastic shell (Regime B, Fig. 4.7b). For a given deflection of the cantilever (the contact force), the total displacement of the AFM piezo transducer is the sum of the cantilever deflection $\delta_k = F/k$, the depth of penetration of the tip inside the wall δ_p, and the deformation (change of curvature) of the wall δ_b. The force and penetration depth are measured, generating loading-unloading curves from which the hardness and elastic response of the material under load is evaluated. Furthermore, data can be collected with high spatial resolution, for example, the investigation of complex tissues and "mapping" of properties [39].

4.3 Applications of Force Spectroscopy on Cell Membranes

Membranes define the cell boundary and surround the various cellular compartments. Such compartmentalization has been essential for the development of life and reflects an important multifunctional tool of organisms. Cellular membranes allow the uptake of nutrients, small molecules and ions, release waste products, bind ligands, transmit signals, convert energy, sense the environment, maintain cell adhesion, control cell migration, and much more while forming a tight barrier [3]. How do the cell membranes fulfill those functions attracts the attention of biologists. Different microscopic approaches for characterizing cellular membranes have been developed. Here, we provide a flavor of the fascinating opportunities offered by the use of AFM-based SMFS as a nanobiotechnological tool in modern membrane biology. The application of SMFS on protein unfolding dynamics, cell adhesion, cell membrane protein, and dynamic endocytosis of living cells would be described.

4.3.1 Protein Unfolding Dynamics

Since the ability of the SMFS based on AFM to measure elasticity was reported in 1997 by Rief et al., the method has proved to stretch a variety of biological polymers, such as polysaccharides [40], DNA [41], and proteins [42]. Hermann E. Gaub used SMFS to investigate the mechanical properties of titin, the giant sarcomeric protein of striated muscle. Individual titin molecules were repeatedly stretched, and the applied force was recorded as a function of the elongation. At large extensions, the restoring force exhibited a sawtooth like pattern, with a periodicity that varied between 25 and 28 nm. Carrion-vazquez et al., compared the process of unfolding the protein by the mechanical SMFS and chemical method. They unfolded I27 domains by SMFS and chemical denaturant, respectively, the unfolding rates obtained by the two methods are same [43]. Furthermore, the transition state for unfolding appears at the same position on the folding pathway when assessed by either method, as shown in Fig. 4.8. They found that mechanical

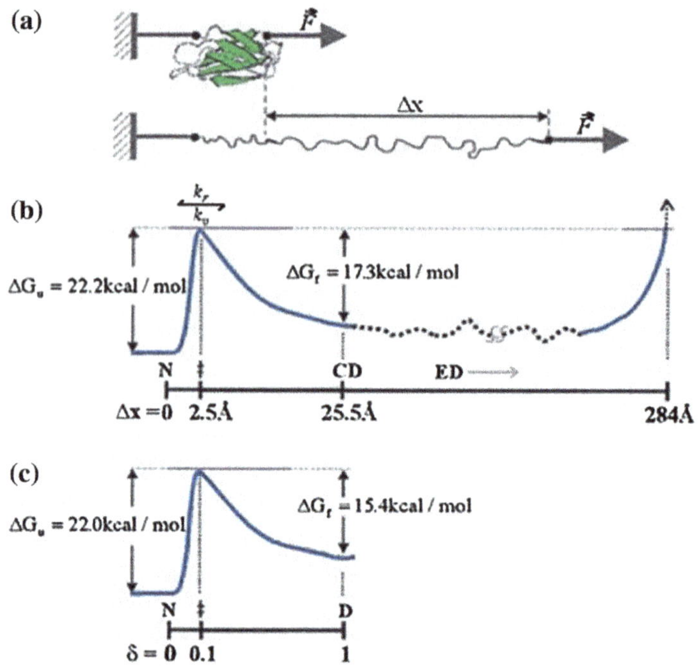

Fig. 4.8 Comparison of the folding pathway of an Ig domain denatured by an applied force or chemical denaturants. **a** Model of the stretching of a single Ig domain. Under an applied force, an Ig domain unravels, causing an increase in the end-to-end length, Δx. **b** Diagram of the folding pathway for an Ig domain as determined using AFM. The changes in free energy (ΔG) are plotted versus the reaction coordinate (end-to-end extension; Δx). Three distinct states are identified: native (N, $\Delta x = 0$), condensed denatured (CD, $\Delta x = 25.5$ Å), and extended denatured (ED, $25.5 < \Delta x < 284$ Å). The transition state, ‡, is located 2.5 Å away from the native state and 23 Å away from the condensed denatured state. **c** The folding pathway determined by chemical denaturants. The changes in free energy (ΔG) between the native, N, and the denatured, D, state are shown vs. the reaction coordinate characterized by a fractional distance δ, where $0 < \delta < 1$ and $\delta_{\ddagger} = m^{\ddagger-N}/m^{D-N} = 0.1$. The putative intermediate is not shown, as its position in the folding coordinate has not been determined. Reproduced from Ref. [43] with permission, Copyright (1999) National Academy of Sciences, U.S.A

unfolding of a single protein by AFM does indeed reflect the same event that is observed in traditional unfolding experiments. The way is now open for the extensive use of AFM to measure folding reactions at the single molecule level. Recently, Hao Yu et al. reported the dynamics of the unfolding individual bacteriorhodopsin proteins [44]. They unfolded bacteriorhodopsin, a model membrane protein, with ultrashort cantilevers optimized for 1-ms SMFS and thereby uncovered previously unobserved protein dynamics and three major intermediates. Furthermore, the elucidated that the unfolding pathway with improved precision resolved a long-standing discrepancy in the size scale of the fundamental structural elements involved in bacteriorhodopsin unfolding, as deduced from experiments or molecular dynamics. By using ultrashort cantilevers optimized for improved spatiotemporal resolution, force spectroscopy allows to reveal a multiplicity of closely spaced and transiently occupied intermediate states, representing small changes in the molecular conformation. These researches initiate the mechanical unfolding of membrane proteins and, more broadly, enable experimental access to previously obscured protein dynamics.

4.3.2 Membrane Protein Mechanics Under Native Conditions

Membrane proteins, often arranged in lipid-protein domains, are involved in basic cellular activities, such as solute and ion transport, energy transduction in respiratory and photosynthetic systems, or sensory stimuli transduction, and information processing; hence, they are important drug targets. SMFS is a powerful tool to address the mechanics of single-membrane proteins in their native environment, i.e., embedded in the lipid bilayer and immersed in a physiological salt solution [45]. When the applied force overcomes the stability of the protein, it unfolds and gives rise to peaks in the force curve that corresponds to elements of secondary structure in the protein [46]. For the investigation of membrane proteins, both strategies of physisorption and covalent bonding via reactive groups have proven to be useful [47, 48]. While imaging the membrane proteins at high resolution, individual proteins were selected, scanning was halted, and the selected protein was then contacted by increasing the force beyond the adhesion threshold, which was found to differ from system to system but it lies typically in the range of 100–200 pN. Upon retracting the tip, a force distance trace was recorded as a protocol for the unfolding and extraction process. A subsequent imaging of the sample allowed the next experiments where a single-membrane protein was extracted would be identified, and the corresponding force distance traces to be selected for further analysis (Fig. 4.9) [49]. By using this technique, Gaub et al. analyzed in depth how the molecular interactions that stabilize the membrane structure of this and other rhodopsins protein are affected by extra- or intracellular unfolding, disulfide bonding, and oligomerization state [50]. SMFS was also used in unfolding and refolding experiments with a bacterial sodium/proton antiporter [51].

4.3.3 Cell Membranes Nanoindentation

The local mechanical properties of cells are frequently probed by force nanoindentation experiments with an AFM-based SMFS. Janshoff Andreas et al. used the typical force nanoindentation measurements to record the stress–strain relationship as well as viscoelastic correspondence principle applied to Hertzian contact

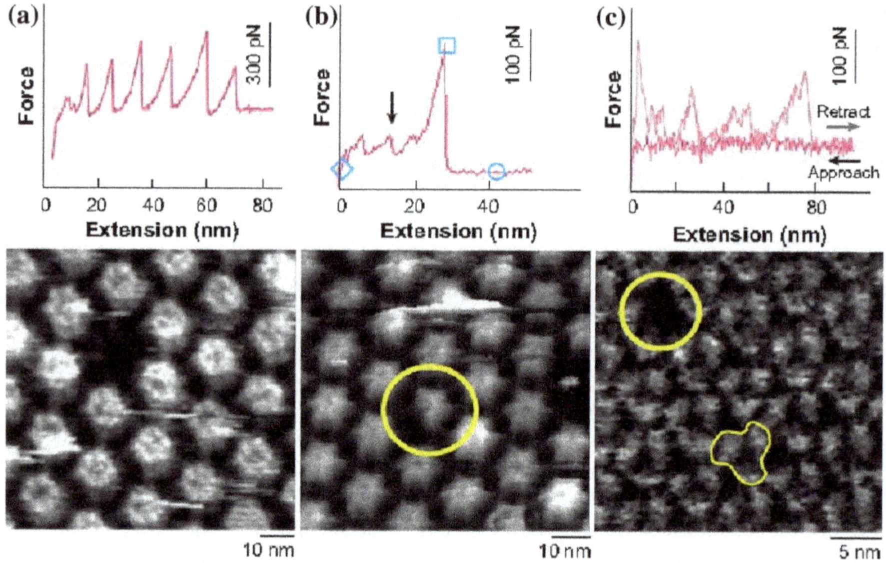

Fig. 4.9 Unzipping individual proteins from a regular array and subsequent high-resolution imaging of the respective vacancy. **a** A distinct relation between the number of force peaks and the related damage is observed on the hexagonally packed intermediate (HPI) layer [52]. Six single peaks and the vacancy corresponding to one hexamer indicate that each force peak corresponds to the removal of one HPI protein. Because one unit contains about 900 residues, only the peptide bridges, holding the HPI proteins together as hexamers, unfold. **b** Similar observations on the S-layer from **c**. glutamicum reveal more complex force peaks: dislodging two subunits required one small (arrow) and one fivefold larger force peak [53]. **c** Unzipping a single bacteriorhodopsin molecule produces a series of force peaks that can unambiguously be related to the seven-transmembrane α-helical structure of bacteriorhodopsin. Reprint from [49], Copyright (2008) by Annual Reviews

mechanics [54]. The cholesterol repletion effect on the nanomechanical properties of human umbilical vein endothelial cell (EA.hy926) was studied using a control-based AFM nanomechanical measurements protocol [55]. Considering that the mechanical properties of the cell could be affected by cell adhesion that induce changes in the cytoskeleton, Samuel T. Souza et al. investigated the influence of the extracellular matrix (ECM) on the elastic properties of fixed macrophage cells using AFM [35]. It is found that there was an increase (about 50%) in the Young's modulus of macrophages adhered to an ECM-coated substrate as compared with an uncoated glass substrate as shown in Fig. 4.10. In addition, cytochalasin D-treated cells have a 1.8-fold reduction of the Young's modulus, indicating the contribution of the actin cytoskeleton to the elastic properties of the cell. Therefore, the cell adhesion influences the mechanical properties of the plasma membrane, providing new information toward understanding the influence of the ECM on elastic alterations of macrophage cell membranes. It also implied that cytoskeleton-mediated cell-matrix interactions directly affect biomechanical events in cells by modifying the physical properties of the cell cytoskeleton.

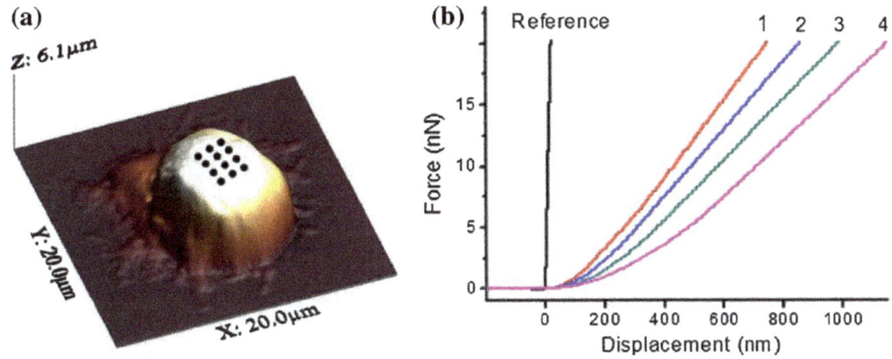

Fig. 4.10 **a** An AFM image of a macrophage showing the indentation area. Twelve black dots represent the indentations performed on the central area of the cell. **b** Force-versus-displacement curves are measured in both the reference substrate and macrophages. (1) Fibronectin, (2) glass, (3) fibronectin + cytochalasin, and (4) glass + cytochalasin. The relation between load force and the indentation depth was calculated as the difference between these curves. Reprinted from Ref. [35], with kind permission from Springer Science + Business Media

4.3.4 Single Molecule Transporting via Transporter

Transporting individual molecules across cell membranes is a fundamental process of cellular metabolism. Although the crystal diffraction technique has greatly contributed to our understanding of the structures of the involved transporters, a description of the dynamic transport mechanism at the single molecule level has been extremely elusive. Using SMFS, the structure and dynamics of glucose cotransporter SGLT were studied by Peter Hinterdorfer et al. [56, 57]. Amino acid transporters across the plasma membranes, which mediate and regulate the flow of ionic nutrients into the cells [58, 59], have been classified into distinct families according to their substrate specificity, transport mechanism, and regulatory properties [60]. SMFS was applied to record the transporting events of single cysteine via amino acid transporters in cell membranes by attaching cysteine onto an AFM tip [29]. Cysteine was coupled to NHS-PEG-MAL and aldehyde-PEG-NHS via a thiol group and an amino group, respectively. For aldehyde-PEG-NHS linked cysteine, only the unbinding force "fu" was detected (Fig. 4.11b), which indicates the unbinding force between the amino acid transporter and a cysteine tethered on the AFM tip. For another NHS-PEG-MAL crosslinker tethered cysteine, both the "fi" and "fu" signals were found. The arrow head "fi" indicates the force signal of transporting the cysteine into the cell, and the arrow head "fu" indicates the force signal of pulling the cysteine out of the cell (Fig. 4.11d). The report also reached the conclusion that there are at least two sites of interaction between the transporter and the substrate, one is involved in the initial binding and another one is related to the translocation pathway [56].

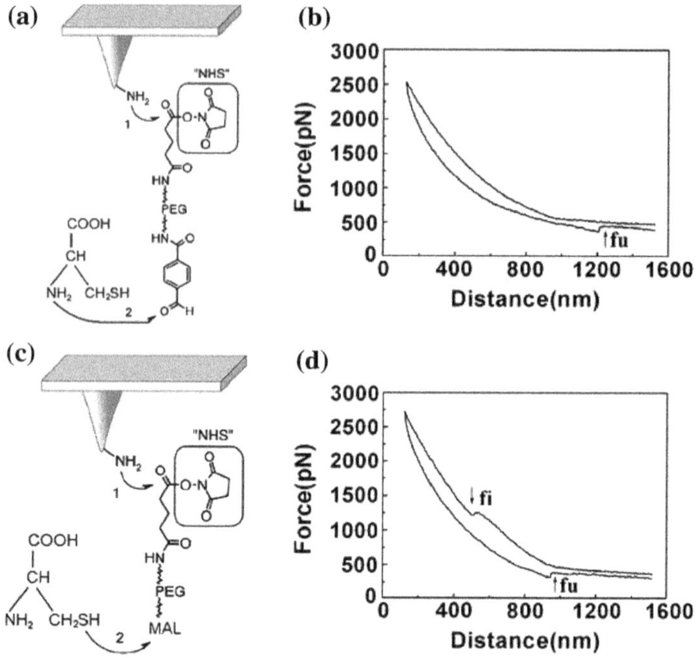

Fig. 4.11 Force spectroscopy of cysteine-modified AFM tip with different crosslinker. **a** Scheme of cysteine conjugation to the AFM tip via aldehyde-PEG-NHS. **b** The typical force-distance cycle of cysteine coated onto the AFM tip via cysteine-aldehyde coupling on the living HeLa cell in HBSS buffer showing the specific force signal in the retrace process. **c** Scheme of cysteine conjugation to the AFM tip via the crosslinker NHS-PEG-MAL. **d** The typical force-distance cycle on HeLa cells with a cysteine coated on the AFM tip via NHS-PEG-MAL. The living cells were pre-incubated in HBSS buffer to remove the extracellular amino acid and then to reduce the intracellular amino acid concentration before performing force spectroscopy. Reproduced from Ref. [29] by permission of The Royal Society of Chemistry

4.3.5 Single Nanoparticle Endocytosis by SMFS

The interactions between nanoparticles and cells determine the entrance of nanoparticles into cells, which is critical for most biomedical applications of nanoparticles [61–64], e.g., drug delivery [65], subcellular sensors [66], photodynamic therapies [67], and cytotoxicity [68]. Endocytosis works as one of the effective way for nanoparticles internalization [69, 70]. Both experimental and theoretical investigations show that the nanoparticle size [71, 72], shape [61], surface chemistry [65], and ligand arrangement affect its active endocytosis. The SMFS was applied to demonstrate the possibility of measuring the endocytosis force of living cells upon the uptake of single semiconductor quantum dots (QDs) at a single particle level under native conditions [73] (Fig. 4.12). In the force-distance cycle, both the force of cellular uptake of single QD and single QD detachment from the cell were recorded. The value of uptake forces “fi” (endocytosis) ranges

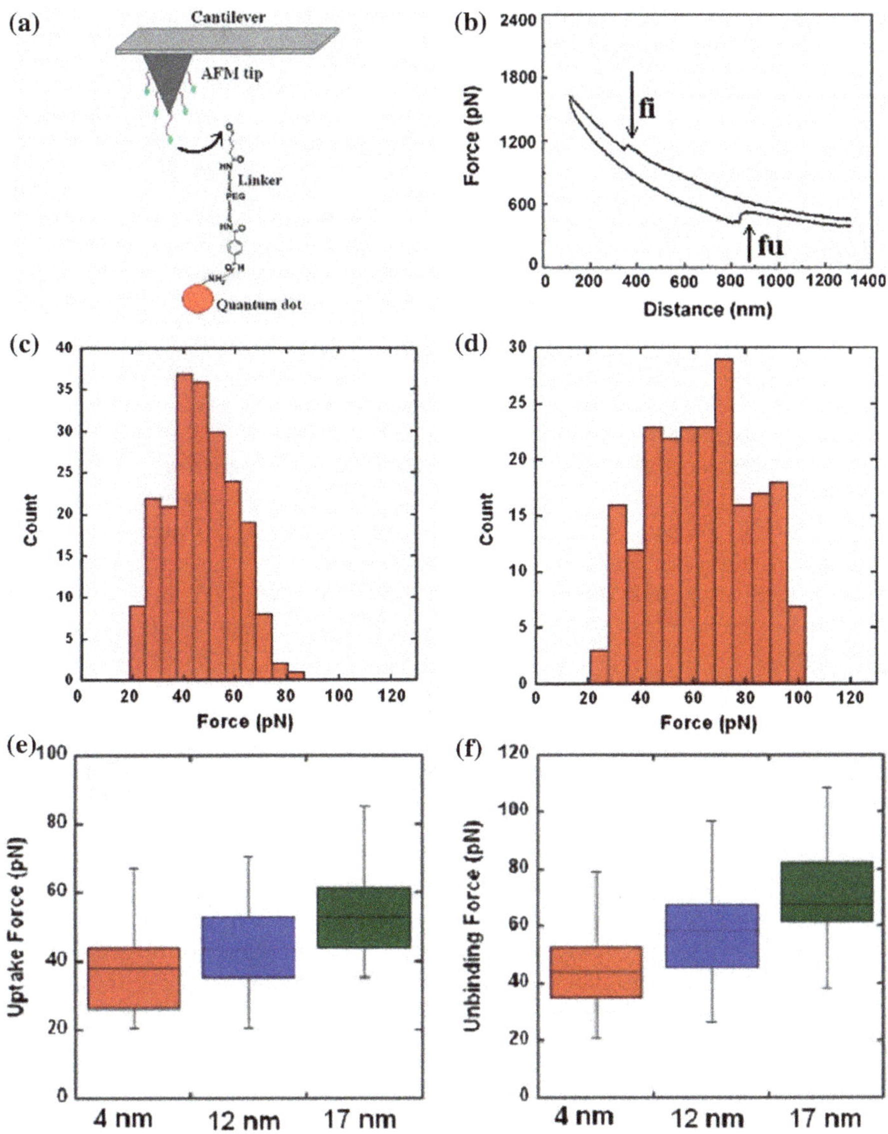

Fig. 4.12 Force measurement of living cells uptaking single quantum dot. **a** Functionalization of AFM tips by QDs. The green dots show that QDs are covalently coupled to AFM tip via a heterobifunctional PEG, and the arrow indicates the details about the PEG and QD. **b** A real typical interaction force curve between single QD and the HeLa cell in DMEM at 37 °C. The long arrow indicates the uptaking force signal "fi." The short arrow indicates the unbinding force signal "fu." **c** The histogram of uptaking forces "fi" ($n > 200$). **d** The histogram of uptaking forces "fi" after blocking with cytochalasin B ($n > 200$). Statistical force values of uptaking **e** and unbinding **f** events for single Au NP with the diameters of 4, 12, and 17 nm. The vertical lines through boxes represent the distribution of interaction force values. The red, blue, and green box mean the main distribution of interaction force values for 4, 12, and 17 nm Au NP, respectively. The black line in every box represents the average of the force distribution, which increases 8–20 pN as the size of Au NP changes from 4 to 12 nm or 12 to 17 nm. Reproduced from Ref. [28, 73] by permission of The Royal Society of Chemistry

from 22 to 80 pN with the loading rate of 1.2×10^5 pN/s, and the maximum of "fi" distribution is about 45 ± 5 pN. The value of unbinding forces ranges from 20 pN to 103 pN with the retracting rate of 1.2×10^5 pN/s, and the maximum of "fu" distribution is about 70 ± 5 pN.

The endocytosis force of nanoparticles is related with the nanoparticle size and surface charge; therefore, it is necessary to detect the endocytosis process with different size particle. With the unique physical and chemical properties, gold nanoparticles (Au NPs) were selected to study the endocytosis process by SMFS. Au NPs with the diameters between 4 and 100 nm have been extensively used in the biomedical applications [74, 75], such as biosensors, drug and gene delivery, and novel photodynamic therapies [61, 71, 76, 77]. It is revealed that both the uptake and unbinding force values were dependent upon the size of gold nanoparticles [28]. In additional, comparing with the endocytosis force of 4 nm QD capped with mercaptoethylamine, 4 nm Au NP capped with cysteine showed a little bit lower endocytic force value. Because that there are net negative charges from the cysteine and the capped citrate ions on the surface of Au NPs, which can repel the negatively charged cell membranes and thus decrease the interaction force. However, on the surface of mercaptoethylamine capped quantum dot, there are lots of positive charges that enhance the interaction force considerably. The SMFS provides a suitable technique to detect the endocytosis force of single nanoparticle on living cell.

4.3.6 Single Organic/Inorganic-Nanoparticle Trans-membrane Dynamic Process

As the first step of cellular uptake, the trans-membrane dynamic process study is really important. Using the ultrafast and sensitive force tracing technique, the trans-membrane dynamic process of single Au NPs was directly recorded, and the crucial role of membrane cholesterol on the dynamic process was discovered [78]. Wang et al. tracked the dynamic process of cellular endocytosis of single Au NP with the average diameters of 5 nm, 10 nm, and 20 nm under physiological conditions, which could reveal new aspects of the dynamic mechanism of NPs entry into living cells. They show that the forces of Au NP endocytosis into the living cells vary from 20 pN to 140 pN, with the average values of 45 ± 14 pN, 67 ± 16 pN, 126 ± 25 pN for 5, 10, and 20 nm Au NPs, respectively. And the duration range between 20 ms and 160 ms, and the average duration for 5, 10, and 20 nm Au NPs are 45 ± 18 ms, 55 ± 18 ms, 81 ± 20 ms, respectively. However the average velocity of Au NP movement during the endocytosis process can be calculated as 0.489 μm s^{-1}, 0.436 μm s^{-1}, and 0.333 μm s^{-1} for 5, 10, and 20 nm Au NPs, respectively. It is noted that the force and duration are increased with the size of Au NP, but the average velocities decrease as the size of Au NPs increases. The case may be related to the size of the Au NP diameters, with the increase of the interaction area between Au NPs and the cell membrane as well as the local membrane curvature at the contact region becoming larger, which result in the increasing endocytosis force and duration from 5, 10 to 20 nm Au NPs.

As one of the organic materials, polymer plays more and more roles in many fields. Since fabricated by Tomalia et al. [79], dendrimers have been considered as one of the most promising molecular structures to elaborate nanostructured materials [80–82]. Szoka et al. reported that polyamidoamine (PAMAM) dendrimers were much superior to other polymeric scaffolds for drug and gene delivery due to their well-defined structure and homogeneously functionalized outer shell [83]. Recent studies demonstrated that dendrimers were capable to enter cells by endocytosis [81] (Fig. 4.13). Utilizing force tracing, the endocytosis process of single G5-PAMAM was studied. The biophysical parameters, including the force, duration, and speed of G5-PAMAM transporting via endocytosis were obtained at the single-particle and microsecond levels. The ~ 19 ms of duration and ~ 1.0 μm s^{-1} of speed unquestionably highlighted the fact that the invagination of the G5-PAMAM nanoparticle was quite a rapid process. Block experiments illustrate that the G5-PAMAM transporting through cell membranes relies on the caveolin-mediated/clathrin-mediated endocytosis, and macropinocytosis pathways with the main pathway being clathrin-mediated endocytosis and macropinocytosis, which is also consistent with the previous literature [84].

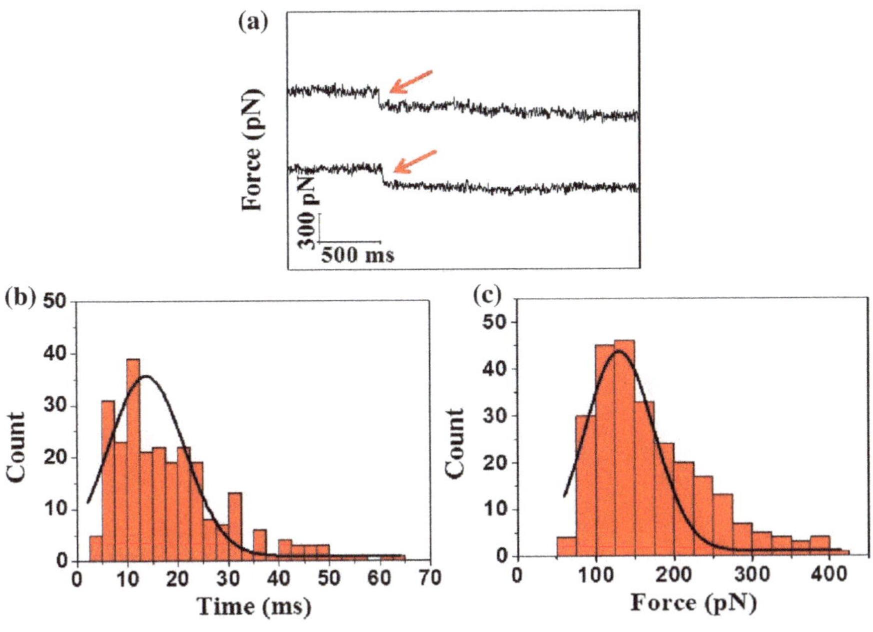

Fig. 4.13 **a** Representative force tracing curves, the red arrows make the endocytic force signals clear. **b** The distribution of endocytic duration, ($n \approx 250$). **c** The distribution of endocytic force, ($n \approx 250$). Reproduced from Ref. [84] by permission of The Royal Society of Chemistry

4.3.7 Single Virus Trans-membrane Dynamic Process

As the smallest Trojan horses, viruses hijack the host cell machinery and then propagate and cause fatal harm to the host [85]. The first step in most viral infections is the penetration of the cell membrane via endocytosis. Endocytosis of a single live virion (Singapore grouper iridovirus, SGIV) through the apical membranes of a host cell was monitored by force tracing [30]. The force tracing test reveals that the maximum velocity during the cell entry of a single SGIV by membrane invagination is approximately 200 nm s^{-1}, the endocytic force is approximately 60.8 ± 18.5 pN. And in the force tracing measurements, two major peaks for the viral displacement were identified at 81.0 ± 6.2 nm and 180.2 ± 21.6 nm. This could be attributed to the SGIV size of 200 nm.

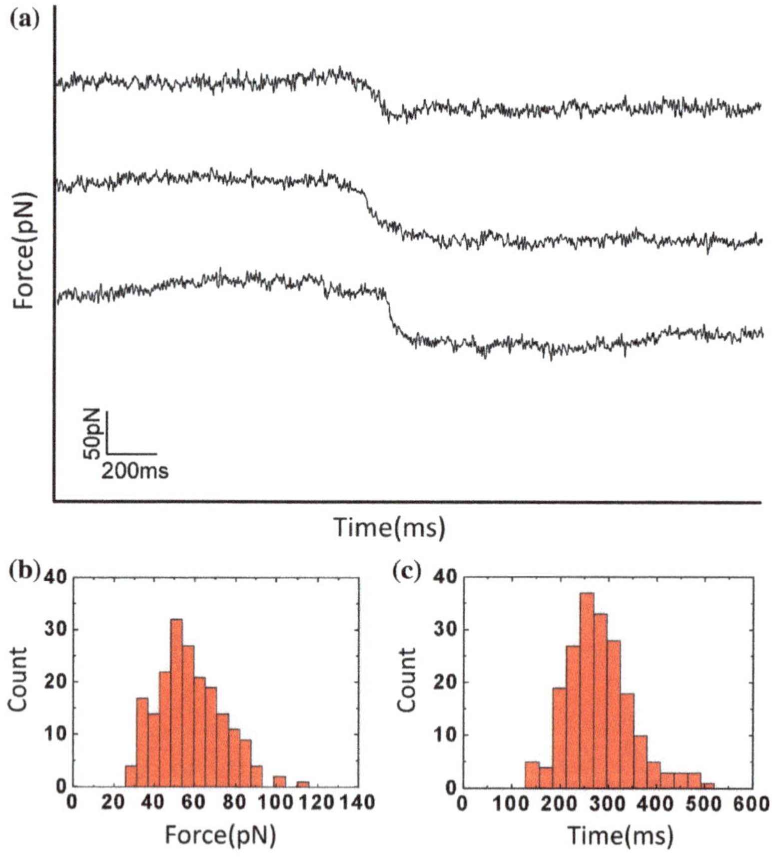

Fig. 4.14 Force tracing curves based on AFM. **a** Typical force tracing curves showing viral invagination via cell membranes. **b** Distribution of force for cellular uptake of virus ($n > 190$ from about 2000 force tracing curves). **c** Distribution of time for viral invagination via cell membranes. Reproduced from Ref. [87] by permission of The Royal Society of Chemistry

The majority of the virions are most likely attached to the side of the AFM tip, which corresponds to the peak in the distribution of the viral displacement at approximately 180.2 ± 21.6 nm. A peak at 81.0 ± 6.2 nm indicates that these virions are attached to the tip apex. The peaks in the time distributions were observed at 0.82 ± 0.06 s and 1.42 ± 0.68 s, corresponding to the two peaks in the displacement distribution.

Recently, Wang et al. used force tracing technique to directly record the endocytic force and wrapping time of single human enterovirus 71 (HEV71) virus-like particles via cell membranes. As shown in Fig. 4.14, while the membrane invaginates to form the vesicle packing virus and finish endocytosis process, the AFM tip will bend downward and a force signal could be detected. The force of virus infection ranges from 40 to 80 pN with a maximum distribution of 58 ± 16 pN. It suggests that the invagination of virus is driven by a force of about 60 pN. The time distribution of viral invagination (duration) is 279 ± 68 ms. The force tracing provides a potent technique to explore the fast dynamic process.

4.4 Methods

For the force spectroscopy experiments, the functionalization of AFM tips with relevant biomolecules is required, which is critical for successfully performing a single-molecule experiment. While the sample preparing is of equal importance, such as preparing the living cell, pre-treatment of cells, and modifying the molecules on the support substrate. The quality and reproducibility of the tip and sample preparation govern the reliability of single molecule force spectroscopy measurements [86].

4.4.1 Conjugation of Molecules/Nanoparticles to AFM Tips

AFM tips modification is the essential procedure for the detection of single molecular force. Usually the biomolecules are conjugated to the AFM tips through bifunctional crosslinker via covalent bond, which is generally obtained using either thiol or silane surface chemistries [10, 87–89]. The biomolecules should be attached strongly at low surface density to achieve single molecule detection without unspecific adsorption. The methods mentioned above described are also available for attaching cells onto AFM cantilevers. The procedure of conjugating bifunctional crosslinker to the AFM tip is described in detail in Chap. 3. The quality of the surface modifications could be assessed using fluorescence or other surface analysis techniques [90, 91]. Most importantly, quantitative force measurements require accurate determination of the cantilever spring constant, which can be achieved using various methods, among which thermal tune is one of the most used approaches.

4.4.2 Preparing Supporting Surfaces and Living Cell Samples

Generally, mica, glass, and silicon could be used as supports for biomolecules and cell culture. For mica, the freshly pealed surface with positive charge can be modified with APTES, and the function of glass and silicon molecules is just the same as preparing AP-AFM tip cantilever (described in Chap. 3). Procedures are also available to immobilize cells on support substrates. The cells, using in force spectroscopy experiments, are sub-cultured every 2 or 3 days when the petri dish is covered by cells achieved 75% confluence [92]. The adherent Vero cells are washed with PBS (phosphate buffer solution) for three times and serum-free medium one time to remove cell debris and unattached cells before used in the force tracing experiments. As a general rule, it is recommended to keep functionalized tips and samples hydrated and in conditions where their functionality remains intact.

4.4.3 Recording Force Curves

Because of the short lifetime of functionalized tip (due to tip contamination or damage) and sample, they should be freshly prepared just before the experiments. Especially, when the sample surface is fragile and modified with loosely bound material, as often in the case for cell surfaces, the "activity" of the biological tip may be lost after recording a single image or a few force curves. In these conditions, it may therefore be useful to visualize the morphology of the sample surface with an unmodified tip to identify a region that is sufficiently smooth, homogeneous, and stable before engaging a functionalized tip on the same region. To measure discrete molecular interaction forces, it may sometimes be useful not only to dilute the surface density of biomolecules on the tip, but also to modulate the contact force and the contact time between tip and sample. To get reliable force data on a given system, users should record several hundred force curves using many independent tips and samples.

4.4.4 Data Analysis

The SMFS data usually are analyzed by the customized software written in MATLAB. By executing a command of "kspec19," the files contain sensitivity of the AFM cantilever (Using the molecules or nanoparticles modified AFM tips to perform force spectroscopy on substrate without sample), and the force-distance cycle data will be loaded by the software. In addition, the spring constant of AFM cantilever used for collecting data is inputted. After then, the every force–distance curve will be showed successively, the force signal would be selected and detected by the software. While the measuring and statistics of data are accomplished.

4.5 Conclusions

AFM has become an invaluable tool in membrane biology research. The high temporal-spatial resolution SMFS allows the interaction between molecules and molecular assemblies on the nanoscale, conformational changes and dynamics of membrane proteins, free energy landscape of a protein including the native folded state and unfolded state, and the dynamic process of single molecule/nanoparticle transporting to be revealed. These features make the AFM unique and outstanding, promising highly exciting results in the future from the microscopic to the nanoscopic scales.

However, the technique of SMFS alone could not fully uncover the cell membrane dynamic process exactly. Developing new AFM-based approaches will contribute more to multifunctional tools that will reveal unique insights into pertinent questions in membrane biology.

Acknowledgements This work was supported by the National Natural Science Foundation of China (No. 31330082, 21773017 and 21673023) and Jilin Provincial Science Research Foundation of China (No. 20160520133JH).

References

1. Frederix P, Bosshart PD, Engel A (2009) Atomic force microscopy of biological membranes. Biophys J 96(2):329–338
2. Goksu EI, Vanegas JM, Blanchette CD et al (2009) AFM for structure and dynamics of biomembranes. Biochim Biophys Acta 1788(1):254–266
3. Muller DJ (2008) AFM: a nanotool in membrane biology. Biochemistry 47(31):7986–7998
4. Hoh JH, Cleveland JP, Prater CB et al (1992) Quantized adhesion detected with the atomic force microscope. J Am Chem Soc 114(12):4917–4918
5. Florin E, Moy V, Gaub H (1994) Adhesion forces between individual ligand-receptor pairs. Science 264(5157):415–417
6. Casuso I, Rico F, Scheuring S (2011) Biological AFM: where we come from—where we are—where we may go. J Mol Recognit 24(3):406–413
7. Kabaso D, Gongadze E, Elter P et al (2011) Attachment of rod-like (BAR) proteins and membrane shape. Mini Rev Med Chem 11(4):272–282
8. Puchner EM, Gaub HE (2009) Force and function: probing proteins with AFM-based force spectroscopy. Curr Opin Struct Biol 19(5):605–614
9. Lulevich V, Kim S, Grigoropoulos CP et al (2011) Frictionless sliding of single-stranded DNA in a carbon nanotube pore observed by single molecule force spectroscopy. Nano Lett 11(3):1171–1176
10. Kienberger F, Costa LT, Zhu R et al (2007) Dynamic force microscopy imaging of plasmid DNA and viral RNA. Biomaterials 28(15):2403–2411
11. Suzuki T, Iwazaki A, Katagiri H et al (1999) Enhanced expression of glucose transporter GLUT3 in tumorigenic HeLa cell hybrids associated with tumor suppressor dysfunction. Eur J Biochem 262(2):534–540
12. Puntheeranurak T, Stroh C, Zhu R et al (2005) Structure and distribution of the *Bacillus thuringiensis* Cry4Ba toxin in lipid membranes. Ultramicroscopy 105(1–4):115–124

13. Muller DJ, Dufrene YF (2011) Atomic force microscopy: a nanoscopic window on the cell surface. Trends Cell Biol 21(8):461–469
14. Clausen-Schaumann H, Seitz M, Krautbauer R et al (2000) Force spectroscopy with single bio-molecules. Curr Opin Chem Biol 4(5):524–530
15. Moy V, Florin E, Gaub H (1994) Intermolecular forces and energies between ligands and receptors. Science 266(5183):257–259
16. Lee G, Chrisey L, Colton R (1994) Direct measurement of the forces between complementary strands of DNA. Science 266(5186):771–773
17. Dammer U, Hegner M, Anselmetti D et al (1996) Specific antigen/antibody interactions measured by force microscopy. Biophys J 70 (5):2437–2441
18. Dufrene YF, Pelling AE (2013) Force nanoscopy of cell mechanics and cell adhesion. Nanoscale 5(10):4094–4104
19. Sharma S, Rasool HI, Palanisamy V et al (2010) Structural-mechanical characterization of nanoparticle exosomes in human saliva, using correlative AFM, FESEM, and force spectroscopy. ACS Nano 4(4):1921–1926
20. Stevenson J, Brown AJ (2009) How essential is cholesterol? Biochem J 420(2):e1–e4
21. Janmey PA, McCulloch CA (2007) Cell mechanics: integrating cell responses to mechanical stimuli. Annu Rev Biomed Eng 9:1–34
22. Butt HJ, Cappella B, Kappl M (2005) Force measurements with the atomic force microscope: technique, interpretation and applications. Surf Sci Rep 59(1–6):1–152
23. Puntheeranurak T, Wimmer B, Castaneda F et al (2007) Substrate specificity of sugar transport by rabbit SGLT1: single-molecule atomic force microscopy versus transport studies. Biochemistry 46(10):2797–2804
24. Czajkowsky DM, Shao Z (1998) Submolecular resolution of single macromolecules with atomic force microscopy. FEBS Lett 430(1–2):51–54
25. Fisher TE, Marszalek PE, Oberhauser AF et al (1999) The micro-mechanics of single molecules studied with atomic force microscopy. J Physiol 520(1):5–14
26. Sarkar A, Robertson RB, Fernandez JM (2004) Simultaneous atomic force microscope and fluorescence measurements of protein unfolding using a calibrated evanescent wave. Proc Nat Acad Sci USA 101(35):12882–12886
27. Marko JF, Siggia ED (1995) Stretching DNA. Macromolecules 28(26):8759–8770
28. Shan YP, Ma SY, Nie LY et al (2011) Size-dependent endocytosis of single gold nanoparticles. Chem Commun 47(28):8091–8093
29. Shang X, Shan YP, Pan YG et al (2013) The force of transporting a single amino acid into the living cell measured using atomic force microscopy. Chem Commun 49(74):8163–8165
30. Pan Y, Wang S, Shan Y et al (2015) Ultrafast tracking of a single live virion during the invagination of a cell membrane. Small 11(23):2782–2788
31. Haque F (2003) Application of nanoindentation development of biomedical to materials. Surf Eng 19(4):255–268
32. Turnbull A, White D (1996) Nanoindentation and microindentation of weathered unplasticised poly-vinyl chloride (UPVC). J Mater Sci 31(16):4189–4198
33. Fang T-H, Kang S-H, Hong Z-H et al (2012) Elasticity and nanomechanical response of Aspergillus niger spores using atomic force microscopy. Micron 43(2–3):407–411
34. Roos WH, Wuite GJL (2009) Nanoindentation studies reveal material properties of viruses. Adv Mater 21(10–11):1187–1192
35. Souza ST, Agra LC, Santos CEA et al (2014) Macrophage adhesion on fibronectin evokes an increase in the elastic property of the cell membrane and cytoskeleton: an atomic force microscopy study. Eur Biophys J Biophys Lett 43(12):573–579
36. Oliver W, Pharr G. (2004) Measurement of hardness and elastic modulus by instrumented indentation: advances in understanding and refinements to methodology. J Mate Res 19(1): 3–20
37. Kurland NE, Drira Z, Yadavalli VK (2012) Measurement of nanomechanical properties of biomolecules using atomic force microscopy. Micron 43(2):116–128

38. Digiuni S, Berne-Dedieu A, Martinez-Torres C et al (2015) Single cell wall nonlinear mechanics revealed by a multiscale analysis of AFM force-indentation curves. Biophys J 108 (9):2235–2248
39. Bushby AJ (2001) Nano-indentation using spherical indenters. Nondestruct Test Eva 17(4–5):213–234
40. Askarova S, Sun Z, GY, Sun GY, Meininger GA, Lee JC-M (2013) Amyloid-b ppeptide on sialyl-lewisx-selectin-mediated membrane tether mechanics at the cerebral endothelial cell surface. PLOS One 0060972
41. Rief M, Clausen-Schaumann H, Gaub HE (1999) Sequence-dependent mechanics of single DNA molecules. Nat Struct Biol 6(4):346–349
42. Rief M, Gautel M, Oesterhelt F et al (1997) Reversible unfolding of individual Titin immunoglobulin domains by AFM. Science 276(5315):1109–1112
43. Carrion-Vazquez M, Oberhauser AF, Fowler SB et al (1999) Mechanical and chemical unfolding of a single protein: a comparison. Proc Nat Acad Sci U S A 96(7):3694–3699
44. Yu H, Siewny MGW, Edwards DT et al (2017) Hidden dynamics in the unfolding of individual bacteriorhodopsin proteins. Science 355(6328):945–950
45. Engel A, Muller DJ (2000) Observing single biomolecules at work with the atomic force microscope. Nat Struct Mol Biol 7(9):715–718
46. Garcia-Saez AJ, Schwille P (2007) Single molecule techniques for the study of membrane proteins. Appl Microbiol Biotechnol 76(2):257–266
47. Kada G, Kienberger F, Hinterdorfer P (2008) Atomic force microscopy in bionanotechnology. Nano Today 3(1–2):12–19
48. Chen J, Liu T, Gao J et al (2016) Variation in carbohydrates between cancer and normal cell membranes revealed by super-resolution fluorescence imaging. Adv Sci 1600270
49. Engel A, Gaub HE (2008) Structure and mechanics of membrane proteins. Annu Rev Biochem 77:127–148
50. Kessler M, Gottschalk KE, Janovjak H et al (2006) Bacteriorhodopsin folds into the membrane against an external force. J Mol Biol 357(2):644–654
51. Kedrov A, Ziegler C, Janovjak H et al (2004) Controlled unfolding and refolding of a single sodium-proton antiporter using atomic force microscopy. J Mol Biol 340(5):1143–1152
52. Chini B, Parenti M (2004) G-protein coupled receptors in lipid rafts and caveolae: how, when and why do they go there? J Mol Endocrinol 32(2):32–338
53. Scheuring S, Stahlberg H, Chami M et al (2002) Charting and unzipping the surface layer of Corynebacterium glutamicum with the atomic force microscope. Mol Microbiol 44(3): 675–684
54. Brueckner BR, Noeding H, Janshoff A (2017) Viscoelastic properties of confluent MDCK II cells obtained from force cycle experiments. Biophys J 112(4):724–735
55. Yan B, Ren J, Liu Y et al. (2017) Study of cholesterol repletion effect on nanomechanical properties of human umbilical vein endothelial cell via rapid broadband atomic force microscopy. J Biomech Eng T Asme 139(3):034501–34501-5
56. Puntheeranurak T, Kasch M, Xia X et al (2007) Three surface subdomains form the vestibule of the Na^+/Glucose cotransporter SGLT1. J Biol Chem 282(35):25222–25230
57. Wimmer B, Raja M, Hinterdorfer P et al (2009) C-terminal Loop 13 of Na^+/Glucose cotransporter 1 contains both stereospecific and non-stereospecific sugar interaction sites. J Biol Chem 284(2):983–991
58. Christensen HN (1990) Role of amino acid transport and countertransport in nutrition and metabolism. Physiol Rev 70(1):43–77
59. Hundal HS, Taylor PM (2009) Amino acid transceptors: gate keepers of nutrient exchange and regulators of nutrient signaling. Am J Physiol Endocrinol Metab 296(4):E603–E613
60. Palacin M, Bertran J, Zorzano (1998) Molecular biology of mammalian plasma membrane amino acid transporters. Physiol Rev 78(4):969–1054
61. Chithrani BD, Ghazani AA, Chan WCW (2006) Determining the size and shape dependence of gold nanoparticle uptake into mammalian cells. Nano Lett 6(4):662–668

62. Dausend J, Musyanovych A, Dass M et al (2008) Uptake mechanism of oppositely charged fluorescent nanoparticles in HeLa cells. Macromol Biosci 8(12):1135–1143
63. Klostranec JM, Chan WCW (2006) Quantum dots in biological and biomedical research: recent progress and present challenges. Adv Mater 18(15):1953–1964
64. Dixit SK, Goicochea NL, Daniel MC et al (2006) Quantum dot encapsulation in viral capsids. Nano Lett 6(9):1993–1999
65. Chavanpatil MD, Khdair A, Panyam J (2007) Surfactant-polymer nanoparticles: a novel platform for sustained and enhanced cellular delivery of water-soluble molecules. Pharmacol Res 24(4):803–810
66. Heller DA, Jeng ES, Yeung TK et al (2006) Optical detection of DNA conformational polymorphism on single-walled carbon nanotubes. Science 311(5760):508–511
67. Hirsch LR, Stafford RJ, Bankson JA et al (2003) Nanoshell-mediated near-infrared thermal therapy of tumors under magnetic resonance guidance. Proc Nat Acad Sci U S A 100(23): 13549–13554
68. Liu N, Mu Y, Chen Y et al (2013) Degradation of aqueous synthesized CdTe/ZnS quantum dots in mice: differential blood kinetics and biodistribution of cadmium and tellurium. Part Fibre Toxicol 10:37
69. Shi X, von dem Bussche A, Hurt RH et al (2011) Cell entry of one-dimensional nanomaterials occurs by tip recognition and rotation. Nat Nano 6(11):714–719
70. Gao H, Shi W, Freund LB (2005) Mechanics of receptor-mediated endocytosis. Proc Nat Acad Sci U S A 102(27):9469–9474
71. Zhang S, Li J, Lykotrafitis G et al (2009) Size-dependent endocytosis of nanoparticles. Adv Mater 21(4):419–424
72. Wang Z, Tiruppathi C, Minshall RD et al (2009) Size and Dynamics of caveolae studied using nanoparticles in living endothelial cells. ACS Nano 3(12):4110–4116
73. Shan Y, Hao X, Shang X et al (2011) Recording force events of single quantum-dot endocytosis. Chem Commun 47(12):3377–3379
74. Shukla R, Bansal V, Chaudhary M et al (2005) Biocompatibility of gold nanoparticles and their endocytotic fate inside the cellular compartment: a microscopic overview. Langmuir 21(23):10644–10654
75. Jin H, Heller DA, Sharma R et al (2009) Size-dependent cellular uptake and expulsion of single-walled carbon nanotubes: single particle tracking and a generic uptake model for nanoparticles. ACS Nano 3(1):149–158
76. Kamei K, Mukai Y, Kojima H et al (2009) Direct cell entry of gold/iron-oxide magnetic nanoparticles in adenovirus mediated gene delivery. Biomaterials 30(9):1809–1814
77. Desai MP, Labhasetwar V, Walter E et al (1997) The mechanism of uptake of biodegradable microparticles in Caco-2 cells is size dependent. Pharm Res 14(11):1568–1573
78. Ding B, Tian Y, Pan Y et al (2015) Recording the dynamic endocytosis of single gold nanoparticles by AFM-based force tracing. Nanoscale 7(17):7545–7549
79. Tomalia DA, Baker H, Dewald J et al (1985) A new class of polymers: starburst-dendritic macromolecules. Polym J 17(1):117–132
80. Jain V, Bharatam PV (2014) Pharmacoinformatic approaches to understand complexation of dendrimeric nanoparticles with drugs. Nanoscale 6(5):2476–2501
81. Albertazzi L, Serresi M, Albanese A et al (2010) Dendrimer internalization and intracellular trafficking in living cells. Mol Pharm 7(3):680–688
82. Zong H, Thomas TP, Lee K-H et al (2012) Bifunctional PAMAM dendrimer conjugates of folic acid and methotrexate with defined ratio. Biomacromol 13(4):982–991
83. Haensler J, Szoka FC (1993) Polyamidoamine cascade polymers mediate efficient transfection of cells in culture. Bioconjug Chem 4(5):372–379
84. Saovapakhiran A, D'Emanuele A, Attwood D et al (2009) Surface modification of PAMAM dendrimers modulates the mechanism of cellular internalization. Bioconjug Chem 20(4): 693–701
85. De Clercq E (2006) Antiviral agents active against influenza A viruses. Nat Rev Drug Discov 5(12):1015–1025

86. Hinterdorfer P, Dufrene YF (2006) Detection and localization of single molecular recognition events using atomic force microscopy. Nat Methods 3(5):347–355
87. Ebner A, Wildling L, Kamruzzahan ASM et al (2007) A new, simple method for linking of antibodies to atomic force microscopy tips. Bioconjug Chem 18(4):1176–1184
88. Kamruzzahan ASM, Ebner A, Wildling L et al (2006) Antibody linking to atomic force microscope tips via disulfide bond formation. Bioconjug Chem 17(6):1473–1481
89. Hinterdorfer P, Baumgartner W, Gruber HJ et al (1996) Detection and localization of individual antibody-antigen recognition events by atomic force microscopy. Proc Natl Acad Sci USA 93(8):3477–3481
90. Ebner A, Hinterdorfer P, Gruber HJ (2007) Comparison of different aminofunctionalization strategies for attachment of single antibodies to AFM cantilevers. Ultramicroscopy 107(101):922–927
91. Diao J, Ren D, Engstrom JR et al (2005) A surface modification strategy on silicon nitride for developing biosensors. Anal Biochem 343(2):322–328
92. Pan Y, Zhang F, Zhang L et al (2017) The process of wrapping virus revealed by a force tracing technique and simulations. Adv Sci 4:1600489

Chapter 5
Super-Resolution Imaging of Membrane Heterogeneity

Jing Gao, Junling Chen and Hongda Wang

5.1 Introduction

The classical fluid mosaic model, proposed in 1972 [1], indicated a homogeneous distribution of membrane molecules throughout the plasma membrane. After several decades, with a number of technical revolutions, many studies have proposed the opposite hypothesis that the plasma membrane is heterogeneous with various kinds of microdomains/clusters. However, inappropriate experimental methods may cause artifacts and incorrect conclusions in the course of characterizing microdomains. For example, some biochemical treatments may induce artificial protein clustering; time consuming and destructive sample preparation in electron microscopy can occasionally lead to low specificity and artifacts. Meanwhile, it cannot be ignored that each technique has its own applicability and limitations. For example, although conventional fluorescence imaging techniques show the capability of directly imaging membranes in vivo, they are suffered from diffraction limit [2] and fail to resolve domains of nanometric sizes. Thus, new imaging methods with higher resolution are required to directly visualize and pick up fine details of the tiny membrane structure. Fortunately, during the last ten years, the proposed super-resolution fluorescence imaging techniques have broken through the optical diffraction limit and greatly improved the spatial resolution to tens of nanometers. With successful applications of these methods, more and more refined organizations

---

Jing Gao, Junling Chen—These two authors contributed equally to this work.

J. Gao · J. Chen · H. Wang (✉)
State Key Laboratory of Electroanalytical Chemistry, Changchun Institute of Applied Chemistry, Chinese Academy of Sciences, No. 5625, Renmin Street, Changchun 130022, Jilin, People's Republic of China
e-mail: hdwang@ciac.ac.cn

H. Wang and G. Li (eds.), *Membrane Biophysics*,
https://doi.org/10.1007/978-981-10-6823-2_5

of membrane constituents have been uncovered to promote a better understanding of the membrane heterogeneity.

In the following sections, we will discuss why and how the development of super-resolution microscopy has fundamentally facilitated the study of biological membrane organization. Firstly, we will briefly introduce the category and imaging principle of these so-called "super-resolution" optical techniques. Secondly, we will focus on the data analysis and quantitative statistic, which will solve many previously inaccessible questions about membrane morphology and dynamics. Finally, we will summarize some new advances in membrane heterogeneity that revealed by super-resolution imaging techniques. We hope this part may help a deeper understanding of membrane structure and open the door to the next wave of discoveries in the membrane field.

5.2 Super-Resolution Imaging Techniques

5.2.1 Principles of Super-Resolution Imaging

5.2.1.1 Stimulated Emission Depletion Microscopy

As proposed by Hell and Wichmann in 1994 [3], stimulated emission depletion fluorescence (STED) microscopy as the first super-resolution fluorescence imaging technology has been widely used in many fields [4–8]. It is a lens-based fluorescence microscope that uses two lasers (excitation laser and de-excitation laser) to overcome the diffraction limit. In brief, during STED microscopy imaging, one laser serves as an excitation laser to excite the fluorophores of a sample; then the other laser (STED beam) is shaped to a toroidal profile to de-excite the fluorophores, not at the precise center of the focal volume, leading to an effective focal volume size with only tens of nanometers (Fig. 5.1) [9]. Theoretically, the improvement of STED microscopy resolution has no physical bottleneck. As long as the intensity of STED beam is sufficiently high, the imaging resolution of the center point can be reduced to the molecular size. When scanning across the sample with this small center spot, an image with a great high resolution will be generated, which can generally achieve to a resolution of less than 20 nm. Moreover, STED microscopy has gradually developed multi-color imaging, living cell imaging, and 3-D imaging. Particularly in studying biology, this technique has been successfully applied to resolve protein organization in cellular microcompartments, such as dendritic spines [10], synapses [11], and plasma membrane microdomains [12].

5.2.1.2 Structured Illumination Microscopy

Structured illumination microscopy (SIM) is a far-field non-scanning imaging technique. In conventional SIM constructions, a periodic diffraction grating with a

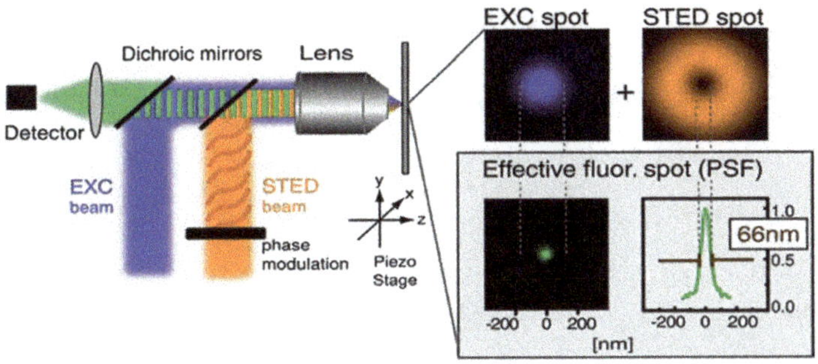

Fig. 5.1 Principle of STED microscopy. With being superimposed by the STED beam (orange) which is a doughnut-shaped depletion beam, all fluorophores excited by the excitation beam (blue) can be de-excited, except those located in the center of the depletion beam. By this way, an effective fluorescent spot of sub-diffraction size (green, below) is yielded. Reprinted by permission from Macmillan Publishers Ltd: Ref. [9], Copyright 2006

known pattern is inserted into the illumination path and projected onto the sample to form an interference pattern (moiré fringes). The periodic illumination pattern can be created through the interference of multiple light sources in the axial direction [13], the lateral direction [14], or both [15]. After a large number of images are collected using different azimuth illumination patterns, the spatial pattern of the samples can be estimated from these generated interference patterns. Since moiré fringes are much coarser than the original illumination pattern and sample information, detection can be facilitated by increasing the spatial frequency in the pattern (Fig. 5.2) [16]. Thus, the improvement of SIM resolution is proportional to the spatial frequency of the illumination mode. Typically, the maximum spatial frequency is the sum of the highest spatial frequency in the mode illumination and the highest detected value. Because the mode itself is still limited by diffraction, the horizontal resolution can only be increased about twice.

The emergence of saturated structured illumination microscopy (SSIM) has broken the barrier. It improves the resolution by applying optical nonlinear near-saturation illumination [17]. When the illumination is strong enough to excite the majority of the fluorescent molecules, the additional increased illumination intensity will no longer increase linearly with the emission intensity, but rather a nonlinear relationship at the near excitation saturation threshold (Fig. 5.2c). Therefore, a nonlinear relationship between excitation and emission can be obtained using saturated illumination. This allows the resulting moiré fringes to have a higher spatial frequency than the initially applied illumination pattern, thus increasing the imaging spatial resolution. The resolution of SSIM is theoretically unrestricted because it depends on the degree of light saturation. However, the photobleaching of dyes and the reduction of signal-to-noise ratio are actually the most decisive factors to limit the resolution. SSIM has demonstrated a 50-nm resolution in two dimensions [17].

Similar to STED, the biggest advantage of SIM and SSIM is the ability to easily achieve multi-color imaging of any bright and photo-stable fluorophores [18].

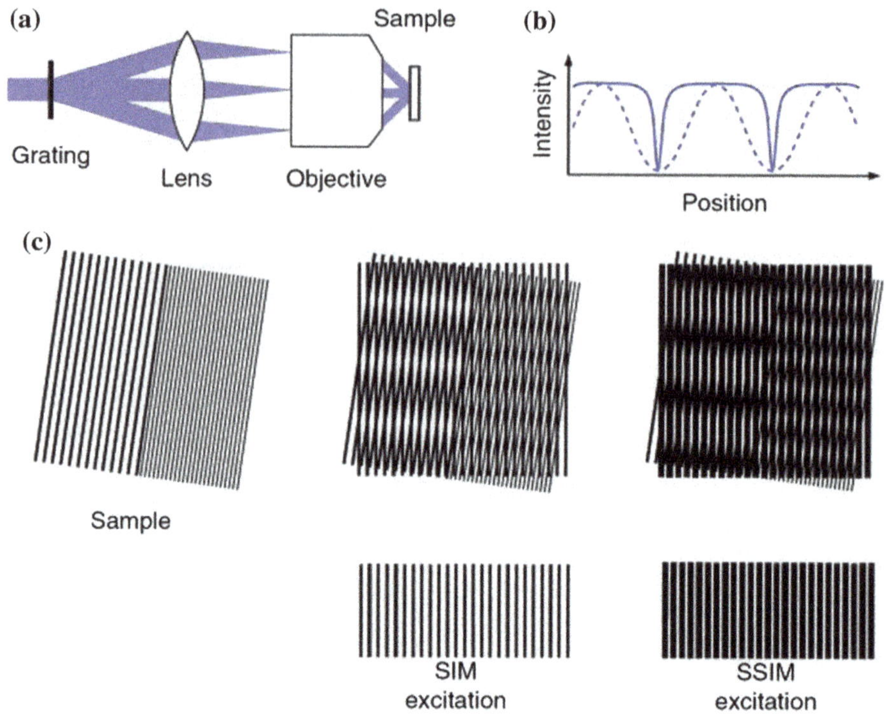

Fig. 5.2 Principle of (S)SIM. **a** A grating in the beam path diffracts the excitation beam, and the diffracted beams are focused to the back focal plane of the objective. The beams interfere in the sample and create a sinusoidal pattern (**b**, dashed line). When the emitted intensity is no longer linearly dependent on the excitation intensity, the effective excitation pattern is no longer sinusoidal (**b**, solid line). **c** Moiré fringers: When two high-frequency patterns are multiplied, Moiré fringers appear. While SIM is capable of resolving structures half the size of the diffraction-limited resolution, SSIM allows features much smaller than this to be resolved. Reproduced from Ref. [16], Hirvonen and Smith (2011), with permission from CSIRO Publishing

Additionally, it is possible for SIM and SSIM to accomplish 3-D high-resolution imaging [19], although SSIM will increase the photobleaching of fluorophores.

5.2.1.3 Single-Molecule Localization Microscopy

Another approach to attain super-resolution is to determine the position of each fluorescent probe in a sample with high precision, which is called single-molecule localization microscopy (SMLM). The major principle of this approach is separating the overlap fluorescence signals by only allowing one molecule per diffraction-limited area to fluoresce at a given time. As activated fluorescent molecules are sparsely distributed each time, it is possible to accurately record their coordinates. By obtaining a series of images until molecules bleach, molecules can

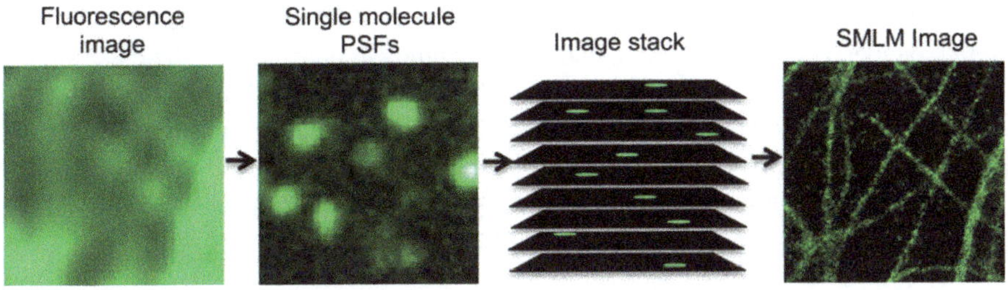

Fig. 5.3 Principle of SMLM. For a sample labeled by fluorescent probes at high density, the diffraction-limited epifluorescence image is generated with all fluorophores emitting photons at the same time (panel 1 from the left). For SMLM, the majority of fluorophores are switched to a dark state so that isolated, spatially separate, single-molecule emissions are stochastically imaged at each time (panel 2). Repeating this cycle and accumulating all single-molecule localizations from thousands of frames (panel 3) can reconstruct the final super-resolution image (panel 4). Reprinted with the permission from Ref. [20]. Copyright 2015 American Chemical Society

be individually imaged, localized, and subsequently deactivated, so that the coordinates of many fluorophores can be mapped and finally a complete reconstructed fluorescence image is built up based on the identified locations of every detected molecule (Fig. 5.3) [20]. In 2006, three laboratories almost simultaneously proposed and implemented this concept. They all utilized the fluorescent molecules with the ability to switch between fluorescent and dark states. According to the fluorescent probe they used, this method is divided into photoactivated localization microscopy (PALM) [21] and fluorescence photoactivated localization microscopy (FPALM) [22] that use fluorescent proteins, and stochastic optical reconstruction microscopy (STORM) [23] that uses organic fluorescent dyes.

The resolution of this technique is limited by the number of photons detected per photoactivation event, which varies from several hundred for fluorescent proteins [21] to several thousand for cyanine dyes [23, 24]. In practice, a lateral resolution of approximately 20 nm has been established experimentally using the cyanine dyes [23, 24]. Much progress is still being made to improve both the spatial resolution and temporal resolution of SMLM, for example, optimizing the photoproperties of fluorescent probes, developing new types of fluorescent probes, exploring the replaceable and better imaging buffer, and improving data algorithms. Thus, SMLM has become an important tool for single-color imaging, multi-color imaging, living cell imaging, and 3-D imaging in various fields [25–27].

5.2.2 Analytical Methods of Super-Resolution Imaging Data

Super-resolution fluorescence images can offer the detailed morphology of membrane components as demonstrated above. To better and accurately understand the information of their spatial arrangement and interactions, we need to apply suitable and credible algorithms to quantitatively analyze their distribution features. So far, there have been several common analytical methods to characterize the clustering of membrane molecules, which are mainly classified into coordinate-based analyses and image-based analyses. Different methods have their individual analytical advantages and applicable scope, so we need to select the optimal method according to the biological characters of research objects. This is a premise for correct interpretation and assessment of super-resolution images.

5.2.2.1 Pair Correlation Function Analysis

Pair correlation function (PCF) analysis, a representative of correlation-based analyses, describes overall clustering of molecules in an image. The peaks arising from a single molecule distribute over a 2-D Gaussian signature, which is a unique spatial signature allowing PCF analysis to identify the peak of a particular molecule. PCF analysis can objectively describe the spatial scales of density fluctuation [28]. It qualifies the probability of one molecule at a given distance from another molecule compared to random distribution [29, 30]. The total PCF of all peaks can be represented as:

$$g(r)^{\text{peaks}} = \left(g(r)^{\text{centroid}} + g(r)^{\text{protein}}\right) * g(r)^{\text{PSF}} \tag{5.1}$$

where $g(r)^{\text{PSF}}$ is the correlation function of the effective point spread function (PSF) of uncertainty in localization, and * represents convolution. The convolution of $g(r)^{\text{centroid}}$ (protein correlation function at $r = 0$) with $g(r)^{\text{PSF}}$ quantifies the correlation arising from multiple appearances of the same molecule and is defined as $g(r)^{\text{stoch}}$. The convolution of $g(r)^{\text{protein}}$ (protein correlation function at $r > 0$) with $g(r)^{\text{PSF}}$ represents the correlation function of the relative spatial distribution of the proteins.

When a protein population is randomly distributed, the correlation function describing its organization, $g(r)^{\text{protein}}$, is ~ 1. The total PCF under these conditions reduces to:

$$g(r)^{\text{peaks}} = g(r)^{\text{stoch}} + 1 \tag{5.2}$$

When proteins are clustering or aggregation, $g(r)^{\text{protein}}$ will be larger than 1 at the given distance.

With defining $g(r)^{\text{protein}} = \text{A} \times \exp(-r/\xi) + 1$ (exp represents exponential, the correlation length, ξ is a measure of the domain size of the protein cluster, and A is roughly the amplitude of the protein correlations extrapolated to $r = 0$), the total correlation will be:

$$g(r)^{\text{peaks}} = g(r)^{\text{stoch}} + (A \times exp(-r/\xi) + 1) * g(r)^{\text{PSF}} \tag{5.3}$$

Through fitting the correlation function with optimal model function, PCF analysis can provide a robust description of spatial distribution of molecules, such as molecule cluster size, number of molecules in clusters, density of molecules in clusters. However, choosing an optimal model function needs to have a priori knowledge of the distribution of the proteins, making interpretation of PCF curves highly model-dependent. Generally, for experimental data, how clusters are organized is difficult to pre-know, so an exponential function is a good starting model.

5.2.2.2 Ripley's K Analysis

Another correlation-based analysis is Ripley's K function [31]. Similar to the PCF analysis, it can also easily discriminate clustered distribution from random distribution by calculating the average number of molecules that exist near another molecule within different radius (r). Conceptually, as the cumulative form of the PCF, it quantifies the number of molecules within a circle of specified radius from the other molecules.

$$K(r) = \frac{1}{n^2} \sum\nolimits_i \sum\nolimits_j \frac{N_r(d_{ij})}{\lambda} \tag{5.4}$$

where n is the total number of localizations within a region of interest (ROI), d_{ij} is the distance between two localizations i and j, N is the number of localizations around localization i within the distance j, and λ is a weighting factor correcting for the area of the ROI.

For a random distribution, $K(r)$ is proportional to πr^2. It can be normalized by the overall density to form Ripley's H function ($H(r)$ = $\sqrt{k(r)/\pi - r}$, which is zero corresponding to a random distribution. For a clustered distribution, $H(r)$ value is larger than zero and related to the average molecule density in clusters.

Both the PCF and Ripley's function have been widely used to quantitatively analyze multi-types of molecule distributions [32–35]. Even so, several points ought to be paid attention to when using these methods: (1) We should confirm that the clustering distribution is uniform in the whole image (cluster sizes and shapes have no significant differences), because both methods are difficult to interpret the heterogeneous distribution of multiple clustering types; (2) we need to exclude the self-clustering effect from the multi-blinking of a single fluorophore and the edge effects; (3) we should better have a prior estimate of the cluster distribution and then choose an optimal model to fit, because these two functions are highly model-dependent.

5.2.2.3 Threshold-Based Analysis

Unlike the coordination-based methods, threshold-based method is model-independent and identifies individual clusters by segmenting grouped localizations from coordinate lists [36]. It has a higher degree of freedom to set a suitable cluster threshold according to the image, such as the distance between molecules, fluorescence intensity of molecules in a ROI, or the local density of localizations. When localizations meet the threshold, they are identified and considered to belong to the same cluster. Once cluster boundaries are determined, a binary map will be generated, in which quantitative properties of clusters can be extracted by Image J (NIH, https://imagej.nih.gov/ij/). In detail, the map of identified clusters is imported and then converted to the binary map by performing "Process" → "Binary" → "Make Binary". Next, performing "Analyze Particles" in "Analyze" will gain a series of parameters, such as cluster area size, cluster perimeter, cluster diameter, cluster circularity, cluster number.

5.3 Applications of Super-Resolution Imaging in Studying Membrane Heterogeneity

5.3.1 Updated Information of Lipid Rafts

The lipid raft hypothesis was firstly proposed in 1997 [37]. It was defined as a mosaic ordered platform enriched in cholesterol, saturated sphingolipid, and phospholipids. Since this concept proposed, the existence and biological function of the lipid raft has always been a controversial issue. With the development of biochemical and biophysical methodology, a deeper understanding and more accurate definition of lipid raft have been updated. For example, Suzuki in 2004 defined the lipid raft as a "molecular complex in the membrane consisting of at least three molecules that include a molecule which plays a critical role in the creation of the complex itself" [38]. Then, the dynamic nature of the lipid raft and the important role of proteins in stabilizing raft were revealed and highlighted. Recently, lipid rafts are generally thought as highly dynamic, small (10–200 nm), heterogeneous, sterol- and sphingolipid-enriched domains [39]. Sometimes, larger platforms are stably formed from small rafts through protein–protein and protein–lipid interactions [40]. Despite many advances in studying lipid raft properties and its important biological functions, there are no suitable tools to observe these raft domains at the required temporal and spatial scales until the emergence of super-resolution fluorescence microscopy. This technique offers a promising platform to update more comprehensive information of lipid rafts at the single-molecule level.

In 2007, Kellner et al. [41] applied STED microscopy to investigate the acetylcholine receptors (AChR, a lipid raft-associated receptor). They found that AChRs aggregated into $\sim$55 nm clusters in the membrane of Chinese hamster ovary (CHO) cells (Fig. 5.4). They also revealed that cholesterol depletion led to the formation of larger AChR clusters. When removing cortical actin skeleton, they observed the disappearance of organization on membrane sheets, suggesting that a lack of cortical actin cytoskeleton could cause the loss of cluster formation. Their work illustrated that besides cholesterol and saturated lipids, proteins (especially those related to the cytoskeleton) play a vital role in membrane organization.

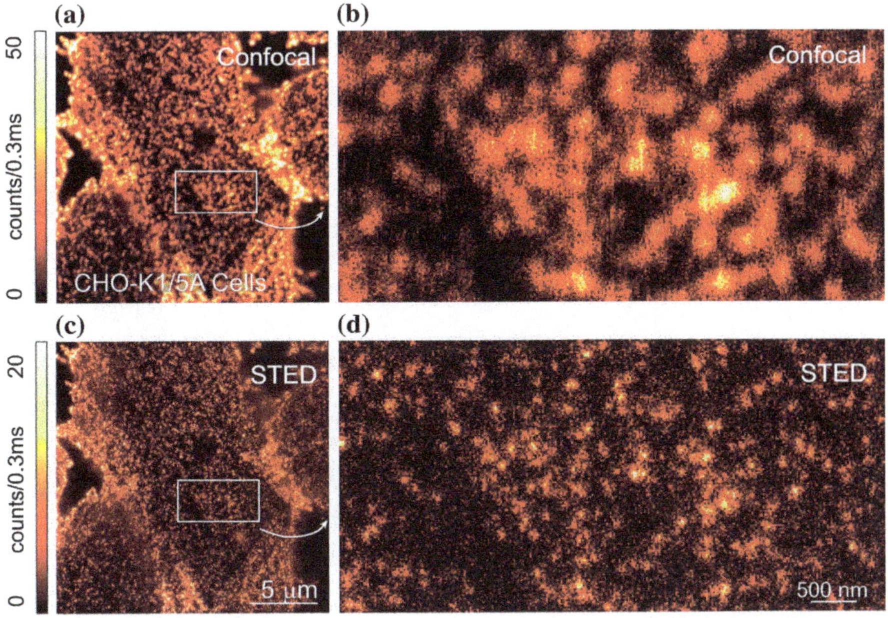

Fig. 5.4 Direct imaging of AChR nanoclusters on the CHO-K1/A5 cell membranes. **a**, **c** Confocal and STED images of AChRs on fixed cell membranes, respectively. **b**, **d** High-magnification views of the boxed areas in (**a**) and (**c**). Magnified STED image shows that AChRs are clustered distributed (**d**), which improves the resolution significantly compared with the confocal image (**b**). Reprinted from Ref. [41], Copyright 2007, with permission from Elsevier

More raft-associated proteins have been studied to demonstrate the link between membrane rafts and the cytoskeleton. For example, super-resolution data demonstrated that hemagglutinins were organized into clusters [42], which were related to lipid rafts [43] and in an action-dependent manner [44]. Tetherin as a raft-associated protein interacting with the cytoskeleton was found to be formed into clusters of 70–90 nm at the plasma membrane by super-resolution microscopy. With PALM and pair correlation analysis, glycosylphosphatidylinositol (GPI)-anchored proteins were revealed to form nanoscale clusters, which was sensitive to the cellular levels of cholesterol and sphingomyelin and cross-correlated with actin after antibody cross-linking (Fig. 5.5) [45].

Besides these studies focused on the role of raft-associated proteins in membrane organization, more effort was put in directly observing lipid rafts with specific fluorescent probes. Mizuno et al. [46] designed probes for cholesterol- and sphingolipid-enriched microdomains dedicated for PALM imaging (Fig. 5.6). They found that cholesterol-enriched microdomains have two types: line-shaped ones with widths of around 150 nm and round ones with an average radius of 118 nm; all sphingomyelin-enriched microdomains were round with an average radius of 124 nm. Moreover, through revealing the different distributed changes of these two types of microdomains with the depletion of cholesterol or sphingomyelin,

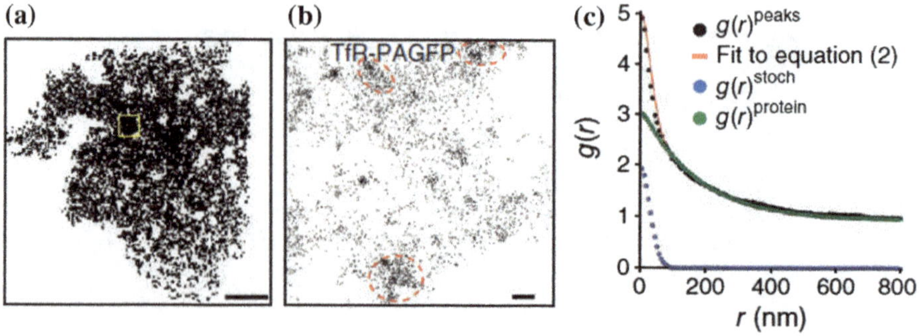

Fig. 5.5 **a** Distribution of peak centers of TfR-PAGFP on the COS-7 cell membrane. A boxed section is selected for correlation analysis. Scale bar is 5 μm. **b** Spatial distribution of peak centers of TfR-PAGFP in the selected region in (**a**). Beside peak clusters associated with individual molecules, there was noticeable grouping of their clusters (red dashed ovals). **c** Measured correlation function of all peaks ($g(r)^{\text{peaks}}$) in (**b**) was well fit to PCF. Correlation protein correlation function ($g(r)^{\text{protein}}$) was evaluated by subtracting the contribution from multiple appearances of single protein ($g(r)^{\text{stoch}}$) from the measured correlation function. Reprinted by permission from Macmillan Publishers Ltd: Ref. [45], Copyright 2011

they concluded that cholesterol- and shingomyelin-enriched domains occupy different regions on the plasma membrane. Their work not only provided the organization of rafts at the single-molecule level, but also illustrated the different localizations of different components of rafts.

With design of appropriate probes of lipid rafts, super-resolution fluorescent imaging can not only analyze the structural components of rafts, but also explore the functions of lipid rafts—regulating the distribution (clustering/exclusion) and diffusion of membrane proteins to mediate protein interactions in cellular signaling and trafficking. Based on super-resolution imaging and Ripley's K-function analysis, the non-random distribution of kinase Lck (a lipid raft marker) was quantified [47], and later it was identified that its conformation states could regulate clustering, thereby linking intramolecular arrangement to intermolecular patterning [48]. The altered clustering distributions of another putative raft marker-LAT (linker for activation of T cells) in resting and activated T cells were imaged and analyzed by super-resolution microscopy and quantitative statistical cluster analysis. They found the increased LAT cluster after T cell stimulation was caused by the translocation of subsynaptic LAT-containing vesicles to the cell surface [49].

About the studies on how rafts affect diffusion, Christian Eggeling et al. combined STED imaging and fluorescence correlation spectroscopy (FCS) to detect single lipid molecules in nanosized areas in the plasma membrane of living cells and found that sphingolipids and glycosylphosphatidylinositol-anchored proteins, unlike phosphoglycerolipids, are transiently trapped in cholesterol-mediated domains with ~20 nm diameter [50]. Their work demonstrated that this non-invasive optical recording of molecular time traces and fluctuation data in nanodomains is a powerful approach to study the dynamics of biomolecules in living cells.

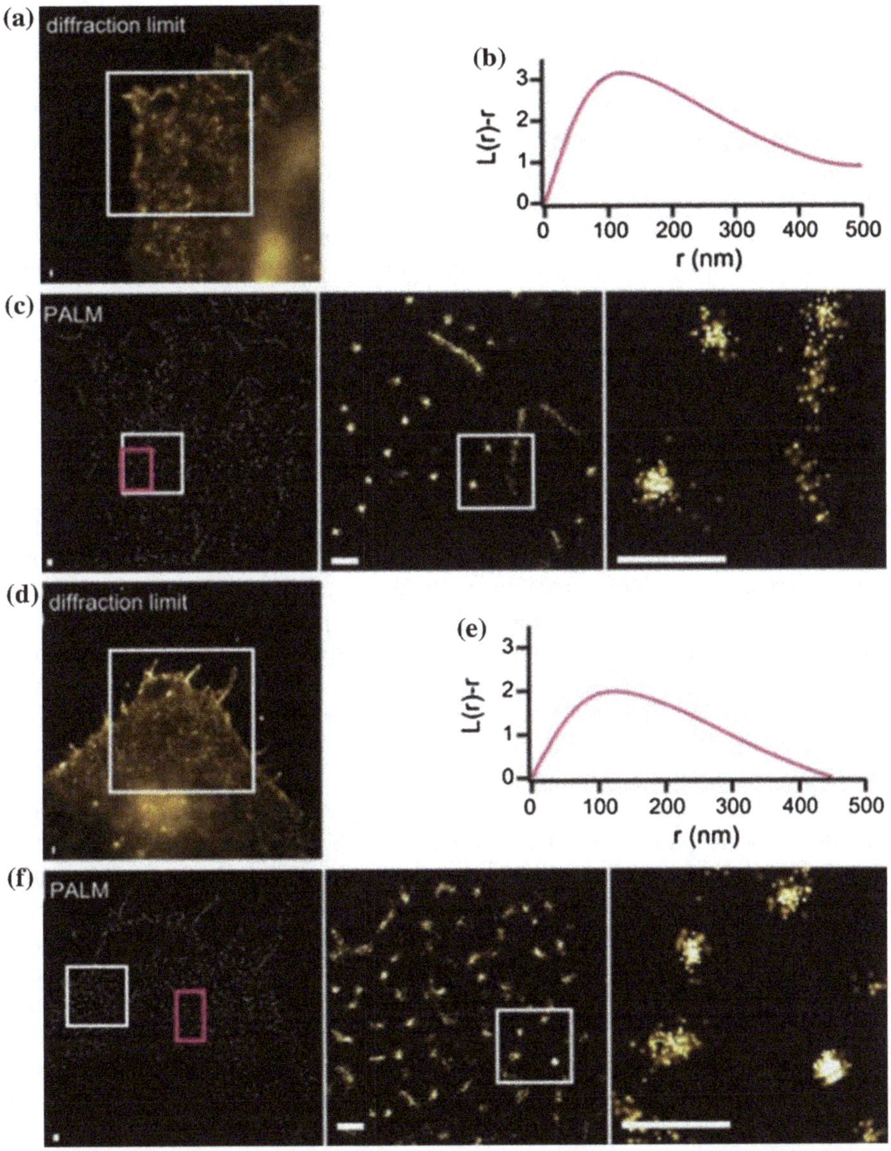

Fig. 5.6 PALM imaging and Ripley's K analysis of the cholesterol—and sphingomyelin-enriched domains on HeLa cell membranes. **a**, **d** Diffraction-limited images of the cholesterol- and sphingomyelin-enriched domains of the plasma membrane, respectively. **b**, **e** Representative Ripley's K-function analysis. **c**, **f** PALM images of the same regions in (**a**) and (**d**). The spots represent a detected single-molecule switching event, and the size of the respective spots corresponds to the precision of the fitting. Reproduced from Ref. [46] by permission of The Royal Society of Chemistry

Indeed, the information of rafts gained from model membranes or fixed cellular membranes is often controversial, because some artificial phenomena are unpredictable but inevitable. Therefore, PALM imaging of raft proteins in living cells has been used. Hess et al. found that HA molecules formed irregular clusters on length scales from $\sim$40 nm up to many micrometers, and an effective diffusion coefficient was determined by quantifying the dynamics of HA within clusters [42]. Later, in 2011, Gunewardene et al. used FPALM to simultaneously observe raft-(HA), non-raft-(transferring receptor, TfR) and cytoskeleton-associated proteins (β-actin) in living fibroblasts [51]. They revealed correlations between the membrane proteins (HA and TfR) and actin skeleton and underlined the role of the actin skeleton in lipid raft formation.

5.3.2 Heterogeneous Distribution of Membrane Proteins

Besides lipids and some raft-associated proteins, nanoscale organizations of other membrane proteins have also been directly observed by super-resolution fluorescence microscopy, including receptor proteins (Fig. 5.7) [35], enzymes (Fig. 5.8) [52], immune proteins [53, 54], adhesions [21, 55], virus proteins [56, 57], and synapses [9, 58]. All these proteins have been found to gather into clusters on both fixed and living cell membranes, verifying that clustering is not an artificial phenomenon caused by fixation [42, 59, 60]. The clustered distribution of proteins in the cell membrane is common, and protein clusters may serve as functional units to participate in various physiological activities. Interestingly, some studies have shown that the clustering properties of membrane proteins (such as epidermal growth factor receptor (EGFR)) are associated with ionic protein–lipid interaction (Fig. 5.9) [61], external environment, cell conditions, and cell types [35], which may be one of factors that lead to the heterogeneous distribution of membrane proteins.

With the ability to directly observe the nanodomains in the living cells, real-time super-resolution fluorescence imaging can provide the unprecedented specific dynamic information of the domain. For example, PALM imaging can not only visualize the different structures of adhesions (such as paxillin) from the elongated structures up to several microns to point clusters of 100–300 nm in size, but also track the growth, fusion, decomposition of the elongated newborn adhesions which happens in a few minutes and the existence of point clusters which keep stable for tens of minutes [59]. Meanwhile, the effect of synaptic movements in different cells on the dynamic characteristics of these structures can be detected as well. Moreover, other membrane microdomains with stronger dynamic behavior have been studied. For instance, by fast, three-dimensional super-resolution imaging, the assembly of endocytic cargo proteins (such as transferrin (Tf)) was investigated, which took only tens of seconds, with a lifetime of only one minute [62]. In addition, the heterogeneity of membrane protein dynamics can be revealed by single-particle tracking. Non-clustered proteins (such as VSVG proteins) exhibited homogeneous diffusion, whereas clustered proteins (such as a viral protein, Gag)

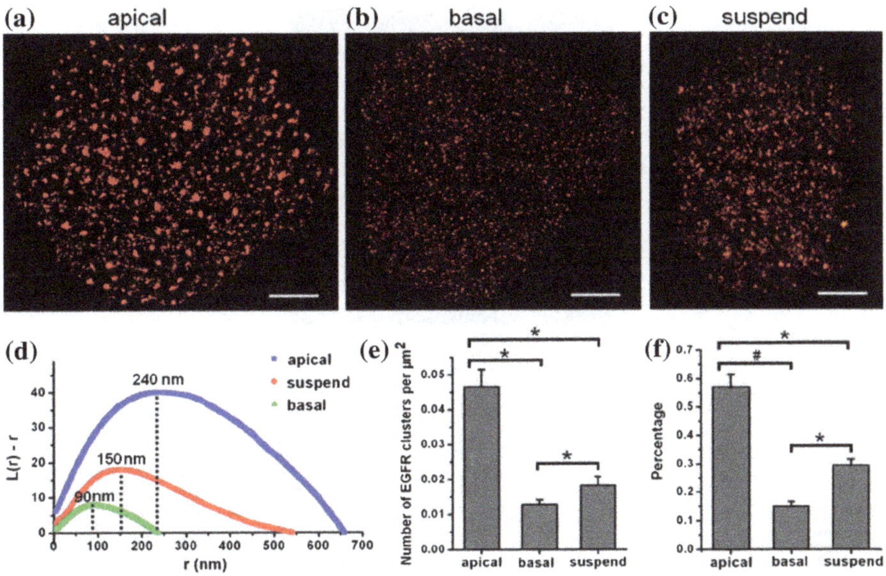

Fig. 5.7 EGFR proteins form clusters on different membrane surfaces of adherent COS-7 cells and suspended cells. **a–c** Reconstructed dSTORM images of EGFR on the apical surface of the adherent COS-7 cell (**a**), on the basal surface of the adherent COS-7 cell (**b**), and on the membrane of the suspended COS-7 cell (**c**), respectively. Scale bars indicate 10 μm. **d** Ripley's K-function analysis of EGFR clustering. The analysis is applied to the stochastic regions of $2 \times 2\ \mu m^2$ in the reconstructed images. The values of r_{max} are 240, 90, and 150 nm on the apical surface, the basal surface of the adherent cell, and on the membrane of the suspended cell, respectively. **e** Normalized average number of EGFR clusters per μm^2 under the different conditions. **f** The percentage of EGFR forming clusters under the different conditions. Data in **e** and **f** are the means ± standard deviation (s.d.), $*p < 0.05$, $\#p < 0.01$, analysis of variance by the two-tailed unpaired t-test

displayed unique free-diffusion and immobilized regions, demonstrating the mechanism of protein capture in microdomains [56] (Fig. 5.10). By this way, some carrier proteins (such as TfR and EGFR) were also recruited by vesicular proteins (such as clathrins) through targeting sequences [63]. Later, it was found that protein capture can be regulated by passive protein–protein interactions in microdomains, but also can be achieved by interacting with potential active actin skeletons [64].

With application of multi-color super-resolution fluorescence imaging, the assembly of different proteins can be simultaneously observed at the nanometer scale. This technique, on one hand, corrects previous mistakes caused by inappropriate experimental methods and, on the other hand, helps to find something new in the field of protein–protein interactions. We briefly introduce some significant examples in the following: (1) Some proteins that were observed to be colocalized at low resolution were found virtually non-overlapping or even interwoven by super-resolution imaging [55]; (2) some microdomains that were observed to be independent by immunoelectron imaging (a method often exists clustering artifact)

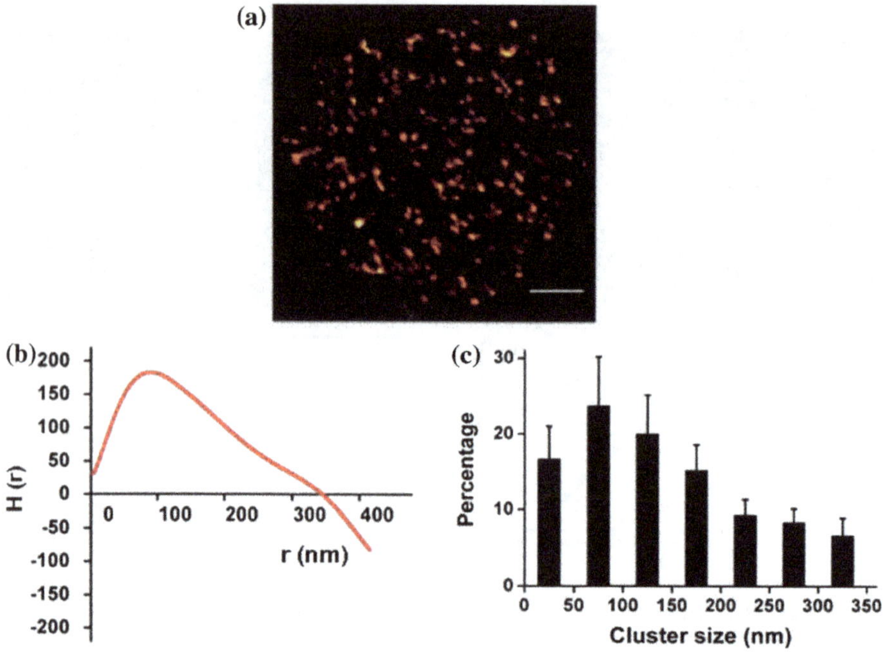

Fig. 5.8 Distribution feature of Na^{+}–K^{+} ATPase clusters on the inside-out membranes of human red blood cells. **a** The dSTORM image of Na^{+}–K^{+} ATPases. Scale bar is 2 μm. **b** A representative of H-function plot in a region of $2 \times 2\ \mu m^2$ to analyze the clustered distribution of Na^{+}–K^{+} ATPases. **c** The percentage of different cluster sizes of Na^{+}–K^{+} ATPases. Data are the means ± standard deviation (s.d.)

were eventually found to seriously overlap depending on the activation of the cell [65]; (3) protein interaction at short distances was directly observed, e.g., the synaptic vesicles docking at reaction sites, the activation of the downstream signal [66, 67]; (4) localizations of different molecules in the protein microdomains have been determined, e.g., the adaptor protein SLP-76 localized at the edge of the immune protein LAT nanodomains (Fig. 5.11) [68]; (5) the molecule density in microdomains can be quantified by calculating the cluster size and the protein number in cluster [69–71], e.g., clustering kinetics of E-cadherins in new forming junctions were detected by PALM [71].

One of the main purposes to study the protein microdomains is to understand their formation and regulation mechanisms. Protein cluster size is an important parameter which indicates the existing mode, density, activation state, or binding rate of proteins. Although measuring cluster size is quite challenging, it can be realized by implementing the molecular counting in PALM imaging. It has been found that some membrane proteins (e.g., antiretroviral protein tetherin, immune protein TCR, G-protein signaling receptor) formed clusters containing a few to a few tens of molecules [53, 54, 57], while other proteins (e.g., chemotaxis receptor and intercellular adhesion protein) formed clusters which follow exponential [72]

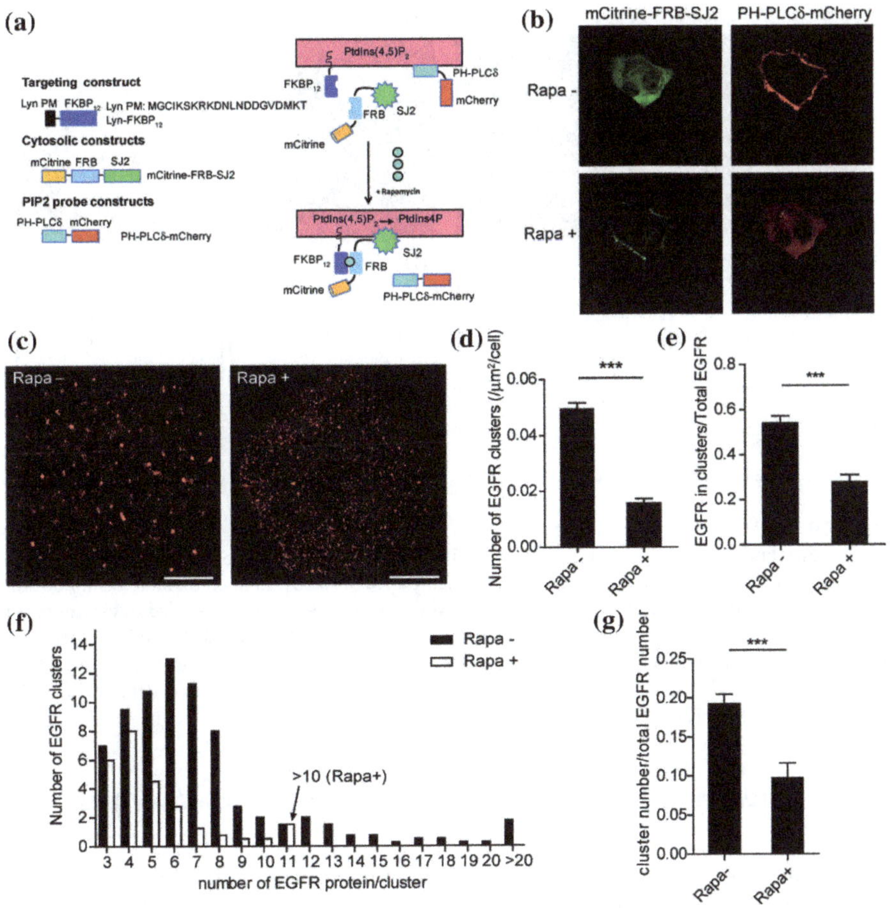

Fig. 5.9 Inducible depletion of PIP2 significantly impairs the EGFR clustering in the plasma membrane. **a** Schematic diagram of the components (left) and principle (right) of inducible PIP2 depletion system. **b** Validation of the system in COS-7 cells visualized by confocal microscope. Scale bar, 20 μm. **c** dSTORM images of EGFR in COS-7 cells transfected with the inducible PIP2 depletion system before (Rapa−) and after (Rapa+) rapamycin treatment. Scale bar, 10 μm. **d** Surface density of EGFR clusters before (Rapa−) and after (Rapa+) rapamycin treatment. **e** Percentage of EGFR proteins clustering. **f** Average number of clusters containing different amount of EGFR proteins. **g** The number of EGFR clusters, which is normalized by corresponding total amount of surface EGFR protein. All error bars denote SD. ***$P < 0.001$, two-tailed unpaired t-test

or power-law distribution [73]. Different cluster sizes imply the distinct mechanisms of cluster formation. Exponential distribution of cluster size suggests that random self-assembly of receptors is accomplished by the random diffusion of molecules and the direct interaction between them [72] (Fig. 5.12), while power-law distribution indicates that the cluster size is regulated by dynamic fusion and decomposition processes, as well as endocytosis [73].

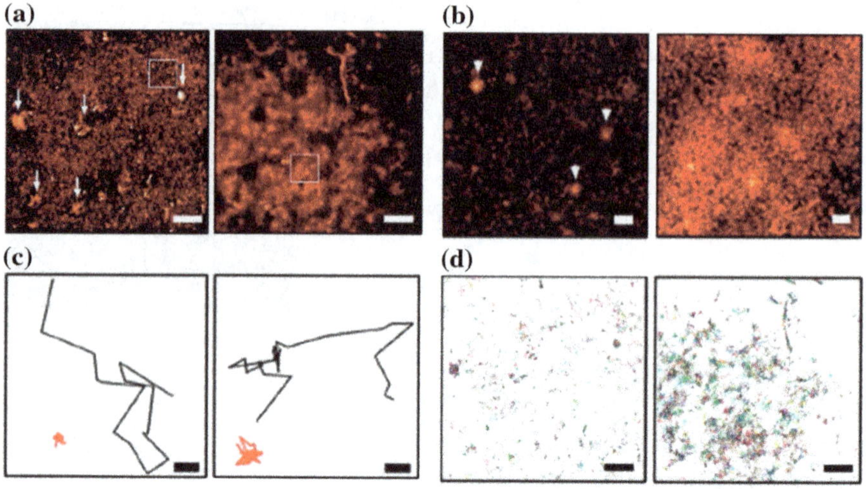

Fig. 5.10 sptPALM imaging of Gag and VSVG expressed in live COS-7 cells. **a** PALM images of Gag-Eos and VSVG-Eos, integrated over 500 s (10,000 images). Scale bars, 2 μm. **b** The corresponding magnified images, showing enriched puncta as arrows indicate. Scale bars, 200 nm. **c** Two representative magnified Gag and VSVG single-molecule trajectories. The tracks represent diffusive (black) and confined (red) movement. Scale bar, 100 nm. **d** Complete sptPALM trajectories of localized Gag and VSVG molecules that are longer than 15 frames (750 ms). Each color indicates a different track. Scale bar, 2 μm. Reprinted by permission from Macmillan Publishers Ltd: Ref. [56], Copyright 2008

5.3.3 Clustering Distribution of Cell Surface Carbohydrates

As an indispensable membrane component, carbohydrates participate in the organization of the plasma membrane and perform diverse functions. However, the traditional structure models of the plasma membrane are proposed without appropriately considering the role of carbohydrates, suggesting that the plasma membrane is laterally compartmentalized with various membrane microdomains and covered by a thick layer of numerous carbohydrates [74–77]. However, recent findings by Wang et al. [78, 79] imply that the proteins on the ectoplasmic side of cell membranes form a dense protein layer on the top of a lipid bilayer, and the carbohydrates stay in microdomains at the ectoplasmic side to participate in various recognition processes. Besides, glycosylated membrane proteins have been reported to aggregate into microdomains on the plasma membranes via super-resolution imaging [52, 76, 80, 81]. Therefore, a deep study of carbohydrates is necessary to reveal the comprehensive organization of plasma membrane.

Recently, with the exploits of the fluorescent probes of carbohydrates and advances fluorescent imaging technology, the fined organization of cell surface carbohydrates has been investigated. In 2015, Letschert et al. [82] combined super-resolution imaging and metabolic labeling to visualize membrane-associated glycoproteins at the nanoscale resolution. They firstly treated cells with

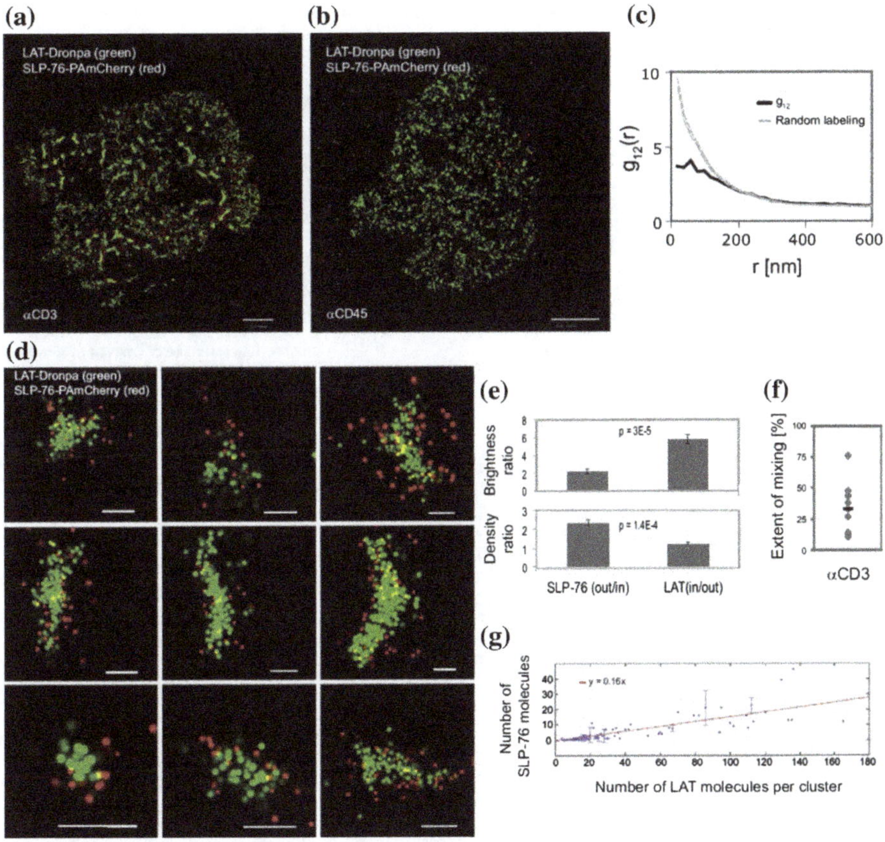

Fig. 5.11 Nanoscale organization of SLP-76 is revealed at the rim of LAT clusters. **a**–**c** Dual-color PALM images of a LAT-Dronpa and SLP-76-PAmCherry expressing Jurkat T cells on **a** αCD3- **b** αCD45-coated coverslips, and **c** bivariate correlation curve. **d** Zoomed images showing that SLP-76 preferred to localize at the rims of LAT clusters. Scale bars are 200 nm. **e** The top shows the ratio of LAT and SLP-76 brightness in the rim versus inner part of the cluster. The bottom shows density ratios of LAT and SLP-76 as calculated by dividing the brightness measurements with the matching areas of the rim and inner part of the cluster. **f** The extent of mixing between LAT and SLP-76. **g** The number of SLP-76 molecules in LAT clusters is plotted as a function of LAT cluster size. Reprinted from Ref. [68], Copyright 2011, with permission from Elsevier

$Ac_4ManNAz$ and $Ac_4GalNAz$ (analogues of biosynthetic precursors ManNAc and GalNAc) which can be incorporated into cell surface carbohydrates upon cell permeation and deacetylation, and then they metabolically labeled Sia and mucin-type O-linked glycans by reacting alkyne-bearing Alexa 647 with catalysis of Cu(I). Similarly, they also labeled the O-GlcNAc-modified plasma membrane proteins by using a GlcNAc analogue ($Ac_4GlcNAz$). Then, by dSTORM imaging, they found that these labeled carbohydrates homogeneously distributed over the entire basal plasma membranes of U2OS and SK-N-MC cells with high localization

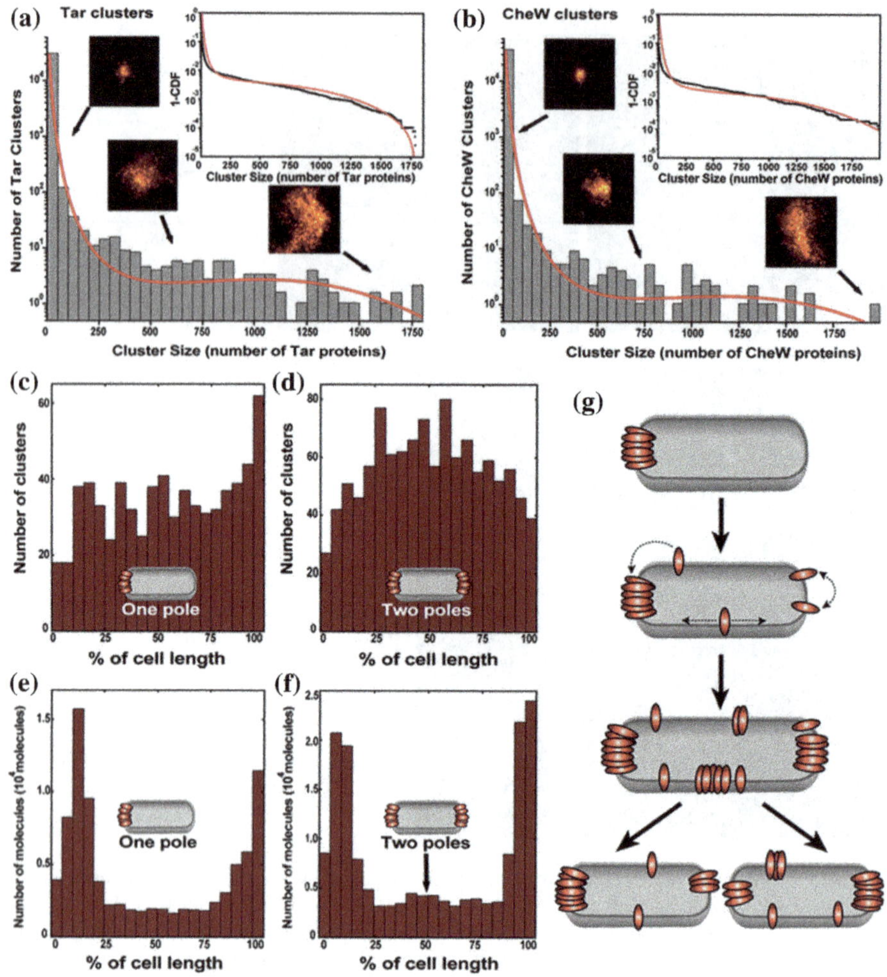

Fig. 5.12 Chemotaxis cluster-size distribution and model. **a**, **b** Histograms of cluster size, measured by the number of closely spaced Eos-labeled Tar (**a**) and CheW (**b**) proteins. Smaller clusters occur much more frequently than larger clusters. Sample images of clusters are shown with arrows that indicate cluster size. To evaluate the fit in a bin-independent representation, the cumulative distribution function (CDF) (insets) is plotted, with fitting the self-assembly model to the data (red). **c**, **d** Cells with one (**c**) or two (**d**) large polar clusters have the highest density of remaining smaller clusters furthest from the existing cluster(s). **e**, **f** Cells with two large polar clusters (**f**) exhibit higher Tar-receptor density at mid-cell (arrow) in comparison to cells with one polar cluster (**e**). **g** Model of receptor self-assembly in which cluster locations are maintained within a population of growing and dividing cells. Cluster nucleation is most likely to occur where receptor density is high, which occurs far from any existing cluster. Dotted arrows denote receptor diffusion within the membrane. Reprinted from Ref. [72]

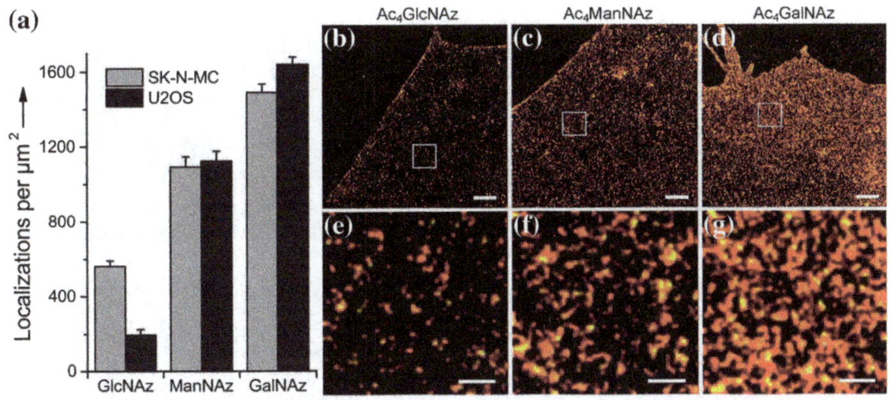

Fig. 5.13 Super-resolution imaging and analysis of cell surface glycoproteins. **a** Localization density of membrane-associated glycans labeled with Alexa Fluor 647 by click chemistry reaction. Error bars represent the standard error of the mean of 12–32 imaged cells. **b–d** dSTORM images of glycoconjugates in the basal membrane of SK-N-MC neuroblastoma cells. **e–g** The enlarged images corresponding to (**b–d**). Scale bars: 1 μm in (**b–d**) and 200 nm in (**e–g**). Reproduced from Ref. [82] by permission of Wiley & Sons Ltd

densities (Fig. 5.13). Their work confirmed that the metabolic labeling approach coupled with bio-orthogonal chemistry could be successfully applied to super-resolution imaging of carbohydrates. However, there is a question needs to be noticed when using this method: The labeled carbohydrates are not the natural ones, and thus, their morphology may be different from the native ones'. Not only this method, but also some synthesized carbohydrate precursors are poorly incorporated into the cell surface carbohydrates [83–85], which may lead to an incomplete localization of carbohydrates.

In the same year of 2015, Chen et al. [36] used lectins to label cell surface carbohydrates and performed dSTORM imaging to visualize their distribution. They labeled GlcNAcs with Alexa 647-linked wheat germ agglutinin (WGA) on Vero cells. They revealed different morphologies of GlcNAcs on the apical and basal membranes of fixed Vero cells with the sub-diffraction resolution (Fig. 5.14). GlcNAcs were mainly distributed into larger clusters on apical membranes with a higher cluster density compared with those on basal membranes. Additionally, most GlcNAc clusters were found to be colocalized with lipid rafts (Fig. 5.15). Meanwhile, most GlcNAc clusters became smaller or even disappeared with disrupting the intact lipid rafts, indicating that stable existence of most GlcNAc clusters was dependent on the intact lipid rafts. Not only one kind of carbohydrate, in 2016, they further systemically investigated the spatial organization of other six representative carbohydrates by dSTORM [86] and found that all of them distributed in clusters on the cell membrane and connected with each other to form a conjoined functional platform.

After observing the clustered distribution of carbohydrates on the cell membrane, the underlying formation mechanism of carbohydrate clusters was also

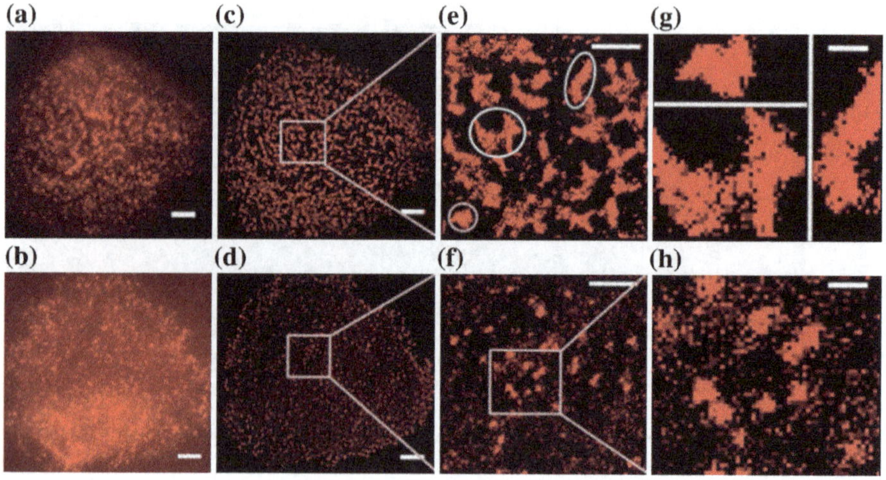

Fig. 5.14 Super-resolution images of N-GlcNAcs on apical and basal membranes of Vero cells. Compared with blurry conventional TIRF images of N-GlcNAcs on the Vero apical (**a**) and basal (**b**) membranes, corresponding dSTORM images (**c**) and (**d**) reveal the distribution features of N-GlcNAcs on cellular membranes at the nanometer resolution. Enlarged images **e** and **f** illustrate the detailed differences in the distribution of N-GlcNAcs. The further enlarged images show three representative clusters of N-GlcNAcs on the apical surface (**g**), and almost uniform and small clusters on the basal membrane (**h**). Scale bars are 5 μm in (**a–d**), 2 μm in (**e**, **f**), 200 nm in (**g**, **h**)

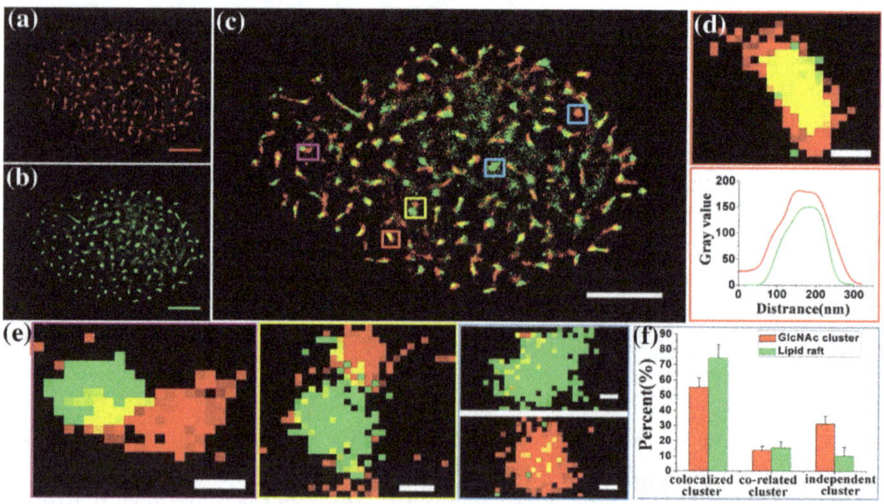

Fig. 5.15 Dual-color dSTORM images revealing the positional relationship between clusters of N-GlcNAcs and lipid raft domains on the Vero apical membrane. **a–c** dSTORM images of N-GlcNAcs (**a**), lipid rafts (**b**), and the merged two channels (**c**). **d** Magnified image of the colocalized domains and the corresponding intensity profiles showing that N-GlcNacs and lipid rafts are perfectly colocalized. **e** Enlarged images in (**c**) showing three kinds of positional relationships: colocalized domains whose overlapping area is >0.02 μm^2 (left), corelated domains which connect with each other but the overlapping area is <0.02 μm^2 (middle), and independent domains which have no overlapping area (right). **f** The percentage of three types of domains on the membrane. Scale bars are 5 μm in (**a–c**), 200 nm in (**d**, **e**)

studied [87]. The clusters of Gal and GlcNAc (as the specific ligands of galectin 3) decreased significantly and got smaller after treating with lactose [88, 89], which can destroy the galectin-related lattices by competing with galectin ligands, whereas Sia (as the non-specific ligands of galectins) clusters only changed a little (Fig. 5.16). The results suggested that the formation of Gal and GlcANc clusters

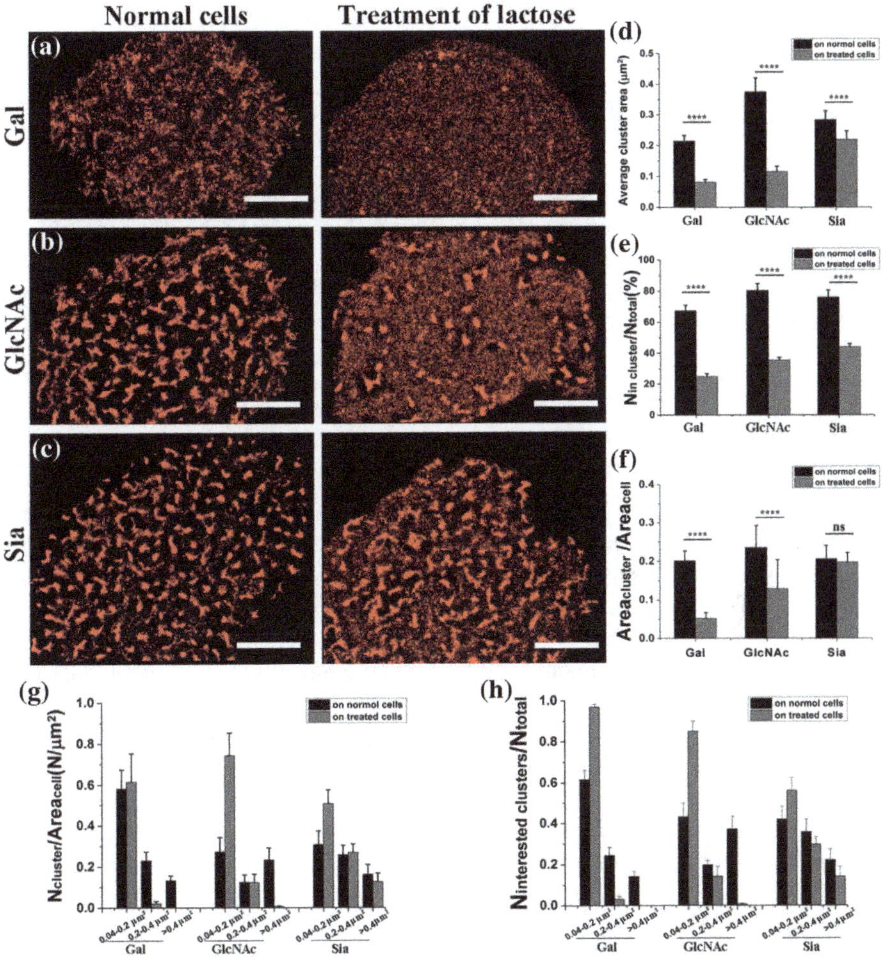

Fig. 5.16 Effect of galectins on the formation of carbohydrates on Vero membranes. **a–c** Comparative dSTORM images of different carbohydrates (**a** for Gal, **b** for GlcNAc) on normal Vero apical membranes (up) and on membranes treated with lactose (lower). Scale bars are 5 μm. **d–f** Histograms of the distribution of average cluster area (**d**), ratio of the number of localizations in clusters to total localizations on the entire cell membrane (**e**), cluster coverage percentage on the entire cell membrane (**f**). **g**, **h** Average cluster density (**g**) and number percentage of different kinds of clusters (**f**) on the normal and treated membranes. All results are the means ± S.D. Statistical significance is given by $^{****}P < 0.0001$; $^{ns}P > 0.05$

mostly depend on the cross-linking of galactin 3. Moreover, when treating cells with methyl β cyclodextrin (MβCD) or cytochalasin B (CB) to destroy lipid rafts or inhibit the network formation of actin filaments, most of carbohydrate clusters dramatically became smaller and fewer, and carbohydrates were dispersed into larger batches on the cell membrane without the specialized domains confined by actin cytoskeleton. Their findings indicate that carbohydrate clusters can stably exist with limited size with the contributions of lipid rafts as the stable factor and actin cytoskeleton as the restrictive factor.

5.4 Protocols for Performing STORM

There are various kinds of super-resolution imaging techniques as we summarized above, and every technique has its own approach. For STORM, the setup of the instruments is relatively simple, and the preparation of cell samples is more typical. Therefore, we only take STORM as an example to introduce how to perform it and what should take care during the imaging.

5.4.1 Antibody Labeling

(1) Dye aliquoting: for Alexa Fluor 647, dissolve 1 mg dye in 1 mL DMSO (final concentration 1 mg/mL) and aliquot into tubes for 10 μL per tube. Store aliquots at −20 °C protected from light.
(2) Set up labeling reaction: 50 μL secondary antibodies (1 mg/mL), 6 μL $NaHCO_3$ (1 mol/L), and 1 μL Alexa Fluor 647 (of 10 μL aliquot). Allow the reaction to proceed for about 30 min at room temperature (RT) in the dark on a rocking platform.
(3) Equilibrate a Nap-5 gel filtration column (GE Healthcare) by running three column volumes of PBS through the column.
(4) After the incubation, add 140 μL PBS to bring the reaction volume up to 200 μL (the minimum column loading volume) and gently vortex. Add the entire volume to the center of the column.
(5) Allow the sample to enter the column, and after the last drip, add sufficient PBS to collect the eluant in 1.5-mL Eppendorf tubes (about 150 μL per tube). For IgG, about 1 ml PBS should be required.
(6) Measure the absorbance at 280 and 650 nm by the UV-spectrophotometer to calculate the antibody concentration and labeling ratios of the eluant in each tube. When the labeling ration of dyes and antibodies is 0.7–1, the eluant is pooled for use.

5.4.2 *Sample Preparation*

(1) Coverslip cleaning: Prior to cell plating, coverslips are sonicated in a tabletop sonicator for ~20 min in 1 mol/L KOH. They are then rinsed thoroughly in Milli-Q water and sterilized in the biosafety cabinet under ultraviolet light for >30 min.
(2) Cell culture: Cells were detached with trypsin/EDTA, plated on pre-cleaned coverslips and cultured to ~75% confluency.
(3) Fix: Add 500 μL of 4% paraformaldehyde (PFA) in one dish. Incubated at RT for 10 min. Rinse cells twice quickly with PBS.
(4) Reduce: Make a fresh solution of 0.1% $NaBH_4$ (1 mg/mL) in PBS immediately before use. Add 500 μL per dish and incubate for 5 min at RT. Rinse cells three times quickly with PBS.
(5) Block: Add 200 μL blocking buffer (3% bovine serum albumin (BSA, m/v)) per dish and place the cells on the rocking platform for 30–60 min.
(6) Primary antibody: Dilute primary antibodies with blocking buffer for addition. The dilution ratio is not constant. It can be adjusted according to the specificity of the antibody. Add ~150 μL of staining solution per dish and incubate for 1 h at RT on the rocking platform. Then rinse with wash buffer (0.2% BSA in PBS) and wash twice. Each wash proceeds for 5 min on the rocking platform.
(7) Secondary antibody: Dilute secondary antibodies-conjugated Alexa Fluor 647 to 1–2 μg/mL (about 1: 100) with PBS. Add ~150 μL per dish and incubate for 1 h at RT on the rocking platform protected from light. Then rinse with wash buffer and wash twice with PBS. Each wash proceeds for 5 min on the rocking platform.
(8) Post-fixation: Add 500 μL of 4% paraformaldehyde (PFA) to fix antibodies. Incubate at RT for 5 min. Rinse cells three times quickly with PBS.
(9) Fiducial markers: 100-nm-diameter TetraSpeck microspheres are embedded to correct the x-y drift of sample and the optical registration between different channels.

5.4.3 *Imaging Buffer*

Two different thiols are used to keep the organic fluorophores photoswitch effectively. β-mercaptoethanol generates bright, long-lived single-molecule photoswitching events, making it ideal for single-color imaging. MEA reduces the photon number slightly but also reduces the non-specific blinking, making it ideal for multi-color imaging. An enzymatic oxygen scavenging system is also used to reduce photobleaching.

Reagents:

(1) Dilution buffer: 10 mM Tris (pH 8.0), 50 mM NaCl.
(2) Oxygen scavenger: Dissolve 14 mg glucose oxidase to 200 μL dilution buffer and vortex to mix. Add 50 μL catalase (20 mg/mL) to glucose oxidase solution. Centrifuge at the speed of 12,000 g for 1 min. Catalase may precipitate out. Use the yellow supernatant for imaging buffers. Store at 4 °C for up to 2 weeks.
(3) MEA: 77 mg MEA dissolve in 1 mL 360 mM hydrochloric acid. Store at 4 °C for up to 1 month.
(4) Imaging buffer base: 10% glucose, 50 mM Tris (pH 8.0), 10 mM NaCl.
(5) Live-cell imaging buffer base: 10 mL DMEM of high glucose without phenol red, 750 μL 1 M HEPES (pH adjusted to 8.0), 400 μL 50% glucose.

Buffer compositions:

(1) STORM buffer using βME: fixed sample 100 × Imaging buffer base, 1 × βME, 1 × Oxygen Scavenger
(2) STORM buffer using MEA: fixed sample 80 × Imaging buffer base, 10 × MEA (1 M), 1 × Oxygen Scavenger
(3) STORM buffer using βME: live sample 100 × Live-cell imaging buffer base, 0.5 × βME, 1 × Oxygen Scavenger
(4) STORM buffer using MEA: live sample 100 × Live-cell imaging buffer base, 0.6 × MEA (1 M), 1 × Oxygen Scavenger. All buffers generally last for about 0.5–1 h.

5.4.4 Using STORM to Image a Sample

(1) Turn on all imaging equipment. Allow the lasers to warm up and the EMCCD camera to cool.
(2) Place the sample on the microscope stage. Use the brightfield lamp and the eyepieces to locate and focus on the cells.
(3) Switch the detection channel to the camera. Ensure that several of microspheres are visible in the field of view. Adjust the microscope to TIRF illumination mode.
(4) Record data using an illumination laser sequence depending on the probes being used and with the imaging speed and camera gain set for optimal signal-to-noise ratio.
(5) Save the data in a format compatible with the data processing software.

5.4.5 Image Reconstruction

There are now a number of different software packages available for obtaining single-molecule coordinates from time series images. A freely available plug-in for

Image J named quickPALM [90] can be used to analyze raw images. Image TIFF stacks are firstly preprocessed via background subtraction. For each frame, points corresponding to single photoemission events are identified with a minimum SNR of 2–4. Then, fluorescence peaks are identified in each frame and fitted a least-squares fit with an elliptical Gaussian function. Individual least-squares fit estimates are performed by a threshold of the peak height and the peak widths in the two lateral dimensions. After rejecting the poor fit and asymmetric PSFs, the coordinates of detected molecules are determined by the centers of gravity of their PSFs. STORM images are reconstructed as a density map using the precise localization data of single fluorescent molecules.

5.4.6 Note

(1) See Invitrogen.com for additional protocols for conjugating amine-reactive compounds to antibodies or amine-modified oligonucleotides.
(2) $NaBH_4$ should be made freshly, and it will produce lots of bubbles.
(3) Oxygen scavenger, MEA, and βME could not store for long time. They all should be prepared in time. Moreover, it is necessary to change the imaging buffer in sample every hour during the acquisition of the images.
(4) Turn on the microscope several hours before starting to image to help reduce stage drift. Sample drift will reduce resolution and lead to artificially large clusters being detected.
(5) Selecting the appropriate laser power and integration time for different samples. High laser and short integration time will increase the time resolution, but typically result in a decrease of the localization precision.
(6) TIRF illumination is important for imaging. The quality of TIRF is determined by the illumination alignment of the microscope. The sample should be placed flat on the microscope stage, otherwise it would also affect TIRF.
(7) 405 nm laser illumination can increase the rate of photoactivation, but high laser power will result in an increase in background. Therefore, it is better to use low levels of 405 nm laser illumination and image more frames in order to acquire sufficient localizations.
(8) Two-channel localization microscopy is possible with sequential acquisitions. Two-color (d)STORM, PALM, or two-color (d)STORM and PALM using fluorescent dyes or proteins are achievable.
(9) No matter which software package is employed, the setup of every parameter should be paid attention in order to optimize the single-molecule detection and localization process. Comparing different methods and results is fully recommended.
(10) MATLAB offers a range of interpolation algorithms for use with the griddata function. However, some algorithms lead to long processing times. Thus, a powerful computer system is suggested for data analysis.

Acknowledgements This work was supported by the National Key R&D Program of China (No. 2017YFA0505300), the National Natural Science Foundation of China (No. 21525314, 21703231 and 21721003).

References

1. Singer S, Nicolson GL (1972) The fluid mosaic model of the structure of cell membranes. In: Day SB, Good RA (eds) Membranes and viruses in immunopathology, pp 7–47
2. Abbe E (1873) Beiträge zur Theorie des Mikroskops und der mikroskopischen Wahrnehmung. Archiv für mikroskopische Anatomie 9(1):413–418
3. Hell SW, Wichmann J (1994) Breaking the diffraction resolution limit by stimulated emission: stimulated-emission-depletion fluorescence microscopy. Opt Lett 19(11):780–782
4. Klar TA, Hell SW (1999) Subdiffraction resolution in far-field fluorescence microscopy. Opt Lett 24(14):954–956
5. Birka H, Katrin IW, Stefan WH (2008) Stimulated emission depletion (STED) nanoscopy of a fluorescent protein-labeled organelle inside a living cell. Proc Natl Acad Sci USA 105(38): 14271–14276
6. Birka H, Willig KI, Wurm CA et al (2010) Stimulated emission depletion nanoscopy of living cells using SNAP-tag fusion proteins. Biophys J 98(1):158–163
7. Mueller V, Ringemann C, Honigmann A et al (2011) STED nanoscopy reveals molecular details of cholesterol- and cytoskeleton-modulated lipid interactions in living cells. Biophys J 101(7):1651–1660
8. Blom H, Brismar H (2014) STED microscopy: increased resolution for medical research? J Intern Med 276(6):560–578
9. Willig KI, Rizzoli SO, Westphal V et al (2006) STED microscopy reveals that synaptotagmin remains clustered after synaptic vesicle exocytosis. Nature 440(7086):935–939
10. Nägerl UV, Willig KI, Hein B et al (2008) Live-cell imaging of dendritic spines by STED microscopy. Proc Natl Acad Sci 105(48):18982–18987
11. Kittel RJ, Wichmann C, Rasse TM et al (2006) Bruchpilot promotes active zone assembly, Ca < sup > 2 + </sup > channel clustering, and vesicle release. Science 312(5776):1051–1054
12. van den Bogaart G, Meyenberg K, Risselada HJ et al (2011) Membrane protein sequestering by ionic protein-lipid interactions. Nature 479(7374):552–555
13. Bailey B, Farkas DL, Taylor DL et al (1993) Enhancement of axial resolution in fluorescence microscopy by standing-wave excitation. Nature 366(6450):44–48
14. Gustafsson MGL (2000) Surpassing the lateral resolution limit by a factor of two using structured illumination microscopy. Journal of Microscopy-Oxford 198:82–87
15. Gustafsson MGL, Shao L, Carlton PM et al (2008) Three-dimensional resolution doubling in wide-field fluorescence microscopy by structured illumination. Biophys J 94(12):4957–4970
16. Hirvonen LM, Smith TA (2011) Imaging on the nanoscale: super-resolution fluorescence microscopy. Aust J Chem 64(1):41–45
17. Gustafsson MG (2005) Nonlinear structured-illumination microscopy: wide-field fluorescence imaging with theoretically unlimited resolution. Proc Natl Acad Sci USA 102(37):13081–13086
18. Reto F, Lin S, Hesper R E et al (2012) Time-lapse two-color 3D imaging of live cells with doubled resolution using structured illumination. Proceedings of the National Academy of Science 109(14):5311–5315
19. Fitzgibbon J, Bell K, King E et al (2010) Super-resolution imaging of plasmodesmata using three-dimensional structured illumination microscopy. Plant Physiol 153(4):1453–1463
20. Whelan DR, Bell TDM (2015) Super-resolution single-molecule localization microscopy: tricks of the trade. J Phys Chem Lett 6(3):374–382

21. Betzig E, Patterson GH, Sougrat R et al (2006) Imaging intracellular fluorescent proteins at nanometer resolution. Science 313(5793):1642–1645
22. Hess ST, Girirajan TPK, Mason MD (2006) Ultra-high resolution imaging by fluorescence photoactivation localization microscopy. Biophys J 91(11):4258–4272
23. Rust MJ, Bates M, Zhuang X (2006) Sub-diffraction-limit imaging by stochastic optical reconstruction microscopy (STORM). Nat Methods 3(10):793–795
24. Bates M, Huang B, Dempsey GT et al (2007) Multicolor super-resolution imaging with photo-switchable fluorescent probes. Science 317(5845):1749–1753
25. Heilemann M, van de Linde S, Mukherjee A et al (2009) Super-resolution imaging with small organic fluorophores. Angew Chemie-Intern Ed 48(37):6903–6908
26. Dani A, Huang B, Bergan J et al (2010) Superresolution imaging of chemical synapses in the brain. Neuron 68(5):843–856
27. Matsuda A, Shao L, Boulanger J et al (2010) Condensed mitotic chromosome structure at nanometer resolution using PALM and EGFP—histones. PLoS ONE 5(9):12
28. Veatch SL, Cicuta P, Sengupta P et al (2008) Critical fluctuations in plasma membrane vesicles. ACS Chem Biol 3(5):287–293
29. Sengupta P, Jovanovic-Talisman T, Skoko D et al (2011) Probing protein heterogeneity in the plasma membrane using PALM and pair correlation analysis. Nat Methods 8(11):969–975
30. Sengupta P, Jovanovic-Talisman T, Lippincott-Schwartz J (2013) Quantifying spatial organization in point-localization super resolution images using pair correlation analysis. Nat Protocols 8(2):345–354
31. Ripley BD (1979) Tests of randomness for spatial point patterns. J R Stat Soc Ser B-Methodol 41(3):368–374
32. Lehmann M, Rocha S, Mangeat B et al. (2011) Quantitative multicolor super-resolution microscopy reveals tetherin HIV-1 interaction. Plos Pathog 7(12)
33. Malkusch S, Muranyi W, Müller B et al (2013) Single-molecule coordinate-based analysis of the morphology of HIV-1 assembly sites with near-molecular spatial resolution. Histochem Cell Biol 139(1):173–179
34. Owen DM, Rentero C, Rossy J et al (2008) PALM imaging and cluster analysis of protein heterogeneity at the cell surface. J Biophotonics 3(7):446–454
35. Gao J, Wang Y, Cai M et al (2015) Mechanistic insights into EGFR membrane clustering revealed by super-resolution imaging. Nanoscale 7(6):2511–2519
36. Chen J, Gao J, Wu J et al (2015) Revealing the carbohydrate pattern on a cell surface by super-resolution imaging. Nanoscale 7(8):3373–3380
37. Simons K, Ikonen E (1997) Functional rafts in cell membranes. Nature 387(6633):569–572
38. Kusumi A, Koyama-Honda I, Suzuki K (2004) Molecular dynamics and interactions for creation of stimulation-induced stabilized rafts from small unstable steady-state rafts. Traffic 5 (4):213–230
39. Klymchenko AS, Kreder R (2014) Fluorescent probes for lipid rafts: from model membranes to living cells. Chem Biol 21(1):97–113
40. Pike LJ (2006) Rafts defined: a report on the keystone symposium on lipid rafts and cell function. J Lipid Res 47(7):1597–1598
41. Kellner RR, Baier CJ, Willig KI et al (2007) Nanoscale organization of nicotinic acetylcholine receptors revealed by stimulated emission depletion microscopy. Neuroscience 144(1): 135–143
42. Hess ST, Gould TJ, Gudheti MV et al (2007) Dynamic clustered distribution of hemagglutinin resolved at 40 nm in living cell membranes discriminates between raft theories. Proc Natl Acad Sci 104(44):17370–17375
43. Takeda M, Leser GP, Russell CJ et al (2003) Influenza virus hemagglutinin concentrates in lipid raft microdomains for efficient viral fusion. Proc Natl Acad Sci 100(25):14610–14617
44. Gudheti Manasa V, Curthoys Nikki M, Gould Travis J et al (2013) Actin mediates the nanoscale membrane organization of the clustered membrane protein influenza hemagglutinin. Biophys J 104(10):2182–2192

45. Sengupta P, Jovanovic-Talisman T, Skoko D et al (2011) Probing protein heterogeneity in the plasma membrane using PALM and pair correlation analysis. Nat Meth 8(11):969–975
46. Mizuno H, Abe M, Dedecker P et al (2011) Fluorescent probes for superresolution imaging of lipid domains on the plasma membrane. Chemical Science 2(8):1548–1553
47. Owen DM, Rentero C, Rossy J et al (2010) PALM imaging and cluster analysis of protein heterogeneity at the cell surface. J Biophotonics 3(7):446–454
48. Rossy J, Owen DM, Williamson DJ et al (2013) Conformational states of the kinase Lck regulate clustering in early T cell signaling. Nat Immunol 14(1):82–89
49. Williamson DJ, Owen DM, Rossy J et al (2011) Pre-existing clusters of the adaptor Lat do not participate in early T cell signaling events. Nat Immunol 12(7):655–662
50. Eggeling C, Ringemann C, Medda R et al (2009) Direct observation of the nanoscale dynamics of membrane lipids in a living cell. Nature 457(7233):1159–1162
51. Gunewardene MS, Subach FV, Gould TJ et al (2011) Super-resolution imaging of multiple fluorescent proteins with highly overlapping emission spectra in living cells. Biophys J 101(6):1522–1528
52. Wu J, Gao J, Qi M et al (2013) High-efficiency localization of Na^{+}-K^{+} ATPases on the cytoplasmic side by direct stochastic optical reconstruction microscopy. Nanoscale 5 (23):11582–11586
53. Lillemeier BF, Mörtelmaier MA, Forstner MB et al (2010) TCR and Lat are expressed on separate protein islands on T cell membranes and concatenate during activation. Nat Immunol 11(1):90–96
54. Scarselli M, Annibale P, Radenovic A (2012) Cell type-specific β2-adrenergic receptor clusters identified using photoactivated localization microscopy are not lipid raft related, but depend on actin cytoskeleton integrity. J Biol Chem 287(20):16768–16780
55. Shroff H, Galbraith CG, Galbraith JA et al (2007) Dual-color superresolution imaging of genetically expressed probes within individual adhesion complexes. Proc Natl Acad Sci 104(51):20308–20313
56. Manley S, Gillette JM, Patterson GH et al. (2008) High-density mapping of single-molecule trajectories with photoactivated localization microscopy. Nature methods 5 (2):155–157
57. Lehmann M, Rocha S, Mangeat B et al (2011) Quantitative multicolor super-resolution microscopy reveals tetherin HIV-1 interaction. PLoS Pathog 7(12):e1002456
58. Rabinovich GA, Toscano MA, Jackson SS et al (2007) Functions of cell surface galectin-glycoprotein lattices. Curr Opin Struct Biol 17(5):513–520
59. Shroff H, Galbraith CG, Galbraith JA et al (2008) Live-cell photoactivated localization microscopy of nanoscale adhesion dynamics. Nat Methods 5(5):417–423
60. Hein B, Willig KI, Wurm CA et al (2010) Stimulated emission depletion nanoscopy of living cells using SNAP-tag fusion proteins. Biophys J 98(1):158–163
61. Wang Y, Gao J, Guo X et al (2014) Regulation of EGFR nanocluster formation by ionic protein-lipid interaction. Cell Res 24(8):959–976
62. Jones SA, Shim SH, He J et al (2011) Fast, three-dimensional super-resolution imaging of live cells. Nat Methods 8(6):499–505
63. Subach FV, Patterson GH, Renz M et al (2010) Bright monomeric photoactivatable red fluorescent protein for two-color super-resolution sptPALM of live cells. J Am Chem Soc 132(18):6481–6491
64. Gudheti Manasa V, Curthoys Nikki M, Gould Travis J et al. Actin mediates the nanoscale membrane organization of the clustered membrane protein influenza hemagglutinin. Biophys J 104(10):2182–2192
65. D'Amico F, Skarmoutsou E (2008) Quantifying immunogold labelling in transmission electron microscopy. J Microsc 230(1):9–15
66. Purbhoo MA, Liu H, Oddos S et al. (2010) Dynamics of subsynaptic vesicles and surface microclusters at the immunological synapse. Sci Signal 3(121):ra36–ra36
67. Williamson DJ, Owen DM, Rossy J et al (2011) Pre-existing clusters of the adaptor Lat do not participate in early T cell signaling events. Nat Immunol 12(7):655–662

68. Sherman E, Barr V, Manley S et al (2011) Functional nanoscale organization of signaling molecules downstream of the T cell antigen receptor. Immunity 35(5):705–720
69. Monks CR, Freiberg BA, Kupfer H et al (1998) Three-dimensional segregation of supramolecular activation clusters in T cells. Nature 395(6697):82–86
70. McGill MA, McKinley RFA, Harris TJC (2009) Independent cadherin-catenin and Bazooka clusters interact to assemble adherens junctions. J Cell Biol 185(5):787–796
71. Binh-An Truong Q, Mani M, Markova O et al (2013) Principles of E-Cadherin Supramolecular Organization In Vivo. Curr Biol 23(22):2197–2207
72. Greenfield D, McEvoy AL, Shroff H et al (2009) Self-organization of the *Escherichia coli* chemotaxis network imaged with super-resolution light microscopy. PLoS Biol 7(6): e1000137
73. Quang B-AT, Mani M, Markova O et al (2013) Principles of E-cadherin supramolecular organization in vivo. Curr Biol 23(22):2197–2207
74. Goñi FM (2014) The basic structure and dynamics of cell membranes: An update of the Singer-Nicolson model. Biochim Biophys Acta 1838(6):1467–1476
75. Nicolson GL (2014) The Fluid—mosaic model of membrane structure: still relevant to understanding the structure, function and dynamics of biological membranes after more than 40 years. Biochim Biophys Acta 1838(6):1451–1466
76. Truong_quang B, Lenne P (2014) Membrane microdomains: from seeing to understanding. Front Plant Sci 5:18
77. Li S, Zhang X, Wang W (2014) Selective aggregation of membrane proteins by membrane-mediated interactions. Sci China Chem 57(12):1683–1689
78. Shan Y, Wang H (2015) The structure and function of cell membranes examined by atomic force microscopy and single-molecule force spectroscopy. Chem Soc Rev 44(11):3617–3638
79. Zhao W, Tian Y, Cai M et al (2014) Studying the nucleated mammalian cell membrane by single molecule approaches. PLoS ONE 9(5):e91595
80. Wang Y, Gao J, Guo X et al (2014) Regulation of EGFR nanocluster formation by ionic protein-lipid interaction. Cell Res 24(8):959–976
81. Saka SK, Honigmann A, Eggeling C et al (2014) Multi-protein assemblies underlie the mesoscale organization of the plasma membrane. Nat commun 5:4509–4522
82. Letschert S, Göhler A, Franke C et al (2014) Super-resolution imaging of plasma membrane glycans. Angew Chem Int Ed 53(41):10921–10924
83. Hang HC, Yu C, Kato DL et al (2003) A metabolic labeling approach toward proteomic analysis of mucin-type O-linked glycosylation. Proc Natl Acad Sci 100(25):14846–14851
84. Boyce M, Carrico IS, Ganguli AS et al (2011) Metabolic cross-talk allows labeling of O-linked *β-N*-acetylglucosamine-modified proteins via the *N*-acetylgalactosamine salvage pathway. Proc Natl Acad Sci 108(8):3141–3146
85. Stairs S, Neves AA, Stöckmann H et al (2013) Metabolic glycan imaging by isonitrile–tetrazine click chemistry. ChemBioChem 14(9):1063–1067
86. Chen J, Gao J, Zhang M et al (2016) Systemic localization of seven major types of carbohydrates on cell membranes by dSTORM imaging. Sci Rep 6:30247
87. Chen J, Gao J, Cai M et al (2016) Mechanistic insights into the distribution of carbohydrate clusters on cell membranes revealed by dSTORM imaging. Nanoscale 8(28):13611–13619
88. Torreno-Pina JA, Castro BM, Manzo C et al (2014) Enhanced receptor–clathrin interactions induced by *N*-glycan–mediated membrane micropatterning. Proc Natl Acad Sci 111(30): 11037–11042
89. Lajoie P, Partridge EA, Guay G et al (2007) Plasma membrane domain organization regulates EGFR signaling in tumor cells. J Cell Biol 179(2):341–356
90. Henriques R, Lelek M, Fornasiero EF et al (2010) QuickPALM: 3D real-time photoactivation nanoscopy image processing in ImageJ. Nat Methods 7(5):339–340

Chapter 6
Analysis and Applications of Single-Molecule Fluorescence in Live Cell Membranes

Hua He, Xiaojuan Wang and Fang Huang

6.1 Introduction

The cell membrane, also called the plasma membrane, is a thin phospholipid bilayer interweaved with cholesterol, proteins, and carbohydrates, where multiple biological processes occur to keep the cell adapted to its environment. As a highly heterogeneous and dynamically organized entity in the cells, the plasma membrane actively participates in numerous cellular functions. Today, deciphering the dynamics of membrane organization on different length scales has become the key to understand biological functions of membranes. In line with its ability to detect the dynamic parameters of molecules in the plasma membrane, fluorescence microscopy is certainly one of the most important techniques. In particular, with the emergence of high-sensitivity detectors, high-quality optics, and diverse fluorescent probes in the past decades, single-molecule fluorescence (SMF) techniques have come into being and provided researchers with unprecedented information by directly detecting single-molecule behaviors in living cells. Because SMF techniques are intrinsically free from ensemble averaging, they can explore properties and behaviors of single molecules and unravel molecular heterogeneities in the live cell membrane. Historically, the detection of single fluorescent molecules was first achieved at cryogenic temperatures by absorption in 1989 [1], and then by emission in 1990 [2]. The cryogenic temperature was used to increase the absorption cross section of a fluorophore as well as to improve its photo-stability. With the advancements of the optics and high-sensitivity detectors, single-fluorescent molecules were observed at room temperature by near-field scanning optical microscopy [3]. However, the aqueous condition was still missing at that time, limiting the application of SMF techniques in the biological field. In 1995, single

H. He · X. Wang · F. Huang (✉)
State Key Laboratory of Heavy Oil Processing and Center for Bioengineering and Biotechnology, China University of Petroleum (East China), Qingdao 266580, China
e-mail: fhuang@upc.edu.cn

H. Wang and G. Li (eds.), *Membrane Biophysics*,
https://doi.org/10.1007/978-981-10-6823-2_6

fluorescently labeled myosin molecules on the glass substrate were directly imaged in aqueous solution by total internal reflection fluorescence (TIRF) microscopy [4], where individual ATP turnover reactions were also visualized. This technique was further employed to image single EGF ligands in the plasma membrane in 2000 [5]. Since then, SMF has become a central technique to probe cellular events in live cell membranes and has been booming in the past decades. As the techniques mature, modern SMF imaging is achieved by various optics, which are commonly divided into two types of configurations: wide-field imaging and confocal detection. SMF imaging, essentially including single-molecule tracking (SMT), is often coupled with different methods such as fluorescence resonance energy transfer (FRET), fluorescence correlation spectroscopy (FCS), and super-resolution fluorescence imaging. All these techniques are revolutionizing our views of the cell membrane. SMF-based super-resolution imaging techniques have been invented to overcome the diffraction barrier and have attracted great attention in recent years because of the ultrahigh resolutions down to 20–50 nm. The super-resolution imaging techniques are specially discussed in Chap. 5.

We here start by describing the basics of three most common SMF techniques used on live cell membrane research including standard SMF imaging (including SMT), single-molecule FRET (sm-FRET) and FCS. Then, the main findings and their contributions to the understanding of live cell membranes are discussed. We finally provide the practical protocols of applying SMF techniques to live cell membranes, including the choice of fluorescent probes, labeling strategies, cell sample preparation, instrumentation setup, and data analysis.

6.2 Principles

In order to detect single molecules in live cells, the key is to achieve a high signal–noise ratio. In addition to bright and photostable fluorophores (see Sect. 6.4.1), high-numerical-aperture (NA) objectives and high-sensitivity detectors, such as avalanche photodiode (APD) and cooled electron multiplying charge-coupled device (EMCCD) camera, are essential to detect weak signals from single-fluorescent molecules. Besides, another critical point is to reduce the background noise by constraining the laser illumination to a shallow depth or small volume with TIRF and confocal microscopies, respectively. These two types of microscopy configurations are commonly employed for SMF imaging, sm-FRET and FCS.

6.2.1 Wide-Field Imaging

The wide-field microscope performs single-molecule imaging by using wide-field illumination and EMCCD as the detector. EMCCD can simultaneously visualize hundreds to thousands of molecules with the best time resolution down to several milliseconds. The epi-illumination is the simplest form of wide-field illumination. Because the excitation volume is large in the z-direction, the fluorescence out of the focal plane is also collected by the objective, resulting in a low contrast in the image. Although the epi-illumination can be applied to single-molecule imaging, a very thin sample is required (e.g., *E. coli*). A more popular form of wide-field imaging is total internal reflection (TIR) illumination as shown in Fig. 6.1. TIR is an optical phenomenon that occurs when light traveling in a medium of high refractive index (e.g., glass) arrives at an interface with a medium of low-refractive index (e.g., water) at an angle larger than a critical angle. The TIR produces an "evanescent field" (or called as "evanescent wave") with the same frequency as the incident light, which exponentially decays into the medium of low-refractive index. This depth can range from 50 to 150 nm by adjusting the incident angle of the light. The evanescent field is capable of exciting fluorophores and allows imaging of molecular events at the cell surface with little or no out-of-focus fluorescence. Currently, TIRF microscopes (TIRFM) are commonly divided into two types of setups: prism-type and objective-type, as shown in Fig. 6.2. Each setup has its own advantages [6, 7]. Prism-type TIRFM employs a prism attached to the coverslip to direct a light beam toward the TIR interface [8]. This configuration is inexpensive to set up and produces a "cleaner" evanescent-field fluorescence than objective-type setup. However, the prism occupies the space above the coverslip, thereby restricting thick sample accessibility. Objective-type TIRFM has the advantage of providing a free space above the coverslip. In this configuration, an objective with very high NA (>1.45) is used to achieve illumination at the critical angle and collects the fluorescence as well [9]. It thus permits the combination of other instrumentation such as micromanipulators, differential interference contrast (DIC), or scanning probe microscopes. Objective-type TIRFM is now more widely used since it represents an ideal tool to investigate the multiple and dynamic events occurring at the plasma membrane of living cells, which minimizes the cellular autofluorescence [10].

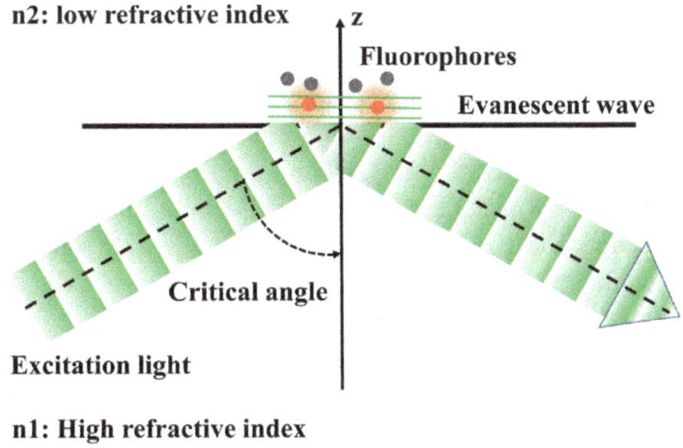

Fig. 6.1 An evanescent wave (or field) is produced at the interface of two media having different refractive indices when a beam of light is reflected away. This wave decays exponentially in the z-direction and will only excite fluorescent molecules at a depth slightly greater than the width of a cell membrane

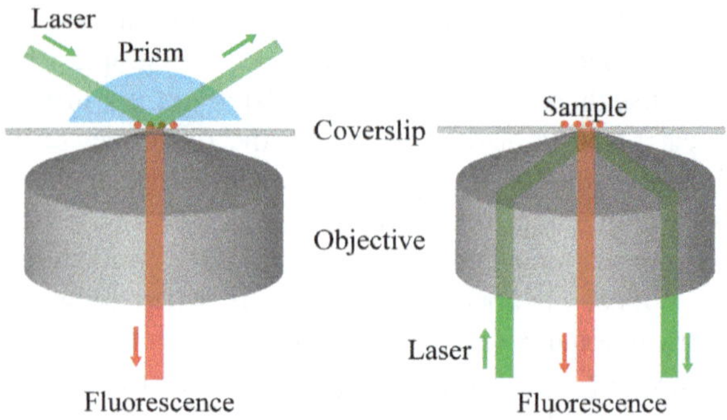

Fig. 6.2 Prism- and objective-type TIRFM

Single-molecule imaging allows tracking of single molecules in living cells by recording image sequences [11, 12]. Analyzing time- and space-dependent fluorescence can obtain certain physical or chemical parameters of single molecules or the local environments where they reside, which are important for understanding their biological relevance in cellular processes at the molecular level [11, 13]. Modern SMT methods are achieving almost the ultimate spatial precision and time resolution for tracking single molecules, of course, which are determined by the currently available fluorescent probes.

6.2.2 *Confocal Imaging*

Confocal laser scanning microscopy is another common approach to reduce out-of-focus fluorescence and improve the image contrast by setting a pinhole in the light path. Unlike wide-field imaging that uses a camera as the detector, confocal imaging is a point-detection-based method [14], as shown in Fig. 6.3. The laser beam is tightly focused by an objective into a small volume which excites and scans across the sample. The fluorescence is collected by the same objective, spatially filtered with a pinhole to remove the out-of-focus fluorescence, and then detected by a point detector (e.g., APD). The advantage of confocal fluorescence microscopy is that it allows to image deeper into a sample and reconstructs a three-dimensional (3D) image by scanning the sample along the z-direction. However, because the point detector essentially tracks one molecule at a time rather than probe multiple regions of the sample simultaneously, this single-imaging modality is less practical for single-molecule imaging and tracking in live cell membranes. Even so, confocal detection is frequently combined with other techniques such as FCS, photon counting, and FRET for single-molecule analysis. These confocal-based detection techniques are of great value for the study of live cells as they afford much higher sensitivity and temporal resolution down to the picoseconds scale than camera-based wide-field imaging.

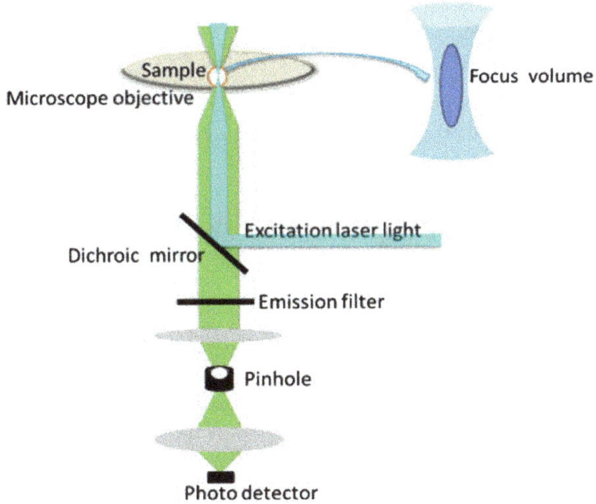

Fig. 6.3 Confocal setup. Reprinted from Ref. [14], Copyright 2014, with permission from Elsevier

6.2.3 *sm-FRET*

FRET is a physical process that relies on the distance-dependent energy transfer from a donor to an acceptor fluorophore via long-range dipole–dipole coupling when they are close to each other within 10 nm [15]. The energy transfer efficiency is given by:

$$E = \frac{1}{1 + (R/R_0)^6}, \tag{6.1}$$

where R is the distance between the pairs and R_0 is the characteristic distance. Due to its sensitivity to distance, FRET is a powerful tool to study conformational distribution and dynamics of fluorophore-labeled biological molecules. Similar to other ensemble techniques by averaging signal over many molecules, ensemble FRET is difficult to probe dynamic events such as relative motion between the donor and acceptor because these events or conformational changes are not synchronized. sm-FRET opens up new opportunities to decipher the kinetics of structural changes without the need for synchronization [16, 17]. Since sm-FRET was first demonstrated on a DNA template [18], it has been widely used to investigate intra- or inter-molecular interactions (Fig. 6.4), such as conformational dynamics of proteins and nucleic acids, and their interactions with each other [19, 20]. When coupled with SMF imaging performed on a confocal or wide-field setup, FRET can be used to map spatial and temporal distributions and functional states of molecules in living cells. Since FRET occurs over distances in the range of 1–10 nm, it can resolve molecular interactions and conformations with a spatial resolution far beyond the optical diffraction limit.

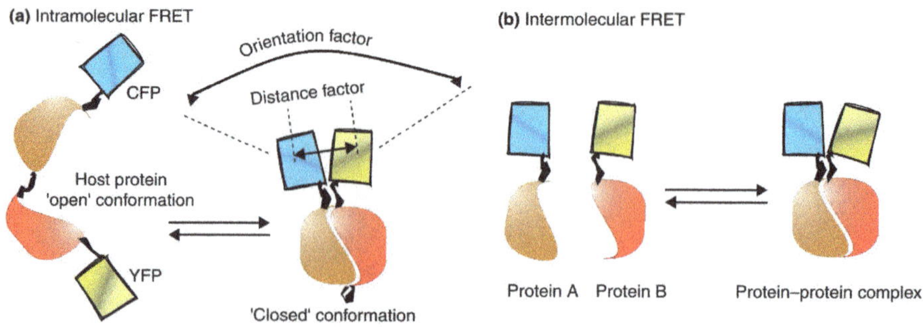

Fig. 6.4 Intramolecular and intermolecular FRET. **a** Intramolecular FRET can occur when both the donor and acceptor chromophores are on the same host molecule, which undergoes a transition, for example, between "open" and "closed" conformations. In each square box corresponding to CFP or YFP (shown in cyan or yellow, respectively), a diagonal line represents the chromophore. The amount of FRET transferred strongly depends on the relative orientation and distance between the donor and acceptor chromophores: the parallel orientation and the shorter distance (<100 Å) generally yield larger FRET. **b** Intermolecular FRET can occur between one molecule (protein A) fused to the donor (CFP), and another molecule (protein B) fused to the acceptor (YFP). When the two proteins bind to each other, FRET occurs. When they dissociate, FRET diminishes. Reprinted from Ref. [20], Copyright 2001, with permission from Elsevier

6.2.4 FCS

FCS was first developed in early 1970s to measure diffusion coefficients of macromolecules in solution by analyzing thermodynamic fluctuations in fluorescence intensity of a system [21–24]. As the technique advanced [25, 26], in particular by its combination with confocal microscopy, FCS has received a boost for cell biology research [27–31]. A scheme of a typical setup is shown in Fig. 6.5 [32]. In FCS, a high-NA objective (water or oil as immersion media) is employed to focus the laser beam. The excited fluorescence within the focal volume is collected by the same objective and then passes through a pinhole to reduce the detection volume in the z-direction. The resultant detection volume is confined to a volume of approximately 10–15 fL. The temporal fluctuations in fluorescence intensity are recorded by an APD detector. This configuration allows extremely low concentrations of fluorophores ranging from 10^{-10} to 10^{-6} M to be used. The fluorescence intensity fluctuations are caused by the diffusion of fluorophores through the tiny detection volume or the changes of their emission properties due to chemical reactions, photodynamic processes, or conformational changes. The resulting data are mathematically analyzed by a tool that relates the fluorescence signal with itself at different lag times. The autocorrelation curve is calculated from the intensity trace as follows:

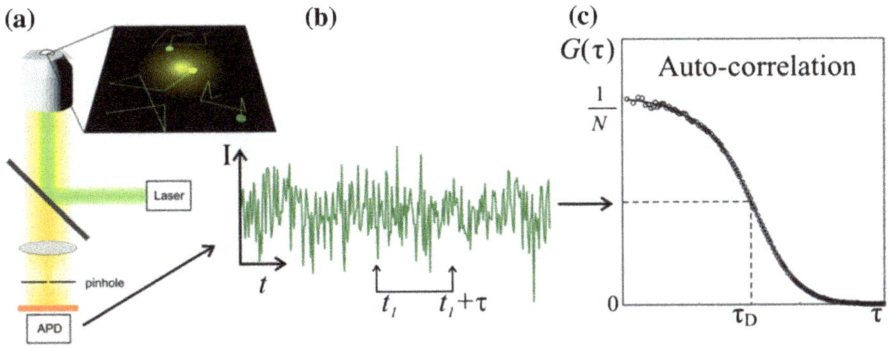

Fig. 6.5 Principles of fluorescence correlation spectroscopy. **a** The sample is illuminated by focusing the laser beam with an objective. The emitted photons are then spectrally filtered and detected with an APD. A pinhole before the detector limits the detection volume in the z-direction so that the light from upper and lower planes is eliminated. **b** The emission of photons diffusing in and out the focal volume induces fluctuations in the intensity signal. The fluorescence trace recorded during a given period of time is autocorrelated. The autocorrelation function measures the self-similarity of the fluorescence signal with time. **c** The graph shows the values obtained for the autocorrelated fluorescence signal in (**b**). The diffusion coefficient and the particle concentration can be determined by fitting the experimental values to the autocorrelation function in (**b**) (solid line). Reprinted from Ref. [32], with kind permission from Springer Science+Business Media

$$G(\tau) = \frac{\langle \delta F(t) \cdot \delta F(t+\tau) \rangle}{\langle F(t) \rangle^2}, \tag{6.2}$$

where G is the autocorrelation function, F is the fluorescence intensity as a function of time t, τ is the correlation time, the angular bracket refers to time averaging, and $\delta F(t) = F(t) - \langle F(t) \rangle$. To obtain dynamic parameters of interest, the autocorrelation curve is fitted with different mathematical model functions [33], which take into account the practical processes that occur in different samples. The fluorescence fluctuations caused by other sources can also affect the autocorrelation curve, such as photo-physical processes, triplet states, and intrinsic blinking behaviors of fluorophores. These effects should be considered in the model function used for fitting. For a single fluorescent species in a homogenous solution without chemical kinetics, the detection volume is approximated by a 3D Gaussian profile. The autocorrelation function describing 3D Brownian diffusion is expressed as follows:

$$G(\tau) = \frac{1}{N} \cdot \frac{1}{\left(1 + \frac{\tau}{\tau_D}\right)} \cdot \frac{1}{\sqrt{1 + \left(\frac{\omega_0}{z_0}\right)^2 \frac{\tau}{\tau_D}}}, \tag{6.3}$$

where ω_0 and z_0 are the lateral and axial radii, respectively, whose ratio defines the shape of the detection volume; N is the average number of fluorescent molecules in the detection volume; τ_D is the characteristic diffusion time, which is related to the diffusion coefficient D through the expression as follows:

$$\tau_D = \omega_0^2/4D. \tag{6.4}$$

For a simple homogeneous solution of a fluorescent species, FCS can determine molecular diffusion coefficient, concentration, and single-molecule brightness [34, 35].

FCS provides a powerful tool to characterize molecular diffusion in living cells, particularly well suited for studying cell membrane events, as the molecular diffusion in the cell membrane is slower than in the cytoplasm. FCS was first used to probe molecular behaviors in the membrane three decades ago [36] and has then been applied to study molecular lateral diffusion in cell membranes [29, 37, 38]. For fluorophore-labeled biomolecules within lipid membranes, the diffusion occurs approximately in two dimensions (2D) in the focal plane. The autocorrelation function for 2D diffusion is expressed as follows:

$$G(\tau) = \frac{1}{N} \cdot \frac{1}{1 + \tau/\tau_D}. \tag{6.5}$$

For membrane measurements, the correct adjustment of the laser focus on the membrane is critical during recording autocorrelation curves [39]. It is generally recommended to combine the FCS setup with a confocal scanning imaging module allowing the measurement at different z-positions of the bilayer [40]. Actually, the thickness of cell membrane is much smaller than the region detected by FCS. If fluorophores are not labeled exclusively to the cell membrane, those fluorophores positioned outside the cell membrane would contribute to the detection signals. Nevertheless, their contributions are low because molecules in the cytoplasm diffuse in and out of the detection volume more rapidly than those in the cell membrane. This problem can be reduced by decreasing the thickness of the excitation region with TIR illumination (described in Sect. 6.2.1). In addition, a variation of FCS is fluorescence cross-correlation spectroscopy (FCCS), which is useful for two-species analysis [41]. In this mode, two species are labeled with spectrally different dyes [42, 43] and excited within the same focal volume by using two lasers. The intensity fluctuations in the two channels give a cross-correlation function whose amplitude is proportional to the relative concentration of the interacting molecules that move together through the detection volume. FCCS is mostly designed to probe interactions or dynamic co-localization by spatiotemporal coincidence [44, 45].

6.3 Applications in Live Cell Membranes

Owing to high signal–noise ratios, SMF techniques now become well-established tools for investigating the complex behaviors of different molecules in live cell membranes [46]. In practical research systems, multiple SMF techniques such as

SMF imaging, FRET, and FCS are used alone or in combination. They can determine the structure, dynamics, and functions of single molecules and thus reveal their information that may lost in ensemble averages. Below we discuss representative examples of SMF techniques to illustrate how to measure physicochemical parameters and uncover critical information on cell membranes. Because of length constraints, we cite only a limited number of works; however, many others in this field that are not introduced here are important as well.

6.3.1 Membrane Receptor Oligomerization and Stoichiometry

An earlier pioneering work of SMF imaging on cell membrane was to image the epidermal growth factor receptor (EGFR) [5], which was lighted up by binding with extracellular epidermal growth factor (EGF) labeled with Cy3. In this work, TIRF imaging was proposed to detect Cy3–EGFs on apical and basolateral plasma membranes, respectively. Due to a slight difference in the refractive indices of the cytoplasm and the culture medium, the TIR on the apical surface of the cell membrane can occur by carefully adjusting the incident angle of the laser beam (Fig. 6.6). The investigators directly observed single molecular events in the plasma membrane by tracking each Cy3 spot. They found that the EGFR dimerization involved the formation of a cell surface complex of one EGF molecule and an EGFR dimer ($EGF - (EGFR)_2$) before the second EGF molecule bound. Meanwhile, the intermolecular interactions in living cells were first probed with sm-FRET by using a mixture of Cy3- and Cy5-labeled EGFs, further confirming that EGF-EGFR complexes indeed formed dimers at the single-molecule level. In the subsequent work, SMF imaging method was used to further study the binding kinetics of EGF to EGFR in live cell membranes [47]. It was concluded that a small number of EGFR on the cell surface formed dimeric binding sites, which interacted with EGF two orders of magnitude faster than the monomeric binding sites. The high association rate of the first binding event and the positive cooperativity of the binding of two EGFs to the predimeric sites facilitated the formation of signaling dimers. These single-molecule results provided valuable information to understand the early events of EGF-signaling pathways.

Because the intensity of each spot in an image is associated with the number of labeled molecules, the number of the bound ligand conjugated with a fluorophore in a cluster can be determined by fitting multiple intensity peaks [48, 49]. The researchers found that the cluster size distribution depended on the concentration of EGF applied to cells [47]. Such size distribution of EGF was used to estimate the association rate constants between EGF and EGFR. By SMF imaging method, it was further demonstrated that the number of extracellular signaling molecules required to induce certain cellular responses was very small. In HeLa cells, 300 EGF molecules were enough to induce an intracellular calcium response [49]. In

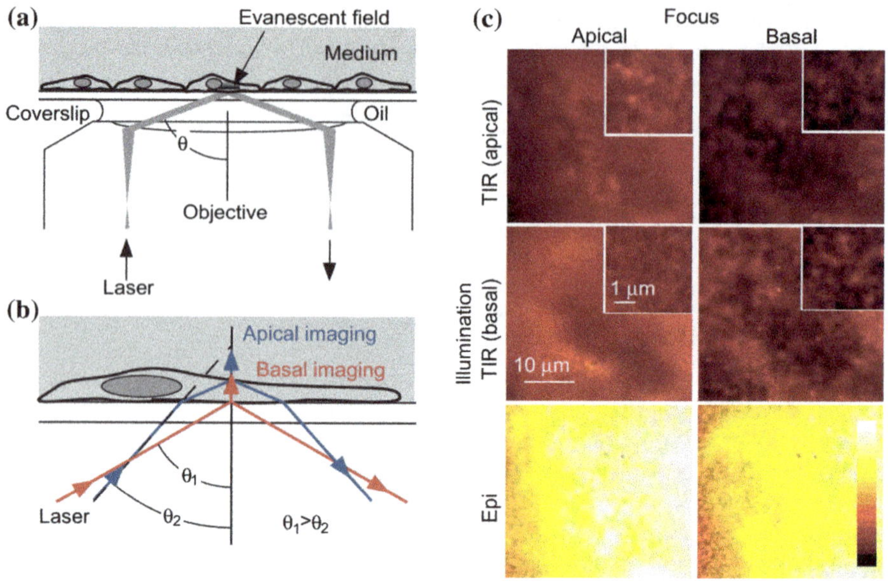

Fig. 6.6 TIR microscopy for visualization of single molecules on the cell surface. **a** TIR microscopy. An objective-type TIR microscope was used. **b** For imaging the basal cell surface, a laser beam was reflected between a coverslip and the culture medium. For imaging the apical plasma membrane, the incident angle of the laser beam was reduced ($\theta_1 > \theta_2$) and the laser beam was reflected between the cytoplasm and culture medium. **c** To demonstrate the effect of the illumination mode, images of A431 cells were taken in the same field under different illumination modes in the presence of 100 ng/mL Cy3–EGF. Cells were 2–3 μm thick, and by changing the focus, we could easily distinguish between the apical surface and the basal surface. Under apical TIR mode, fluorescent spots were shown on both apical and basal surfaces. In this case, spots on the basal surface would be excited by epifluorescence illumination. Because of excitation of Cy3–EGF in the basolateral surface and free Cy3–EGF in the evanescent-field, background fluorescence was higher on the apical surface than on the basal surface. Even in this condition, individual Cy3–EGF molecules could be visualized on the apical surface (inset). Under basal TIR mode, spots were only observed on the basal surface because of the limited excitation depth of TIR. Under the epifluorescence configuration, background from fluorescent EGF in the culture medium was high and no fluorescent spots on the cell surface were detected. Reprinted by permission from Macmillan Publishers Ltd: Ref. [5], Copyright 2000

addition, only 40 molecules of nerve growth factor per single growth cone were required to trigger the extension of the growth cone of chick dorsal root ganglion cells [50]. Also, a photobleaching-based method was developed to count single molecules in live cells [51]. Single step photobleaching generally implies a single emitter. The number of fluorophores can be counted by observing the number of discrete photobleaching steps. This method provides a straight way to reveal the stoichiometry of membrane receptors. As a demonstration, the authors characterized the photobleaching behavior of a four-emitter system, i.e., a four-unit molecular complex with each unit fused with a green fluorescence protein (GFP) molecule. By using this method, they resolved the subunit composition of

GFP-tagged NR1:NR3 NMDA receptor in live cell membranes [51]. Similar membrane protein complexes, such as transforming growth factor-β (TGF-β) receptors, were also identified [52–55]. It was demonstrated that GFP-labeled TGF-β receptor, Type II (TβRII), existed as monomers at low expression level in the resting cells, dimerized upon TGF-β stimulation for activation, and mainly formed high-order oligomers when the expression level was high [55]. G protein-coupled receptors (GPCRs), the largest family of membrane receptors in eukaryotes, were found to act through two models: monomers versus dimers and the monomers were capable of signaling [56]. Although GPCR homo- and heterodimers had been reported in many literature [56], the strengths of the evidence for dimer formation in living cells varied greatly among different research methods and experimental systems [57, 58]. SMF studies revealed that the monomers and dimers existed together and were in dynamic equilibrium for different membrane proteins [59–62], e.g., M1 muscarinic receptors, formyl-peptide (FP) receptors (FPRs) and β1- and β2-adrenergic receptors (Fig. 6.7). Additionally, the 2D monomer–dimer equilibrium constant as well as the association and dissociation rate constants were able to be determined. SMF studies demonstrated that monomers continually converted into dimers and dimers dissociated into monomers within 91 ms, and at any given moment, 2500 and 3500 receptor molecules participated in transient dimers and monomers, respectively [59].

6.3.2 *Membrane Receptor Dynamics and Interaction*

SMF imaging of membrane proteins for their stoichiometric analysis often involves the use of SMT technique since proteins are highly dynamic in living cells. SMT has been widely used to probe the spatiotemporal complexity of single-molecular behaviors of proteins including multi-subunit proteins and protein complexes for understanding their dynamics and deciphering the molecular mechanisms of protein functions. In order to characterize the dynamic properties of molecules in the plasma membrane, a key parameter is the lateral diffusion coefficient [50, 63]. Lino et al. identified GFP-fused E-cadherin (E-cad-GFP) clusters and measured their diffusion constants by SMT with 33-ms time resolution and determined cluster stoichiometry by using their quantized fluorescence intensity [52]. It was found that the translational diffusion coefficient of E-cad-GFP was reduced by a factor of 10–40 upon oligomerization. According to this, an oligomerization-induced trapping model was proposed. Similar characterization was carried out for other proteins such as Ca^{2+} channel proteins [53], H-Ras [64], cyclic adenosine 3',5'-monophosphate (cAMP) receptors [65], and peptide-binding proteins [66]. Taking advantage of the better photo-stability compared to dyes and fluorescent proteins, photostable quantum dots (QDs) were used to track glycine receptors over extended periods (>20 min) and to analyze their lateral dynamics in the neuronal membrane of living cells [67]. Chung et al. studied the dimerization dynamics of individual EGFRs in living cells with QD-conjugated anti-EGFR antibody [68]. By analyzing the local

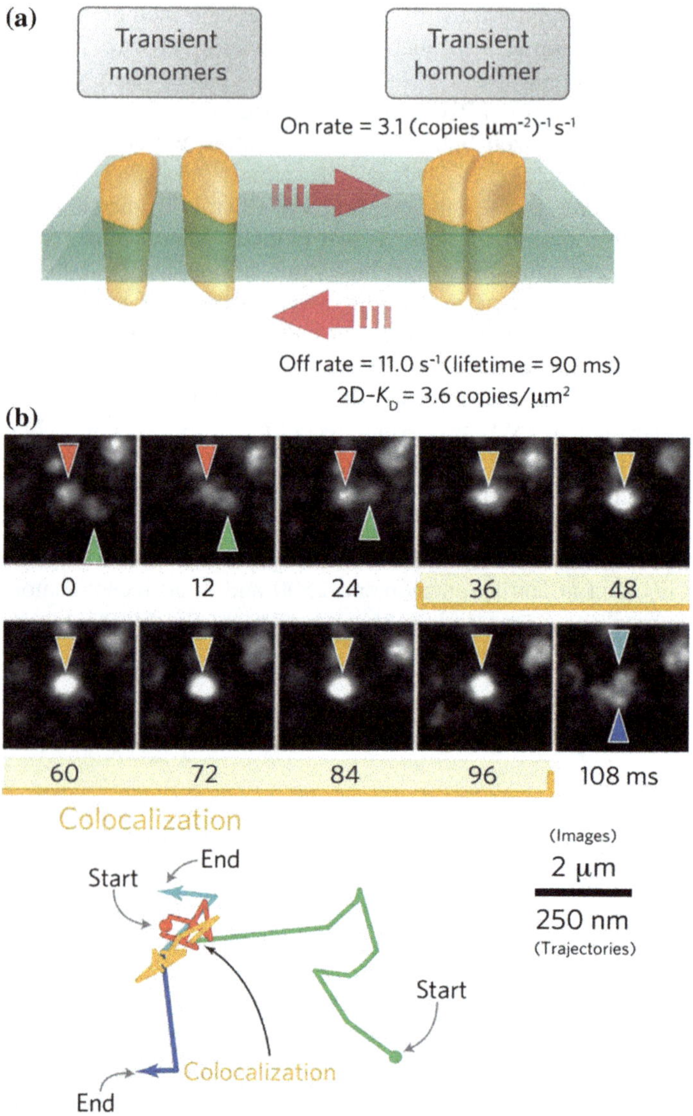

Fig. 6.7 **a** In the case of the FPR, the equilibrium has been fully characterized, with a two-dimensional equilibrium constant of 3.6 copies per mm^2 and dissociation and two-dimensional association rate constants of 11.0 s^{-1} and 3.1 (copies mm^{-2})$^{-1}$ s^{-1}, respectively. Agonist binding to FPR did not affect the dynamic monomer–dimer equilibrium. **b** A representative image sequence and trajectories of two diffusing FPR molecules. The receptor molecules were bound by Alexa 594-conjugated FPs (ligand), and their behaviors in the PM were recorded at 4-ms resolution (shown every 12 ms; i.e., every third frame). The two molecules (red and green arrowheads and trajectories) first became co-localized in the tenth frame (36 ms), diffused together for ~68 ms (indicated as "co-localization" in the image sequence; see also the orange trajectory) and then separated into monomers (dark and light blue arrowheads and trajectories). Reprinted by permission from Macmillan Publishers Ltd: Ref. [62], Copyright 2014

diffusivity of single receptors from each trajectory, they identified both fast and slow diffusion states, corresponding to monomers and dimers, respectively. It was found that EGFRs fluctuated continuously between monomer and dimer states in the absence of EGF. Upon ligand addition, they exhibited a very slow diffusivity state correlated with kinase activation. Taking advantage of narrow emission spectra of QDs, two-color QD tracking was performed to visualize the dimerization process of human erbB1 in real time, which demonstrated that erbB1 dimerization was promoted by domain co-confinement and stabilized by ligand binding [69]. The dynamic interactions of two molecules can be monitored in detail through sm-FRET method, as FRET occurs at the length scale of only a few nanometers. For example, Murakoshi et al. employed sm-FRET to observe the activation of the small G protein Ras at single molecule level [54]. It was found that on activation the diffusion of Ras was greatly suppressed/immobilized, suggesting the formation of large and activated Ras-signaling complexes. This approach was also applied to probe the inter-EGF distances within pairs of EGF molecules on the A431 cell surface. Distance measurements from EGF to the plasma membrane showed that high-affinity EGFR ectodomains were oriented flat on the membrane, whereas low-affinity ectodomains stood proud from it [70].

6.3.3 *Diffusion Dynamics and the Density of Membrane Receptors Probed by FCS*

FCS is a powerful tool to probe protein dynamics and interactions in live cell membrane. By measuring the diffusion of molecules, FCS identified some slowly diffusing populations of membrane-associated proteins or peptides, suggesting their binding to more ordered domains in the plasma membrane [71, 72]. Also, the binding constants of fluorescently labeled ligands to the γ-amino butyric acid A (GABAA) and to the β2-adrenergic receptors were determined in living cells [73, 74]. By analyzing the mobility of ligand–receptor complexes, two different diffusion coefficients were revealed, corresponding to the fast and hindered mobility, respectively. This approach was also used to investigate the interactions of T cell receptor with CD8 during T cell activation [75]. It demonstrated the presence of two TCR and CD8 subpopulations with different lateral diffusion rate constants. It was proposed that the most efficient route of the ligand binding started with CD8–pMHC interaction. In addition to the variation in diffusion dynamics, the fluctuations in the density of membrane proteins can reveal the dynamics of membranes. For example, Hao et al. measured the density of Arabidopsis respiratory burst oxidase homolog D (RbohD) fused with GFP at the plasma membrane by FCS [76]. In control seedlings, the mean density of GFP–RbohD was 29.84 ± 4.12 molecules/μm^2. When treated with Tyrphostin A23 (TyrA23) and methyl-β-cyclodextrin (mβCD), the GFP–RbohD densities were increased to 49.97 ± 6.43 and 35.16 ± 4.87 molecules/μm^2, respectively. It suggested that clathrin-dependent endocytosis was the main pathway for internalization of GFP–RbohD. By contrast, the density of GFP–RbohD decreased dramatically to 17.43 ± 3.59 molecules/μm^2 in the presence of NaCl, supporting the NaCl-induced internalization of GFP–RbohD. These results suggested that the density of signaling proteins on the plasma membrane was closely related to signal transduction in plants. Also, FCS-based approaches can be used to study the interactions of protein pairs in or near the plasma membrane of living cells (Fig. 6.8) [77]. The membrane proteins in different domains have different diffusion coefficients because of the interaction of proteins with microdomain-associated molecules [78, 79]. Using FCCS, the investigators demonstrated the close correlation between the localization of membrane proteins and clathrin in vivo [80].

6.3.4 Conformational Dynamics of Individual Proteins

sm-FRET has been used to investigate the conformational dynamics of membrane proteins in living cells, which allows the quantification of intramolecular conformational changes by attaching the donor and acceptor to two different sites in the same protein. The first demonstration of sm-FRET applying to monitor conformational changes of proteins in living cells has been reported in 2010 [81]. By a combination of sm-FRET with SMT, the authors monitored individual SNARE (soluble *N*-ethylmaleimide-sensitive factor attachment protein receptor) proteins entering into membrane-tethered complexes in living cells. The individual SNARE protein was specifically labeled with both a FRET donor and an acceptor, which performed only low-efficient FRET when the protein was free whereas yielded FRET with low or high efficiency depending on dye location when folded into a SNARE complex (Fig. 6.9). It was found that individual proteins rapidly incorporated into folded complexes at the cell membrane after injection. It should be noted that, in spite of its advances, this method still remains a great challenge in living cells owing to the difficulty in designing labeled molecules and suboptimal cell culture conditions.

6.3.5 Membrane Organization

The cell membrane is a complex and dynamic organization with a variety of components and functions. Quantifying these components and movements provides insights into the local microstructures and molecular interactions in the plasma membrane such as membrane microdomains (e.g., lipid rafts) [82]. Microdomains are tiny regions of tightly packed lipids and are also highly dynamic. They can affect protein movement, localization, and turnover. The motion pattern of proteins varies inside or outside membrane microdomains, which is closely related to its

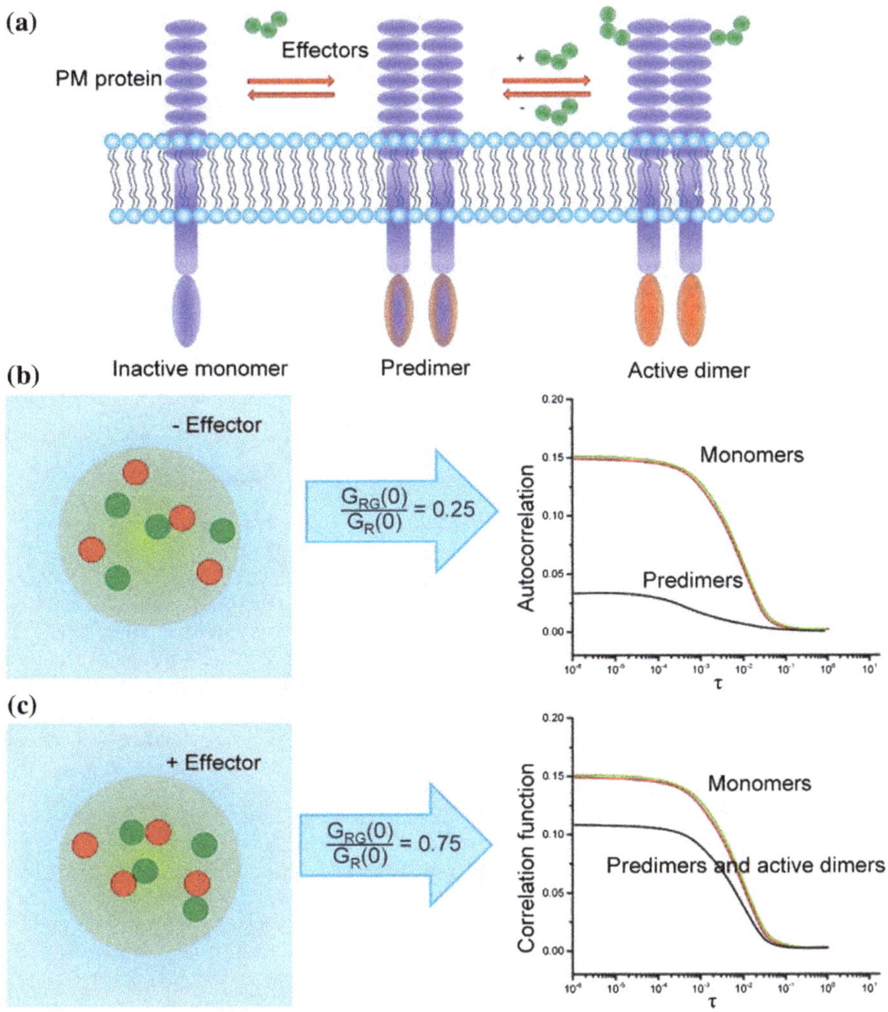

Fig. 6.8 FCCS approach used in the study of protein oligomerization. **a** Schematic showing PM protein dynamic equilibrium of the monomer, predimer, and active dimer. **b**, **c** Theoretical FCCS analysis predicts the proportion of monomeric and dimeric PM protein in living cells. The green circles represent the observation volume in the FCCS experiment. For the resting state case (**b**), the ratio of $G_{RG}(0)$ (cross-correlation curves shown in black) to $G_R(0)$ or $G_G(0)$ should theoretically be zero but shows a finite minimum value resulting from the predimers without stimulation by effectors. For the effector stimulation case (**c**), the ratio of $G_{RG}(0)$ to $G_R(0)$ or $G_G(0)$ increases dramatically due to the induced hetero-oligomerization of red and green fluorescently tagged proteins. Thus, the oligomerization or co-localization equilibrium of PM protein is resolved by calculating the minimum and maximum ratios for the set of resting and stimulated experimental conditions, respectively. Reprinted from Ref. [77], Copyright 2016, with permission from Elsevier

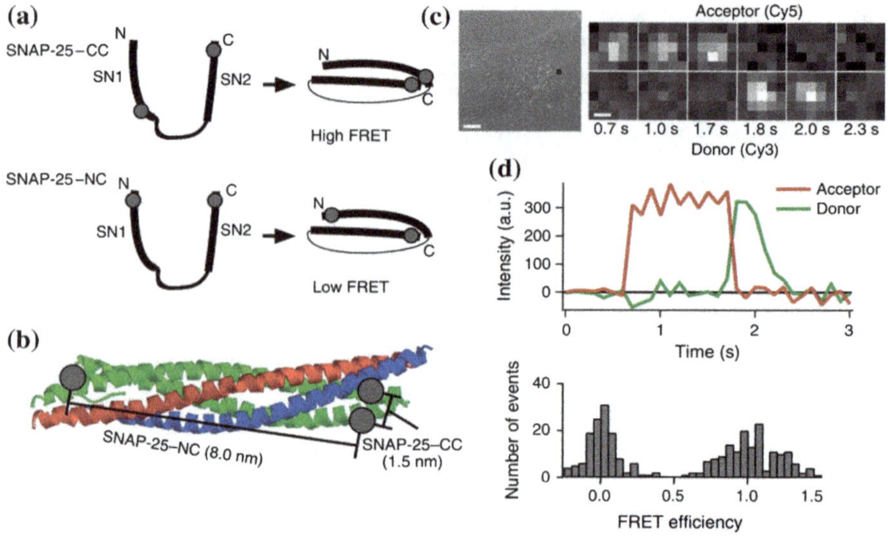

Fig. 6.9 sm-FRET detection in live cells. **a** Schematic of SNAP-25–CC and SNAP-25–NC label site locations. Isolated SNAP-25(left) is unstructured, resulting in low FRET because of the large separation of the label attachment sites. When SNAP-25 enters SNARE complex (right; synaptobrevin and syntaxin are not shown) the SNAP-25–CC and SNAP-25–NC constructs yield high- and low-FRET states, respectively. Gray circles represent the approximate label sites. **b** Location of label sites (gray circles) in assembled SNARE complex (Protein Data Bank (PDB) identifier: 1sfc; syntaxin, red; synaptobrevin, blue; SNAP-25, green). SNAP-25–CC construct was designed such that dye distances are less than 2 nm apart in the final SNARE complex and give high FRET. SNAP-25–NC was designed so dyes are 8 nm apart in parallel complex and give low FRET. **c** Details of a movie of fluorescence emission in the donor (bottom) and acceptor (top) channel taken from the boxed location indicated in the differential interference contrast image for SNAP-25–CC in a BS-C-1 cell. Time after injection is indicated. Scale bars, 10 μm (left), 0.5 μm (right). **d** Intensity time course for sm-FRET example from (**c**). Green illumination was active for entire interval plotted. **e** FRET efficiency histogram for single-molecule events using Cy3–Cy5 labeled SNAP-25–CC in a BS-C-1 cell accumulated for 0.2 s (FRET < 0.5) and 1.0 s (FRET > 0.5) to sufficiently populate both peaks. Scale bars: 10 μm. Reprinted by permission from Macmillan Publishers Ltd: Ref. [81], Copyright 2010

function [76, 83]. The detection of membrane microdomains remains challenging by using traditional imaging methods, but single-molecule microscopy allows direct visualization of lipid microdomains labeled with fluorescence dyes containing saturated acyl chains [84]. Single lipid dye molecules were tracked on a millisecond time scale and used to search for lipid microdomains or rafts in the plasma membrane of human coronary artery smooth muscle (HASM) cells. With about 15 probe molecules localized in each domain, the location, size, shape, and motion of single domain became clearly visible. It showed that the domains had a size of 0.7 μm (0.2–2 μm), covering about 13% of total membrane area. Douglass and Vale used SMT to demonstrate that the adaptor protein LAT and the tyrosine kinase Lck co-clustered in discrete microdomains in the plasma membrane of signaling T cells

[85]. The properties of these microdomains depended on protein–protein interactions but not on actin or lipid rafts. It was demonstrated that the protein microdomains not only affected the free diffusion of molecules in the membrane but also trapped and immobilized specific proteins. Similar SMT technique was employed to demonstrate the presence of microdomains in the cytoplasmic leaflet of the cell membrane [64]. In this work, a lipid-anchored yellow fluorescent protein (YFP) was produced, which was targeted to the cytoplasmic leaflet of the plasma membrane in living cells. By recording and analyzing trajectories of individual YFP molecules diffusing in the cytoplasmic leaflet of the plasma membrane, the researcher found that the diffusion of 30–40% of the lipid-anchored YFP molecules was constrained in domains. These domains had a typical size of 200 nm, not depending significantly on the actin cytoskeleton or cholesterol.

Membrane organization has been well characterized by FCS and the derived approaches. FCS measurements on various spatial scales were applied to identify lipid-dependent (microdomain model) or cytoskeleton-based (meshwork model) processes in living cells with a high temporal resolution [78]. It was found that putative raft markers were dynamically compartmented into small microdomains that were sensitive to the cholesterol and sphingomyelin levels, while actin-based cytoskeleton barriers were responsible for the confinement of the transferrin receptor protein. When such two organizations were disrupted, a free-like diffusion was observed. The results suggested that two main compartmentalizing forces were at work in the plasma membrane. In addition, FCS combined with stimulated-emission-depletion (STED) illumination allows accurate quantitative analysis of nanodomains and their influence on the diffusion of molecules in live cell membranes [86]. The authors used STED–FCS to measure and compare the diffusion of various fluorescent lipid analogs in live cell membranes and demonstrated details of the transient formation of molecular complexes. This combination technique was also employed to reveal a constant level of confined diffusion of raft lipid analogs that varied greatly in their partitioning behavior [87] and to quantify the nanoscale dynamics of membrane lipids in a living cell owing to a fast gathering of statistically accurate data [88].

6.4 Methods

Despite their great advances, applying SMF techniques to live cell membranes remains many practical challenges due to the weak detective signal and the complex membrane composition and structure. The design of SMF experiments involves several practical considerations, including the choice of fluorescent probes, the strategies for labeling the targeting biomolecules, the cell culture and sample preparation, the method and instrumentation for signal detection, and the data analysis.

6.4.1 Choice of Probes

Generally, native biomolecules in the cell membrane are invisible under the optical microscope; a fluorescent probe has to be attached to them for optical detection [89, 90]. Improvements in fluorescent probes will enhance the utility of fluorescence detection for studying molecular events in living cells. An ideal fluorescent probe should have (a) high extinction coefficient, (b) high quantum yield, (c) good photo-stability under intense illumination, (d) easy conjugation, and (e) no perturbation to the targeting molecule. Frequently used fluorescent probes can be categorized into three groups: fluorescent proteins (FPs, such as GFP and YFP), organic dyes (such as Cy3 and Cy5), and fluorescent nanoparticles (such as QDs and noble-metal nanocluster). Each group has its advantages and disadvantages, and the choice of probes depends on the design of SMF experiments and the practical systems under investigation.

FPs are a class of genetically encoded proteins that can emit fluorescence upon light excitation [91]. By fusing its gene into the target gene, FPs can be expressed in many organisms, such as bacteria, mammalian cells, yeast, and human cells. Because such labeling is achieved at the genetic level, nonspecific labeling is intrinsically avoided [92, 93]. FPs are more biocompatible and less phototoxic than organic dye molecules. The disadvantage of FPs is that they are not as bright as many organic dye molecules and also prone to photobleach. Albeit these, FPs have been successfully used in many SMF studies in living cells by using TIR or confocal optics to provide high signal–noise ratios [51, 54, 55, 94, 95]. For sm-FRET experiments, FPs are generally not suggested for use as acceptors as they can be partly excited by donor excitation, limiting quantitative sm-FRET analysis to some extent. Instead, an organic dye molecule can be chosen as the acceptor by pairing with a donor FP for sm-FRET detection [54]. Note that FPs such as GFP, CFP, and YFP tend to form dimers [96]. If possible, it is preferable to use monomeric GFP (mGFP) containing the A206K mutation to prevent artificial oligomerization. Other monomeric variants of FPs (e.g., mEosFP, mRFP, mKate, and mCherry) can also be chosen for SMF experiments in living cells.

Organic dyes are indispensable for SMF detection in cell biology owing to their small sizes, high brightness, and a wide available spectral range. However, attaching dye molecules to biomolecules needs additional and tedious labeling procedures such as coupling chemical reactions and adsorption. The similar coupling chemistries are also required for fluorescent QDs. Currently, well-established QDs are semiconductor nanocrystals with a core–shell architecture such as CdSe/ZnS. In addition to the higher extinction coefficients, QDs are more photostable and much brighter than organic dyes and FPs [97]. The disadvantage of QDs in single-molecule application is the large size. Typically, the size of fully functionalized QDs can reach 50 nm, which may cause the effects of steric hindrance and crosslinking. Nevertheless, QDs do not pass through the cell membrane and thus are readily used to target proteins, lipids, and other cell surface molecules by conjugation to a specific ligand or antibody [98, 99]. QDs are also good candidate probes for SMT experiments in living cells due to their high resistance to photobleaching [100, 101]. For sm-FRET measurements, QDs cannot work as acceptors because of their broad absorption spectra but this feature along with their narrow fluorescence emission makes it easy to identify and simultaneously track multiple QDs, each with a different color emission.

6.4.2 *Labeling Strategy*

The FP-based method is straightforward and relatively easy to perform by properly constructing the plasmid. Upon expression, FPs will be covalently linked to the target protein and the specificity is absolute, which is well suited for living cell experiments. For organic dyes and QDs, suitable functional groups such as peptides, proteins, or oligonucleotides are required for modifications by covalent binding or noncovalent attachment. In this regard, organic dyes are more advantageous than QDs because a toolbox of functionalized dyes is commercially available along with established labeling protocols, purification, and characterization as well as information on the site-specificity of the dye bio-conjugates. For QD bio-conjugates, there are many excellent reviews on the methods and labeling strategies [102–105], and readers interested in this topic are encouraged to refer to them. We here focus on FP labeling by fusing enhanced green fluorescent protein (EGFP) to chemokine receptor CXCR4 as an example. The detailed protocol, cell growth, sample preparation, SMF experiments, and data analysis are discussed in the next sections. All these procedures are applicative to other surface molecules on cell membranes. Cell culture conditions, instrumentation methods, parameter settings, and data analysis can also be followed by other SMF experiments with dyes and QD bio-conjugates as probes.

6.4.3 *Protocol*

6.4.3.1 Materials and Reagents

1. T-REx-293 cell line containing the stable expression of pcDNA6/TR that encodes the Tet repressor protein (TetR), Lipofectamine 2000, Dulbecco's Modified Eagle Medium (DMEM), Blasticidin, fetal bovine serum, GlutaMAX, and pcDNA4/TO plasmid were purchased from Invitrogen.
2. Tetracycline (Sigma-Aldrich).
3. PBS buffer, pH 7.4.
4. Ethanol, 70% (v/v) for various cleaning and sterilization tasks.
5. Chambered coverslip with eight wells (Lab-Tek®).
6. Immersion oil with a refractive index (n = 1.518, Olympus).

6.4.3.2 Equipments

In this section, we list the components that have been performed for SMF imaging of EGFP–CXCR4 in live cell membranes. However, other components with comparable specification should also work.

1. TIRF microscope (Nikon Ti series). Note that any TIRF microscope that is capable of visualizing fluorescent markers will work for this protocol.
2. 100 × TIR objective, NA = 1.49.
3. EMCCD (iXon DU-897, Andor).
4. Laser (488 nm, Cobolt).
5. Nikon filter cubes (DM 505 nm/BA520 nm).

6.4.3.3 Procedures

CXCR4–EGFP Plasmid Construction

1. CXCR4–EGFP fusion construct was generated by overlap PCR using the appropriate primers.
2. The PCR product was double digested with DNA restriction enzymes *Hin*dIII and *Eco*RI and then cloned into the pcDNA4.0/TO vector digested with the same two DNA restriction enzymes.
3. All resulting constructs were confirmed by DNA sequencing.

Cell Culture

1. Cells were grown in DMEM/F12 with GlutaMAX and supplemented with 10% fetal bovine serum, HEPES (15 mM), non-essential amino acids (0.1 mM), penicillin (100 μg/mL), streptomycin (100 μg/mL) and grown at 37 °C under 5% (v/v) CO_2.
2. The pcDNA4/TO/CXCR4–EGFP vector was transfected into T-Rex-HEK293 cells using Lipofectamine 2000.
3. After transfection, cells were further cultured for 48 h, and then selective medium containing 5 μg/mL Blasticidin and 60 μg/mL Zeocin was added.
4. Cells were continuously cultured in the selective medium for another 4 weeks.
5. Colonies stably integrating the recombinant vectors were picked up and screened for better receptor expression.

Sample and Microscope Preparation

1. 1×10^4 cells/mL were seeded into Lab-Tek eight-well chambers. Note that the constructed stable T-Rex-293 cell line can produce recombinant EGFP–CXCR4 by the induction of tetracycline.
2. Stable cell lines were induced by 1 μg/mL tetracycline at 37 °C under 5% (v/v) CO_2 for a specified time and washed by PBS. Note that the expression level of EGFP–CXCR4 increases as the induction time is prolonged.

3. Turn on the lasers and laser shutters on the TIRF microscope. Turn on the computer that is connected to the EMCCD, which has the NIS-elements microscope imaging software (Nikon).
4. In the control panel of the capture software, set preliminary parameters. Laser intensity: 10%, exposure time: 100 ms, EM gain: 300, conversion gain: 1×.
5. Drop immersion oil on the 100× objective lens.
6. Place the samples on the stage of the TIRF microscope and adjust the lens position.
7. Open the shutter for the 488-nm laser and adjust the focus position to observe EGFP signals.
8. Adjust the laser beam angle to achieve TIR beyond the critical angle.
9. Find a target area containing EGFP–CXCR4 of interest and optimize the focus, laser intensity, incidence angle of TIR, exposure time, and EM gain. Note that optimization should be performed as soon as possible to minimize photobleaching of EGFP as well as obtain an optimal signal-to-background ratio.
10. Movies of 100–500 frames were acquired for each sample. Note that crop the field of view to minimize the file size and acquire an image stream within a 35-ms exposure time.

Data Analysis

Image sequences can be analyzed using the acquisition software available from the microscope or by using the public domain NIH Image program ImageJ/FIJI software with lots of plug-ins, which is freely available at http://fiji.sc/Fiji. Many plug-ins and other open source codes provide powerful tools to extract and analyze fluorescence signal of single molecules. We recommend performing the quantification analysis using suitable plug-ins. For example, the background fluorescence can be subtracted using the rolling ball method in Image J software. Fluorescent spots analysis can be done using Speckle TrackerJ plugin.

6.4.4 Note

1. Coverslips should be cleaned in $K_2Cr_2O_7/H_2SO_4$ solution or Piranha solution (H_2SO_4: H_2O_2 = 3:1). Coverslips should be checked to make sure that fewer or no fluorescent spots of impurities under the imaging conditions used in SMF experiments before use.
2. To improve cell adhesion, coverslips can be modified (incubation for 60 min to overnight) with, e.g., collagen or poly-l-lysine (PLL) before use.
3. Optimization should be performed to minimize photobleaching of fluorescent probes, e.g., fluorescence proteins or organic dyes. For EMCCD, the gain should be used to improve the detection signals so that SMF experiments are performed at a low laser output power, but too high gains decrease the dynamic range of the detectors.

4. For APD detection in a confocal system, the output power of the laser should be strictly controlled. Too many photons may damage the detector.
5. TIR should be carefully adjusted for SMF imaging using a TIRF microscopy. A larger incidence angle leads to the thinner excitation depth and minimizes the interference from the autofluorescence of cells, but also may decrease single-molecule brightness.
6. Semiconductor QDs, e.g., CdSe/ZnS, are ideal for real-time and long-time (ranging from seconds to minutes) tracking of single molecules due to the fact that they do not fade when exposed to light. Those with suppressed blinking should be chosen. In addition, QDs emitted at >600 nm or NIR region are favorable for SMF imaging with high signal–noise ratios.
7. The filters should be optimized according to the fluorescence probes used in the experiments. "Green" probes, such as Alexa Fluor 488, Atto-488, or EGFP are used in green fluorescence channel excited at 488 nm. The band-pass filter with a transmission range of ~500–550 nm or long-pass filter with a transmission range of >500 nm is suitable. "Red" probes, such as Alexa Fluor 594, mCherry, and red quantum dots are used in red fluorescence channel having a band-pass filter with a transmission range of ~600–700 nm or a long-pass filter with a transmission range of >580 nm.
8. Image processing. The digital image can be enhanced by adjusting the contrast values. The image filtering processing should be cautiously used as the original data is modified in this process.

Acknowledgements This work was supported by the National Natural Science Foundation of China (No. 21573289) and the Natural Science Foundation of Shandong Province (No. ZR2014BM028).

References

1. Moerner WE, Kador L (1989) Optical detection and spectroscopy of single molecules in a solid. Phys Rev Lett 62(21):2535–2538
2. Orrit M, Bernard J (1990) Single pentacene molecules detected by fluorescence excitation in a p-terphenyl crystal. Phys Rev Lett 65(21):2716–2719
3. Betzig E, Chichester RJ (1993) Single molecules observed by near-field scanning optical microscopy. Science 262(5138):1422–1425
4. Funatsu T, Harada Y, Tokunaga M, Saito K, Yanagida T (1995) Imaging of single fluorescent molecules and individual ATP turnovers by single myosin molecules in aqueous solution. Nature 374(6522):555–559
5. Sako Y, Minoghchi S, Yanagida T (2000) Single-molecule imaging of EGFR signalling on the surface of living cells. Nat Cell Biol 2(3):168–172
6. Tokunaga M, Kitamura K, Saito K, Iwane AH, Yanagida T (1997) Single molecule imaging of fluorophores and enzymatic reactions achieved by objective-type total internal reflection fluorescence microscopy. Biochem Biophys Res Commun 235(1):47–53
7. Axelrod D (2001) Selective imaging of surface fluorescence with very high aperture microscope objectives. J Biomed Opt 6(1):6–13

8. Axelrod D (1981) Cell-substrate contacts illuminated by total internal reflection fluorescence. J Cell Biol 89(1):141–145
9. Stout AL, Axelrod D (1989) Evanescent field excitation of fluorescence by epi-illumination microscopy. Appl Opt 28(24):5237–5242
10. Sako Y, Yanagida T (2003) Single-molecule visualization in cell biology. Nat Rev Mol Cell Biol:SS1–5
11. Schütz G, Schindler H, Schmidt T (1997) Single-molecule microscopy on model membranes reveals anomalous diffusion. Biophys J 73(2):1073–1080
12. Ide T, Yanagida T (1999) An artificial lipid bilayer formed on an agarose-coated glass for simultaneous electrical and optical measurement of single ion channels. Biochem Biophys Res Commun 265(2):595–599
13. Schmidt T, Schütz G, Baumgartner W, Gruber H, Schindler H (1996) Imaging of single molecule diffusion. Proc Natl Acad Sci U S A 93(7):2926–2929
14. Shen C, Knapp M, Puchinger MG, Shahzad A, Gaubitzer E, Shen AD, Koehler G (2014) Using fluorescence correlation spectroscopy (FCS) for IFN-g detection: a preliminary study. J Immunol Methods 407:35–39
15. Stryer L (1978) Fluorescence energy transfer as a spectroscopic ruler. Annu Rev Biochem 47 (1):819–846
16. Weiss S (1999) Fluorescence spectroscopy of single biomolecules. Science 283 (5408):1676–1683
17. Ishii Y, Yoshida T, Funatsu T, Wazawa T, Yanagida T (1999) Fluorescence resonance energy transfer between single fluorophores attached to a coiled-coil protein in aqueous solution. Chem Phys 247(1):163–173
18. Ha T, Enderle T, Ogletree D, Chemla D, Selvin P, Weiss S (1996) Probing the interaction between two single molecules: fluorescence resonance energy transfer between a single donor and a single acceptor. Proc Natl Acad Sci U S A 93(13):6264–6268
19. Padilla-Parra S, Tramier M (2012) FRET microscopy in the living cell: different approaches, strengths and weaknesses. Bioessays 34(5):369–376
20. Truong K, Ikura M (2001) The use of FRET imaging microscopy to detect protein–protein interactions and protein conformational changes in vivo. Curr Opin Struct Biol 11(5): 573–578
21. Magde D, Elson E, Webb WW (1972) Thermodynamic fluctuations in a reacting system—measurement by fluorescence correlation spectroscopy. Phys Rev Lett 29(11):705–708
22. Elson EL, Magde D (1974) Fluorescence correlation spectroscopy. I. Conceptual basis and theory. Biopolymers 13(1):1–27
23. Magde D, Elson EL, Webb WW (1974) Fluorescence correlation spectroscopy. II. An experimental realization. Biopolymers 13(1):29–61
24. Webb WW (1976) Applications of fluorescence correlation spectroscopy. Q Rev Biophys 9 (01):49–68
25. Chen H, Farkas ER, Webb WW (2008) In vivo applications of fluorescence correlation spectroscopy. Methods Cell Biol 89:3–35
26. Haustein E, Schwille P (2007) Fluorescence correlation spectroscopy: novel variations of an established technique. Annu Rev Biophys Biomol Struct 36:151–169
27. Eigen M, Rigler R (1994) Sorting single molecules: application to diagnostics and evolutionary biotechnology. Proc Natl Acad Sci U S A 91(13):5740–5747
28. Rauer B, Neumann E, Widengren J, Rigler R (1996) Fluorescence correlation spectrometry of the interaction kinetics of tetramethylrhodamin α-bungarotoxin with Torpedo californica acetylcholine receptor. Biophys Chem 58(1):3–12
29. Haupts U, Maiti S, Schwille P, Webb WW (1998) Dynamics of fluorescence fluctuations in green fluorescent protein observed by fluorescence correlation spectroscopy. Proc Natl Acad Sci U S A 95(23):13573–13578
30. Schwille P, Haupts U, Maiti S, Webb WW (1999) Molecular dynamics in living cells observed by fluorescence correlation spectroscopy with one-and two-photon excitation. Biophys J 77(4):2251–2265

31. Singh AP, Wohland T (2014) Applications of imaging fluorescence correlation spectroscopy. Curr Opin Chem Biol 20:29–35
32. García-Sáez AJ, Schwille P (2007) Single molecule techniques for the study of membrane proteins. Appl Microbiol Biotechnol 76(2):257–266
33. García-Sáez AJ, Schwille P (2008) Fluorescence correlation spectroscopy for the study of membrane dynamics and protein/lipid interactions. Methods 46(2):116–122
34. Dertinger T, Pacheco V, von der Hocht I, Hartmann R, Gregor I, Enderlein J (2007) Two-focus fluorescence correlation spectroscopy: a new tool for accurate and absolute diffusion measurements. Chem Phys Chem 8(3):433–443
35. Schwille P (2001) Fluorescence correlation spectroscopy and its potential for intracellular applications. Cell Biochem Biophys 34(3):383–408
36. Fahey P, Koppel D, Barak L, Wolf D, Elson E, Webb W (1977) Lateral diffusion in planar lipid bilayers. Science 195(4275):305–306
37. Schwille P, Korlach J, Webb WW (1999) Fluorescence correlation spectroscopy with single-molecule sensitivity on cell and model membranes. Cytometry 36(3):176–182
38. Schlessinger J, Koppel D, Axelrod D, Jacobson K, Webb W, Elson E (1976) Lateral transport on cell membranes: mobility of concanavalin A receptors on myoblasts. Proc Natl Acad Sci U S A 73(7):2409–2413
39. Kahya N (2006) Targeting membrane proteins to liquid-ordered phases: molecular self-organization explored by fluorescence correlation spectroscopy. Chem Phys Lipids 141(1):158–168
40. Benda A, Beneš M, Marecek V, Lhotský A, Hermens WT, Hof M (2003) How to determine diffusion coefficients in planar phospholipid systems by confocal fluorescence correlation spectroscopy. Langmuir 19(10):4120–4126
41. Schwille P, Meyer-Almes F-J, Rigler R (1997) Dual-color fluorescence cross-correlation spectroscopy for multicomponent diffusional analysis in solution. Biophys J 72(4):1878–1886
42. Bacia K, Kim SA, Schwille P (2006) Fluorescence cross-correlation spectroscopy in living cells. Nat Methods 3(2):83–89
43. Haustein E, Schwille P (2004) Single-molecule spectroscopic methods. Curr Opin Struct Biol 14(5):531–540
44. Lillemeier BF, Mörtelmaier MA, Forstner MB, Huppa JB, Groves JT, Davis MM (2010) TCR and Lat are expressed on separate protein islands on T cell membranes and concatenate during activation. Nat Immunol 11(1):90–96
45. Larson DR, Gosse JA, Holowka DA, Baird BA, Webb WW (2005) Temporally resolved interactions between antigen-stimulated IgE receptors and Lyn kinase on living cells. J Cell Biol 171(3):527–536
46. Xie XS, Yu J, Yang WY (2006) Living cells as test tubes. Science 312(5771):228–230
47. Teramura Y, Ichinose J, Takagi H, Nishida K, Yanagida T, Sako Y (2006) Single-molecule analysis of epidermal growth factor binding on the surface of living cells. EMBO J 25 (18):4215–4222
48. Nguyen AH, Nguyen VT, Kamio Y, Higuchi H (2006) Single-molecule visualization of environment-sensitive fluorophores inserted into cell membranes by staphylococcal γ-hemolysin. Biochemistry 45(8):2570–2576
49. Uyemura T, Takagi H, Yanagida T, Sako Y (2005) Single-molecule analysis of epidermal growth factor signaling that leads to ultrasensitive calcium response. Biophys J 88(5):3720–3730
50. Tani T, Miyamoto Y, Fujimori KE, Taguchi T, Yanagida T, Sako Y, Harada Y (2005) Trafficking of a ligand-receptor complex on the growth cones as an essential step for the uptake of nerve growth factor at the distal end of the axon: a single-molecule analysis. J Neurosci 25(9):2181–2191
51. Ulbrich MH, Isacoff EY (2007) Subunit counting in membrane-bound proteins. Nat Methods 4(4):319–321

52. Iino R, Koyama I, Kusumi A (2001) Single molecule imaging of green fluorescent proteins in living cells: E-cadherin forms oligomers on the free cell surface. Biophys J 80(6):2667–2677
53. Harms GS, Cognet L, Lommerse PH, Blab GA, Kahr H, Gamsjäger R, Spaink HP, Soldatov NM, Romanin C, Schmidt T (2001) Single-molecule imaging of L-type Ca^{2+} channels in live cells. Biophys J 81(5):2639–2646
54. Murakoshi H, Iino R, Kobayashi T, Fujiwara T, Ohshima C, Yoshimura A, Kusumi A (2004) Single-molecule imaging analysis of Ras activation in living cells. Proc Natl Acad Sci U S A 101(19):7317–7322
55. Zhang W, Jiang Y, Wang Q, Ma X, Xiao Z, Zuo W, Fang X, Chen Y-G (2009) Single-molecule imaging reveals transforming growth factor-β-induced type II receptor dimerization. Proc Natl Acad Sci U S A 106(37):15679–15683
56. Maurice P, Kamal M, Jockers R (2011) Asymmetry of GPCR oligomers supports their functional relevance. Trends Pharmacol Sci 32(9):514–520
57. James JR, Oliveira MI, Carmo AM, Iaboni A, Davis SJ (2006) A rigorous experimental framework for detecting protein oligomerization using bioluminescence resonance energy transfer. Nat Methods 3(12):1001–1006
58. Meyer BH, Segura J-M, Martinez KL, Hovius R, George N, Johnsson K, Vogel H (2006) FRET imaging reveals that functional neurokinin-1 receptors are monomeric and reside in membrane microdomains of live cells. Proc Natl Acad Sci U S A 103(7):2138–2143
59. Kasai RS, Suzuki KG, Prossnitz ER, Koyama-Honda I, Nakada C, Fujiwara TK, Kusumi A (2011) Full characterization of GPCR monomer–dimer dynamic equilibrium by single molecule imaging. J Cell Biol 192(3):463–480
60. Hern JA, Baig AH, Mashanov GI, Birdsall B, Corrie JE, Lazareno S, Molloy JE, Birdsall NJ (2010) Formation and dissociation of M1 muscarinic receptor dimers seen by total internal reflection fluorescence imaging of single molecules. Proc Natl Acad Sci U S A 107(6): 2693–2698
61. Calebiro D, Rieken F, Wagner J, Sungkaworn T, Zabel U, Borzi A, Cocucci E, Zürn A, Lohse MJ (2013) Single-molecule analysis of fluorescently labeled G-protein–coupled receptors reveals complexes with distinct dynamics and organization. Proc Natl Acad Sci U S A 110(2):743–748
62. Kusumi A, Tsunoyama TA, Hirosawa KM, Kasai RS, Fujiwara TK (2014) Tracking single molecules at work in living cells. Nat Chem Biol 10(7):524–532
63. Shibata SC, Hibino K, Mashimo T, Yanagida T, Sako Y (2006) Formation of signal transduction complexes during immobile phase of NGFR movements. Biochem Biophys Res Commun 342(1):316–322
64. Lommerse PH, Blab GA, Cognet L, Harms GS, Snaar-Jagalska BE, Spaink HP, Schmidt T (2004) Single-molecule imaging of the H-Ras membrane-anchor reveals domains in the cytoplasmic leaflet of the cell membrane. Biophys J 86(1):609–616
65. Ueda M, Sako Y, Tanaka T, Devreotes P, Yanagida T (2001) Single-molecule analysis of chemotactic signaling in Dictyostelium cells. Science 294(5543):864–867
66. Vrljic M, Nishimura SY, Brasselet S, Moerner W, McConnell HM (2002) Translational diffusion of individual Class II MHC membrane proteins in cells. Biophys J 83(5): 2681–2692
67. Dahan M, Levi S, Luccardini C, Rostaing P, Riveau B, Triller A (2003) Diffusion dynamics of glycine receptors revealed by single-quantum dot tracking. Science 302(5644):442–445
68. Chung I, Akita R, Vandlen R, Toomre D, Schlessinger J, Mellman I (2010) Spatial control of EGF receptor activation by reversible dimerization on living cells. Nature 464 (7289):783-U163
69. Low-Nam ST, Lidke KA, Cutler PJ, Roovers RC, en Henegouwen PMVB, Wilson BS, Lidke DS (2011) ErbB1 dimerization is promoted by domain co-confinement and stabilized by ligand binding. Nat Struct Mol Biol 18(11):1244–1249
70. Webb SE, Roberts SK, Needham SR, Tynan CJ, Rolfe DJ, Winn MD, Clarke DT, Barraclough R, Martin-Fernandez ML (2008) Single-molecule imaging and fluorescence

lifetime imaging microscopy show different structures for high-and low-affinity epidermal growth factor receptors in A431 cells. Biophys J 94(3):803–819
71. Zhang D, Manna M, Wohland T, Kraut R (2009) Alternate raft pathways cooperate to mediate slow diffusion and efficient uptake of a sphingolipid tracer to degradative and recycling compartments. J Cell Sci 122(20):3715–3728
72. Gerken M, Krippner-Heidenreich A, Steinert S, Willi S, Neugart F, Zappe A, Wrachtrup J, Tietz C, Scheurich P (2010) Fluorescence correlation spectroscopy reveals topological segregation of the two tumor necrosis factor membrane receptors. Biochim Biophys Acta 1798(6):1081–1089
73. Hegener O, Jordan R, Häberlein H (2004) Dye-labeled benzodiazepines: development of small ligands for receptor binding studies using fluorescence correlation spectroscopy. J Med Chem 47(14):3600–3605
74. Meissner O, Häberlein H (2003) Lateral mobility and specific binding to GABAA receptors on hippocampal neurons monitored by fluorescence correlation spectroscopy. Biochemistry 42(6):1667–1672
75. Gakamsky DM, Luescher IF, Pramanik A, Kopito RB, Lemonnier F, Vogel H, Rigler R, Pecht I (2005) CD8 kinetically promotes ligand binding to the T-cell antigen receptor. Biophys J 89(3):2121–2133
76. Hao H, Fan L, Chen T, Li R, Li X, He Q, Botella MA, Lin J (2014) Clathrin and membrane microdomains cooperatively regulate RbohD dynamics and activity in Arabidopsis. Plant Cell 26(4):1729–1745
77. Li X, Xing J, Qiu Z, He Q, Lin J (2016) Quantification of Membrane protein dynamics and interactions in plant cells by fluorescence correlation spectroscopy. Mol Plant 9(9): 1229–1239
78. Lenne PF, Wawrezinieck L, Conchonaud F, Wurtz O, Boned A, Guo XJ, Rigneault H, He HT, Marguet D (2006) Dynamic molecular confinement in the plasma membrane by microdomains and the cytoskeleton meshwork. EMBO J 25(14):3245–3256
79. Leutenegger M, Ringemann C, Lasser T, Hell SW, Eggeling C (2012) Fluorescence correlation spectroscopy with a total internal reflection fluorescence STED microscope (TIRF-STED-FCS). Opt Express 20(5):5243–5263
80. Fan L, Hao H, Xue Y, Zhang L, Song K, Ding Z, Botella MA, Wang H, Lin J (2013) Dynamic analysis of Arabidopsis AP2 σ subunit reveals a key role in clathrin-mediated endocytosis and plant development. Development 140(18):3826–3837
81. Sakon JJ, Weninger KR (2010) Detecting the conformation of individual proteins in live cells. Nat Methods 7(3):203–205
82. Owen DM, Williamson D, Rentero C, Gaus K (2009) Quantitative microscopy: protein dynamics and membrane organisation. Traffic 10(8):962–971
83. Demir F, Horntrich C, Blachutzik JO, Scherzer S, Reinders Y, Kierszniowska S, Schulze WX, Harms GS, Hedrich R, Geiger D (2013) Arabidopsis nanodomain-delimited ABA signaling pathway regulates the anion channel SLAH3. Proc Natl Acad Sci U S A 110 (20):8296–8301
84. Schütz GJ, Kada G, Pastushenko VP, Schindler H (2000) Properties of lipid microdomains in a muscle cell membrane visualized by single molecule microscopy. EMBO J 19(5): 892–901
85. Douglass AD, Vale RD (2005) Single-molecule microscopy reveals plasma membrane microdomains created by protein-protein networks that exclude or trap signaling molecules in T cells. Cell 121(6):937–950
86. Mueller V, Ringemann C, Honigmann A, Schwarzmann G, Medda R, Leutenegger M, Polyakova S, Belov V, Hell S, Eggeling C (2011) STED nanoscopy reveals molecular details of cholesterol-and cytoskeleton-modulated lipid interactions in living cells. Biophys J 101(7):1651–1660
87. Sezgin E, Levental I, Grzybek M, Schwarzmann G, Mueller V, Honigmann A, Belov VN, Eggeling C, Coskun Ü, Simons K (2012) Partitioning, diffusion, and ligand binding of raft

lipid analogs in model and cellular plasma membranes. Biochim Biophys Acta 1818(7):1777–1784
88. Eggeling C, Ringemann C, Medda R, Schwarzmann G, Sandhoff K, Polyakova S, Belov VN, Hein B, von Middendorff C, Schönle A (2009) Direct observation of the nanoscale dynamics of membrane lipids in a living cell. Nature 457(7233):1159–1162
89. Shaner NC, Steinbach PA, Tsien RY (2005) A guide to choosing fluorescent proteins. Nat Methods 2(12):905–909
90. Giepmans BN, Adams SR, Ellisman MH, Tsien RY (2006) The fluorescent toolbox for assessing protein location and function. Science 312(5771):217–224
91. Shimomura O, Johnson FH, Saiga Y (1962) Extraction, purification and properties of aequorin, a bioluminescent protein from the luminous hydromedusan, Aequorea. J Cell Comp Physiol 59(3):223–239
92. Tsien RY (1998) The green fluorescent protein. Annu Rev Biochem 67(1):509–544
93. Griesbeck O (2004) Fluorescent proteins as sensors for cellular functions. Curr Opin Neurobiol 14(5):636–641
94. Hibino K, Shibata T, Yanagida T, Sako Y (2009) A RasGTP-induced conformational change in C-RAF is essential for accurate molecular recognition. Biophys J 97(5):1277–1287
95. Elf J, Li GW, Xie XS (2007) Probing transcription factor dynamics at the single-molecule level in a living cell. Science 316(5828):1191–1194
96. Zacharias DA, Violin JD, Newton AC, Tsien RY (2002) Partitioning of lipid-modified monomeric GFPs into membrane microdomains of live cells. Science 296(5569):913–916
97. Resch-Genger U, Grabolle M, Cavaliere-Jaricot S, Nitschke R, Nann T (2008) Quantum dots versus organic dyes as fluorescent labels. Nat Methods 5(9):763–775
98. Jaiswal JK, Goldman ER, Mattoussi H, Simon SM (2004) Use of quantum dots for live cell imaging. Nat Methods 1(1):73–78
99. Jaiswal JK, Simon SM (2004) Potentials and pitfalls of fluorescent quantum dots for biological imaging. Trends Cell Biol 14(9):497–504
100. Pinaud F, Clarke S, Sittner A, Dahan M (2010) Probing cellular events, one quantum dot at a time. Nat Methods 7(4):275–285
101. Liu SL, Zhang ZL, Sun EZ, Peng J, Xie M, Tian ZQ, Lin Y, Pang DW (2011) Visualizing the endocytic and exocytic processes of wheat germ agglutinin by quantum dot-based single-particle tracking. Biomaterials 32(30):7616–7624
102. Alivisatos AP, Gu W, Larabell C (2005) Quantum dots as cellular probes. Annu Rev Biomed Eng 7:55–76
103. Medintz IL, Uyeda HT, Goldman ER, Mattoussi H (2005) Quantum dot bioconjugates for imaging, labelling and sensing. Nat Mater 4(6):435–446
104. Michalet X, Pinaud FF, Bentolila LA, Tsay JM, Doose S, Li JJ, Sundaresan G, Wu AM, Gambhir SS, Weiss S (2005) Quantum dots for live cells, in vivo imaging, and diagnostics. Science 307(5709):538–544
105. Jiang X, Zhu M, Narain R (2014) Quantum Dots Bioconjugates. In: Narain R (ed) Chemistry of bioconjugates: synthesis, characterization, and biomedical applications. Wiley, Hoboken, New Jersey, pp 315–326

Chapter 7
Lipid Cubic Phase for Membrane Protein X-ray Crystallography

Jialu Zha and Dianfan Li

7.1 Introduction

Integral membrane proteins (IMPs), including membrane-embedded enzymes, channels, linkers, structural anchor proteins, transporters, and receptors, play key roles in all living cells. Approximately one-third of the most genomes encode IMPs, and half of the current drugs target this important class of macromolecule [1]. Despite their importance, structural biology of IMPs lags far behind their soluble counterparts due to their hydrophobic nature and the lack of membrane mimetic that preserves IMP stability upon membrane disrupture and solubilization. Currently, structures of 698 unique IMPs were solved (http://blanco.biomol.uci.edu/mpstruc/), with a total of 2232 entries in the Protein Databank (PDB), accounting for only 1. 7% of total entries.

Thus far, X-ray crystallography is the most used technique for membrane protein structure determination, owing to its capability of yielding atomic resolution information that is often crucial for understanding mechanism details and structure-based drug discovery applications. Essential to this approach is crystallization, a process that IMPs pack together in an ordered manner to form solid crystals. Crystals can cause incident X-ray to diffract. X-ray diffraction data are then collected on a detector. These data reflect the three-dimensional position of atoms because they are self-organized in a periodic array inside crystals. The process from crystal to structure determination of IMP is no different than soluble proteins, and

J. Zha · D. Li
National Center for Protein Science Shanghai, Shanghai Science Research Center, CAS Center for Excellence in Molecular Cell Science, Shanghai Institute of Biochemistry and Cell Biology, Chinese Academy of Sciences, 333 Haike Road, Shanghai 201210, China

J. Zha · D. Li (✉)
University of Chinese Academy of Sciences, 333 Haike Road, Shanghai 201210, China
e-mail: dianfan.li@sibcb.ac.cn

H. Wang and G. Li (eds.), *Membrane Biophysics*,
https://doi.org/10.1007/978-981-10-6823-2_7

the detailed principle is well explained in a number of books [2, 3]. This chapter will focus on the crystallization process.

Crystallization of proteins is generally challenging and even more so for IMPs because of their hydrophobicity, high dynamics, and instability [4, 5]. According to statistics in the literature, obtaining diffraction quality of IMPs presents the biggest bottleneck in IMP X-ray crystallography, while overexpression and purification also hurdle the whole pipeline [6]. Traditionally, the crystallization of IMPs is performed in micelles (in surfactant, *in surfo*) [7] formed by detergents. Detergents are wide-used and often unavoidable IMP-solubilizing amphiphiles, but are meanwhile problematic for the crystallization process [8]. Generally, IMPs are more stable in long-chain detergents, but these micelles will mask regions for crystal contacts. On the other hand, short-chain detergents leave more space for crystal contacts but increase the risk of non-specific hydrophobic contacts that lead to aggregation [8, 9]. Therefore, successful *in surfo* crystallization often relies on high stability of IMPs in short-chained detergents, such as *n*-decyl-β-**D**-maltopyranoside (DM), *n*-nonyl-β-**D**-glucopyranoside (NG), and *n*-octyl-β-**D**-glucopyranoside (OG), which is achieved as a result of either broad screening for naturally stable homologs [10, 11] or, by extensive and systematic mutagenesis [12–21].

Reconstitution of IMPs from micelles to more lipid bilayer-alike membrane mimetic offers the advantage of restoring or maintaining the structure integrity of IMPs and therefore has been one of the driving forces in advancing the field of IMP structural biology. They include lipid bicelles [22, 23] and the lipid cubic phase (LCP) [7].

LCP, a transparent, viscous, and sticky material, is a lyotropic crystal containing a three-dimensionally continuous lipid bilayer that separates two continuous water channels [7]. LCP is formed by mixing water with certain neutral lipids such as monoacylglycerols (MAGs) [24]; it can tolerate high amount of native lipids such as phospholipids and cholesterol [25], two major components for biomembranes. This makes LCP more membrane-alike than micelles, in both composition and structure (Fig. 7.1). The lipid environment is believed to be favorable in maintaining

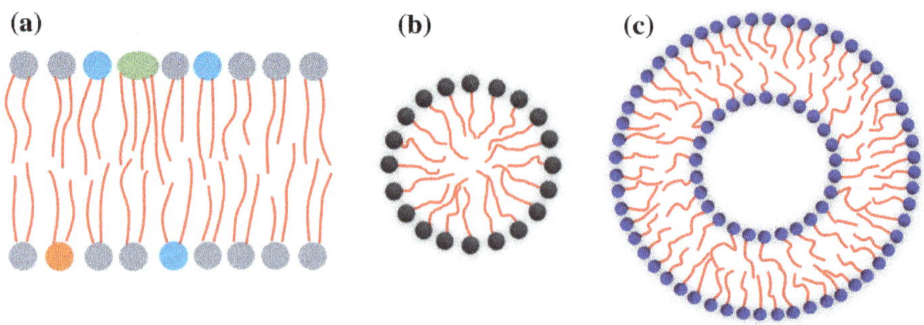

Fig. 7.1 Schematic of biomembranes, micelles, and LCP bilayer. **a** Native lipid bilayer that is mostly composed of phospholipids with different head groups. **b** Micelle monolayer. **c** LCP lipid bilayer

IMP stability for crystallization. Moreover, the packing of LCP crystals is exclusively type I, which generally characterizes less solvent content than type II *in surfo* crystals, and therefore typically diffracted to higher resolutions [7].

Two decades ago, Landau and Rosenbusch reported the first application of LCP for membrane protein crystallization [26]. Hexagonal crystals of a red-colored proton pump called bacteriorhodopsin were obtained in LCP that later yielded a 2.5 Å structure [27]. Since then, the method has been advanced in many aspects. Toolsets have been developed to facilitate simple and efficient LCP mixing, high-throughput crystallization and imaging [28, 29]. The phase diagrams of several LCP-forming lipids were mapped out, as well as the compatibility with natural lipids [25], detergents [30], salts [31], precipitants [32], and many other chemicals often used in biochemistry. Such results serve as guidelines in designing protocols, strategies, and precipitant screens for IMP crystallization. Vigorous efforts have been made in exploring LCP for the crystallization of non-bacteriorhodopsin proteins, including β-barrel membrane proteins [33, 34], single-pass peptides [35], complexes [36, 37], enzymes [38–40], and transporters [41]. The success of these projects, together with frequent workshops [7], encouraged the community in using this technology. Optimization strategies of LCP microcrystals using rationally designed host lipids were also proposed [42], tested [35, 43], and applied to several IMPs with great success [40, 41, 44, 45]. Special procedures in the manipulation of LCP crystals, including harvesting and heavy atom soaking for experimental phasing, have been established [46–48]. Ultra-thin plates for in situ X-ray diffraction data collection [49, 50], as well as the combination of LCP technology and serial femtosecond crystallography (SFX) using X-ray free electron laser (XFEL), have been developed to speed up structure determination [51].

With such advances, the LCP method is becoming increasingly popular in IMP X-ray crystallography. At the time of this chapter, LCP method has been responsible for 447 PDB entries, accounting for 20% of total IMP entries.

7.2 Principle

7.2.1 Phase Diagram and LCP-Forming Lipids

Lipid cubic phase, as its name implies, is a type of phase that belongs to cubic space groups and has lipid component. When monoolein is mixed with water, different phases spontaneously form, depending on the hydration and temperature [24, 52] (Fig. 7.2a). Two common cubic phases exist, the *Pn3m* and *Ia3d*. According to the phase diagram, LCP of monoolein forms at a broad range of temperatures, as well as hydration, featuring a very forgiving system. At room temperature conditions (20 °C) at which most of the biochemical experiments and crystallization were carried out, the monoolein LCP forms with 28–40% of hydration. Beyond that, LCP coexists with excess water (Fig. 7.2a).

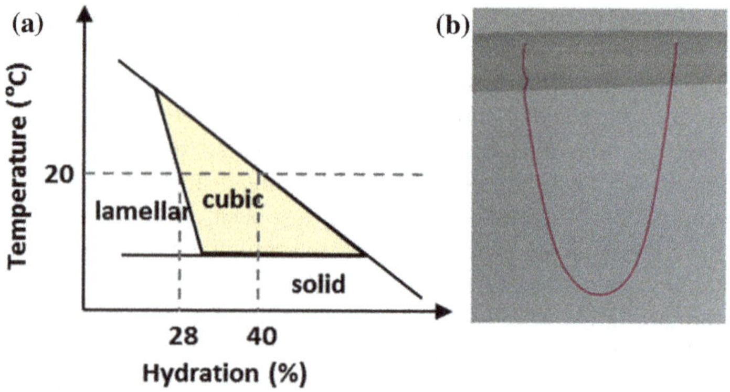

Fig. 7.2 Phase diagram of monoolein LCP, and the demonstration of the stickiness of LCP. **a** A simplified phase diagram of monoolein–water mixture under different hydration and temperature conditions. A detailed diagram is available in reference [24]. The drawing is modified from reference [53]. **b** A string of LCP (0.4 mm in diameter) attached to a pipette tip, showing its stickiness

The monoolein mentioned above is a monoacylglycerol (MAG). Many other MAGs, such as 6.9, 7.7, 7.8, 7.9, 8.7, 8.8, 8.9, and 9.7 MAG (N.T MAGs, Fig. 7.3a), are also known to form LCP [7]. Common to all these MAGs is the presence of a *cis* double bond in the acyl chain, which introduces a kink in the tail region, giving the characteristic curvature of LCP bilayer (Fig. 7.1c).

Apart from the host lipids mentioned above, newly emerging LCP-forming lipids are worth noting and perhaps provide new dimensions for crystal optimization. Landau and co-workers [54] designed a MAG named monodihydrosterculin (Fig. 7.3b). Compared to monoolein, the *cis* double bond was replaced with a cyclopropyl, making a more rigid 'kink' in the acyl chain. Unlike monoolein, this lipid forms stable LCP at 4 °C (and so for the 7.9 MAG), making it attractive for low-temperature crystallization. Another family of lipids are the isoprenyl lipids, such as phytantriol [55] (Fig. 7.1d) and β-XylOC$_{(16+4)}$ [56] (Fig. 7.1e), and GlyNCOC$_{15+4}$ and GlyNMeCOC$_{15+4}$ [57] (Fig. 7.1f, g). These isoprenyl lipids are chemically more stable than the MAGs, therefore more suitable for crystallization experiments at extreme pHs or high temperatures.

7.2.2 Compatibility

Biochemical and biophysical studies, in particular crystallization of IMPs, require the use of a range of chemicals and are carried out under different temperatures. To study IMPs in LCP, therefore, requires phase stability of LCP with respect to these chemical and physical changes. Thus far, the monoolein-based LCP is the most characterized for phase stability in this regard [24, 31, 36, 58]. Earlier efforts in the characterization of monoolein LCP showed extraordinary stability. It can be formed

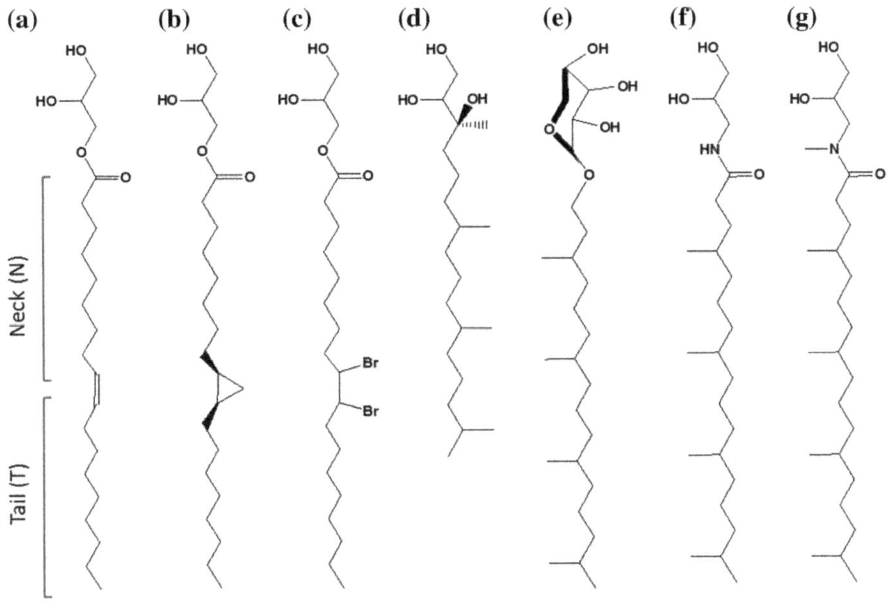

Fig. 7.3 LCP-forming lipids. **a** MAGs. The MAGs contain a glycerol head group and a hydrocarbon tail with a *cis* double bond. The two parts separated by the double bond are named as Neck (N) and Tail (T) in the N.T MAG nomenclature. According to this, monoolein is a 9.9 MAG. **b** Monodihydrosterculin. **c** Bromo-MAG. **d** Phytantriol. **e** b-XylOC$_{(16+4)}$. **f** GlyNCOC$_{15+4}$. **g** GlyNMeCOC$_{15+4}$

at temperatures between 17 and 95 °C [24]. Once fully hydrated, adding water to LCP does not 'dilute' or change its phase property. For crystallization, this means the IMP concentration in LCP remains the same with excess of precipitant solution during crystallization. Compared to detergent micelles, this is a unique advantage.

LCP is also remarkably stable regarding chemical reagents for IMP purification and crystallization. For salts, it can accommodate up to 5 M NaCl [31] and presumably to high concentrations of other salts too. Interestingly, LCP is stable even in the presence of 8 M urea or 6 M GuHCl [36], two chaotropes for protein unfolding–refolding studies. This has made it possible for IMP refolding from these chaotropes into LCP [59].

The LCP bilayer can also accommodate native lipids, which are often co-purified with or supplemented to IMP for stabilization and function purposes [25]. Owing to their low solubility in detergent micelles, including native lipids such as phospholipids and cholesterol at high concentrations for crystallization, it is usually technically challenging. By contrast, solubility is not an issue for LCP because of its dense lipid bilayer (2 molar concentration). Instead, it is the phase transition boundary that limits the amount of these lipids in the system. Fortunately, as noted, LCP can accommodate large amounts of several native lipids. These include dioleoyl phosphatidyl choline (DOPC, ~20 mol%), dioleoyl phosphatidyl ethanolamine (DOPE, ~20 mol%), dioleoyl phosphatidyl serine (DOPS, ~7 mol%), diacylglycerol and triacylglycerol (~5 mol%), fatty acids (~10 mol%) [60], cardiolipin (6 mol%), and cholesterol (~28 mol%) [25].

Another very important component associated with IMPs is detergents, which can destabilize or even solubilize LCP at high enough concentrations. Similar to native lipids, the compatibility of usual detergents for LCP has been characterized [30, 61]. It should be noted that the phase characterization in these works was carried out using 9.9 MAG, water, and detergents. In a real crystallization, the components are normally more complex because of the presence of salts, the membrane protein itself, buffer, reducing agents, etc. The phase property in such experiments might respond to the detergent loading differently than reported. However, they should not deviate much from the reported values. Therefore, the data provide very informative guidelines for maximum detergent loading. For monoolein LCP, the concentrations for cubic–lamellar phase boundary are as follows: 8–10% (w/v) *n*-dodecyl-β-**D**-maltopyranoside (DDM) [30], ~6% (w/v) OG, ~6% (w/v) NG [61]. For recently developed detergents, such as maltose–neopentyl glycol (MNG-DDM) amphiphiles [62, 63], the compatibility data are not available and await for future characterization. In the absence of phase diagram data, one can check the phases by examining birefringence of the mesophase; the lamellar phase characterizes bright birefringence, while the LCP does not [64, 65] (Fig. 7.4a, c, and d).

Crystallization generally requires incubation of protein with precipitant solutions at various pH values. The pH stability of LCP depends on the host lipids. LCP made of MAGs is only stable in pH ranging from ~3.5 to 9.5 because MAGs can be hydrolyzed to form free fatty acid at very acidic and alkali conditions. The use of phytantriol, or modification of the ester bond of the MAGs [57], leads to LCP that is less sensitive to extreme pH.

Apart from pH and salts, polymers and alcohols are major ingredients in precipitant formula. Polyethylene glycols (PEGs) are one of the most used families of polymers in protein crystallization. For the low molecular weight PEGs, such as PEG 400, the monoolein LCP can tolerate to as high as 48% (v/v) without flipping the phase [66, 67]. At above 48%, the LCP undergoes a transition to a more fluid, oily mesophase called lipid sponge phase [36, 68]. Compared to LCP, lipid sponge phase features smooth edges (Fig. 7.4b). It is also less sticky and viscous and hence can move relatively easily on a surface. The exact boundary for phase transition regarding concentrations of high molecular weight PEGs remains to be determined. But, generally, as the molecular weight increases, the concentration boundary decreases [69]. Interestingly, low molecular weight PEGs, such as PEG 400 [70], triethylene glycol [71], PEG 550, are more often reported in LCP crystallization conditions than their high molecular weight PEGs. Another common reported reagent for LCP crystallization is 2-methyl-2,4-pentanediol (MPD) [36]. Like PEG 400, MPD is also a spongifier agent with a cubic–sponge phase transition concentration of about 15% (v/v) [36]. Other reagents known to induce sponge phase include N-methylpyrrolidinone [72], DMSO [73], Jeffamine and pentaerythritol propoxylate (PPO), KSCN, butanediol [36, 67, 72, 74].

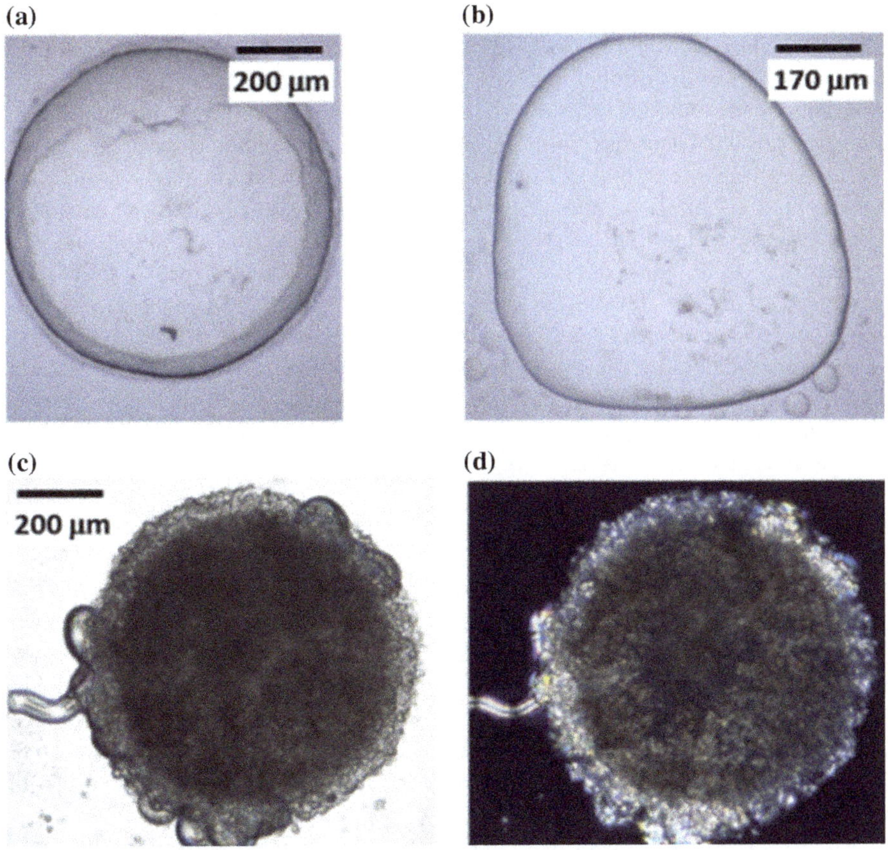

Fig. 7.4 Different lipid phases. **a** Lipid cubic phase. **b** Lipid sponge phase. Compared to the LCP, lipid sponge phase has smooth edges. **c** Typical lamellar phase under normal light. **d** Lamellar phase viewed with polarized light showing characteristic of birefringence

Similar to that of detergents, MAGs with shorter chain are generally less tolerant to the polymers. According to our experience, for example, the 7.8 MAG LCP will transit to sponge phase in the presence of 30% PEG 400 or 8% MPD.

7.2.3 Reconstitution and Functionality

LCP crystallization of IMPs is meaningful only if IMPs are reconstituted into the LCP bilayer and remain functionally relevant. Experiments have been designed and carried out to test both reconstitution and IMP functions in LCP.

For reconstitution, the experiment design involved chemical modification of monoolein [33, 34, 38, 75]. Two bromines are added to either side of the *cis* double bond of monoolein (Fig. 7.3c). This lipid, termed bromo-MAG, can also form LCP. The introduction of bromines at the middle of the acyl chain quenches fluorescence of Trp residues that are generally near the membrane interface by proximity effects. Should IMPs be reconstituted to the LCP bilayer, the intrinsic Trp fluorescence will decrease in a bromo-MAG concentration-dependent manner (Fig. 7.5). Both the α-helical membrane protein such as diacylglycerol kinase (DgkA) and β-barrel membrane proteins (OpcA and ButB) were shown to be reconstituted into the LCP bilayer using this elegant experiment [33, 34, 38].

That the LCP can support membrane protein function was conveniently demonstrated by the enzymatic assay of DgkA [38, 39]. As mentioned above, fluorescence quenching assay shows that DgkA is reconstituted into LCP. Coincidently, LCP-forming MAGs are also DgkA's substrate which gets phosphorylated by MgATP. Therefore, DgkA reconstituted into LCP is surrounded by a vast amount of lipid substrate (~2 M). The assay took advantage of the viscous and sticky nature of LCP (Fig. 7.2b) and a coupled assay through which the ATP-to-ADP conversion is coupled to the oxidation of NADH by the enzymatic reactions of the pyruvate kinase and lactate dehydrogenase, a process that causes decrease of absorbance at 340 nm (Fig. 7.6a).

LCP reconstituted with DgkA is dispensed at the side of a microplate well, a geometry that avoids interference of subsequent spectrophotometric measurements (Fig. 7.6b). Being sticky (Fig. 7.2b), LCP stays at where it is put, with DgkA enzyme in its lipid bilayer. Assay mix containing water-soluble MgATP is then added to the LCP. MgATP diffuses into LCP, initiates the kinase activity, produces MgADP, which releases to the assay mix where it meets the two coupled enzymes (Fig. 7.6b), oxidizes NADH, and causes the decrease of A_{340} as the phenotype for DgkA function (Fig. 7.6c).

An additional experiment with the bacterial phosphatidylglycerol phosphate synthase (PgsA) [38] also demonstrates that LCP supports IMP enzymatic activity. PgsA transfers the phosphatidyl group from cytidine diphosphate–diacylglycerol

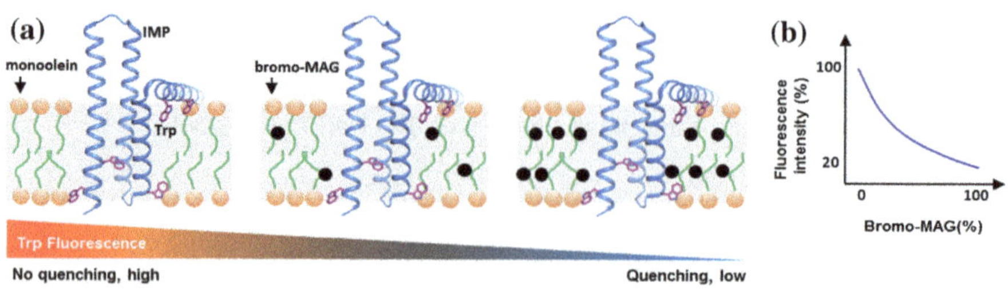

Fig. 7.5 Fluorescence quenching of IMPs in LCP as evidence for reconstitution. **a** Principle. A DgkA monomer is shown as an example for IMPs. The Trp residues (magenta sticks) are buried in the bilayer and can only be quenched by quenchers in the membrane because of proximity requirements. Upon the addition of bromo-MAG, the bromines quench Trp fluorescence of DgkA in a dose-dependent manner. **b** Schematic drawing of a typical fluorescence quenching result. The original curve for this experiment is in reference [38]

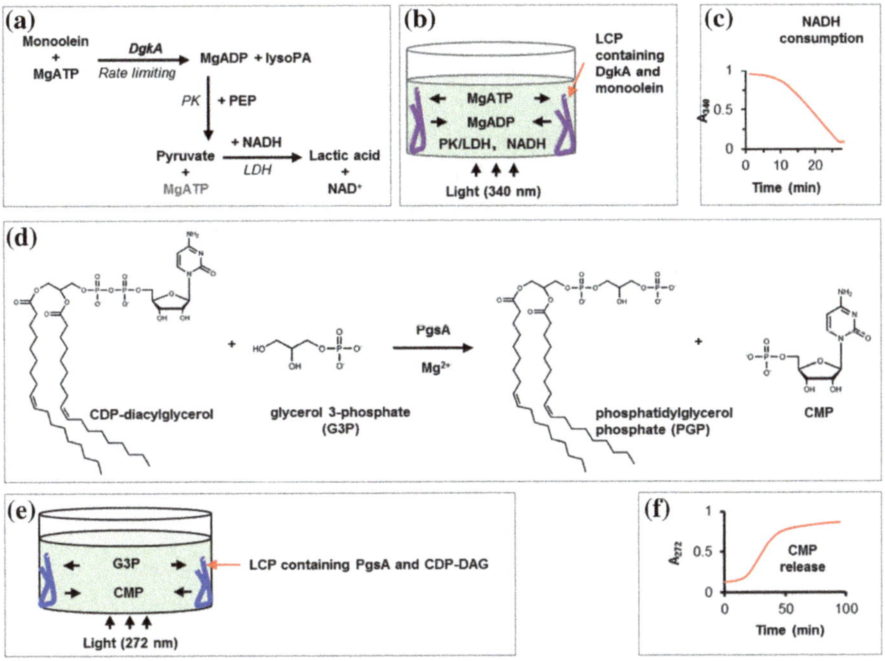

Fig. 7.6 Functional assay of two IMP enzymes in LCP. **a** The principle of the coupled assay for DgkA. The ATP–ADP conversion of DgkA reaction is coupled to the oxidation of the A_{340}-absorbing NADH to the non-absorbing NAD^+, by the two coupled enzymes, pyruvate kinase (PK) and lactate dehydrogenase (LDH). The coupled enzymes are in excess to ensure the rate of NADH oxidation reflects the DgkA activity. **b** Schematic drawing of the DgkA assay in LCP. In a microplate well, LCP-containing DgkA (and monoolein as the host lipid) is dispensed at the side, avoiding interference with the absorbance measurement. Note that LCP is sticky so it stays at where it is put, even with vigorous shaking. DgkA is hydrophobic so it will stay in the LCP bilayer. MgATP is added to the assay mix that soaks LCP. The water-soluble MgATP diffuses into the nano-porous LCP, initiates the reaction, produces MgADP which then releases from LCP to the assay mix, triggering the oxidation of NADH mediated by the coupled enzyme pre-loaded in the assay mix. **c** Schematic drawing of the progress curve of DgkA assay in LCP. The initial lag phase reflects the time required for equilibration of MgATP between the assay mix and LCP. **d** Reaction catalyzed by PgsA. Note that the UV-absorbing cytidine moiety is tethered to a hydrophobic molecule in the substrate CDP-DAG, but is water-soluble as in the product form. This means that LCP can separate the substrate and product without further manipulation. **e** Schematic drawing of the setup for the PgsA direct assay. LCP containing the enzyme PgsA and its lipid substrate CDP-DAG is dispensed in a microplate well similar to the DgkA assay. Although the CDP-DAG absorbs UV light, the geometry ensures that the molecule is not seen by the beam. Upon the addition of G3P into the assay mix, the water-soluble substrate diffuses into LCP, initiates the activity, and produces CMP. Again the water-soluble CMP diffuses from LCP to the assay mix and gets detected by the UV light. **f** Schematic drawing of the progress curve of PgsA assay in LCP. Real experimental data for both DgkA and PgsA assay are in reference [38]

(CDP-DAG) to glycerol 3-phosphate (G3P) to form phosphatidylglycerol phosphate and cytidine monophosphate (CMP) (Fig. 7.6d). The principle of assaying this enzyme takes advantage of the phase separation feature of LCP so that substances in LCP are naturally separated based on its hydrophobicity, without the need for extraction with organic solvents. The membrane-embedded PgsA and the substrate CDP-DAG are reconstituted into LCP, which is dispensed in the microplate as for the DgkA (Fig. 7.6e). Because of the diacyl chain, the CDP-DAG anchors to LCP and does not release from LCP. Again, the LCP is deposited at the side of the microplate well; therefore, the UV light does not see the UV-absorbent cytosine moiety in CDP-DAG. Upon the addition of assay mix containing Mg^{2+} and G3P, the water-soluble substrate and co-factor diffuse into LCP and initiate activity, producing CMP which is released from LCP and detected by UV light (Fig. 7.6f). Using this assay, the dependency of PgsA activity on substrate and co-factor is revealed.

Recently, our laboratory successfully used LCP to characterize an integral membrane enzyme which produces inorganic phosphate for its reaction (unpublished), adding one more example for LCP as a membrane mimetic in supporting membrane protein activity.

7.2.4 Crystallization

The mechanism of LCP crystallization has been probed by both molecular dynamics simulation and experiments [76–78]. The most accepted hypothesis, with convincing supporting experimental evidence, is the following. Lamellar phase, which is a multilayered bilayer, forms locally under precipitant conditions. IMPs concentrate on this region and form nuclei. The nuclei grow to crystals with proteins fed from the LCP reservoir where IMPs can diffuse three-dimensionally. Consistent with this mechanism is the type I packing (layered) observed in all LCP crystals to date. In addition, lamellar-phase signature patterns were observed in regions edging LCP crystals using small-angle X-ray scattering (SAXS) analysis.

According to the mechanism, ideally the mesophase should be cubic phase or sponge phase during the crystallization. Care should be taken to avoid conditions that flip LCP to lamellar phase (see Sect. 7.2.2). Examples of optimized crystals grown in the cubic phase and sponge phase are shown in Fig. 7.7. Occasionally, lamellar phase can change to LCP overtime, upon equilibration with the precipitant solution. In a few of these cases, we have observed crystal growth. An example of this is shown in Fig. 7.8. Therefore, it is important to monitor the crystallization drop even if the initial phases are not LCP or sponge phase.

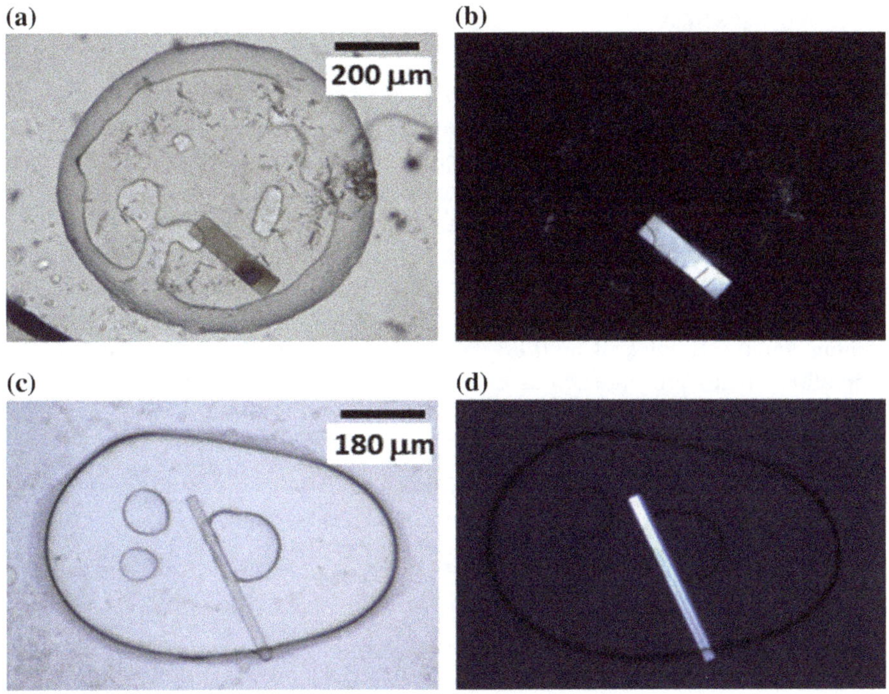

Fig. 7.7 Optimized crystals grown in the mesophases. **a** A crystal in LCP under normal light. **b** Same crystal viewed under polarized light. **c** Crystal grown in the sponge phase. **d** Same crystal in c viewed under polarized light

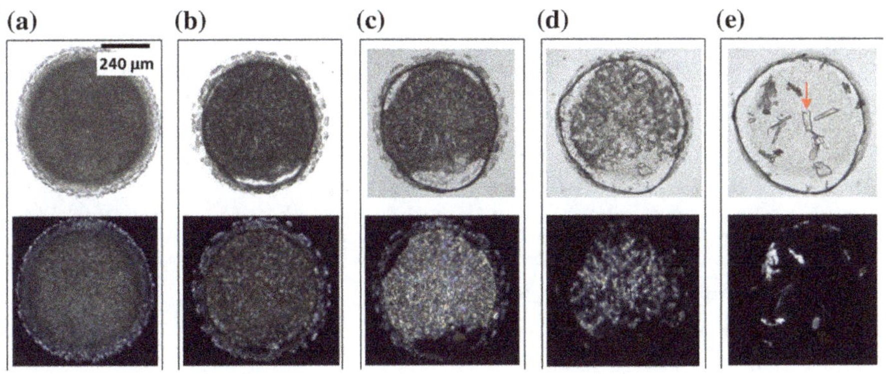

Fig. 7.8 An example of crystal growth under conditions where lamellar phases are originally formed but later converted to LCP. Top and bottom panels show images taken with normal light and polarized light, respectively. **a–e** Images taken on days 0, 3, 5, 7, and 14, respectively. A red arrow in e indicates a crystal that appeared between day 7 and day 14

7.3 Application

In the past two decades, the LCP method has been successfully applied to the crystallization of a broad range of IMPs, as listed below and as shown in Fig. 7.9 as a gallery. For many of these projects, the use of LCP was key to the success of structure determination, which in turn provides deep insights into functions.

As an interesting note, in addition to supporting the crystallization of IMPs, LCP can also be used for the crystallization of soluble proteins, such as lysozyme [79] and insulin [49]. Because the lysozyme and insulin are readily available, the crystallization of these two proteins can be used for the purposes of teaching/ training, for the practice of both crystallization setup and harvesting, and for the calibration of instruments such as crystallization robots, imagers, and ultraviolet (UV) microscopes.

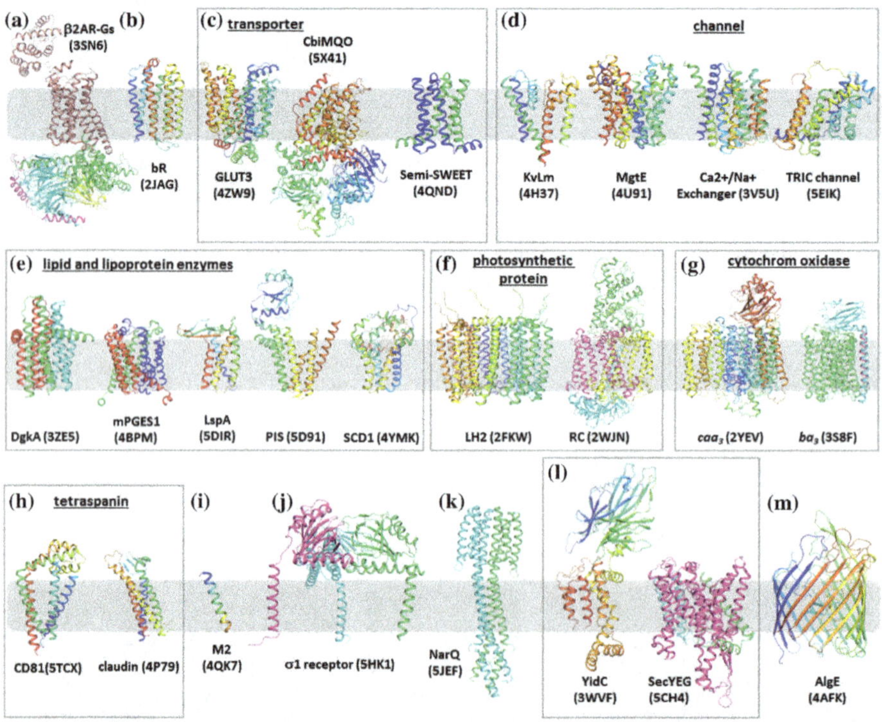

Fig. 7.9 Structure gallery of representative structures solved using LCP crystals. Protein names are shown for each structure with the PDB entry ID in brackets. **a** GPCR–Gs complex. **b** Bacteriorhodopsin (bR). **c** Transporters, including secondary transporters, semi-SWEET transporters, and an ABC transporter. **d** Ion channels. **e** Membrane-embedded enzymes involved in lipid metabolism and lipoprotein maturation. **f** Light-harvesting complex II (LH2) and the photosynthetic reaction center (RC). **g** Two cytochrome *C* oxidases. **h** Two tetraspannins, CD81 and claudin. **i** Single-TM peptide M2 channel. **j** Single-TM protein σ1 receptor. **k** Two-component system NarQ. **l** Membrane insertases. **m** Outer membrane protein AlgE

7.3.1 G-protein Coupled Receptors (GPCRs) and Its Complexes

The GPCR family and associated complex are probably the most visible and prominent cases in LCP crystallization. GPCR superfamily contains more than 800 members, and mediates most cellular responses to environmental stimuli (such as smell and taste), hormones, bioactive lipids, metabolites, and neurotransmitters [80]. A third of all current drugs are targeted through GPCRs [81]. GPCRs all share a similar transmembrane domain composed of seven transmembrane (TM) helices. Therefore, they are also called 7TM receptors. Upon ligand binding at the extracellular side, the TM helices rearrange, allowing the G protein or arrestin to bind at the cytosolic pocket, which in turn triggers a series of downstream cellular events.

In 2007, 10 years after the first report for LCP crystallization, the first high-resolution structure of a non-rhodopsin GPCR, the β2-adrenergic receptor (β2AR) [82, 83], was solved using LCP crystals of an extensively engineered construct. This landmark event prompted the use of this technology in the GPCR field and opened a new era in GPCR biology. So far, 46 unique GPCR structures, including the GPCR–Gs complex [84] (Fig. 7.9a) and the rhodopsin GPCR–arrestin complex [85], have been solved using crystals grown in LCP. Among 190 entries for GPCRs, only a few are solved using crystals from detergent micelles, such as the thermostabilized β1AR [86–88], the A_{2A} receptor with an agonist bound [89], the A_{2A} in detergent lauryl maltose neopentyl glycol [90], and the recent thermostabilized β1AR mini-GS complex [19].

The various GPCR structures show the striking diversity of binding pockets for ligands ranging from small molecules to large peptides. Structures with antagonists and agonists revealed the ground and activated states of the receptors and explained the activation mechanism and constitutive activity. Structures with allosteric ligands or ions reveal complex allosteric modulation. The GPCR–Gs and GPCR–arrestin complex structures capture the activated GPCR with G proteins or arrestin bound, providing atomic details of transmembrane signaling, as well as basis for biased signaling. All these successes associated with LCP dramatically advanced our understanding of GPCR signaling and provided rational basis for the design of new or better drugs against this largest family of drug targets.

7.3.2 Microbial Rhodopsin

Bacteriorhodopsin, for which the first LCP structure was reported, is an archaea origin proton pump that shares significant homology and the 7-TM topology with the GPCRs (Fig. 7.9b), as noted [91]. Being highly stable, colored, and readily crystallizable, this protein had been the benchmark sample for LCP crystallization, for various purposes including LCP technology development such as serial crystallography, crystal delivery devices [92], and validation of newly synthesized

LCP-forming lipids, such as the cyclopropyl analogue of monoolein, monodihydrosterculin [54], MAGs with different chain length and *cis* double bond positions [93], the β-XylOC$_{16+4}$ [56], GlyNCOC$_{15+4}$, and GlyNMeCOC$_{15+4}$ [57].

Apart from the bacteriorhodopsin, a number of other microbial homologs such as channelrhodopsin [94] and sensory rhodopsin [95] were also crystallized using LCP. Note that the sensory rhodopsin was crystallized in monovaccenin (11.7 MAG), rather than the most commonly used monoolein (9.9 MAG).

The work of channelrhodopsin is worth noting here. Channelrhodopsin is originally identified in green algae. What makes it interesting is its light-gated cation (H^+, Ca^{2+}, Na^+, K^+) channel activity, meaning a way in modulating intracellular acidity, calcium influx, and many other processes by light at a particular wavelength. This makes the channelrhodopsin a key tool in optogenetics where they activity of vertebrate neurons is controlled by light through this channel. Although the protein has been used for decades in this field, little was known for its working mechanism, until in 2012 when the crystal structure of the channelrhodopsin was solved at 2.3 Å using LCP crystals. The structure showed an electronegative pore composed of residues in four of its seven TMs, revealing a convincing cation-conducting pathway. Further, the structure shows two constrictions in the pathway as potential gates for cation conductance. This structure provides a rational basis for the design of channelrhodopsin variants with useful properties in optogenetics, such as new excite wavelength and selective pore for ion conductance.

7.3.3 Transporters

So far, secondary transporters, more specifically, the major facilitator proteins, are the most reported with respect to LCP crystallization among different types of transporters. These secondary carriers include the bile acid transporter [96]; the glucose transporter, GLUT3 complexed with **D**-glucose [97] (Fig. 7.9c); the peptide transporter, PepT [41]; two multidrug and toxic compound extrusion (MATE) transporters, *pf*MATE [98, 99] and VcmM [100]; the lipid flippase, MurJ [101]; and the nitrate/nitrite antiport, NarK [102]. It is noted that the PepT structure solved using LCP crystals (2.3 Å) [41] is of much higher resolution than the *in surfo* structure (3.3 Å) [103].

Apart from revealing architecture and ligand-binding details of the mentioned transporters, the LCP structures, together with structures determined in detergents, collectively capture high-resolution information of the secondary transporters in different conformations necessary for the transport function [104]. These states, including inward open, outward open, or occluded, support the 'alternating access' hypothesis [105] in transporting matters across the biomembrane.

ATP-binding cassette (ABC) transporters, unlike the secondary transporters, use energy from ATP hydrolysis to transport substance across the biomembrane. The ability to use ATP-hydrolyzing energy means that ABC transporters can move ligands against their concentration gradient. ABC transporters contain multiple subunits, usually one or two ATPase subunits, one or two transmembrane subunit, and sometimes additional units for substrate specificity. Recently, an energy-coupling factor transporter was crystallized in LCP that yielded a 3.5 Å structure [106] (Fig. 7.9c). This is the first in this family that is crystallized in LCP, demonstrating the suitability of LCP for crystallizing ABC transporters, which includes many physiologically and pharmaceutically important proteins such as multidrug resistant complexes that impede cancer chemotherapy [107].

The use of LCP has also led to the structure determination of a relatively new class of transporters, the SWEET transporters. SWEET transporters are responsible for sugar uptake and are originally found in plants as 7-TM transporters with two structural repeats, TM1-3 and TM5-7. In prokaryotes, the 'half' version was found instead and thus was called 'semi-SWEET.' Recently, the structure of such semi-SWEET transporters has been reported by three independent groups, to as high as 1.7 Å resolutions [108–110]. The structures showed that semi-SWEET transporters are composed of only three TM helices (Fig. 7.9c), representing the smallest known transporters. The semi-SWEET transporters function as a dimer, mimicking the eukaryotic monomer with two internal repeats connected by an inversion TM linker. The structures identified a cavity with conserved residues, as most likely the sugar-binding pocket [111]. Moreover, the outward open, outward occluded, and inward open states captured in these structures suggest an alternative access model [111] similar to the secondary transporters.

7.3.4 Channels and Exchangers

Ion channels play important roles in ion homeostasis in all living cells. The LCP method has also shown promises in structure elucidation of this type of IMPs. The channels crystallized in LCP include the K^+ channel, KvLm [112]; the bacterial Mg^{2+} channel, MgtE [113]; the Ca^{2+}/Na^+ exchanger [114, 115]; Ca^{2+}/H^+ exchanger [116]; and the trimeric intracellular cation (TRIC) channel [117]. It is noted that although these channels share the same classification here, they have distinctly different topologies (Fig. 7.9d), further enriching the diversity of structure folds associated with the LCP technology.

KvLm (Fig. 7.9d) is a voltage-gated K^+ channel and functions as a tetramer. Importantly, the tetramer state as revealed by the LCP structure was occupied with ion and ready to conduct through the selectivity filter, thus representing the first closed state of this type of channel and adding an important piece of structure information in its functional conformation cycle.

The structure of Ca^{2+}/Na^+ exchanger solved using LCP crystal was the first high-resolution structure of this class of protein. The structure revealed a 10-TM topology that shows inverted symmetry between the two 5-TM bundles (Fig. 7.9d). As mentioned, LCP crystals generally diffracted better than those grown in detergents. This was the case for this channel. The LCP structure was determined at 1.9 Å resolution and was used to solve the 3.6 Å *in surfo* structure. The high-resolution LCP structure allowed the authors to identify a Ca^{2+}-binding site and three convincing Na^{+}-binding sites. These structures, together with mutagenesis data and biochemical and biophysical data in the literature, suggest an alternative access model similar to transporters and 'a progressive antagonistic effect of multiple Na^{+} binding on Ca^{2+} affinity' mechanism for the Na^{+}/Ca^{2+} exchange through the reaction cycle.

TRIC channels are trimeric intracellular cation channels localized on the membrane of sarcoplasmic and endoplasmic reticulum. This counter-cation family of protein is responsible for maintaining K^{+}/Ca^{2+} homeostasis. Using crystals grown in LCP, Chen and co-workers solved the high-resolution structure of TRIC channels to 1.6 Å resolution [117] (Fig. 7.9d) that provides higher resolution and complementary information to *in surfo* structures [118, 119].

7.3.5 Membrane-Embedded Lipid or Lipoprotein Enzymes

Many IMP enzymes participate in synthesis and degradation of membrane lipids such as phospholipids, cholesterol, glycolipids, and sphingolipids. Because lipids are essential components of all membranes, revealing how they are metabolized at a molecular level is a key in understanding the biochemical mechanism and in controlling these enzymes for therapeutic reasons [120]. LCP crystallization method has advantages over other methods in studying this type of IMP. Lipid substrates are usually difficult to solubilize with detergents, but can be reconstituted into LCP at as much as 20 mol% without phase change [25], as noted. This is of particular importance for kinetic assays, where high concentrations of substrates are required, and for crystallization with saturating concentrations of lipid substrates.

Recently, many IMP enzymes involved in lipid metabolism and lipoprotein maturation have been crystallized in LCP and have the structure solved. Such enzymes include DgkA [39], the human prostaglandin E2 synthase 1 (mPGES1) [40], the signal peptidase LspA [121], the CDP-alcohol phosphotransferase AF2299 [122], the phosphatidylinositol synthase PIS [123], and the mouse stearoyl-CoA desaturase 1 (SCD1) [124] (Fig. 7.9e).

The application of LCP was key to the success for the crystal structure determination of DgkA because many years of crystallographic efforts using detergents yielded big crystals but no structures [43, 125]. By contrast, the LCP crystals provided a high-resolution structure at 2.05 Å (Fig. 7.9e). The structure reveals a homo-trimer composed of three structurally heterogeneous monomers. The active site is shared between two adjacent subunits, explaining biochemical data in the literature [39, 126]. An ATP analogue-bound structure [127] supports the direct phosphotransfer mechanism [128] that shares features with common kinases despite no sequence or structure homology found between them.

The mPGES1 enzyme catalyzes the isomerization of prostaglandin H2 to E2 which is a potent bioactive lipid that regulates many physiological and pathophysiological functions such as inflammation and pain. As such, it is a potential drug target for inflammation, pain, and cancer [129]. The LCP structure (2.08 Å) of mPGES1 reveals a homo-trimer (Fig. 7.9e), and the active site where a functionally necessary co-factor glutathione is hosted is shared between two adjacent subunits. In addition, a high-affinity inhibitor is observed in this active site [40]. It is interesting to note that the structure was determined using sulfur SAD [130], the first report of this method with LCP crystals. Moreover, the crystallization of mPGES1 employed optimization of DOPC concentrations in LCP [40], which was also the first of such applications with this technology.

The signal peptidase LspA was crystallized in LCP with an antibiotic [121], the cyclic peptide globomycin. This structure (Fig. 7.9e) therefore provides a basis for the design of new or better antibiotics by inhibiting LspA, an essential enzyme in maturation of lipoproteins that are essential to bacteria growth. An interesting note is that the protein was solubilized in Fos-choline detergent prior to reconstitution. Fos-choline detergents are generally deemed as harsh detergent and unsuitable for crystallization. In this case, it is possible that the vast excess amount of monoolein in LCP masks the effect of Fos-choline on destabilizing the membrane protein.

The LCP structure of the phosphatidylinositol synthase revealed a 6-TM fold. The substrate CDP-DAG was crystallized in the active site of the enzyme, together with essential magnesium ions (Fig. 7.9e).

The SCD1 crystal structure (Fig. 7.9e) was long sought because of the pharmaceutical importance in inhibiting the enzyme for treating obesity [131]. SCD1 introduces a double bond to the acyl chain of stearoyl-CoA forming oleoyl-CoA, an essential precursor for the synthesis of phospholipids, cholesterol, and triglycerides. The 2.6 Å resolution structure showed a 4-TM domain that is capped by a cytosolic domain. With the substrate stearoyl-CoA bound, the active site is revealed, as well as a putative pathway for substrate and product exchange through a plausible lateral gate. This structure should allow rational design of inhibitors as potential drugs for obesity.

7.3.6 Photoreaction Proteins

In 2006, in exploring the LCP for suitability for the crystallization of membrane protein complexes, and to investigate the application of lipid sponge phase for membrane protein crystallization, Cherezov et al. crystallized the light harvesting complex II (LH2) from *Rhodoblastus acidophilus* in the sponge phase that yielded a structures at 2.45 Å resolution (Fig. 7.9f). The LH2 structure is composed of nice small peptides each containing two transmembrane helices. The structure forms two concentric rings, along with its bacteriochlorophyll chromophores. Observed in the structure are also the carotenoid chromophores. This result was a then important message; LCP supports the crystallization of membrane complexes that contain a large number of TM helices, as well as chromophores. The fact that LH2 shows characteristic absorption spectra in LCP and that the LCP structure is overall similar to that obtained in detergent micelles also gives confidence of the then relatively new crystallization method in maintaining functionally relevant conformations.

In addition to the LH2, two photoreaction centers from *Rhodobacter sphaeroides* and *Blastochloris viridis* are crystallized in the lipid sponge phase. These crystals yield two structures to 2.20 and 1.86 Å resolutions [132] (Fig. 7.9f). Similar to the case of LH2, the crystallization conditions contain spongifier reagents, and these complexes are composed of amino acid residues, as well as chromophore pigments, providing more evidence that lipid sponge phase is suited for the crystallization of multicomponent complexes.

Cytochrome oxidase complexes: The application of the LCP method to the ba_3 cytochrome C oxidase gave a higher resolution structure (1.8 Å) (Fig. 7.9g) than the already existing *in surfo* structures, again showing the advantage of the bilayer method (*121*). The fact that the complex was crystallized in a bilayer allowed the disclosure of the interaction profile between lipids and the complex. As well, the atomic resolution of the active site provided the mechanism of electron transfer and the diffusion path of oxygen into the active site.

Similarly, the LCP method was applied to another cytochrome *C* oxidase, the caa_3-type, or the complex IV in the respiratory chain (Fig. 7.9e). Extensive screens of crystals grown in detergent micelles failed to yield diffraction data with high enough resolution for structure determination. By contrast, the LCP crystal structure was solved at 2.36 Å resolution (*27*). The structure revealed a novel integral membrane subunit and a native lipid that was co-purified with the complex. As well, the structure provided a different electron transfer mechanism for the cytochrome *C* that is covalently tethered to the oxidase compared to the soluble and free cytochrome *C*, suggesting a new site for electron entry and exit. It is noted that the caa_3 oxidase contains 24 TM helices, representing the largest TM domain among all LCP structures to date.

7.3.7 Tetraspanins

Tetraspanin superfamily features a 4-TM topology and contains four or more cysteine residues in the extracellular loop 2 region that contains a highly conserved CCG motif [133]. In addition, the TM region usually contains a CXXC motif for palmitoylation. This superfamily has more than 30 members, with functions ranging from cell adhesion and proliferation, and immunity. The structures of two tetraspanin proteins were recently solved, with the tight junction claudin at 2.4 Å resolution [134], and the TAPA-1 (Target of the Antiproliferative Antibody 1, also known as CD81) protein that is involved in viral infection and regulation of cell growth and development [135]. Although sharing the same membrane topology, the two tetraspanin structures displayed a very different arrangement of the TM helices (Fig. 7.9h). The claudin adapts a tight packing of the helices, whereas the TMs of CD81 are separated into two bundles [133, 134]. Given the successes, LCP is likely to continue in contributing to the elucidation of more tetraspanin structures.

7.3.8 Single TM Peptides

While LCP supports the crystallization of aforementioned complexes that have a large number of TMs, it has also been proven successful in the crystallization of single TM peptides. The first such peptides are the antibiotic gramicidin [35], which is a 16-residue β-helical protein produced through non-ribosome biosynthesis. The ultra-high-resolution (1.08 Å) crystal structure revealed an intertwined dimer complex. The other three, solved a few years later, are zinc transporter by protein design [136], the glycophorin A transmembrane peptide [137], and the M2 channel [138] (Fig. 7.9i), at 2.7, 2.8, and 1.1 Å resolutions, respectively. The zinc transporter was the first de novo-designed membrane protein. The LCP structure at atomic resolution echoes the rationale in the design of the ion conductance pore through which metals and protons are counter-transported. The structure also represents the smallest protein that is capable of conducting metal ions through an alternative access mechanism similar to transporters.

Overall, the success of obtaining high-resolution structures of four single-cross peptides should encourage the use of LCP for the crystallization of small transmembrane peptides, many of which play important roles in anchoring, regulation of complexes, signal transduction, and oligomerization processes.

7.3.9 Single TM Proteins

Unlike single TM peptides that lack soluble part, the single TM proteins have a large soluble domain at either N or C terminus, or both. This type of protein was long proven very challenging to crystallize because of the lack of transmembrane mass and rigidity (the TM can tilt more freely than the multipass IMPs) for crystal packing, the flexibility between the TM and soluble domain, and for predicted high solvent contents which are usually associated with poor diffraction quality. Therefore, in the past, these bitopic proteins are generally crystallized with truncation of the TM helix, whereas the TM region is usually studied by nuclear magnetic resonance (NMR) spectroscopy. Importantly, single TM proteins are naturally abundant, with functions ranging from important types of receptors such as Toll-like receptors [139], immunoglobulin-like receptors [140], tumor necrosis factor receptors [141], receptor-type kinase [142], regulators [143], enzymes (such as P450 enzymes) [144, 145], and many more. In *Escherichia coli*, yeast, and pig, over half (55%) of IMPs only cross the membrane once [146]; therefore, this is the most abundant topology for IMPs.

Recently, a non-classical receptor termed σ1 receptor that contains a single TM at the N terminus was crystallized in LCP, and the crystals yielded a 2.51 Å structure [147]. The receptor forms a trimer (Fig. 7.9j). The single TM forms most of the crystal contacts. The soluble domain adopts a cupin-like β-barrel with a ligand-binding site that can host chemically diverse small molecules. The successful crystallization of the σ1 receptor will surely be the first of many LCP structures of the single-TM membrane proteins with a broad range of functions.

7.3.10 Two-Component System

Bacteria sense environmental signals via a so-called two-component system, with which the extracellular (or periplasmic) receptor domain recognizes small ligands such as nutrients and repellants, and the TM region transduces signal to the intracellular domain and triggers downstream cascades by phosphorylation- or methylation-mediated events [148]. These proteins have large domains in both sides of the membrane; the cytoplasmic domain along is ~60 Å long or more [149]. By contrast, the monoolein LCP water channel has a diameter of 40 Å. As a simple comparison, it might appear that LCP crystallization of this type of protein is not feasible. The recent structures of a two-component system, the sensor histidine kinase NarQ from *E. coli* (Fig. 7.9k), with and without the nutrients bound [150], however, demonstrate that LCP is flexible enough to accommodate such large domains for crystallization.

The structure revealed that the protein exists as a symmetric dimer in the apo form and as an asymmetric dimer with a ligand. Regarding the sensor mechanism, the binding of ligand to the sensor domain causes a conformational change that translates a piston-like shift of the TM domain, which in turn is converted to lever-like helical rotation into the coiled-coil HAMP domain (Histidine kinase, Adenylate cyclases, Methyl-accepting proteins, and Phosphotases domain). The high-resolution structure (1.9 Å) should attract the use of LCP for the elucidation of other important two-component systems such as methyl-accepting proteins in understanding bacterial behavior in responses to chemo-attractant and repellent.

7.3.11 Membrane Protein Insertase and Translocon

The LCP crystallization technique has also been applied to membrane protein translocon with great success; the structures of YidC and SecYEG complex were solved at 2.4, 2.7 Å (*apo*), and 3.6 Å (with peptide bound) resolutions [151–153], respectively. The YidC insertase adapts a 5-TM topology (Fig. 7.9i), forming a positively charged hydrophilic patch that is opened sideward to the bilayer and to the cytoplasm, and closed on the extracellular side. The structure suggested an insertion mechanism for single-TM membrane proteins that have an acidic N-terminal extracellular domain. The SecYEG complex structures (Fig. 7.9i) reveal that the protein-producing pore is either occupied by the peptide substrate or blocked by the cytoplasmic loop of SecG, avoiding a constant open pore which would impair the membrane permeability.

7.3.12 *Outer Membrane β-Barrels*

The LCP crystallization method has also been used in structure determination of a number of β-barrel proteins with good success. The vitamin B12 transporter, BtuB [34], and adhesion protein, OpcA [33], were the first ones that demonstrate the feasibility of LCP crystallization for this type of IMP. More recently, the structures of the OmpF porin [154], intimin and invasion [155], and the alginate transporter AlgE [49] have been solved using LCP crystals. All the membrane proteins except intimin were solved by both LCP and the micelle method, and the overall structures were similar, suggesting that LCP also supports functionally relevant forms of β-barrel proteins. In fact, the BtuB still binds vitamin B12, and the OpcA binds sialic acid in LCP provided functional evidence that LCP should preserve functionality of this class of membrane proteins. The AlgE structure (Fig. 7.9m) was solved in LCP with many different forms, and these structures collectively display a conformational landscape through which the alginate substrate is transported across the outer membrane.

Other than being found in the outer membrane of Gram-negative bacteria, β-barrel membrane proteins also exist in the outer membranes of mitochondria of eukaryotes and chloroplasts of plants. Thus far, all LCP-based β-barrel structures have been of bacterial origin.

7.4 Methods

Structure determination through LCP crystallization involves making LCP, setting up crystallization either manually or robotically, monitoring crystal growth, optimizing crystals, manipulating crystal (soaking and harvesting), and collecting diffraction data. This chapter will describe general methods for each step. As well, new methods that are somewhat unconventional, i.e., the in situ and LCP-SFX, are briefly described. Finally, the process from making LCP to growing crystals is laid out step-by-step using DgkA as the example.

7.4.1 *Making LCP*

As noted, LCP is composed of water and lipids (Fig. 7.2a). Thus, the process is basically the hydration of lipid by mixing the two effectively. For the purpose of crystallization, IMPs are included in the aqueous solution as in detergent micelles.

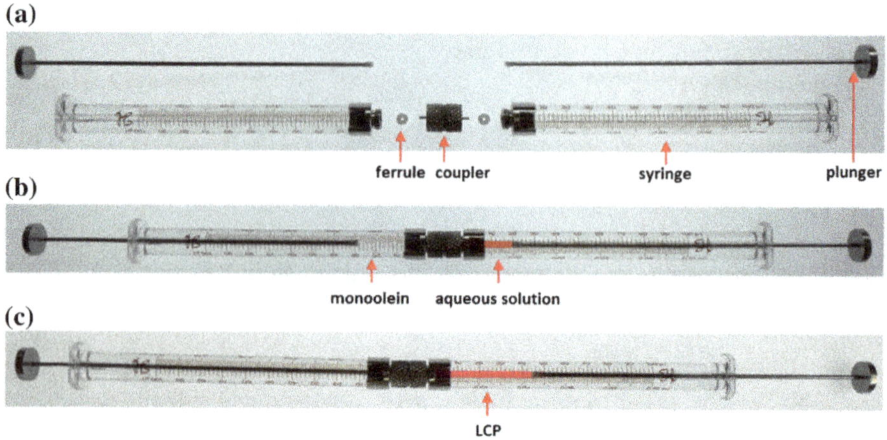

Fig. 7.10 Making LCP with the coupled syringe device. **a** Coupled device. The coupler, fitted with the ferrules at both ends, couples the two syringes. **b** Two parts of aqueous solution and three parts of monoolein are, respectively, filled into the two syringes, which are then connected through the coupler and ready to mix. **c** Mixing the two thoroughly formed LCP. Ponceau dye was added to the aqueous solution for visualization purposes

Alternatively but rarely, membrane proteins that are tolerant to organic solvents can be co-dissolved with lipids. After the removal of solvents, LCP can be made using the lipid–protein mixture with water [35].

The mixing was originally achieved through centrifugation forces, as described in the first report of LCP crystallization [26]. Monoolein, protein, salts, detergents, and precipitants are all added to an Eppendorf tube. The mixture is subjected to centrifugation at 10,000 g for 2.5 h with periodical inversion of the tube for effective mixing. This low-throughput method is revolutionized by a coupled devices developed in the Caffrey group in 1998 [156]. This coupled syringe device is composed of two Hamilton syringes, two Teflon ferrules, and a homemade metal coupler that connects the two syringes through a halo needle (Fig. 7.10a). To make LCP, lipids and water (or protein-containing detergent solution) are filled separately to the two syringes (Fig. 7.10b). Air is carefully extruded by pushing either lipids or protein solution before connecting to the coupler. After assembly, lipids and water are mixed through the needle by pushing the two plungers back and forth. A heavily turbid mixture is first formed, followed by the transparent, homogenous, and clear LCP (Fig. 7.10c) that gives resistance in mixing.

7.4.2 *Crystallization Plates*

The process of crystallization normally begins with a broad initial screen typically with thousands of conditions. Should initial hits be identified, many rounds of subsequent optimization are often required. To save precious membrane protein materials, miniaturization is required. LCP crystallization is generally carried out by depositing 30–50 nL of LCP bolus containing 12–20 nL of protein onto a crystallization well, followed by covering the bolus with 0.8–1 μL of precipitant solution. At this scale, a crystallization trial with 1000 conditions (10- of 96-well plates) requires only 0.12–2 mg of protein, given the protein concentration in the range of 10–100 mg/mL. Compared to *in surfo* crystallization experiments that dispense drops at a 100-nL scale, LCP crystallization uses 5–10-fold less material. For IMPs that are generally produced at low yields, this is a highly attractive feature.

The extreme low volume presents practical challenges too, however. Conventionally, crystallization trials are set up in plastic plates, which are water permeable. Because the water loss happens at very low rate, it is not a noticeable problem for sitting drop or hanging drop experiments [157] where tens or hundreds of microliters of precipitant solution are contained in a well. This is different in the case of LCP crystallization where only sub-microliters of precipitant solution are contained. Thus, an absolute seal is required. This is achieved by using a glass sandwich plate system [158] (Fig. 7.11). A 1-mm glass plate that is salinized to prevent aqueous solution from spreading forms the bottom layer of the sandwich. A double-sticky tape that is pre-punctuated to have 96 holes with the diameter of 5–6 mm is the middle layer of the sandwich. The depth of the individual wells is 0.12 mm. It is in these wells that the LCP and precipitant solution are hosted. The wells are sealed with a 0.1-mm cover glass for crystal growth.

The glass sandwich plate not only provides tightness required for reproducible crystallization results as noted, but also provides excellent optic features for observing crystals grown in LCP. Crystals, especially initial hits, from LCP experiments are usually small, as 5–20 μm needles as exemplified in Fig. 7.12. Glass offers much better optic quality than the plastics. Moreover, unlike plastic, glass does not show birefringence under polarized light, thus avoiding interferences during observation of birefringent microcrystals.

7.4.3 *Robotics*

Crystallization using robots offers the advantage of automation, speed, reproducibility, and miniaturization. Because LCP is a gel-like material, it was robot-unfriendly for conventional liquid-handling robots. In overcoming this problem, in 2004, the first robot for LCP crystallization was developed [29]. The main difference between the LCP robot and other liquid-handling robots is the use of a positive-displacement arm for delivering LCP bolus onto the glass sandwich plate. It turns out the stickiness of LCP (Fig. 7.2b), once thought troublesome for manipulation, works in favor for miniaturization. To set up the crystallization experiment, a syringe filled with LCP is mounted to the motorized pump which pushes the plunger down via a rotational motion under controls of a microchip which in turn can be controlled by a computer [28]. Each pulse squeezes out a tiny

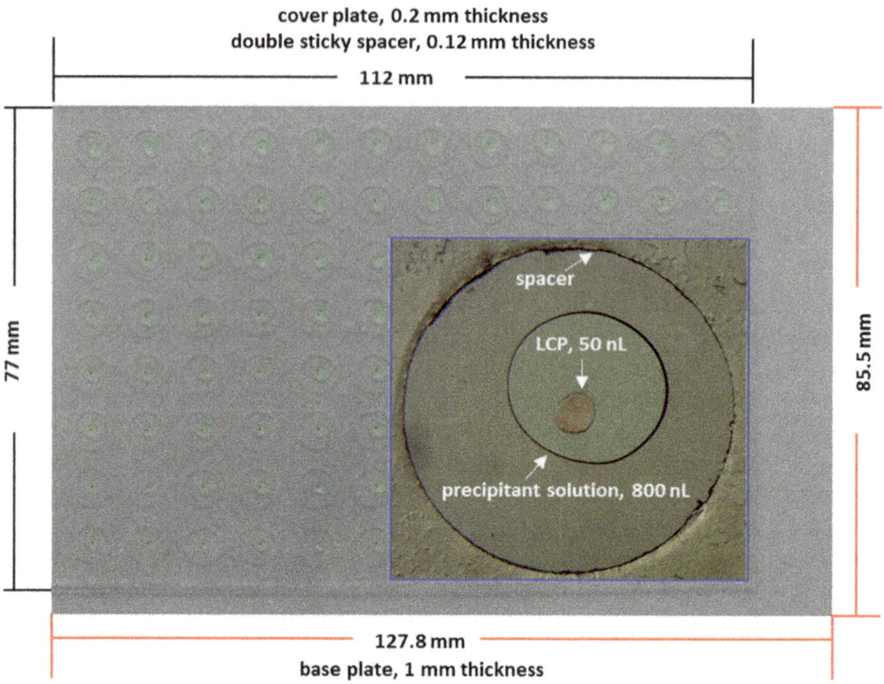

Fig. 7.11 A glass sandwich plate. The sandwich assembly contains a glass bottom plate, a double-sticky spacer, and a glass cover plate. The dimensions are shown appropriately. The spacer has the same length and width with the cover plate. Each well contains 50 nL of LCP and 800 nL precipitant solution. LCP was made with the Ponceau red dye solution with monoolein. The precipitant solution contains green nickel sulfate, for better visibility. The inset shows a zoomed well, with various components indicated by arrows

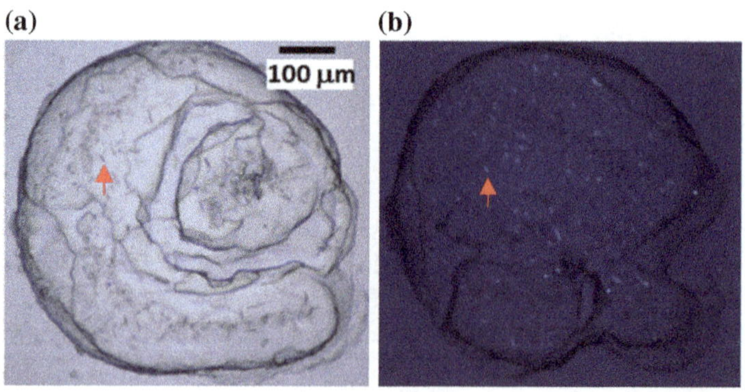

Fig. 7.12 Typical small crystals obtained from an initial LCP crystallization screen. **a** Crystallization drop viewed with normal light. **b** Same drop viewed with polarized light. Red arrows indicate a birefringent crystal

LCP bolus, which touches the sandwich plate. Withdrawal of the pump leaves the LCP bolus on the plate, and so it continues. The 'touch-and-deliver' mode, as well as the stickiness of LCP, makes it possible to deposit LCP reproducibly even at pico-liter volumes [159].

To set up crystallization robotically, a good calibration is required. This is even so in the case of LCP crystallization. First, the deposit should be at the center of the well. Too close to the edge will cause loss of precipitant solution into the spacer due to capillarity either when the cover glass is applied with pressure for sealing, or during well-opening for harvesting. Second, it is key to calibrate the distance between the needle and the glass plate. A distance of about 0.2 mm (a little larger than the double-sticky spacer) is recommended. Larger values may result in no dispense (no touch), while the smaller values may cause uneven LCP surface as reasoned below. An LCP bolus, when deposited onto any substrate, in this case the bottom glass plate, is microscopically rough at the surface because of its gel-like texture, and the roughness feature will remain without pressing from top (Fig. 7.13a). The surface gets flattened (Fig. 7.13b) when the cover glass touches the LCP surface and shapes it as an approximate cylinder between the two glass plates. For this to happen, the bolus needs to be a little higher than the double-sticky spacer so that the cover glass can touch and press the LCP. This is a key step because otherwise the uneven surface will reflect light and interfere with observation of crystals, especially microcrystals.

Another important consideration is to control and to record the temperature, as well as the humidity around the plate during set up. Because the LCP bolus is extremely small, any dehydration would change the composition of the mesophase which could mean phase change, protein stress, or even denaturation and, at least, irreproducibility. If no accurate humidity control is available, a domestic humidifier can be used to increase the humidity around the plate.

There are now a number of commercially available LCP robots, such as the FLEXUS Crystal IMP, TTP Mosquito, Art Robbins Gryphon, and Formulatrix NT8, and all share the feature of using the microsyringe pump as originally

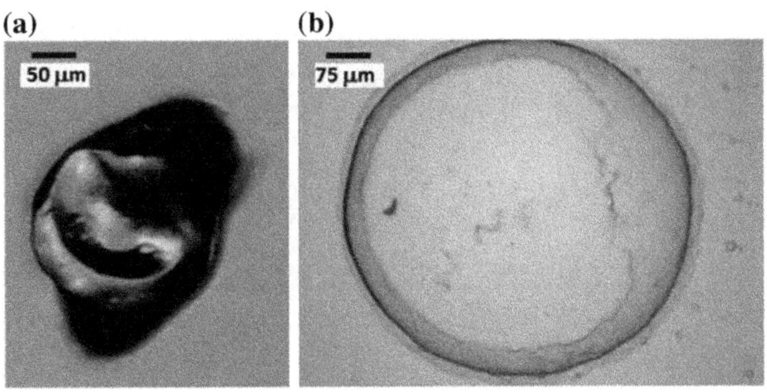

Fig. 7.13 Rough and smooth LCP. **a** Gel-like LCP is rough on the surface after depositing on a glass plate. **b** Surface gets flattened when the cover glass touches and presses on the LCP

reported. The liquid-handling part (8 or 96 at a time) and the status of the deck (move or stay during setup) are the major differences between these robots. Using these robots, setting up a plate typically finishes in 4–6 min. For the small volume of precipitant solution (0.8 μL), evaporation can significantly change the concentration of precipitant solution between first wells and last wells. For projects that are sensitive to subtle change of precipitant concentrations, it may be beneficial to set up crystallization using a small part (a quarter for example) at a time or to set up crystallization in a closure with controlled, high humidity.

It should be noted that the LCP crystallization process can be set up manually too [53], in cases that robots are not available, or setting up crystallization with larger volume under limited known conditions. To do so, LCP is filled into a 10-uL syringe which is mounted to a 50-step repetitive dispenser (Fig. 7.14a). Each click moves the ratchet to dispense 200 nL of LCP. Precipitant solution can be added using micropipettes. Wells are sealed with square cover slides.

The next step in the process is to monitor crystal growth. This can be done under microscope manually, with bright field or polarized light. Crystals with small sizes

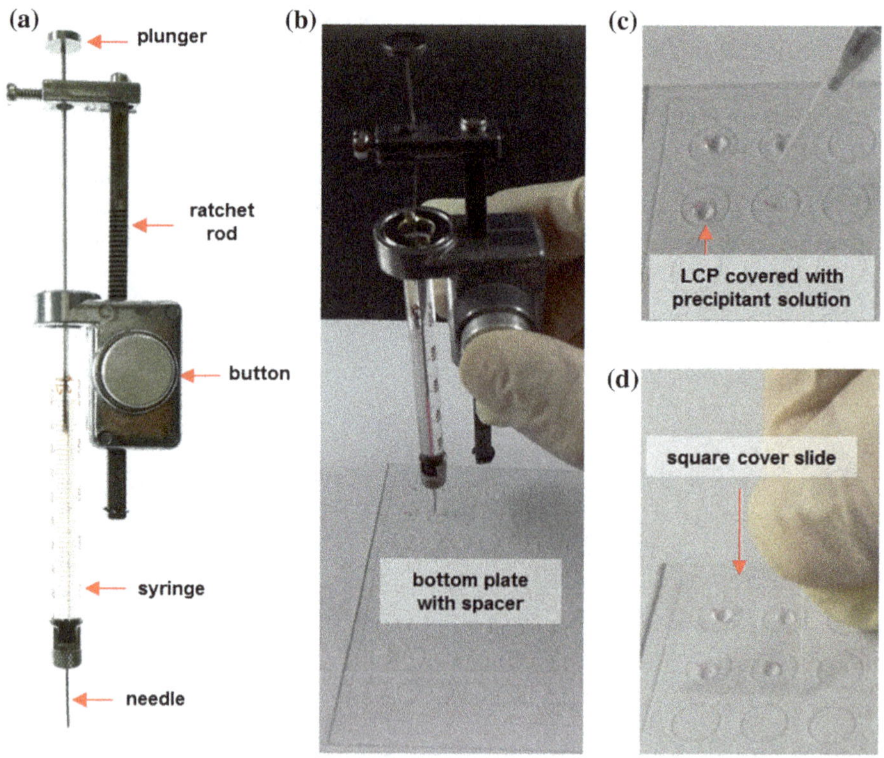

Fig. 7.14 Setting up LCP crystallization manually. **a** Repeat dispenser. Various parts are appropriately labeled. **b** Dispensing LCP onto the bottom plate. **c** Covering LCP with precipitant solution using a micropipette. **d** Sealing the wells with a square cover slide

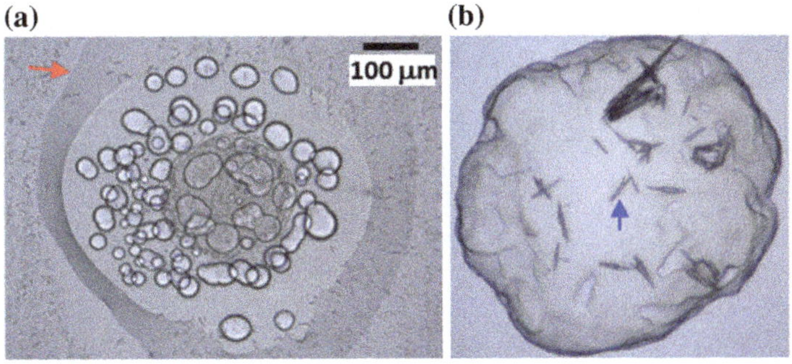

Fig. 7.15 Salt crystals from LCP crystallization trials. **a** Salt crystals that appear in both the mesophase and the precipitant solution. The red arrow points to the mesophase boundary. **b** Salt crystals that only appear in mesophase. A blue arrow indicates such a crystal

such as 1–2 µm can be observed with bright microscopes. The first question to ask, when crystallization hits are obtained, is usually that whether the crystals are of protein or of other components in the rather complex crystallization mixture. Such salt, detergent, and lipid crystals are rare in LCP crystallization trials, but they do appear, especially at high concentrations of salts. Usually, the salt crystals appear both in LCP and in precipitant solution. Therefore, the co-appearance of similarly shaped crystals in both LCP and surrounding precipitant solution most certainly points to salt crystals (Fig. 7.15a). Sometimes, salt crystals only appear in LCP, probably because the LCP lipids compete with salts for water. We have seen this type of crystals in conditions containing sodium fluoride (Fig. 7.15b). In these cases, if synchrotron resources are limited, UV microscope can be very useful for the identification of protein crystals as most membrane proteins contain UV fluorescent tryptophan residues. Because glass attenuates UV light, better results are obtained by facing the UV light with the thin cover glass. Recently, a technic, termed Second Order Nonlinear Imaging of Chiral Crystals (SONNIC) that makes use of chirality of macromolecules is developed to monitor crystal growth in LCP. This SONNIC device is now available through Formulatrix. Using this laser-based technology, microcrystals at sub-micron sizes can be visualized [160].

Robotic systems have also been developed for taking drop images of LCP plates automatically. A prototype imager custom-designed for the glass sandwich plate was developed in 2004 [29]. Imaging a 96-well plate takes only 4.5 min with this prototype. Since then, more features have been implemented. The current models have the ability to find LCP bolus, to take images under bright field, between cross-polarizers, under UV light, and with SONNIC lasers, at user-defined schedules, typically at 0, 1, 3, 5, 7, 14, 21, and 30 days post-setup. The automated imager has the advantage of not missing snapshots that could have been missed with manual inspection if the timing is not right. This is particularly important for crystals that appear and disappear in a very short time window. For example, the initial hit of the human prostaglandin E2 synthase 1 crystal appeared in day 3 but disappeared in day 5. This was the only hit and the very hit that, after optimization, lead to a 2.08 Å crystal structure [40]. It could have been missed for manual inspection which is normally performed at a less frequent schedule.

7.4.4 Optimization of LCP Crystals

Optimization of crystals is often required for crystallography. Optimization process usually involves the systematic change of pH, salt type, salt concentration, protein concentration, precipitant concentration, temperature, and sometimes protein constructs. Apart from these common methods, a unique strategy for LCP optimization is worth discussion—the host lipid screening.

As mentioned, the most used lipid in LCP crystallization is monoolein, a type of MAG that is readily available. As revealed by SAXS, monoolein LCP contains a lipid bilayer with a thickness of 32 Å and a water channel with a radius of 40 Å [24]. Such microstructure features are coupled to the acyl chain length and position of the *cis* double bond. For example, LCP of 7.7 MAG shows very different characteristics as that of the monoolein; the bilayer thickness of the 7.7 MAG LCP is 6 Å less, whereas the water channel radius is $\sim$17 Å more. In addition, the lipid bilayer of 7.7 MAG LCP is less curved than that of 9.9 MAG [93]. The bilayer thickness and curvature change are important environment variables for the transmembrane portion, whereas the size of the water channels could mean different restraints for the hydrophilic part of membrane proteins, especially for those with large soluble domains. These parameters were reasoned to have an impact on the crystallization of membrane proteins within.

The first systematic screen of host lipid was from the crystallization study of a β-barrel porin, OprB, a glucose porin from *Pseudomonas aeruginosa* [42]. Long efforts in optimizing OprB crystals in monoolein LCP did not yield big enough crystals for X-ray diffraction. Interestingly, a host lipid screen that involves the use of 7.7, 6.9, 7.8, 8.7, and 8.8 MAG proves to be useful. Large crystals (300-μm rods) were obtained with 7.8 MAG that diffracted to 2.8 Å. The host lipid screen was applied later with two α-helical membrane proteins, the microbial DgkA and the human mPGES1 [40, 43]. In both cases, initial crystals were obtained in monoolein LCP. Optimization was not successful until the use of short-chained MAGs, 7.8 MAG for DgkA, and 8.8 MAG for mPGES1. The use of non-monoolein MAGs has also been critical for many other membrane protein crystallization projects. The β2-adrenegic receptor–G-protein complex (β2AR-Gs) failed to crystallize in monoolein but grew crystals in 7.7 MAG which yielded a 3.6 Å structure. For the respiratory complex IV, crystals grown in 7.7 MAG LCP yielded a 2.4 Å structure. For the peptide transporter PepT, crystals grown in detergent micelles diffracted to 3.3 Å. Microcrystals (20 μm) grown in 7.8 MAG LCP yielded a 2.3 Å high-resolution structure. The great success motivated the commercialization of such MAGs (www.avantilipids.com). It is noted that, like all other crystallization conditions, choosing host lipid is still empirical. To save costly short-chained lipids (ten times more expensive than monoolein), it might be wise to set up initial broad screening using monoolein and run optimization with the hit condition in combination with the host lipid screening.

Another optimization strategy is to include native lipids during LCP crystallization. As noted, the LCP made of monoolein can tolerate phospholipids and cholesterol to as high as 20 mol%. By contrast, the solubility of these lipids in mild detergents is usually very poor, a limiting factor for additive lipids in crystallization. The strategy of adding native lipids to LCP is widely used for GPCRs. It was found that cholesterol stabilizes GPCR in LCP [161], and cholesterol was essential for LCP crystallization of β2AR [82]. Indeed, cholesterol was found in the β2AR structure, stabilizing the dimer [82, 83]. Since then, cholesterol is always included for GPCR LCP crystallization, and very likely, it had evolutionarily become as a default condition. However, other native lipids such as phospholipids should not be dismissed if cholesterol alone does not yield any satisfactory results. Indeed, in the case of mPGES1 crystallization [40], phosphatidylcholine, not cholesterol, was required for crystal growth in LCP with the optimal concentration of 0.25 mol%.

7.4.5 Crystal Harvesting

Generally, LCP crystals need to be cryo-cooled for X-ray diffraction at synchrotron sources. Crystals grown in wells are exposed before mounting with microloops. For crystals grown in glass sandwich plates, opening the glass well without damaging the crystals is the first challenge. Wells are cut open by cutting gently either near the edge (Fig. 7.16a–d) [47] or at the center of the drop [162] using a sharp and pointy glass cutter, usually under a microscope. A fine tweezer then picks up the cover glass (Fig. 7.16c), exposing the LCP (Fig. 7.16d), which may stick to either the cover, or the bottom glass, or both. It is not necessary to add precipitant solution if the crystals can be harvested quick enough before phase change caused by evaporation. Exposed crystals are then harvested under a microscope using MiTeGen loops or equivalents. Care must be taken to scoop as little LCP as possible to avoid high background during X-ray diffraction data collection. If the crystals are small, it can be very difficult to spot them under a microscope after opening the well because the surface of LCP is no longer flat. In this case, using a polarizer (but do not rotate fully to darkness) is helpful in spotting crystals by their birefringence [47].

In some precipitant conditions, the LCP turns into sponge phase. Such conditions include high concentrations of PEG 400, MPD, and Jeffamine, as noted. Opening wells containing sponge phase can be more difficult than the LCP because sponge phase is very fluidic so it moves around during the opening. The problem is that with such small volume (generally 30–50 nL), when the oily fluidic sponge phase meets cracked gap of the cover, or glass dust caused by cutting, or the spacer, the capillary effect will absorb the sponge phase and crystals are instantaneously lost [47]. In this case, a different strategy in opening the glass well is needed (Fig. 7.16e–h). Apart from the two concentric rings, two horizontal cuts are made (Fig. 7.16e). This gap prevents propagation of cutting forces which often cause cracking of the wells. A window is then created (Fig. 7.16f). The precipitant solution is then reduced to about 0.2 μL using a piece of tissue paper (Fig. 7.16g) through this window. This step should be carried out under a microscope because otherwise the co-migrating crystals can be drawn out with the sponge phase without being noticed. The reduced volume is not enough for the drop to travel through its way to the spacer. However, it also means that the drop is now more prone to evaporation. Therefore, the subsequent cutting and opening need to be done promptly and swiftly. Harvesting in a cold room (if the crystals are grown at this temperature), temporarily avoiding air flow/circulation, increased humidity, and keeping a cool base of the microscope (use a remote light source, or constantly cool the base by tap water) are good practices to minimize evaporation.

Once harvested, crystals are directly plunged into liquid nitrogen without any further treatment with cryo-protectant.

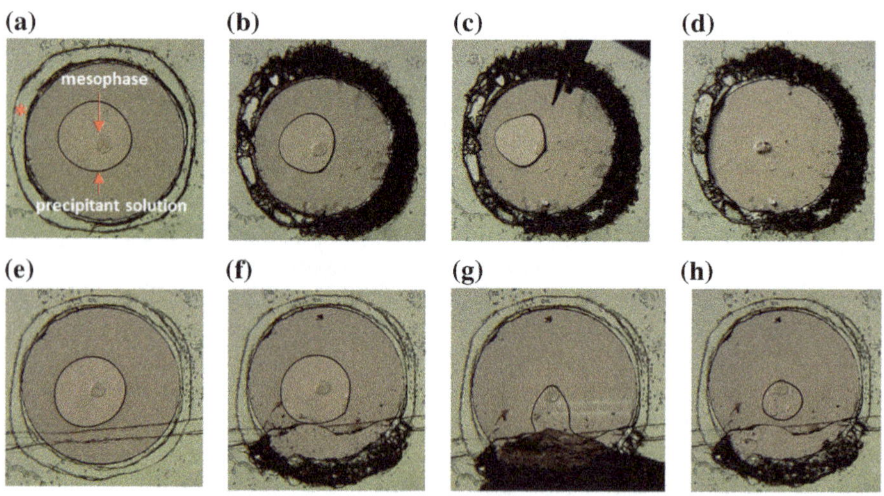

Fig. 7.16 Protocols for opening LCP crystallization wells for crystal harvesting. **a–d** are for cubic phases and **e–h** are for sponge phases. **a** Two concentric rings were cut using a glass cutter. The inner ring should be as close to the well edge as possible. The area labeled with '*' is then cut to loosen the cover glass. **b** Well after cutting the area labeled with '*'. **c** Lifting the cover glass using a fine twizzle. **d** Well with exposed mesophase, ready for harvesting. Fresh precipitant solution should be added in this case as the opening process wiped away all the precipitant solution in the well. **e** Well containing sponge phase was cut similarly as in **a**, with two additional horizontal cuts. The inner one should be close to the precipitant solution. **f** A window is created by cutting this region. **g** The precipitant solution was largely removed using tissue paper. **h** Well after removing most of the precipitant solution. The cover can then be lifted the same way as in **c** and **d**

7.4.6 Crystal Soaking

To determine structures without known structure homology, experimental phasing is required. Natural heavy atoms such as sulfur in Met and Cys residues, as well as naturally bound metals such as zinc, could provide anomalous signal. With high enough resolution and redundancy, the weak phases could be used to solve structures [163]. More often, additional heavy atoms that have stronger anomalous signal need to be introduced to the protein prior to (SeMet labeling), during the process of (co-crystallization with heavy atom), or post crystallization (heavy atom soaking). Technically, LCP crystallization for SeMet protein, or co-crystallization follow standard procedures and hence are not discussed here. For crystal soaking, however, requires special manipulations [46].

The cover glass of desired wells is cut carefully to form a window (same as in Fig. 7.16f) through which the heavy atom solution is injected into the well under a microscope. The liquid drop is then pushed away from the edge of the window. The cover class is wiped dry and sealed with a piece of tape. After incubation with desired time, the heavy atom solution (now about 2 μL) is withdrawn from the well using a small piece of tissue paper. Back-soaking was then performed by repeating the injection process with a heavy atom-free precipitant solution to remove non-specific binding of heavy metals [46].

Alternatively, the soaking can be done with more ease with the hanging drop version of LCP. In this setup, LCP bolus and precipitant solution are deposited onto the hanging drop cover slide. The cover is then flipped and attached to the well which is greased for sealing and filled with precipitant solution. For heavy atom or ligand soaking, the cover is simply detached, and the precipitant solution is replaced with heavy atom-containing solutions. During the soaking and back-soaking, the cover can be detached and reattached multiple times without the trouble of breaking the wells like in the glass sandwich plates. The success of this method relies on the reproducibility of the crystallization conditions that is adapted from glass sandwich plate to the hanging drop setup. In some cases, re-optimizing the hanging drop conditions is required to get crystals with satisfactory quality for soaking experiments.

7.4.7 Data Collection

Data collection for LCP crystals is performed similar to standard crystals, with a few possible exceptions. The surface of LCP becomes rough during the opening of the well and the harvesting process. Upon cryo-cooling, this rough bonus becomes opaque, rendering the small crystals inside invisible at synchrotron [164]. Because the crystals are invisible, the centering is normally done by grid-scanning the whole LCP with an attenuated beam, followed by evaluating each grid based on the diffraction quality of corresponding images. The so-called rastering process [164] has been automated in beamlines GM/CA CAT 23ID-B (APS), I24 (Diamond, UK), PXII (X10SA, SLS, Switzerland), and BL18U (SSRF, Shanghai, China).

Once centered, the data collection is performed as normal. Note that microcrystals in LCP might be very sensitive to radiation damage. An attenuated beam is desirable for data collection in such cases.

Unlike lipid cubic phase, crystals grown in lipid sponge phase are often visible at synchrotron. Crystal centering and data collection in this case are no different than those grown in solution.

7.4.8 *In Meso in Situ*

The process of harvesting LCP crystals presents a few challenges. To open the glass without damaging the crystals requires lots of practice and patience. Even so, the success rate can be low. Fishing crystals without bringing too much LCP is a delicate process which can last for minutes, because one may need to do microsurgery to remove LCP sticking to crystals by scoping away excess LCP using another harvesting loop. The long manipulation can be problematic if there are many crystals to be harvested in a single well, because evaporation can increase precipitant concentration, which in turn can cause the phase change and crystal damage. Unfavorably, the small volume (less than 1 μL) makes the LCP crystallization wells prone to dehydration. Usually, the maximum number one can harvest from a single well is less than twenty. For projects that require data merging from multiple small crystals, the low harvesting efficiency can be frustrating and delays the progress for structure determination.

To overcome these challenges, Huang et al. developed the *in meso* in situ method, avoiding the harvesting process [49, 50]. Crystals were grown in similar setups as in the glass sandwich plate, except for the following differences. First, the bottom and cover glasses are replaced with 25-μm cyclic olefin copolymer (COC) films. Second, the double-sticky spacer is thinner than that used in standard glass sandwich plates (50 vs. 120 μm). In addition, two glasses, one at the top and one at the bottom, sandwich the COC film-spacer assembly to compensate the lack of rigidity for the thin films and to prevent the water loss. Therefore, the COC plates can still be handled the same way as for the glass plates during the crystallization setup and imaging process.

For X-ray data collection, the glass is removed, exposing the COC wells. The well is then cut and mounted to a holder which is compatible with the goniometer. An in-line camera can then detect target crystals in the well for X-ray shooting without liquid nitrogen flow for room temperature data collection. Alternatively, to avoid severe radiation damage at room temperatures, crystals can be cryo-cooled prior to X-ray diffraction. For that, the mounted well is cut open with a scissor and removed free of precipitant solution using a cotton wool or tissue paper before plunged into liquid nitrogen [49]. Data collection is performed the same way as the room temperature set up, except that the cryo-stream is applied. The key success to this method is the application of X-ray transparent material COC. By avoiding the troublesome harvesting process, this method is expected to be welcomed by the membrane protein crystallography community.

7.4.9 Serial Femtosecond Crystallography (SFX) with LCP Crystals

Recently, the SFX gained considerable attention in macromolecule structure biology. It takes advantage of X-ray free electron laser (XFEL) beams which deliver high flux, at 10^{12} photons per pulse, that is, femtoseconds (typically 50 fs) long. Upon hit by XFEL, the temperature of crystal rises to thousands of degrees, in timescales that are much longer than the pulse. Thus, diffraction data are obtained before the crystal-destroying event ('diffraction before destruction') [165, 166]. The beam size of the XFEL is about 1–2 μm in diameter, making the method ideal for microcrystals. Obviously, only one diffraction image can be obtained from one shot. Feeding the beam with fresh crystals for each pulse is essential. A method of doing so was developed by Weierstall et al. using a liquid jet [167, 168]. When pressure is applied, crystals in suspension in the reservoir of the jet are shot out as a 5-μm-diameter flow traveling intersection with the beam direction at a speed of 10 m s^{-1}. XFEL randomly shoot the crystals with a hit rate of approximately 0.004%. Owing to the low efficiency, one has to use lots of crystals which are grown from tens of milligrams of protein [169]. For membrane proteins that are generally difficult to overexpress, purify, and crystallize, this is a great limitation.

The large consumption of crystal for the liquid jet is caused by the requirement for high speed in order to maintain a straight line of liquid. If the flow direction of the crystals can be maintained at a slower rate, less material would be needed. LCP features unique characteristics for such applications. Being viscous and sticky, it can keep its shape and position even without movement. Therefore, one only needs to push it slowly to feed the XFEL beam with fresh samples.

With this basic principle, Weierstall, Cherezov, and co-workers designed a sophisticated LCP injector to facilitate the XFEL diffraction experiments [51]. Briefly, LCP is filled into a reservoir. The front of the reservoir is connected to a nozzle fitted with a glass capillary (30–50 μm in diameter) through which the LCP is extruded. The back of the reservoir is blocked by two Teflon balls. Pressure is applied to this chamber using an HPLC pump. Gas was flowed in a controlled manner to steer the LCP to the beam center. For data collection, the LCP injector is inserted into a vacuum chamber where the LCP is extruded and meet the XFEL beam.

The vacuum was desirable in reducing the X-ray scattering background, but it also creates problems. Under vacuum conditions, evaporation not only dehydrates LCP, but also decreases the temperature of the LCP by evaporative cooling. Both the composition and temperature change have impacts on the phase behavior of

LCP. In LCP made of monoolein, the process causes the change of LCP to lamellar phase, which produces 'finger-printing' type of strong diffraction rings. Apart from masking diffraction spots in the ranges, the strong diffraction exceeds the linear range of the detector and can damage the detector.

The 7.9 MAG, which forms stable LCP at 4 °C, was then used to overcome this problem. Crystals can be grown in 7.9 MAG and then fed into the LCP injector, or, alternatively, crystals grown in monoolein were doped with 7.9 MAG LCP at a ratio of approximately 9:1 (v/v). It was found that the process of mixing and dilution does not usually cause dissolving of membrane protein crystals [166, 170].

Another critical step for the LCP-SFX is to grow crystals microliters volume for the LCP injector. This is done in the microsyringes. Detailed protocols can be found in the literature [170]. Briefly, approximately 10 μL of LCP is injected to a 100-μL syringe pre-filled with 80 μL of precipitant solution. The injection process typically produces a helical bundle of LCP. The syringe is either de-coupled and sealed with Parafilm, or left as an intact assembly in an incubator at a desired temperature for crystal growth. This set up mimics the glass sandwich plate, such that the LCP is surrounded by precipitant solution whose composition is not changed during the incubation period. Crystal growth profile is normally similar to that in the glass sandwich plate, e.g., growth rates and final sizes. It is therefore recommended to set up a crystallization plate using the same LCP but in glass sandwich plate, to follow the crystal growth more conveniently. When crystals grow to the desired sizes in the plates, one could take crystals from the syringes for observation under a microscopes. To do this, a small LCP bolus is extruded carefully from the coupler onto a clean glass plate. A cover glass is then applied on top of the LCP to make a sandwich for crystal examination under light and cross-polarized microscope.

For LCP-SFX, excess precipitant solution needs to be removed for the formation of a homogenous LCP sample. This is done by carefully pushing the precipitant solution from the syringe to an empty one. By pushing slowly, LCP is retained by the coupler, while the precipitant solution goes through. The process does not remove excess precipitant solution completely. Therefore, more monoacylglycerols are added back to the mixture to 'consume' the residual precipitant solution. After this step, a homogenous LCP sample will form.

The power of LCP-SFX method is demonstrated by several successful structures, including several GPCRs [171–174], a GPCR–arrestin complex [85], and the DgkA [127]. The case of DgkA offers a direct comparison of the LCP-SFX with synchrotron radiation. With the third-generation synchrotron, these microcrystals only diffracted to no better than 20 Å [166]. With XFEL-LCP, the structure was solved at 2.18 Å resolution using similarly sized crystals [127]. Regarding the time required, the optimization of the microcrystals to big rectangular crystals that diffracted to 2.18 Å took 3 years, which involves systematic screening of precipitant solution, host lipids, temperature, and mutants [43, 46]. By contrast, the 'optimization' for getting small crystals for XFEL-LCP only took several weeks [166]. Therefore, the XFEL-LCP method offers attractive advantages for crystallization projects that are difficult to optimize.

7.4.10 Protocol—Crystallization of DgkA

The protocol below is for manual setup of LCP crystallization. Note, if crystallization trials are to be set up using robots, the same protocol can be followed except that the steps 'Transfer' and 'Dispense' are carried out as per instructions from the manufacturer for the robots.

7.4.10.1 Materials

1. DgkA mutant that contains 7 point mutations [39], solubilized in DM at 12 mg/mL.
2. 7.8 MAG (Avanti Polar Lipids, Cat. 850531).
3. Precipitant solution, 3–6% MPD, 0.1 M NaCl, 0.1 M sodium citrate pH 5.6 (Hampton Research). Make MPD concentrations as 0.3% increments.
4. Glass sandwich plate (Hampton Research Cat. HR3-151 or equivalent).
5. Cover slides, 18 × 18 mm, 0.1 mm thickness.

7.4.10.2 Equipment

1. Two 0.1-mL Hamilton syringes (Cat. 81065).
2. One 10-μL Hamilton microsyringe (Cat. 80065).
3. Coupler (custom made, see literature [156]).
4. A micropipette that can deliver 0.8–1 μL of solution.
5. Needle with 1 cm length, 22 gauge (Cat. 7787-02, Hamilton). Specify the length while ordering.
6. Repeat dispenser (Cat. PB600-1, Hamilton).
7. Microscope.
8. Incubator at 4 °C.

7.4.10.3 Procedure

Loading Protein. Add 10–15 μL of protein solution to one of the 0.1-mL syringes, with the Teflon ferrule attached. Care must be taken to avoid air bubbles. This can be done by the injection of the protein solution from the back (without the plunger) or from the front (with the plunger in place). If the protein solution is injected from the back, the syringe should be held slightly tilted so that the protein solution can flow slowly without flowing through the front. The plunger can be inserted into place once protein solution reached to the marks of 90–100 μL inside the syringe. An air gap will form. To drive out this air gap, the protein solution is slowly pushed up toward the front of the syringe which now should be in a vertical position. The plunger is slowly pulled down, and then pushed up quickly. By repeating this several times, the protein solution should settle down, driving out the air gap. Care must be taken not to overshoot the plunger. The syringe should not be left at 4 °C or on ice because the Teflon plunger will shrink more than the glass syringe, causing leaking of the protein solution.

Loading Lipid. Bring the 7.8 MAG to room temperature. The solid material will become oily. Use a pipette to add 7.8 MAG (equal volume to the protein) to the other 0.1-mL syringe. To do this, insert the pipette tip to the ferrule, then pull the plunger as well as push the pipette so the viscous 7.8 MAG can flow into the syringe by both suction and injection forces. To measure the volume of the lipid, push the plunger to the very top till no air gap can be seen. Attach the coupler to the syringe once the desired volume is reached. The dead volume of the coupler, generally 1.5–2 μL, is then filled up by carefully and slowly pushing the plunger.

Mixing. The two syringes are then connected together through the coupler, after extruding air bubbles. The lipid and protein solution is mixed by pushing the plungers back and forth. The initial turbid mixture will become clear, indicating of LCP formation.

Transfer. The 10-μL syringe is mounted onto the repeat dispenser (Fig. 7.14a). Ten microliters of LCP is transferred from the 0.1-mL syringe to the 10-μL syringe through the coupler. The coupler is then replaced with the needle. The plunger is fixed to the gripper. LCP can be dispensed stepwise by clicking the button.

Dispense. LCP is dispensed onto the crystallization wells (Fig. 7.14b). Each click delivers 200 nL of LCP. To reduce evaporation, four wells are filled as a group before moving onto the next dispenser. LCP boluses are immediately covered with 0.8–1 μL of precipitant solution. The four wells as a square are then covered with the coverslip. Pressure can be applied to ensure tight sealing.

Monitoring Crystal Growth. The plate is then transferred to a 4 °C incubator for crystal growth. Crystallization process can be monitored using a microscope in a cold room. Crystals should appear after 1 day and continue to grow for 1–2 weeks. Under optimal conditions, rectangular crystals sized at 100–200 μm in the longest dimension can be obtained.

Acknowledgements This work was supported by the 1000 Young Talent Program, the Shanghai Pujiang Talent Program (15PJ1409400), the National Natural Science Foundation of China (No. 31570748 and U1632127), the CAS Shanghai Science Research Center (CAS-SSRC-YJ-2015-02), and Key Program of CAS Frontier Science (QYZDB-SSW-SMC037).

References

1. Yildirim MA, Goh KI, Cusick ME, Barabasi AL, Vidal M (2007) Drug-target network. Nat Biotechnol 25(10):1119–1126
2. Waseda Y, Matsubara E, Shinoda K (2011) X-ray diffraction crystallography: introduction, examples and solved problems. Springer Science & Business Media

3. Rupp B (2009) Biomolecular crystallography: principles, practice, and application to structural biology
4. Carpenter EP, Beis K, Cameron AD, Iwata S (2008) Overcoming the challenges of membrane protein crystallography. Curr Opin Struct Biol 18(5):581–586
5. Parker JL, Newstead S (2016) Membrane protein crystallization: current trends and future perspectives. Adv Exp Med Biol 922:61–72
6. Kloppmann E, Punta M, Rost B (2012) Structural genomics plucks high-hanging membrane proteins. Curr Opin Struct Biol 22(3):326–332
7. Caffrey M, Li D, Dukkipati A (2012) Membrane protein structure determination using crystallography and lipidic mesophases: recent advances and successes. Biochemistry 51 (32):6266–6288
8. Privé GG (2007) Detergents for the stabilization and crystallization of membrane proteins. Methods 41(4):388–397
9. Loll P (2014) Membrane proteins, detergents and crystals: what is the state of the art? Acta Crystallogr F 70(12):1576–1583
10. Cao Y, Jin X, Huang H, Derebe MG, Levin EJ, Kabaleeswaran V, Pan Y, Punta M, Love J, Weng J, Quick M, Ye S, Kloss B, Bruni R, Martinez-Hackert E, Hendrickson WA, Rost B, Javitch JA, Rajashankar KR, Jiang Y, Zhou M (2011) Crystal structure of a potassium ion transporter. TrkH. Nature 471(7338):336–340
11. Mancia F, Love J (2010) High-throughput expression and purification of membrane proteins. J Struct Biol 172(1):85–93
12. Abdul-Hussein S, Andrell J, Tate CG (2013) Thermostabilisation of the serotonin transporter in a cocaine-bound conformation. J Mol Biol 425(12):2198–2207
13. Magnani F, Serrano-Vega MJ, Shibata Y, Abdul-Hussein S, Lebon G, Miller-Gallacher J, Singhal A, Strege A, Thomas JA, Tate CG (2016) A mutagenesis and screening strategy to generate optimally thermostabilized membrane proteins for structural studies. Nat Protoc 11 (8):1554–1571
14. Coleman JA, Green EM, Gouaux E (2016) X-ray structures and mechanism of the human serotonin transporter. Nature 532(7599):334–339
15. Lebon G, Bennett K, Jazayeri A, Tate CG (2011) Thermostabilisation of an agonist-bound conformation of the human adenosine A(2A) receptor. J Mol Biol 409(3):298–310
16. Doré AS, Robertson N, Errey JC, Ng I, Hollenstein K, Tehan B, Hurrell E, Bennett K, Congreve M, Magnani F, Tate CG, Weir M, Marshall FH (2011) Structure of the adenosine A(2A) receptor in complex with ZM241385 and the xanthines XAC and caffeine. Structure 19(9):1283–1293
17. Carpenter B, Nehmé R, Warne T, Leslie AGW, Tate CG (2016) Structure of the adenosine A (2A) receptor bound to an engineered G protein. Nature 536(7614):104–107
18. Miller JL, Tate CG (2011) Engineering an ultra-thermostable β(1)-adrenoceptor. J Mol Biol 413(3):628–638.
19. Carpenter B, Tate CG (2016) Engineering a minimal G protein to facilitate crystallisation of G protein-coupled receptors in their active conformation. Protein Eng Des Sel 29(12):583–594
20. Miller-Gallacher JL, Nehmé R, Warne T, Edwards PC, Schertler GFX, Leslie AGW, Tate CG (2014) The 2.1 Å resolution structure of cyanopindolol-bound β(1)-Adrenoceptor identifies an intramembrane Na(+) ion that stabilises the ligand-free receptor. PLoS One 9 (3):e92727
21. Carpenter B, Nehmé R, Warne T, Leslie AGW, Tate CG (2016) Structure of the adenosine A2A receptor bound to an engineered G protein. Nature 536(7614):104–107
22. Faham S, Bowie JU (2002) Bicelle crystallization: a new method for crystallizing membrane proteins yields a monomeric bacteriorhodopsin structure. J Mol Biol 316(1):1–6
23. Sanders CR, Prosser RS (1998) Bicelles: a model membrane system for all seasons? Structure 6(10):1227–1234
24. Qiu H, Caffrey M (2000) The phase diagram of the monoolein/water system: metastability and equilibrium aspects. Biomaterials 21(3):223–234

25. Cherezov V, Clogston J, Misquitta Y, Abdel-Gawad W, Caffrey M (2002) Membrane protein crystallization in meso: lipid type-tailoring of the cubic phase. Biophys J 83 (6):3393–3407
26. Landau EM, Rosenbusch JP (1996) Lipidic cubic phases: a novel concept for the crystallization of membrane proteins. Proc Natl Acad Sci U S A 93(25):14532–14535
27. Pebay-Peyroula E, Rummel G, Rosenbusch JP, Landau EM (1997) X-ray structure of bacteriorhodopsin at 2.5 angstroms from microcrystals grown in lipidic cubic phases. Science 277(5332):1676–1681
28. Li D, Boland C, Walsh K, Caffrey M (2012) Use of a robot for high-throughput crystallization of membrane proteins in lipidic mesophases. J Vis Exp 67:e4000
29. Cherezov V, Peddi A, Muthusubramaniam L, Zheng YF, Caffrey M (2004) A robotic system for crystallizing membrane and soluble proteins in lipidic mesophases. Acta Crystallogr D 60(Pt 10):1795–1807
30. Ai X, Caffrey M (2000) Membrane protein crystallization in lipidic mesophases: detergent effects. Biophys J 79(1):394–405
31. Caffrey M (1987) Kinetics and mechanism of transitions involving the lamellar, cubic, inverted hexagonal, and fluid isotropic phases of hydrated monoacylglycerides monitored by time-resolved X-ray diffraction. Biochemistry 26(20):6349–6363
32. Cherezov V, Fersi H, Caffrey M (2001) Crystallization screens: compatibility with the lipidic cubic phase for in meso crystallization of membrane proteins. Biophys J 81(1):225–242
33. Cherezov V, Liu W, Derrick JP, Luan B, Aksimentiev A, Katritch V, Caffrey M (2008) In meso crystal structure and docking simulations suggest an alternative proteoglycan binding site in the OpcA outer membrane adhesin. Proteins 71(1):24–34
34. Cherezov V, Yamashita E, Liu W, Zhalnina M, Cramer WA, Caffrey M (2006) In meso structure of the cobalamin transporter, BtuB, at 1.95 Å resolution. J Mol Biol 364 (4):716–734
35. Hofer N, Aragao D, Caffrey M (2010) Crystallizing transmembrane peptides in lipidic mesophases. Biophys J 99(3):L23–25
36. Cherezov V, Clogston J, Papiz MZ, Caffrey M (2006) Room to move: crystallizing membrane proteins in swollen lipidic mesophases. J Mol Biol 357(5):1605–1618
37. Lyons JA, Aragao D, Slattery O, Pisliakov AV, Soulimane T, Caffrey M (2012) Structural insights into electron transfer in caa3-type cytochrome oxidase. Nature 487(7408):514–518
38. Li D, Caffrey M (2011) Lipid cubic phase as a membrane mimetic for integral membrane protein enzymes. Proc Natl Acad Sci U S A 108(21):8639–8644
39. Li D, Lyons JA, Pye VE, Vogeley L, Aragao D, Kenyon CP, Shah ST, Doherty C, Aherne M, Caffrey M (2013) Crystal structure of the integral membrane diacylglycerol kinase. Nature 497(7450):521–524
40. Li D, Howe N, Dukkipati A, Shah ST, Bax BD, Edge C, Bridges A, Hardwicke P, Singh OM, Giblin G, Pautsch A, Pfau R, Schnapp G, Wang M, Olieric V, Caffrey M (2014) Crystallizing membrane proteins in the lipidic mesophase. experience with human prostaglandin E2 synthase 1 and an evolving strategy. Cryst Growth Des 14(4):2034–2047
41. Lyons JA, Parker JL, Solcan N, Brinth A, Li D, Shah STA, Caffrey M, Newstead S (2014) Structural basis for polyspecificity in the POT family of proton-coupled oligopeptide transporters. EMBO Rep 15(8):886–893
42. Li D, Lee J, Caffrey M (2011) Crystallizing membrane proteins in lipidic mesophases. a host lipid screen. Cryst Growth Des 11(2):530–537
43. Li D, Shah ST, Caffrey M (2013) Host lipid and temperature as important screening variables for crystallizing integral membrane proteins in lipidic mesophases. Trials with diacylglycerol kinase. Cryst Growth Des 13(7):2846–2857
44. Tan J, Rouse SL, Li D, Pye VE, Vogeley L, Brinth AR, El Arnaout T, Whitney JC, Howell PL, Sansom MSP, Caffrey M (2014) A conformational landscape for alginate secretion across the outer membrane of Pseudomonas aeruginosa. Acta Crystallogr D 70(Pt 8):2054–2068

45. Rasmussen SGF, DeVree BT, Zou Y, Kruse AC, Chung KY, Kobilka TS, Thian FS, Chae PS, Pardon E, Calinski D, Mathiesen JM, Shah STA, Lyons JA, Caffrey M, Gellman SH, Steyaert J, Skiniotis G, Weis WI, Sunahara RK, Kobilka BK (2011) Crystal structure of the β(2)Adrenergic receptor-Gs protein complex. Nature 477(7366):549–555
46. Li D, Pye VE, Caffrey M (2015) Experimental phasing for structure determination using membrane-protein crystals grown by the lipid cubic phase method. Acta Crystallogr D 71(Pt 1):104–122
47. Li D, Boland C, Aragao D, Walsh K, Caffrey M (2012) Harvesting and cryo-cooling crystals of membrane proteins grown in lipidic mesophases for structure determination by macromolecular crystallography. J Vis Exp 67:4001
48. Liu W, Cherezov V (2011) Crystallization of membrane proteins in lipidic mesophases. J Vis Exp 49:2501
49. Huang CY, Olieric V, Ma P, Howe N, Vogeley L, Liu X, Warshamanage R, Weinert T, Panepucci E, Kobilka B, Diederichs K, Wang M, Caffrey M (2016) In meso in situ serial X-ray crystallography of soluble and membrane proteins at cryogenic temperatures. Acta Crystallogr D 72(Pt 1):93–112
50. Huang C-Y, Olieric V, Ma P, Panepucci E, Diederichs K, Wang M, Caffrey M (2015) In meso in situ serial X-ray crystallography of soluble and membrane proteins. Acta Crystallogr D 71(Pt 6):1238–1256
51. Weierstall U, James D, Wang C, White TA, Wang D, Liu W, Spence JCH, Doak RB, Nelson G, Fromme P, Fromme R, Grotjohann I, Kupitz C, Zatsepin NA, Liu H, Basu S, Wacker D, Han GW, Katritch V, Boutet S, Messerschmidt M, Williams GJ, Koglin JE, Seibert MM, Klinker M, Gati C, Shoeman RL, Barty A, Chapman HN, Kirian RA, Beyerlein KR, Stevens RC, Li D, Shah STA, Howe N, Caffrey M, Cherezov V (2014) Lipidic cubic phase injector facilitates membrane protein serial femtosecond crystallography. Nat Commun 5:3309
52. Clogston J, Rathman J, Tomasko D, Walker H, Caffrey M (2000) Phase behavior of a monoacylglycerol: (Myverol 18–99 K)/water system. Chem Phys Lipids 107(2):191–220
53. Caffrey M, Porter C (2010) Crystallizing membrane proteins for structure determination using lipidic mesophases. J Vis Exp 45:e1712
54. Salvati Manni L, Zabara A, Osornio YM, Schoppe J, Batyuk A, Pluckthun A, Siegel JS, Mezzenga R, Landau EM (2015) Phase behavior of a designed cyclopropyl analogue of monoolein: implications for low-temperature membrane protein crystallization. Angew Chem Int Ed Engl 54(3):1027–1031
55. Barauskas J, Landh T (2003) Phase behavior of the phytantriol/water system. Langmuir 19 (23):9562–9565
56. Borshchevskiy V, Molseeva E, Kuklin A, Buldt G, Hato M, Gordeliy V (2010) Isoprenoid-chained lipid beta-XylOC(16 + 4)—a novel molecule for in meso membrane protein crystallization. J Cryst Growth 312(22):3326–3330
57. Ishchenko A, Peng L, Zinovev E, Vlasov A, Lee SC, Kuklin A, Mishin A, Borshchevskiy V, Zhang Q, Cherezov V (2017) Chemically stable lipids for membrane protein crystallization. Cryst Growth Des 17(6):3502–3511
58. Broecker J, Keller S (2013) Impact of urea on detergent micelle properties. Langmuir 29 (27):8502–8510
59. Li D, Caffrey M (2014) Renaturing membrane proteins in the lipid cubic phase, a nanoporous membrane mimetic. Sci Rep 4:5806
60. Clogston J, Craciun G, Hart DJ, Caffrey M (2005) Controlling release from the lipidic cubic phase by selective alkylation. J Control Release 102(2):441–461
61. Misquitta Y, Caffrey M (2003) Detergents destabilize the cubic phase of monoolein: implications for membrane protein crystallization. Biophys J 85(5):3084–3096
62. Chae PS, Rasmussen SGF, Rana RR, Gotfryd K, Chandra R, Goren MA, Kruse AC, Nurva S, Loland CJ, Pierre Y, Drew D, Popot J-L, Picot D, Fox BG, Guan L, Gether U, Byrne B, Kobilka B, Gellman SH (2010) Maltose-neopentyl glycol (MNG) amphiphiles for

solubilization, stabilization and crystallization of membrane proteins. Nat Methods 7 (12):1003–1008
63. Champeil P, Orlowski S, Babin S, Lund S, le Maire M, Moller J, Lenoir G, Montigny C (2016) A robust method to screen detergents for membrane protein stabilization, revisited. Anal Biochem 511:31–35
64. Caffrey M (2015) A comprehensive review of the lipid cubic phase or in meso method for crystallizing membrane and soluble proteins and complexes. Acta Crystallogr F 71(Pt 1):3–18
65. Caffrey M, Cherezov V (2009) Crystallizing membrane proteins using lipidic mesophases. Nat Protoc 4(5):706–731
66. Ridell A, Ekelund K, Evertsson H, Engström S (2003) On the water content of the solvent/monoolein/water sponge (L3) phase. Colloids Surf A Physicochem Eng Asp 228(1):17–24
67. Engström S, Alfons K, Rasmusson M, Ljusberg-Wahren H (1998) Solvent-induced sponge (L3) phases in the solvent-monoolein-water system. In: Lindman B, Ninham BW (eds) The colloid science of lipids: new paradigms for self-assembly in science and technology. Steinkopff, Darmstadt, pp 93–98. doi:10.1007/BFb0117965
68. Wadsten P, Wohri AB, Snijder A, Katona G, Gardiner AT, Cogdell RJ, Neutze R, Engstrom S (2006) Lipidic sponge phase crystallization of membrane proteins. J Mol Biol 364(1):44–53
69. Wöhri AB, Johansson LC, Wadsten-Hindrichsen P, Wahlgren WY, Fischer G, Horsefield R, Katona G, Nyblom M, Öberg F, Young G, Cogdell RJ, Fraser NJ, Engström S, Neutze R (2008) A lipidic-sponge phase screen for membrane protein crystallization. Structure 16 (7):1003–1009
70. Cherezov V, Liu J, Griffith M, Hanson MA, Stevens RC (2008) LCP-FRAP assay for pre-screening membrane proteins for in meso crystallization. Cryst Growth Des 8(12): 4307–4315
71. Kenworthy AK, Simon SA, McIntosh TJ (1995) Structure and phase behavior of lipid suspensions containing phospholipids with covalently attached poly (ethylene glycol). Biophys J 68(5):1903–1920
72. Imberg A, Evertsson H, Stilbs P, Kriechbaum M, Engström S (2003) On the self-assembly of monoolein in mixtures of water and a polar aprotic solvent. J Phys Chem B 107(10): 2311–2318
73. Takahashi H, Matsuo A, Hatta I (2000) Effects of chaotropic and kosmotropic solutes on the structure of lipid cubic phase: monoolein-water systems. Mol Cryst Liq Cryst Sci Technol Sect A Mol Cryst Liq Cryst 347(1):231–238
74. Evertsson H, Stilbs P, Lindblom G, Engström S (2002) NMR self diffusion measurements of the Monooleoylglycerol/Poly ethylene glycol/water L3 phase. Colloids Surf B Biointerfaces 26(1):21–29
75. Liu W, Caffrey M (2005) Gramicidin structure and disposition in highly curved membranes. J Struct Biol 150(1):23–40
76. Cherezov V, Caffrey M (2007) Membrane protein crystallization in lipidic mesophases. A mechanism study using X-ray microdiffraction. Faraday discussions 136:195–212
77. Qutub Y, Reviakine I, Maxwell C, Navarro J, Landau EM, Vekilov PG (2004) Crystallization of transmembrane proteins in cubo: mechanisms of crystal growth and defect formation. J Mol Biol 343(5):1243–1254
78. Caffrey M (2008) On the mechanism of membrane protein crystallization in lipidic mesophases. Cryst Growth Des 8(12):4244–4254
79. Aherne M, Lyons JA, Caffrey M (2012) A fast, simple and robust protocol for growing crystals in the lipidic cubic phase. J Appl Crystallogr 45(Pt 6):1330–1333
80. Kobilka B (2013) The structural basis of G-protein-coupled receptor signaling (Nobel Lecture). Angew Chem Int Ed Engl 52(25):6380–6388

81. Santos R, Ursu O, Gaulton A, Bento AP, Donadi RS, Bologa CG, Karlsson A, Al-Lazikani B, Hersey A, Oprea TI, Overington JP (2017) A comprehensive map of molecular drug targets. Nat Rev Drug Discov 16(1):19–34
82. Cherezov V, Rosenbaum DM, Hanson MA, Rasmussen SGF, Thian FS, Kobilka TS, Choi H-J, Kuhn P, Weis WI, Kobilka BK, Stevens RC (2007) High resolution crystal structure of an engineered human β(2)-Adrenergic G protein-coupled receptor. Science 318 (5854):1258–1265
83. Rosenbaum DM, Cherezov V, Hanson MA, Rasmussen SG, Thian FS, Kobilka TS, Choi HJ, Yao XJ, Weis WI, Stevens RC, Kobilka BK (2007) GPCR engineering yields high-resolution structural insights into beta2-adrenergic receptor function. Science 318 (5854):1266–1273
84. Rasmussen SG, DeVree BT, Zou Y, Kruse AC, Chung KY, Kobilka TS, Thian FS, Chae PS, Pardon E, Calinski D, Mathiesen JM, Shah ST, Lyons JA, Caffrey M, Gellman SH, Steyaert J, Skiniotis G, Weis WI, Sunahara RK, Kobilka BK (2011) Crystal structure of the beta2 adrenergic receptor-Gs protein complex. Nature 477(7366):549–555
85. Kang Y, Zhou XE, Gao X, He Y, Liu W, Ishchenko A, Barty A, White TA, Yefanov O, Han GW, Xu Q, de Waal PW, Ke J, Tan MH, Zhang C, Moeller A, West GM, Pascal BD, Van Eps N, Caro LN, Vishnivetskiy SA, Lee RJ, Suino-Powell KM, Gu X, Pal K, Ma J, Zhi X, Boutet S, Williams GJ, Messerschmidt M, Gati C, Zatsepin NA, Wang D, James D, Basu S, Roy-Chowdhury S, Conrad CE, Coe J, Liu H, Lisova S, Kupitz C, Grotjohann I, Fromme R, Jiang Y, Tan M, Yang H, Li J, Wang M, Zheng Z, Li D, Howe N, Zhao Y, Standfuss J, Diederichs K, Dong Y, Potter CS, Carragher B, Caffrey M, Jiang H, Chapman HN, Spence JC, Fromme P, Weierstall U, Ernst OP, Katritch V, Gurevich VV, Griffin PR, Hubbell WL, Stevens RC, Cherezov V, Melcher K, Xu HE (2015) Crystal structure of rhodopsin bound to arrestin by femtosecond X-ray laser. Nature 523(7562):561–567
86. Serrano-Vega MJ, Magnani F, Shibata Y, Tate CG (2008) Conformational thermostabilization of the β1-adrenergic receptor in a detergent-resistant form. Proc Natl Acad Sci U S A 105(3):877–882
87. Warne T, Serrano-Vega MJ, Baker JG, Moukhametzianov R, Edwards PC, Henderson R, Leslie AGW, Tate CG, Schertler GFX (2008) Structure of a [bgr]1-adrenergic G-protein-coupled receptor. Nature 454(7203):486–491
88. Lebon G, Warne T, Tate CG (2012) Agonist-bound structures of G protein-coupled receptors. Curr Opin Struct Biol 22(4):482–490
89. Lebon G, Warne T, Edwards PC, Bennett K, Langmead CJ, Leslie AG, Tate CG (2011) Agonist-bound adenosine A2A receptor structures reveal common features of GPCR activation. Nature 474(7352):521–525
90. Sun B, Bachhawat P, Chu ML-H, Wood M, Ceska T, Sands ZA, Mercier J, Lebon F, Kobilka TS, Kobilka BK (2017) Crystal structure of the adenosine A2A receptor bound to an antagonist reveals a potential allosteric pocket. Proc Natl Acad Sci U S A 114(8):2066–2071
91. Lanyi JK (2004) Bacteriorhodopsin. Annu Rev Physiol 66(1):665–688
92. Nogly P, Panneels V, Nelson G, Gati C, Kimura T, Milne C, Milathianaki D, Kubo M, Wu W, Conrad C, Coe J, Bean R, Zhao Y, Bath P, Dods R, Harimoorthy R, Beyerlein KR, Rheinberger J, James D, DePonte D, Li C, Sala L, Williams GJ, Hunter MS, Koglin JE, Berntsen P, Nango E, Iwata S, Chapman HN, Fromme P, Frank M, Abela R, Boutet S, Barty A, White TA, Weierstall U, Spence J, Neutze R, Schertler G, Standfuss J (2016) Lipidic cubic phase injector is a viable crystal delivery system for time-resolved serial crystallography. Nat Commun 7:12314
93. Misquitta LV, Misquitta Y, Cherezov V, Slattery O, Mohan JM, Hart D, Zhalnina M, Cramer WA, Caffrey M (2004) Membrane protein crystallization in lipidic mesophases with tailored bilayers. Structure 12(12):2113–2124
94. Kato HE, Zhang F, Yizhar O, Ramakrishnan C, Nishizawa T, Hirata K, Ito J, Aita Y, Tsukazaki T, Hayashi S, Hegemann P, Maturana AD, Ishitani R, Deisseroth K, Nureki O (2012) Crystal structure of the channelrhodopsin light-gated cation channel. Nature 482 (7385):369–374

95. Gordeliy VI, Labahn J, Moukhametzianov R, Efremov R, Granzin J, Schlesinger R, Buldt G, Savopol T, Scheidig AJ, Klare JP, Engelhard M (2002) Molecular basis of transmembrane signalling by sensory rhodopsin II-transducer complex. Nature 419(6906):484–487
96. Zhou X, Levin EJ, Pan Y, McCoy JG, Sharma R, Kloss B, Bruni R, Quick M, Zhou M (2014) Structural basis of the alternating-access mechanism in a bile acid transporter. Nature 505(7484):569–573
97. Deng D, Sun P, Yan C, Ke M, Jiang X, Xiong L, Ren W, Hirata K, Yamamoto M, Fan S, Yan N (2015) Molecular basis of ligand recognition and transport by glucose transporters. Nature 526(7573):391–396
98. Taniguchi R, Kato HE, Font J, Deshpande CN, Wada M, Ito K, Ishitani R, Jormakka M, Nureki O (2015) Outward- and inward-facing structures of a putative bacterial transition-metal transporter with homology to ferroportin. Nat Commun 6:8545
99. Tanaka Y, Hipolito CJ, Maturana AD, Ito K, Kuroda T, Higuchi T, Katoh T, Kato HE, Hattori M, Kumazaki K, Tsukazaki T, Ishitani R, Suga H, Nureki O (2013) Structural basis for the drug extrusion mechanism by a MATE multidrug transporter. Nature 496(7444): 247–251
100. Kusakizako T, Tanaka Y, Hipolito CJ, Kuroda T, Ishitani R, Suga H, Nureki O (2016) LCP crystallization and X-ray diffraction analysis of VcmN, a MATE transporter from Vibrio cholerae. Acta Crystallogr F 72(Pt 7):552–557
101. Kuk ACY, Mashalidis EH, Lee S-Y (2017) Crystal structure of the MOP flippase MurJ in an inward-facing conformation. Nat Struct Mol Biol 24(2):171–176
102. Fukuda M, Takeda H, Kato HE, Doki S, Ito K, Maturana AD, Ishitani R, Nureki O (2015) Structural basis for dynamic mechanism of nitrate/nitrite antiport by NarK. Nat Commun 6:7097
103. Solcan N, Kwok J, Fowler PW, Cameron AD, Drew D, Iwata S, Newstead S (2012) Alternating access mechanism in the POT family of oligopeptide transporters. EMBO J 31 (16):3411–3421
104. Quistgaard EM, Low C, Guettou F, Nordlund P (2016) Understanding transport by the major facilitator superfamily (MFS): structures pave the way. Nat Rev Mol Cell Biol 17 (2):123–132
105. Kaback HR, Smirnova I, Kasho V, Nie Y, Zhou Y (2011) The Alternating access transport mechanism in LacY. J Membr Biol 239(1–2):85–93
106. Bao Z, Qi X, Hong S, Xu K, He F, Zhang M, Chen J, Chao D, Zhao W, Li D, Wang J, Zhang P (2017) Structure and mechanism of a group-I cobalt energy coupling factor transporter. Cell Res 27(5):675–687
107. Moitra K (2015) Overcoming multidrug resistance in cancer stem cells. Biomed Res Int 2015:635745
108. Wang J, Yan C, Li Y, Hirata K, Yamamoto M, Yan N, Hu Q (2014) Crystal structure of a bacterial homologue of SWEET transporters. Cell Res 24(12):1486–1489
109. Lee Y, Nishizawa T, Yamashita K, Ishitani R, Nureki O (2015) Structural basis for the facilitative diffusion mechanism by SemiSWEET transporter. Nat Commun 6:6112
110. Xu Y, Tao Y, Cheung LS, Fan C, Chen LQ, Xu S, Perry K, Frommer WB, Feng L (2014) Structures of bacterial homologues of SWEET transporters in two distinct conformations. Nature 515(7527):448–452
111. Feng L, Frommer WB (2015) Structure and function of SemiSWEET and SWEET sugar transporters. Trends Biochem Sci 40(8):480–486
112. Santos JS, Asmar-Rovira GA, Han GW, Liu W, Syeda R, Cherezov V, Baker KA, Stevens RC, Montal M (2012) Crystal structure of a voltage-gated K+ channel pore module in a closed state in lipid membranes. J Biol Chem 287(51):43063–43070
113. Takeda H, Hattori M, Nishizawa T, Yamashita K, Shah ST, Caffrey M, Maturana AD, Ishitani R, Nureki O (2014) Structural basis for ion selectivity revealed by high-resolution crystal structure of Mg^{2+} channel MgtE. Nat Commun 5:5374

114. Liao J, Marinelli F, Lee C, Huang Y, Faraldo-Gómez JD, Jiang Y (2016) Mechanism of extracellular ion exchange and binding-site occlusion in the sodium-calcium exchanger. Nat Struct Mol Biol 23(6):590–599
115. Liao J, Li H, Zeng W, Sauer DB, Belmares R, Jiang Y (2012) Structural insight into the ion-exchange mechanism of the sodium/calcium exchanger. Science 335(6069):686–690
116. Waight AB, Pedersen BP, Schlessinger A, Bonomi M, Chau BH, Roe-Zurz Z, Risenmay AJ, Sali A, Stroud RM (2013) Structural basis for alternating access of a eukaryotic calcium/proton exchanger. Nature 499(7456):107–110
117. Su M, Gao F, Yuan Q, Mao Y, Li DL, Guo Y, Yang C, Wang XH, Bruni R, Kloss B, Zhao H, Zeng Y, Zhang FB, Marks AR, Hendrickson WA, Chen YH (2017) Structural basis for conductance through TRIC cation channels. Nat Commun 8:15103
118. Kasuya G, Hiraizumi M, Maturana AD, Kumazaki K, Fujiwara Y, Liu K, Nakada-Nakura Y, Iwata S, Tsukada K, Komori T, Uemura S, Goto Y, Nakane T, Takemoto M, Kato HE, Yamashita K, Wada M, Ito K, Ishitani R, Hattori M, Nureki O (2016) Crystal structures of the TRIC trimeric intracellular cation channel orthologues. Cell Res 26(12):1288–1301
119. Yang H, Hu M, Guo J, Ou X, Cai T, Liu Z (2016) Pore architecture of TRIC channels and insights into their gating mechanism. Nature 538(7626):537–541
120. Shi Y, Burn P (2004) Lipid metabolic enzymes: emerging drug targets for the treatment of obesity. Nat Rev Drug Discov 3(8):695–710
121. Vogeley L, El Arnaout T, Bailey J, Stansfeld PJ, Boland C, Caffrey M (2016) Structural basis of lipoprotein signal peptidase II action and inhibition by the antibiotic globomycin. Science 351(6275):876–880
122. Sciara G, Clarke OB, Tomasek D, Kloss B, Tabuso S, Byfield R, Cohn R, Banerjee S, Rajashankar KR, Slavkovic V, Graziano JH, Shapiro L, Mancia F (2014) Structural basis for catalysis in a CDP-alcohol phosphotransferase. Nat Commun 5:4068
123. Clarke OB, Tomasek D, Jorge CD, Dufrisne MB, Kim M, Banerjee S, Rajashankar KR, Shapiro L, Hendrickson WA, Santos H, Mancia F (2015) Structural basis for phosphatidylinositol-phosphate biosynthesis. Nat Commun 0:8505
124. Bai Y, McCoy JG, Levin EJ, Sobrado P, Rajashankar KR, Fox BG, Zhou M (2015) X-ray structure of a mammalian stearoyl-CoA desaturase. Nature 524(7564):252–256
125. Lorch M, Faham S, Kaiser C, Weber I, Mason AJ, Bowie JU, Glaubitz C (2005) How to prepare membrane proteins for solid-state NMR: a case study on the alpha-helical integral membrane protein diacylglycerol kinase from E. coli. Chembiochem A Eur J Chem Biol 6 (9):1693–1700
126. Lau FW, Chen X, Bowie JU (1999) Active sites of diacylglycerol kinase from Escherichia coli are shared between subunits. Biochemistry 38(17):5521–5527
127. Li D, Stansfeld PJ, Sansom MS, Keogh A, Vogeley L, Howe N, Lyons JA, Aragao D, Fromme P, Fromme R, Basu S, Grotjohann I, Kupitz C, Rendek K, Weierstall U, Zatsepin NA, Cherezov V, Liu W, Bandaru S, English NJ, Gati C, Barty A, Yefanov O, Chapman HN, Diederichs K, Messerschmidt M, Boutet S, Williams GJ, Marvin Seibert M, Caffrey M (2015) Ternary structure reveals mechanism of a membrane diacylglycerol kinase. Nat Commun 6:10140
128. Badola P, Sanders CR 2nd (1997) Escherichia coli diacylglycerol kinase is an evolutionarily optimized membrane enzyme and catalyzes direct phosphoryl transfer. J Biol Chem 272 (39):24176–24182
129. Koeberle A, Werz O (2015) Perspective of microsomal prostaglandin E2 synthase-1 as drug target in inflammation-related disorders. Biochem Pharmacol 98(1):1–15
130. Weinert T, Olieric V, Waltersperger S, Panepucci E, Chen L, Zhang H, Zhou D, Rose J, Ebihara A, Kuramitsu S, Li D, Howe N, Schnapp G, Pautsch A, Bargsten K, Prota AE, Surana P, Kottur J, Nair DT, Basilico F, Cecatiello V, Pasqualato S, Boland A, Weichenrieder O, Wang B-C, Steinmetz MO, Caffrey M, Wang M (2015) Fast native-SAD phasing for routine macromolecular structure determination. Nat Methods 12 (2):131–133

131. Dobrzyn A, Ntambi JM (2005) Stearoyl-CoA desaturase as a new drug target for obesity treatment. Obes Rev Off J Int Assoc Study Obes 6(2):169–174
132. Katona G, Andreasson U, Landau EM, Andreasson LE, Neutze R (2003) Lipidic cubic phase crystal structure of the photosynthetic reaction centre from Rhodobacter sphaeroides at 2.35 Å resolution. J Mol Biol 331(3):681–692
133. Charrin S, Jouannet S, Boucheix C, Rubinstein E (2014) Tetraspanins at a glance. J Cell Sci 127(Pt 17):3641–3648
134. Suzuki H, Nishizawa T, Tani K, Yamazaki Y, Tamura A, Ishitani R, Dohmae N, Tsukita S, Nureki O, Fujiyoshi Y (2014) Crystal structure of a claudin provides insight into the architecture of tight junctions. Science 344(6181):304–307
135. Zimmerman B, Kelly B, McMillan BJ, Seegar TC, Dror RO, Kruse AC, Blacklow SC (2016) Crystal structure of a full-length human tetraspanin reveals a cholesterol-binding pocket. Cell 167(4):1041–1051 e1011
136. Joh NH, Wang T, Bhate MP, Acharya R, Wu Y, Grabe M, Hong M, Grigoryan G, DeGrado WF (2014) De novo design of a transmembrane Zn(2+)-transporting four-helix bundle. Science 346(6216):1520–1524
137. Trenker R, Call ME, Call MJ (2015) Crystal structure of the glycophorin a transmembrane dimer in lipidic cubic phase. J Am Chem Soc 137(50):15676–15679
138. Thomaston JL, Alfonso-Prieto M, Woldeyes RA, Fraser JS, Klein ML, Fiorin G, DeGrado WF (2015) High-resolution structures of the M2 channel from influenza A virus reveal dynamic pathways for proton stabilization and transduction. Proc Natl Acad Sci U S A 112(46):14260–14265
139. Takeda K, Kaisho T, Akira S (2003) Toll-like receptors. Annu Rev Immunol 21:335–376
140. Barclay AN (2003) Membrane proteins with immunoglobulin-like domains—a master superfamily of interaction molecules. Semin Immunol 15(4):215–223
141. Waters JP, Pober JS, Bradley JR (2013) Tumour necrosis factor and cancer. J Pathol 230 (3):241–248
142. Hirayasu K, Arase H (2015) Functional and genetic diversity of leukocyte immunoglobulin-like receptor and implication for disease associations. J Hum Genet 60 (11):703–708
143. Dolan J, Walshe K, Alsbury S, Hokamp K, O'Keeffe S, Okafuji T, Miller SF, Tear G, Mitchell KJ (2007) The extracellular Leucine-Rich repeat superfamily; a comparative survey and analysis of evolutionary relationships and expression patterns. BMC Genom 8(1):320
144. Klug L, Daum G (2014) Yeast lipid metabolism at a glance. FEMS Yeast Res 14(3): 369–388
145. Zanger UM, Schwab M (2013) Cytochrome P450 enzymes in drug metabolism: regulation of gene expression, enzyme activities, and impact of genetic variation. Pharmacol Ther 138 (1):103–141
146. Lomize AL, Lomize MA, Krolicki SR, Pogozheva ID (2017) Membranome: a database for proteome-wide analysis of single-pass membrane proteins. Nucleic Acids Res 45 (Database issue):D250–D255
147. Schmidt HR, Zheng S, Gurpinar E, Koehl A, Manglik A, Kruse AC (2016) Crystal structure of the human sigma1 receptor. Nature 532(7600):527–530
148. Bourret RB, Silversmith RE (2010) Two-component signal transduction. Curr Opin Microbiol 13(2):113–115
149. Bhate Manasi P, Molnar Kathleen S, Goulian M, DeGrado William F (2015) Signal transduction in histidine kinases: insights from new structures. Structure 23(6):981–994
150. Gushchin I, Melnikov I, Polovinkin V, Ishchenko A, Yuzhakova A, Buslaev P, Bourenkov G, Grudinin S, Round E, Balandin T, Borshchevskiy V, Willbold D, Leonard G, Buldt G, Popov A, Gordeliy V (2017) Mechanism of transmembrane signaling by sensor histidine kinases. Science 356(6342):1043–1049
151. Kumazaki K, Kishimoto T, Furukawa A, Mori H, Tanaka Y, Dohmae N, Ishitani R, Tsukazaki T, Nureki O (2014) Crystal structure of Escherichia coli YidC, a membrane protein chaperone and insertase. Sci Rep 4:7299

152. Kumazaki K, Chiba S, Takemoto M, Furukawa A, K-i Nishiyama, Sugano Y, Mori T, Dohmae N, Hirata K, Nakada-Nakura Y, Maturana AD, Tanaka Y, Mori H, Sugita Y, Arisaka F, Ito K, Ishitani R, Tsukazaki T, Nureki O (2014) Structural basis of Sec-independent membrane protein insertion by YidC. Nature 509(7501):516–520
153. Tanaka Y, Sugano Y, Takemoto M, Mori T, Furukawa A, Kusakizako T, Kumazaki K, Kashima A, Ishitani R, Sugita Y, Nureki O, Tsukazaki T (2015) Crystal structures of SecYEG in lipidic cubic phase elucidate a precise resting and a peptide-bound state. Cell Rep 13(8):1561–1568
154. Efremov RG, Sazanov LA (2012) Structure of Escherichia coli OmpF porin from lipidic mesophase. J Struct Biol 178(3):311–318
155. Fairman JW, Dautin N, Wojtowicz D, Liu W, Noinaj N, Barnard TJ, Udho E, Przytycka TM, Cherezov V, Buchanan SK (2012) Crystal structures of the outer membrane domain of intimin and invasin from enterohemorrhagic E. coli and enteropathogenic Y. pseudotuberculosis. Structure 20(7):1233–1243
156. Cheng A, Hummel B, Qiu H, Caffrey M (1998) A simple mechanical mixer for small viscous lipid-containing samples. Chem Phys Lipids 95(1):11–21
157. McPherson A, Gavira JA (2014) Introduction to protein crystallization. Acta Crystallogr F 70(Pt 1):2–20
158. Cherezov V, Caffrey M (2003) Nano-volume plates with excellent optical properties for fast, inexpensive crystallization screening of membrane proteins. J Appl Crystallogr 36(6): 1372–1377
159. Cherezov V, Caffrey M (2006) Picolitre-scale crystallization of membrane proteins. J Appl Crystallogr 39(4):604–606
160. Kissick DJ, Wanapun D, Simpson GJ (2011) Second-order nonlinear optical imaging of chiral crystals. Annu Rev Anal Chem 4:419–437
161. Liu W, Hanson MA, Stevens RC, Cherezov V (2010) LCP-Tm: an assay to measure and understand stability of membrane proteins in a membrane environment. Biophys J 98 (8):1539–1548
162. Cherezov V, Abola E, Stevens RC (2010) Toward drug design: recent progress in the structure determination of GPCRs, a membrane protein family with high potential as pharmaceutical targets. Methods Mol Biol 654:141–168
163. Olieric V, Weinert T, Finke AD, Anders C, Li D, Olieric N, Borca CN, Steinmetz MO, Caffrey M, Jinek M, Wang M (2016) Data-collection strategy for challenging native SAD phasing. Acta Crystallogr D 72(3):421–429
164. Cherezov V, Hanson MA, Griffith MT, Hilgart MC, Sanishvili R, Nagarajan V, Stepanov S, Fischetti RF, Kuhn P, Stevens RC (2009) Rastering strategy for screening and centring of microcrystal samples of human membrane proteins with a sub-10 μm size X-ray synchrotron beam. J R Soc Interface 6(Suppl 5):S587–S597
165. Chapman HN, Caleman C, Timneanu N (2014) Diffraction before destruction. Philos Trans R Soc B 369(1647):20130313
166. Caffrey M, Li D, Howe N, Shah STA (2014) 'Hit and run' serial femtosecond crystallography of a membrane kinase in the lipid cubic phase. Philos Trans R Soc B 369 (1647):20130621
167. Weierstall U, Spence JC, Doak RB (2012) Injector for scattering measurements on fully solvated biospecies. Rev Sci Instrum 83(3):035108
168. Weierstall U (2014) Liquid sample delivery techniques for serial femtosecond crystallography. Philos Trans R Soc B 369(1647):20130337
169. Chapman HN, Fromme P, Barty A, White TA, Kirian RA, Aquila A, Hunter MS, Schulz J, DePonte DP, Weierstall U, Doak RB, Maia FRNC, Martin AV, Schlichting I, Lomb L, Coppola N, Shoeman RL, Epp SW, Hartmann R, Rolles D, Rudenko A, Foucar L, Kimmel N, Weidenspointner G, Holl P, Liang M, Barthelmess M, Caleman C, Boutet S, Bogan MJ, Krzywinski J, Bostedt C, Bajt S, Gumprecht L, Rudek B, Erk B, Schmidt C, Homke A, Reich C, Pietschner D, Struder L, Hauser G, Gorke H, Ullrich J, Herrmann S, Schaller G, Schopper F, Soltau H, Kuhnel K-U, Messerschmidt M, Bozek JD, Hau-Riege SP, Frank M,

Hampton CY, Sierra RG, Starodub D, Williams GJ, Hajdu J, Timneanu N, Seibert MM, Andreasson J, Rocker A, Jonsson O, Svenda M, Stern S, Nass K, Andritschke R, Schroter C-D, Krasniqi F, Bott M, Schmidt KE, Wang X, Grotjohann I, Holton JM, Barends TRM, Neutze R, Marchesini S, Fromme R, Schorb S, Rupp D, Adolph M, Gorkhover T, Andersson I, Hirsemann H, Potdevin G, Graafsma H, Nilsson B, Spence JCH (2011) Femtosecond X-ray protein nanocrystallography. Nature 470(7332):73–77

170. Liu W, Ishchenko A, Cherezov V (2014) Preparation of microcrystals in lipidic cubic phase for serial femtosecond crystallography. Nat Protoc 9(9):2123–2134

171. Liu W, Wacker D, Gati C, Han GW, James D, Wang D, Nelson G, Weierstall U, Katritch V, Barty A, Zatsepin NA, Li D, Messerschmidt M, Boutet S, Williams GJ, Koglin JE, Seibert MM, Wang C, Shah STA, Basu S, Fromme R, Kupitz C, Rendek KN, Grotjohann I, Fromme P, Kirian RA, Beyerlein KR, White TA, Chapman HN, Caffrey M, Spence JCH, Stevens RC, Cherezov V (2013) Serial femtosecond crystallography of G protein-coupled receptors. Science 342(6165):1521–1524

172. Zhang H, Han GW, Batyuk A, Ishchenko A, White KL, Patel N, Sadybekov A, Zamlynny B, Rudd MT, Hollenstein K, Tolstikova A, White TA, Hunter MS, Weierstall U, Liu W, Babaoglu K, Moore EL, Katz RD, Shipman JM, Garcia-Calvo M, Sharma S, Sheth P, Soisson SM, Stevens RC, Katritch V, Cherezov V (2017) Structural basis for selectivity and diversity in angiotensin II receptors. Nature 544(7650):327–332

173. Zhang H, Unal H, Gati C, Han GW, Liu W, Zatsepin NA, James D, Wang D, Nelson G, Weierstall U, Sawaya MR, Xu Q, Messerschmidt M, Williams GJ, Boutet S, Yefanov OM, White TA, Wang C, Ishchenko A, Tirupula KC, Desnoyer R, Coe J, Conrad CE, Fromme P, Stevens RC, Katritch V, Karnik SS, Cherezov V (2015) Structure of the angiotensin receptor revealed by serial femtosecond crystallography. Cell 161(4):833–844

174. Zhang H, Unal H, Desnoyer R, Han GW, Patel N, Katritch V, Karnik SS, Cherezov V, Stevens RC (2015) Structural basis for ligand recognition and functional selectivity at angiotensin receptor. J Biol Chem 290(49):29127–29139

Chapter 8
Electron Microscopic Analysis of the Plasma Membrane and Cell Surface Molecules

Haishuang Chang, Longxing Cao and Yongning He

8.1 Introduction

Cell membrane is a super complex system composed of various molecules such as proteins, lipids, and carbohydrates, and the organization of biological macromolecules on cell membrane plays important roles in the functional activities of membrane systems. Although many biophysical techniques have been developed to characterize cell membrane and the associated molecules, direct visualization of the macromolecules on cell surface at high resolution is still a big challenge. Electron microscopy (EM) is one of the major techniques for visualizing biological structures at different levels. An advantage of EM over other biophysical methods is that it can reach high resolution and provide atomic details of the target molecules. For purified proteins, cryo-electron microscopy is now able to determine the structures at atomic or near-atomic resolution [1, 2]. However, the structural characterization for membrane systems is much more difficult because cell membrane and the associated molecules are not homogeneous and could be extremely dynamic as well. One of the major difficulties is how to preserve cellular samples well enough to retain the structural details. Since cell or tissue samples are usually much thicker than 1–2 μm, they could not be visualized directly by transmission electron microscopy (TEM) [3]. Therefore, the samples need to be immobilized and sectioned into thin slices before EM imaging. The immobilization can be done by either chemical fixation or cryo-fixation. The conventional chemical fixation is commonly used for cellular or tissue samples to visualize structural features at relative low resolution, as it may induce artifacts such as aggregation of proteins,

---

H. Chang · L. Cao · Y. He (✉)
State Key Laboratory of Molecular Biology, National Center for Protein Science Shanghai, Shanghai Science Research Center, CAS Center for Excellence in Molecular Cell Science, Shanghai Institute of Biochemistry and Cell Biology, Chinese Academy of Sciences, 333 Haike Road, Shanghai 201210, China
e-mail: he@sibcb.ac.cn

H. Wang and G. Li (eds.), *Membrane Biophysics*,
https://doi.org/10.1007/978-981-10-6823-2_8

collapse of highly hydrated glycans, and loss of lipids [4–10]. To explore the high-resolution details of a specimen close to its living state, cryo-fixation is admittedly to be a better immobilization method [11]. Cryo-fixation can be done in two different ways. One is relatively straightforward by plunging the specimen on an EM grid directly into liquid ethane cooling down to liquid nitrogen temperature and freezing in milliseconds [12, 13]. However, in the best cases, plunge freezing can only vitrify specimen to a few microns from their surfaces [14–16] and is suitable for the purified proteins, viruses, liposomes, or small cells which are thin enough for vitrifying. The other cryo-fixation method is called high-pressure freezing (HPF) achieved by freezing the specimens with liquid nitrogen under a pressure of 2100 bar. HPF can freeze non-cryoprotected samples up to 0.5 mm in thickness without significant ice crystal damage [17], therefore suitable for the high-quality preservation of cellular and tissue samples.

Here, we will discuss the basic principles and sample preparation methods of TEM imaging and reconstruction for biological samples, and also provide protocols for visualizing the plasma membrane and surface molecules.

8.2 TEM Imaging, Data Collection, and 3D Reconstruction for Biological Samples

8.2.1 TEM Imaging of Biological Samples

As an important imaging technique, TEM has been widely used in physics, material sciences, and biological sciences. Unlike inorganic or metallic materials, biological samples, which are essentially composed of proteins, lipids, carbohydrates, nucleic acids, and water, are more sensitive to temperature, vacuum, and electron dose and give lower imaging contrast. Therefore, the sample preparation procedures, microscope settings, and imaging conditions need to be adapted and optimized for biological samples in different cases. Up-to-date, two imaging strategies are commonly used for biological samples, one is called single particle analysis, and the other is called tomography. Single particle analysis is used for the three-dimensional reconstruction of macromolecules and able to reveal high-resolution details for the purified homogenous samples through the averaging among the selected particles [18]. By contrast, tomography is mostly used for reconstructing 3D structures of relatively larger objects with or without averaging, thus can be applied to heterogeneous samples [19]. Over the past few decades, the hardware and software of TEM have been developed quickly toward the high-resolution imaging of biological samples. A number of protein structures have been determined at atomic resolution by the single particle method in recent years, suggesting that it has become a powerful tool to visualize the high-resolution details of macromolecules as X-ray crystallography [20, 21]. In fact, EM reconstruction has several advantages in structural determination. It usually requires much less sample than crystal screening, and the images can be taken under physiological conditions instead of the crystallization buffers. The structural determination procedure by EM is relatively straightforward without the phase problem in crystallography [22]. However, EM reconstruction could be time-consuming depending on the quality of the samples and the images. As biological samples usually have low contrast under EM, small proteins would be more challenging to achieve high resolution than large proteins.

The structural determination of membrane proteins by EM is similar to the water-soluble proteins. A major problem for membrane protein imaging is that they are usually purified in the presence of detergents, which may affect the sample preparation and give high imaging background. In recent years, amphipathic polymers such as amphipols are utilized to stabilize membrane proteins in aqueous solutions instead of detergents, which improve the imaging quality significantly and are able to achieve high resolution in many cases [23–25]. Meanwhile, nanodisks are also used for stabilizing membrane proteins in solution by modeling the native membrane environment, which may help to avoid artifacts in structural determination [26–28]. Overall, sample preparation and imaging quality play key roles in the TEM analysis of biological samples, which need to be optimized in different cases for obtaining high-resolution information.

8.2.2 *Principle and Application of Single Particle Analysis*

The basic idea of single particle 3D reconstruction is to build a 3D model through a series of 2D projections of the target object based on the central section theorem which states that the Fourier transformation of a 2D projection of a 3D object in real space along a certain angle is equal to the central section that is perpendicular to the projection vector of the Fourier transformation of the 3D object. For TEM imaging, electron beams penetrate the specimen and generate 2D projections perpendicular to the direction of the electron beam; therefore, the 3D structure of the target macromolecules can be reconstructed from the 2D projections with different orientations (Fig. 8.1). In recent years, the single particle analysis has become an important technique for obtaining high-resolution structures of biological macromolecules due to the development of both EM equipment and reconstruction algorithms, and can be applied to the structural characterization of large complexes of macromolecules or membrane proteins, which are usually difficult to crystallize.

Single particle reconstruction is generated by averaging thousands of images of identical particles with different orientations; therefore, sample homogeneity is important for achieving high-resolution reconstruction (Fig. 8.2). Protein samples are commonly expressed and purified from different systems, such as bacteria, yeast, insect cells, or mammalian cells. The complexes with multi-subunits can also be isolated from cells or tissues. In some cases, a large complex sample can be a mixture of sub-complexes, and the compositionally homogeneous complexes may have different conformations; therefore, protein samples need to be analyzed by gel-filtration chromatography and SDS-PAGE for the homogeneity and purity, and sample heterogeneity should be minimized to simplify the reconstruction. An

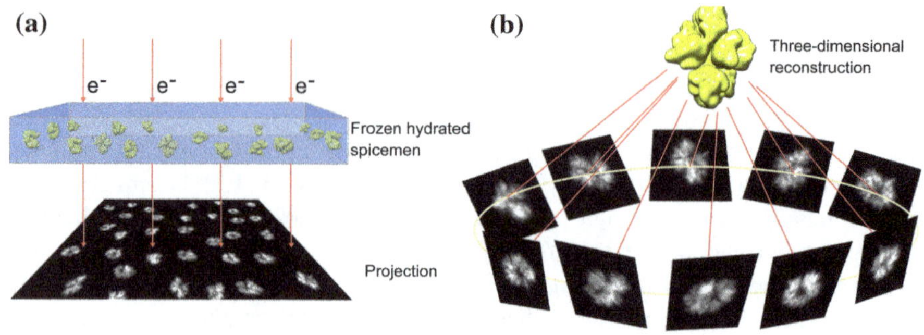

Fig. 8.1 Principle of cryoEM single particle reconstruction. 2D projections of particles are collected by TEM (**a**). A 3D model is reconstructed from the 2D projections (**b**)

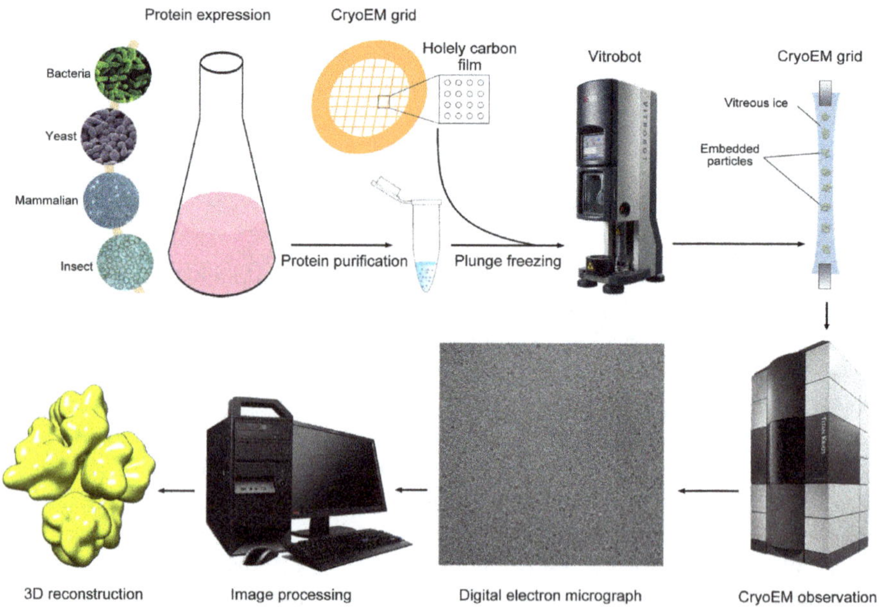

Fig. 8.2 A flowchart of cryoEM single particle reconstruction

alternative way to judge the quality of a protein sample is to visualize it by negative-stain EM, which can provide the images of the single particles with high contrast (Fig. 8.3). Although current image processing software is capable of differentiating sub-complexes or conformational states of the samples, improving homogeneity during sample preparation will reduce the effort for reconstructing structures at high resolution.

Since hydrated biological samples can be damaged under high vacuum or by exposing to the high-energy electrons, they need to be fixed before TEM imaging.

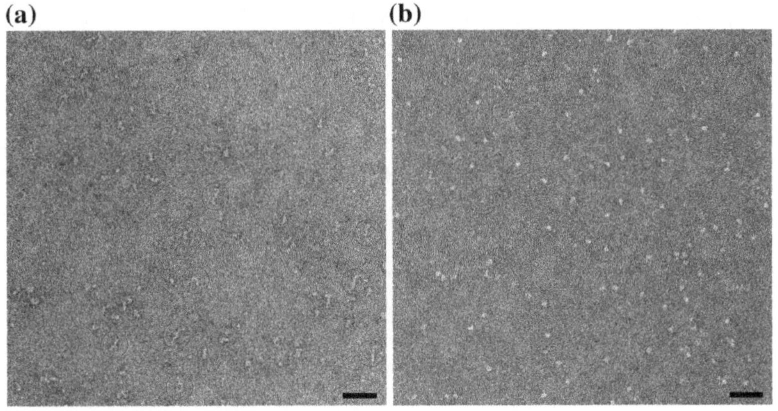

Fig. 8.3 Negatively stained EM images of a cell surface protein (CD205, blue or red circle) showing the different conformations at basic (**a**) and acidic (**b**) pH (bar, 50 nm)

For protein samples, cryo-fixation is the most commonly used preparation technique for single particle analysis. Purified proteins are loaded onto a holey carbon film supported by a copper grid, which is usually glow-discharged to generate a hydrophilic surface. Then, after sample loading, the excess sample is removed by blotting with filter papers. Blotting time is critical for the thickness of ice layer in the holes. Ideally, the ice needs to be thick enough to keep particles in one layer with the same focus, but not too thick, which will reduce the imaging contrast. After blotting, the grid is frozen rapidly by plunging into liquid ethane at liquid nitrogen temperature. This procedure can be done on semi-automated plungers, for example, Vitrobot from FEI, which makes the freezing process much easier and more reproducible. Then the grids can be transferred to cryo-boxes for storage or ready for loading onto an EM cryo-holder for imaging.

Before data collection, the microscope should be aligned with care and the imaging conditions need be chosen properly, including coma-free alignment, condenser aperture, spot size, objective aperture. On the other hand, biological samples usually produce low contrast against background, thus imaging conditions such as defocus and total dose must be optimized to improve the image contrast. A recent option for enhancing image contrast is the phase plate technique, which has been shown to able to improve the image contrast significantly and reduce the number of particles that have to be averaged during single particle reconstruction [29]. Since biological samples are vulnerable to electron beams, cryoEM data collection is usually performed under the "low dose mode" (typically <20 electrons/Å^2), which can be divided into three steps: searching, focusing, and exposure. The searching step is to locate an area of interest on the grid with good particle distribution and ice thickness, and it is typically done at low magnification. During the focusing step, a nearby area is chosen and used for focusing at high magnification. CryoEM images are typically recorded with a small defocus value (1.5–3.0 μm for large proteins or complexes, 2.0–4.5 μm for small proteins) to improve the image contrast and facilitate the subsequent image processing procedures. After focusing, images can be recorded in the exposure mode with the desired magnification. Recently, the direct electron detectors, such as Falcon (FEI) and K2 (Gatan), which have high detective quantum efficiency (DQE), can take movies during the exposure, thus allowing the observation of radiation damage and correction of the beam-induced movement during image processing. Therefore, higher dose can be applied during exposure if these detectors are used. Typically, hundreds or thousands of images are taken for 3D reconstructions, and current microscopes are usually equipped with high-throughput data collection software for image recording.

Single particle analysis is a multi-step process includes motion correction (if applicable), CTF correction, particle picking, 2D classification, initial model generation, 3D classification, 3D refinement, resolution assessment, and interpretation of the final map. A number of software packages have been developed over the past years for single particle reconstruction. Among them, EMAN2 [30], FREALIGN [31], RELION [32], and SPARX [33] are the most popular ones. EMAN2 has extensive options and flexibilities. RELION is widely used for high-resolution structural refinement. Due to the development of direct electron detectors, more and more cryoEM images are recorded in movie mode, where the total electron dose is fractionated into a series of image frames that can be aligned to compensate for specimen drift and beam-induced movement. For images collected in movie mode, the motion correction should be carried out, and the images with extremely low contrast or high contamination should be discarded. Since TEM images are distorted by the systematic aberrations of the microscope and defocus aberration, a contrast transfer function (CTF) needs to be calculated to correct the perturbation. After CTF correction, the particles in each micrograph should be picked out, which might be labor-intensive if picking is done manually. But automated particle picking is also available in software packages such as EMAN2. The quality of the selected particles can affect the subsequent data analysis, as poor particles may introduce higher noise to the structural refinement. The selected particles are aligned and averaged for 2D classification, which is based on k-means or maximum likelihood algorithm, and can be applied to remove poor particles as well as assess the conformational heterogeneity of the sample. To calculate 3D reconstructions, an initial model is required for refinement, which can be generated either from existing structures or form picked particles by common line or random conical tilt method. For heterogeneous samples, 3D classification might be applied to extract more homogeneous subsets, and particles are assigned to different 3D classes representing different conformations, thus allowing the subsequent reconstruction of the 3D models based only on a subset of the particles, and it could be time-consuming in the whole procedure. The 3D model refinement is done by iterative optimization of the rotational and translational parameters of the selected particles, and a new 3D map is calculated after each refinement cycle (Fig. 8.4). To evaluate the resolution of the final model, the selected particles are split into two halves, and the refinement is performed for each half independently. Then the resulting 3D reconstructions are compared in Fourier space by the Fourier shell correlation (FSC) curve, which represents the cross-correlation coefficient between the two structures as a function

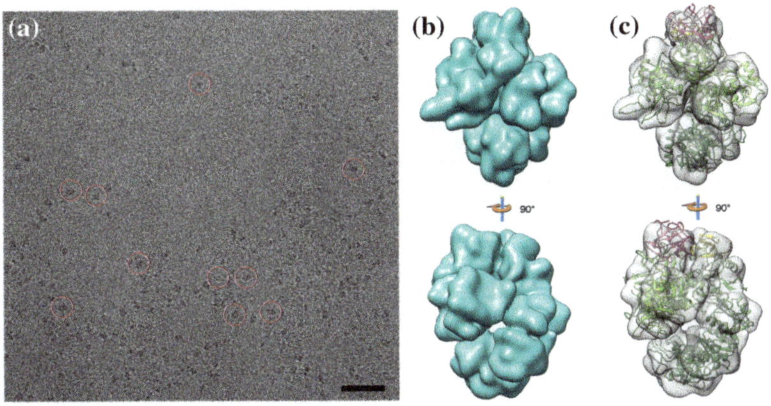

Fig. 8.4 CryoEM image (**a**) of a cell surface protein (CD205, red circle) and the 3D reconstruction of the protein (**b**). The fitting model of the cryoEM reconstruction is shown in (**c**). (bar, 50 nm)

of spatial frequency. The resolution of the model is usually claimed by choosing a threshold value on the FSC curve. The most popular value used for many cryoEM structures determined recently is 0.143. However, the resolution of a cryoEM structure could be anisotropic and the resolution of the flexible regions can be much lower than the rigid core of the macromolecules. The validation of 3D reconstructions can be done by fitting existing crystal structures or by tilt-pair analysis [34]. For atomic or near-atomic-resolution reconstructions, the validation can be done by fitting the atomic structure of amino acid residues into the cryoEM volumes.

8.2.3 Electron Tomography for Biological Samples

As described above, electron tomography (ET) can be applied for 3D structural reconstruction of objects without averaging. The principle of ET has been proposed nearly a century ago [35, 36]. Briefly, by collecting a series of 2D projections of an object at different tilt angles, the 3D structure of the object can be reconstructed from the 2D images based on the central slice theorem [35, 37]. Although the 3D structure of the object can be reconstructed perfectly from 2D images in theory, it is difficult to achieve in practice. For ET data collection, the biological sample on EM grid is loaded onto a specimen holder and inserted into the EM column, and a series of images of projection can be collected by tilting the specimen holder step by step with a certain increment (typically 0.5°–2°). The regular specimen holder cannot tilt up to 90°, therefore would lead to a "missing wedge" in the sampling space [38, 39]. For most of the samples such as thin sections or cryo-samples, the thickness of the sample for the electron beam is also increasing with the tilt angles and may reduce the signal–noise ratio of the images. The tilting angles usually range from

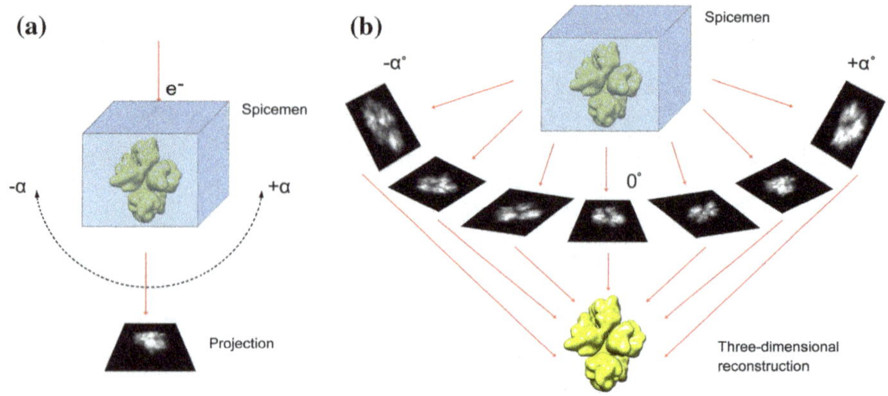

Fig. 8.5 Principle of electron tomography. Tile images are collected by rotating the specimen between $-\alpha$ and $+\alpha$ degree (**a**). A 3D model is reconstructed from the 2D tilting images (**b**)

$-65°$ to $65°$, and the electron dose also needs to be adjusted depending on the sample (cryo-samples or thin sections) as well as the number of images in the tilt series (Fig. 8.5) [40]. To improve the alignment of the tilt images, gold particles (5–10 nm) could be added during sample preparation [41]. ET data collection is usually software-controlled and can be carried out for either cryo-samples or thin sections. If necessary, a second tilt series can be collected by rotating the EM grid on the specimen holder by 90°, which could reduce the "missing wedge" in the sampling space and achieve better tomograms in some cases [42, 43]. A number of software, for example, IMOD, have been developed for the alignment of the tile images and 3D tomogram generation [44].

Membrane samples such as the cell membrane or organelles are usually heterogeneous and difficult to reconstruct by single particle analysis. Therefore, ET would be an appropriate method to generate 3D models for membrane samples at nanometer resolution (2–10 nm). In fact, the macromolecules in 3D tomogram can be averaged in some cases to improve resolution, thus revealing the details of these molecules close to the physiological state. Nevertheless, the quality of final tomograms is largely determined by sample preparation, which will be discussed in the following sections.

8.3 Principle and Application of the Cryo-Fixation Methods for Membrane Samples

8.3.1 Plunge Freezing

Although most of the biological samples including membranes are thicker than 1–2 μm, there are thin samples such as liposomes, organelles, and small cells.

In these cases, plunge freezing can be applied for ice vitrification and is much easier than the high-pressure freezing method described in the next section [45]. The procedure of plunge freezing for membrane samples is quite straightforward by submerging the samples on EM grids into the liquid ethane cooled at liquid nitrogen temperature [13], which is similar to the procedure that has been used routinely in the cryoEM single particle analysis for protein samples. Plunge freezing can be performed with either custom-made manually controlled plunge freezer or with automatic plunge freezers that are commercially available. The freezing conditions such as blotting time, humidity can be adjusted for optimizing freezing conditions with automatic plunge freezers. Several types of EM grids can be used for cryo-imaging or tomographic data collection. Among them, quantifoil grids, lacey carbon grids, C-flat carbon grids, or homemade holey carbon grids are the most common ones. The choice of grids depends on the size of the samples as well as the imaging conditions. Since membrane samples are usually larger than protein samples, grids with larger holes might be preferred for data collection. For example, grids with smaller holes are suitable for membrane proteins, small vesicles, liposomes, while grids with larger holes might be used for large objects such as organelles or cells.

8.3.2 High-Pressure Freezing and Freeze Substitution

For cryo-fixation of cells or living tissues, freezing at low cooling rates (e.g., in a freezer) may lead to the formation of ice crystals in the cytoplasm which would be disastrous to the integrity of cellular structure [46–48]. In the worst cases, the growing ice crystals may poke holes into cellular membranes and lead to the leakage of the cytoplasm and the organelles [49]. To preserve ultrastructure close to the living state, samples need to be frozen fast enough and turned the intrinsic water into vitreous ice and prevented the formation of ice crystals [50, 51]. Due to the low heat conductivity of water, plunge freezing in liquid ethane is an option for water vitrification for very thin samples and has been used successfully in the cryo-single particle analysis and cryo-electron tomography [12, 13, 52]. But plunge freezing cannot generate satisfactory preservation of cells or tissues whose thickness is more than 20 μm [53]. To overcome the limitation, a rapid freezing method under a high pressure (2100 bar) has been developed. At about 2045 bar, the melting point of water is depressed to about −22 °C and the super cooling zone is reduced to −90 °C (Fig. 8.6) [54], thus reducing the rate of ice crystal formation and also slowing down the growth of ice crystals [53, 55–58]. Since air bubbles in specimens may collapse or explode under high pressure, intercellular space should be filled with proper fillers such as 1-hexadecene, 20% ficoll, 20% dextran, or 20% BSA before freezing [8, 41]. Once the whole procedure is performed carefully, HPF can generate good preservation for samples with thickness up to 0.5 mm [59].

After HPF, freeze substitution (FS) needs to be carried out, which is a hybrid method that combines the fine structural preservation of cryo-fixation with resin embedding [60, 61]. During the substitution, the cellular water is replaced by an

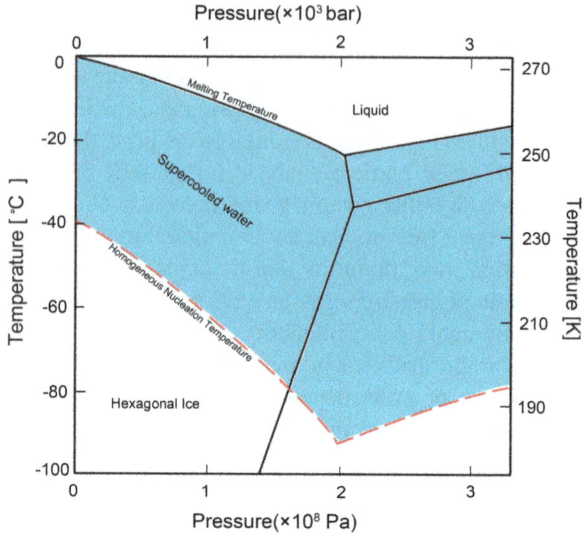

Fig. 8.6 The pressure/temperature phase diagram of water. At about 2045 bar, the melting point of water is −22 °C and the super cooling zone is −90 °C. The blue area shows the region of the super-cooled water [54]

organic solvent such as acetone or methanol at temperatures ranging from −78 to −90 °C. In the meantime, the fixatives, including osmium tetroxide and glutaraldehyde, are dissolved in the solvent and diffused into the samples while they are still frozen under low temperature. When the samples start warming up, the proteins, lipids, and other constituents will be cross-linked by osmium tetroxide and glutaraldehyde at −70 and −40 °C, respectively [62–65]. Finally, the dehydrated samples are embedded in resin and thin-sectioned by a microtome for TEM imaging (Fig. 8.7).

8.3.3 Preservation of the Cell Membrane by HPF

As described above, HPF/FS can preserve cellular ultrastructure close to the living state. The membrane systems in the cytosol, for example, the mitochondria membrane, the nuclear membrane, and the membrane of the Golgi apparatus can be well preserved and visualized clearly (Fig. 8.8a, c). However, the lipid bilayer of the plasma membrane would be damaged and become poorly visible after freeze substitution using conventional HPF protocols in many cases (Fig. 8.8b) [66].

It has been shown that the visualization of the lipid bilayer can be improved by adding of 1–5% water in the substitution solution [66]. For example, the addition of 5% water in the substitution solution can increase the membrane contrast significantly (Fig. 8.9a–c), and the bilayer of the plasma membrane and the mitochondria membrane are all highlighted (Fig. 8.9b, c). Therefore, this protocol may be helpful for monitoring the morphology of membrane systems. However, the addition of

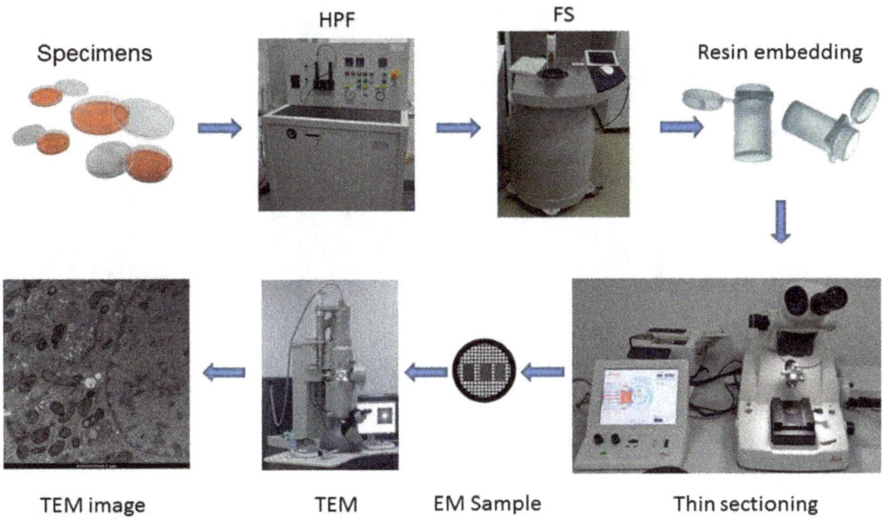

Fig. 8.7 A flowchart of the specimen preparation procedure by the high-pressure freezing and freeze substitution method. HPF: high-pressure freezing; FS: freeze substitution; TEM: transmission electron microscope

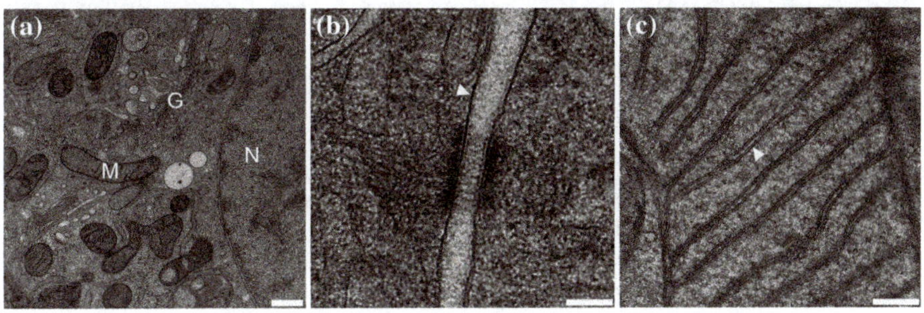

Fig. 8.8 Thin sections prepared by the conventional substitution solution. **a** A low magnification image of a cell. The mitochondria (M), the Golgi apparatus (G), the nucleus (N) are visualized clearly; Bar = 500 nm. **b** A high magnification image of a cell junction between two plasma membranes (arrow head); Bar = 100 nm. **c** A high magnification image of the mitochondria. The lipid bilayer of the mitochondria crista membranes are visualized (arrow head); Bar = 100 nm

water could affect the preservation of the membrane-associated proteins and the cytoplasm may also be damaged by the recrystallization of water.

Alternatively, the pretreatment of specimens with tannic acid for 2 days before the freeze substitution [67, 68] could also enhance the visualization of the membrane bilayers. For example, a specimen pretreated with 0.1% tannic acid for 48 h before the freeze substitution shows that all membrane systems are highlighted significantly (Fig. 8.10a–c). Both the plasma membrane (Fig. 8.10b) and membrane of the organelles are visualized clearly. Although this protocol is more

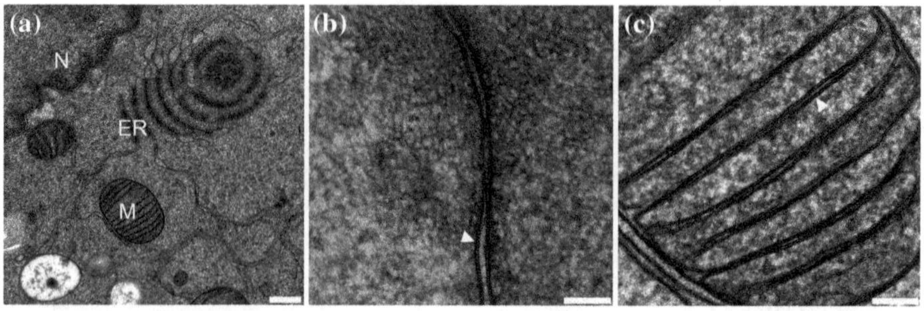

Fig. 8.9 Thin sections prepared by adding 5% water in the substitution solution after HPF. **a** An overview of the cellular structures at low magnification. The mitochondria (M), endoplasmic reticulum (ER), and the nucleus are visualized clearly; Bar = 500 nm. **b** A high magnification view of the plasma membrane (arrowhead). **c** A high magnification view of the mitochondria membrane (arrowhead); Bars = 100 nm

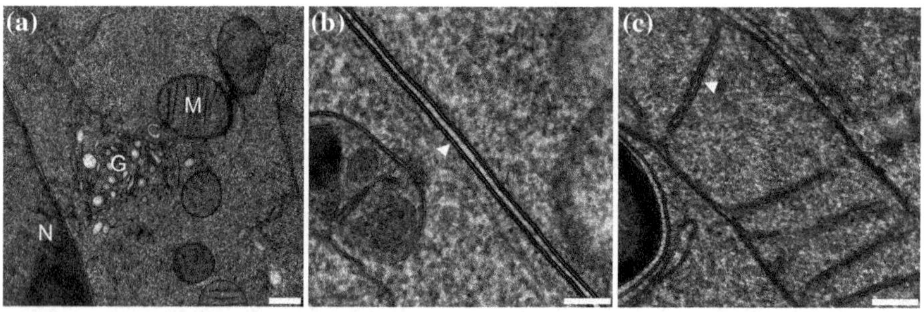

Fig. 8.10 Thin sections prepared by pretreating the specimen with 0.1% tannic acid before substitution. **a** An overview of the cellular structures at low magnification. The mitochondria (M), Golgi apparatus (G), and the nucleus are visualized clearly; Bar = 500 nm. **b** A high magnification view of the plasma membrane (arrowhead). **c** A high magnification view of the mitochondria membrane (arrowhead); Bars = 100 nm

time-consuming and involved extra steps in the whole procedure, it could be a good option for those who need a careful characterization of the membrane systems.

Instead of adding extra reagents, a modified freeze-substitution method has also been shown to be able to preserve membranes reasonably well [69, 70]. The substitution can be carried out in a Styrofoam box for 1.5 h, which is much faster than the conventional substitution procedure. This protocol could generate good preservation results for the plasma membrane and the membrane-associated proteins. The lipid bilayer of the mitochondria in the cytoplasm can also be seen clearly (Fig. 8.11b). Therefore, it might be a fast option for monitoring the plasma membrane and the associated proteins. However, since the rate of substitution is much faster in this case, the preservation of the cytoplasm is not as good as the conventional protocol (Fig. 8.11a).

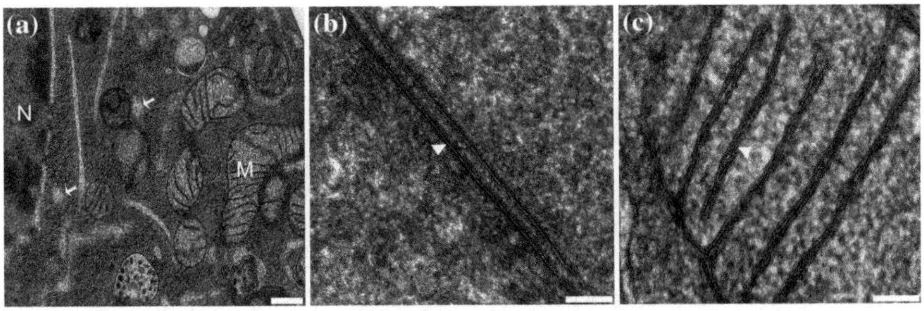

Fig. 8.11 Thin sections prepared by the fast substitution method after HPF. **a** An overview of the cellular structures. The mitochondria (M) and the nucleus (N) are visualized, the cytoplasm is disturbed (arrow). Bar = 500 nm. **b** The plasma membranes of two cells. The lipid bilayer (arrow head) is visualized; Bar = 100 nm. **c** The lipid bilayer (arrow head) of the mitochondrial cristae is visualized; Bar = 100 nm

8.3.4 Visualization of the Cell Membrane and Surface Molecules

After HPF/FS, the specimens are embedded in resins, then thin-sectioned on a microtome. The thin sections are usually stained by heavy metal reagents to increase the contrast of lipid membranes and proteins. Compared to the sample preservation procedure, the imaging by TEM is relatively easy. Before loading the sample grids, the microscope needs to be aligned properly. The thin sections can be screened under TEM at low magnification to check the preservation quality and locate the area of interest, and the visualization of the lipid bilayer could be a practical indicator for the preservation quality of the specimens. Once the target area is found, the magnification can be increased for high-resolution imaging. The cell membranes are usually easy to recognize even the preservation is not as good as expected. By contrast, the cell surface molecules are difficult to identify since there are numerous molecules on cell surface. A typical way of visualizing the cell surface molecules is to label the molecules by the electron-dense markers such as gold particles through antibodies (Fig. 8.12a, b). Since a thin section only contains a small portion of cell surface area, the search of surface molecules may be time-consuming even though the target molecules are supposed to be highly expressed on cell surface. Nevertheless, the labeled molecules could be identified if the specimens are preserved properly. If the target molecules are large enough, its conformation can also be visualized. To obtain the three-dimensional view of the plasma membrane and surface molecules, a series of tilt images can be collected for tomographic reconstruction (Fig. 8.12c, d). The three-dimensional tomograms usually give better contrast than the two-dimensional images, therefore revealing more structural details of the target area.

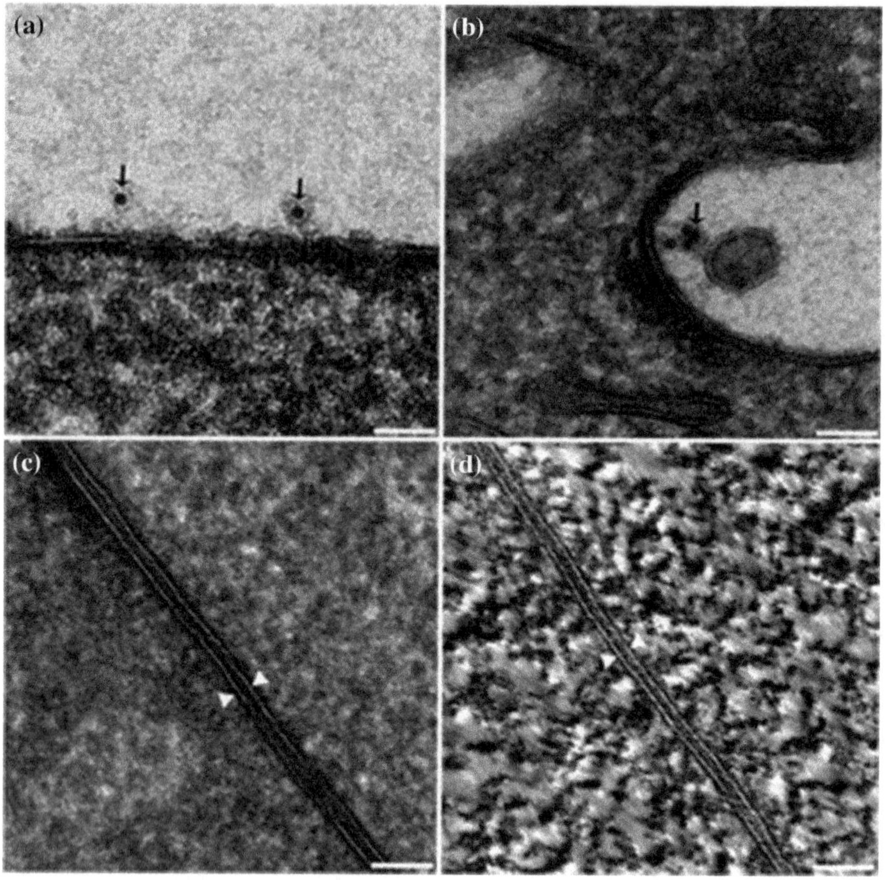

Fig. 8.12 Visualization of the plasma membrane and membrane-associated proteins. **a**, **b** Visualization of the gold-labeled proteins (arrow) on cell surface (**a**) and endosome (**b**). **c** The 2D image of a cell junction between cell membranes (arrow). **d** A tomographic section of a cell junction between cell membranes (arrow)

8.4 Principle and Application of Tokuyasu Method

8.4.1 Principle of Tokuyasu Method

Tokuyasu method was initially developed for eliminating the artifacts during the dehydration and embedding procedures of ultramicrotomy, which combines chemical fixation and cryo-sectioning for cellular and tissue samples [71]. The specimens are chemically fixed first, then infiltrated with sucrose to improve the plasticity of the specimens during sectioning, and sucrose also acts as a cryo-protectant to reduce the ice crystal damage during freezing [71]. The frozen specimens are then cryo-sectioned and transferred carefully onto EM grids after the sections are thawed at room temperature. The sections can be immunolabeled or directly dried and embedded for TEM analysis. To prevent the potential damage by air-drying, the sections are usually embedded in materials such as methyl cellulose and uranyl acetate to protect the ultrastructure and also increase contrasting effects [72–76]. Since neither organic solvents nor lipid-solubilizing embedding materials are applied during the process, the proteins in the specimens are well preserved and can be detected by antibodies efficiently [77]. Therefore, this method is commonly used for the immunolabeling of the target molecules, especially the proteins associated with membranes.

8.4.2 Visualization of the Cell Membrane by Tokuyasu Method

Although Tokuyasu method is commonly used for the immunolabeling of macromolecules [71, 77], it usually gives better contrast for the membranes, therefore can be applied to visualize the morphology of cellular membranes. If the sections are negatively stained with uranyl acetate (UA), the membranes are highlighted in white and other cellular areas are shown in gray (Fig. 8.13a, b). The preservation quality of Tokuyasu method may not be as good as HPF, but it is still useful in monitoring the membrane-associated activities such as membrane fusion, endocytosis, and vesicle transportation. In fact, HPF can be also used in the freezing step of Tokuyasu method, which could reduce the freezing damage and achieve better resolution in the images [58].

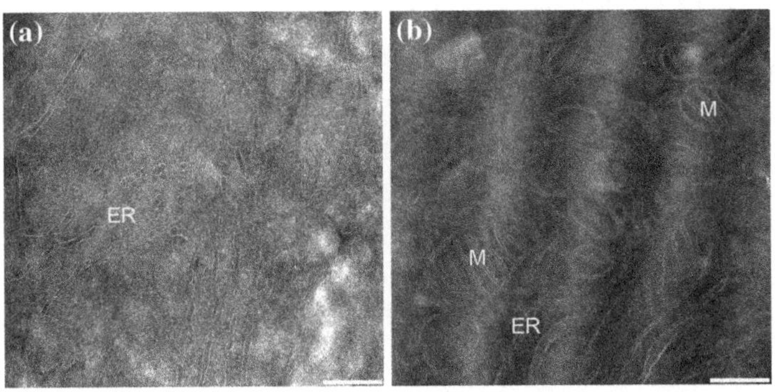

Fig. 8.13 Thin sections prepared by Tokuyasu method. **a**, **b** The cellular membranes (white) are highlighted in the thin sections prepared by Tokuyasu method. M: mitochondria; ER: endoplasmic reticulum; Bar = 250 nm

8.5 Materials and Methods

8.5.1 Materials

8.5.1.1 Materials for HPF and Freeze Substitution

(1) HPF machine and corresponding specimen carriers.
(2) Freeze-substitution machine (e.g., Leica AFS2).
(3) A metallic heater block with holes for 2 mL centrifuge tubes.
(4) 45° fine-tipped tweezers (Ted Pella, Inc.).
(5) Straight fine-tipped tweezers (Ted Pella, Inc.).
(6) 1-Hexadecene (Merck).
(7) 2 mL cryogenic vials (Biologic Research Co.).
(8) Crystalline osmium tetroxide (EMS).
(9) Acetone (>99.7%).
(10) Uranyl acetate (Ted Pella, Inc.).
(11) Tannic acid (EMS).
(12) Glass Pasteur pipette.
(13) 1% OsO_4–0.1% UA freeze-substitution solution in acetone.

8.5.1.2 Materials for Embedding and Sectioning

(1) Ultramicrotome UC7/FC7 (Leica Microsystem).
(2) A regulated ionizer (Diatome, Bienne, Switzerland).
(3) Flat bottom polyethylene capsules (Ted Pella, Inc.).
(4) Pyramitome (Leica EM TRIM2).
(5) Razor blade.
(6) Diamond knife (Diatome, Bienne, Switzerland).
(7) No. 1 Superfine eyelash (Ted Pella, Inc.).
(8) 150 mesh copper grids with formvar films (Ted Pella, Inc.).
(9) Grid storage box (Ted Pella, Inc.).
(10) Epon 812 resin.

8.5.1.3 Materials for Staining

(1) PELCO®SynapTekGridStick™ Kit (Ted Pella, Inc.).
(2) Anhydrous lead citrate $Pb(C_6H_5O_7)_2$.
(3) Lead nitrate $Pb(NO_3)$.
(4) Lead acetate $Pb(CH_3COO)_2 \cdot 3H_2O$.
(5) Sodium citrate $Na_3(C_6H_5O_7) \cdot 2H_2O$.
(6) Syringe filter with 0.22 μm pore size (PALL).
(7) 50 mL syringe.

8.5.1.4 Materials for Tokuyasu Method

(1) 25% glutaraldehyde solution (EMS).
(2) Paraformaldehyde.
(3) Methyl cellulose (Sigma M-6385 Methyl Cellulose 25 centipoises).
(4) Gelation.
(5) Na-Azide 10% (1 g/10 mL distilled water).
(6) 4% Uranyl acetate in water.
(7) Methylcellulose (Sigma-M6385).

8.5.1.5 Electron Microscopy

(1) Electron microscope with stage and computer-controlled goniometer.
(2) Specimen holders.
(3) CCD or DDD camera for digital imaging.

8.5.2 Methods

8.5.2.1 Methods for High-Pressure Freezing

HPF can freeze large specimen (e.g., cells or tissues) with thickness up to 500 μm [53, 78]. To get high-quality biological specimens by HPF, the preceding preparation is critical and could be the limiting factor in many cases [78]. The general principle for specimen preparation includes healthy cells, work as quickly as possible, minimal buffer but never dry out, no air space, right filler, smallest specimen volume, and shallowest carrier (**Note 1**). Here, we outline the general procedures for high-pressure freezing cell and tissue specimens. (The Wohlwend High-Pressure Freezer compact 2 with corresponding specimen carriers are used in this protocol.)

(1) Connect the LN_2 tank to the high-pressure freezing machine.
(2) Start up the high-pressure freezing machine and wait until the system is ready for freezing.
(3) Preclean the specimen carriers (e.g., 0.1 mm/0.2 mm, 0.04 mm/0.26, 0.3 mm aluminum carriers) with acetone and dry in a fume chamber.
(4) Fill the specimen-unloading foam box with LN_2.
(5) For different specimens, use different loading methods (Fig. 8.14).
(6) Specific samples (e.g., bacteria, yeast, Drosophila embryos, specific animal tissues) might need specific loading methods to achieve high-quality freezing result [79–82].
(7) Put the specimen holder with samples into the high-pressure freezing chamber, lock the pin, and freeze the specimen.
(8) Transfer the frozen specimen sandwich immediately to the foam box with LN_2 and unload the specimen under LN_2.

(9) Open the specimen sandwich by moving the cap off with a precooled forceps under LN_2.
(10) Transfer the frozen specimen with a precooled tweezers to the frozen substitution solutions in a 2 mL cryogenic vials under LN_2.
(11) Cover the cap of the specimen-unloading foam box immediately to avoid ice contamination.
(12) Dry the specimen holder with a hair drier.
(13) Repeat steps 5–10 to freeze the next specimen.
(14) Shut down the machine after frozen all of the specimens.

8.5.2.2 Methods for Freeze Substitution

In the freeze-substitution procedure, every step should be performed carefully to avoid the recrystallization of water in the specimen. Here, we outline the three different protocols of the freeze substitution.

(1) Conventional freeze-substitution protocol.

(a) Start up the Leica AFS2 freeze-substitution machine, make sure the freeze-substitution chamber is dry and put the tube holder into the chamber.
(b) Fill ~35 L LN_2 into the Leica AFS2 LN_2 container, precool the chamber, and transfer the 2 mL tubes with specimens and freeze-substitution solution to the freeze-substitution chamber quickly with a precooled long forceps.
(c) Lock the lid of the freeze-substitution chamber.
(d) Set a AFS2 program for a freeze-substitution protocol and start the program (Table 8.1).

(2) Two–step freeze-substitution method with tannic acid.

The procedure for preparing the freeze substitution is similar to the conventional freeze-substitution procedures as shown above. The major difference is the addition

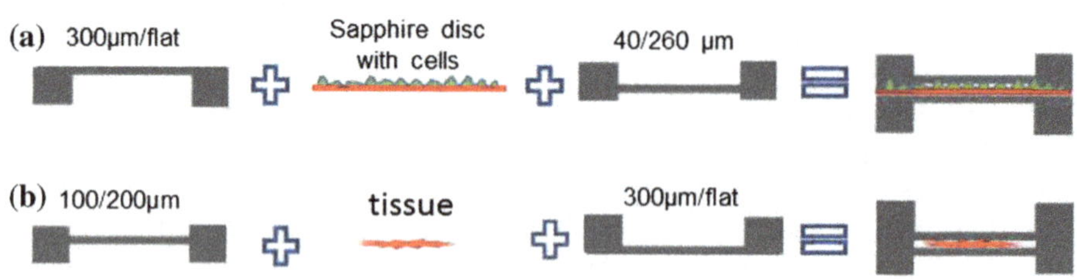

Fig. 8.14 The schematic diagram of specimen loading for high-pressure freezing. **a** For cells samples growing on the sapphire disks. Put the sapphire disk with cells upside on the top of the flat side of a 300 μm carrier, and then put a 40/260 μm carrier onto the sapphire disk as a cap. **b** For tissue samples, put the tissue in the 100 μm cave of a 100/200 μm specimen carrier and then put a 300 μm/flat carrier onto the tissue as a cap. The space between the two aluminum specimen carriers is filled with fillers (for example, 1-hexadecene, 20% ficoll, 20% dextran, 20% BSA, or other fillers), and the sandwich is ready for freezing

Table 8.1 Program for the conventional freeze-substitution method

Step	*T* start (°C)	*T* end (°C)	Duration	Notes
1	−90	−90	48 h	The duration time can be changed within the range of 8–72 h depend on the size or nature of the tissue
2	−90	−60	6 h	
3	−60	−60	24 h	
4	−60	−30	8 h	
5	−30	−30	8 h	
6	−30	−20	4 h	
7	−20	−20	8 h	
8	−20	4	2 h	Prepare pure acetone in 2 ml centrifuge tube and precool them in the chamber for flowing wash
9	4	4	45 min	3 × 15 min washes with precooled acetone to wash away the OsO_4 (**Note 2**)

Warm up to room temperature in 15 min for flowing resin infiltration

of tannic acid before the conventional protocol. The tannic acid treatment protocol is as the following chart (Table 8.2).

(3) Rapid freeze-substitution method.

(a) The frozen specimens are put into the 2 mL cryogenic vials with 1% OsO_4–0.1% UA in acetone after high-pressure freezing.
(b) Immerse the heater block into the LN_2 in a foam box until it is completely cooled down.
(c) Transfer the 2 mL tubes with specimens and substitution solution to the foam box and insert into the heater block immediately.
(d) Pour off LN_2 carefully and put the foam box with the specimen on the shaker.
(e) Shake for 1.5 h with a speed of 100 rpm/min in a fume. Wash the specimen three times in 15 min with acetone in a fume at room temperature.

Table 8.2 Program for the two-step freeze-substitution method

Step	*T* start (°C)	*T* end (°C)	Duration (h)	Notes
1	−90	−90	24	0.05–0.1% tannic acid in acetone >99.7%. Prepare acetone in 2 ml centrifuge tube and precool them in the chamber for flowing wash. Precool the substitution solution with 1% OsO_4–0.1% UA in 2 ml cryogenic vials
2	−90	−90	1	3 × 15 min washes with precooled acetone to wash away the tannic acid (**Note 3**). Then suck off the acetone and add 1.5 ml precooled 1% OsO_4–0.1% UA into the cryogenic vials with specimen

Then substituted as the conventional substitution protocol as above

8.5.2.3 Methods for Resin Infiltration, Embedding, and Polymerization

(1) Conventional resin infiltration.

After freeze substitution, the specimen needs to be further infiltrated with resin, embedded, and polymerized. EPON 812 resin is a widely used embedding resin for electron microscopy and is suitable for most specimens. The infiltration procedures are shown in the following table (Table 8.3).

(2) Fast resin infiltration.

Fast resin infiltration protocol is shown as follows (Table 8.4).

Table 8.3 Procedure of the conventional resin infiltration

Step	Medium	Concentration	Duration	Notes
1	Resin/acetone	1:3	1 h	Pour the acetone carefully and do not pour out the specimen with the acetone
2	Resin/acetone	1:1	1 h	Do not suck out the specimen and let the specimen stay in the bottom of the tube
3	Resin/acetone	3:1	2 h	
4	Resin	100%	2 h	
5	Resin	100%	Overnight	
6	Resin	100%	1 h	

Ready for embedding

Table 8.4 Procedure of the fast resin infiltration

Step	Medium	Concentration	Duration	Notes
1	Resin/acetone	1:3	Centrifuge for 30 s at 6000 rpm	Pour the acetone carefully and do not pour out the specimen with the acetone
2	Resin/acetone	1:1	Centrifuge for 30 s at 6000 rpm	Do not pick out the specimen and let the specimen stay in the bottom of the tube
3	Resin/acetone	3:1	Centrifuge for 30 s at 6000 rpm	Do not pick out the specimen
4	Resin	100%	Centrifuge for 30 s at 6000 rpm	Do not pick out the specimen
5	Resin	100%	Centrifuge for 30 s at 6000 rpm	Pick out the specimen and put it into a new tube with resin
6	Resin	100%	Centrifuge for 30 s at 6000 rpm	

Ready for embedding

(3) Embedding and Polymerization.

(a) For the cells on the sapphire disks, the flat bottom capsules are used for embedding. Put the sapphire disks with the cells upside in the bottom of the capsules (**Note 4**). Then fill the capsules with resin and cap the capsules carefully to avoid air bubbles.
(b) For the tissues, dispense the resin to the flat rubber embedding mold, put the tissues at one end of the cavity and make sure the specimen is placed on the bottom and oriented properly (**Note 5**).
(c) Transfer the embedded specimens to an oven, polymerize at 37 °C for 12 h and 65 °C for 24–48 h.

8.5.2.4 Methods of Chemical Fixation for Tokuyasu Method

(1) Methods for fixing tissues.

(a) Prepare 2% paraformaldehyde-0.2% glutaraldehyde fixation solution.
(b) Excise the tissues as small as possible and immerse into the fixation solution quickly at room temperature (**Note 6**).
(c) Transfer the specimens to fresh fixation after 0.5 h and fix for another 2 h at 4 °C.

(2) Methods for fixing suspension cells.

(a) Add equal double-strength fixative (4% formaldehyde-0.4% glutaraldehyde) to the cell culture and fix 10 min at 37 °C.
(b) Centrifuge 2 min with 800 rgp/min.
(c) Discard the supernatant, add fresh single strength fixative and fix for another 1 h at room temperature.
(d) Centrifuge 5 min with 5000 rgp/min to form a cell pelt.

(3) Methods for fixing sedentary cells.

(a) Grow cells in Petri dishes. Replace the culture medium with fresh medium gently the day before fixation.
(b) Add equal double-strength fixative at room temperature and mix gently.
(c) After 5 min, replace with single strength fixative and fix for another 1 h at room temperature.
(d) Scrape the cells from the dish and transfer to an eppendorf tube.
(e) Centrifuge 5 min with 5000 rgp/min to form a cell pelt.

8.5.2.5 Methods for Gelatine Embedding

(1) Gelatine embedding of tissue.

(a) The fixed tissues are cut into small blocks (<1 mm^3) with a sharp razor blade.
(b) Rinse the tissues in 0.15% glycinein PBS for 10 min.

(c) Wash the tissues with PBS for three times.
(d) Infiltrate with 10% gelatine in 0.1 M phosphate buffer at 37 °C for 15–30 min.
(e) Solidify the gelatin with specimen on ice for 30 min.

(2) Gelatine embedding of cell pelts.

(a) Move the supernatant fixative of the cell pelts.
(b) Rinse twice with PBS and three times with 0.15% glycine in PBS.
(c) Resuspend the cells with 10% gelatin and infiltrate for 10 min at 37 °C.
(d) Centrifuge 5 min with 10,000 rpf/min to form a cell pelt.
(e) Discard the supernatant and solidify the gelatin on ice for 30 min.

8.5.2.6 Cryo-Protection and Specimen Freezing of Tokuyasu Method

(1) Cut the solidified gelatin with tissues or cells into small blocks (<1 mm^3) with a sharp razor blade at 4 °C (**Note 7**).
(2) Transfer the blocks of tissues or cell pellets to 1 mL 2.3 M sucrose in a 1.5 mL centrifuge tube, seal the cap, and leave on a rotating wheel at 4 °C. Change the sucrose after 2 h and infiltrate overnight.
(3) Transfer each of the infiltrated specimen block onto a clean specimen stub, suck off excess sucrose with a filter paper and immerse the specimen stub in LN_2 until freezing is complete.

8.5.2.7 Methods for Thin Sectioning

(1) Thin Sectioning for Plastic-Embedded Specimen.

(a) For the cells on the sapphire disk, the excess resin around the sapphire disks was trimmed off to expose the sapphire disks. Dip the resin block into LN_2 for a few seconds and remove the sapphire disk off with forceps. Trim the surface of the resin block with cells to a ladder shaped face with a razor blade.
(b) For the tissues or other specimens, trim the specimen end of the block to a ladder shaped face with a razor blade to expose the specimen.
(c) Mount the trimmed block on the ultramicrotome, align properly, and use a fresh glass knife to trim off the front face.
(d) Change the glass knife to an ultra 45° diamond knife and align the block properly.
(e) Cut 70–150 nm thin sections.
(f) Pick a formvar-/carbon-coated EM grid by a fine tweezers. Insert the grid into the water of the trough, make the central region of the grid gently touch the edge of a section, and slowly lift the grid up.
(g) Put the grid on a filter paper, wait until the grid is dry, and store the grid into a grid box.

(2) Cryo-sectioning of Tokuyasu method.

(a) Fill ~25 L LN_2 into the tank of the Ultramicrotome UC7/FC7.
(b) Put trim 20 diamond trimming and cryo immune 35° diamond knife in the cryo-chamber.
(c) Start the FC Mont and precool the cryo-chamber of the ultramicrotome to −80 °C (**Note 8**).
(d) Transfer the specimen stub into the precooled cryo-chamber, quickly place the specimen stub into the specimen holder and lock it in place.
(e) Turn on the antistatic line to full power and trim off the face of the block.
(f) Trim the block to a rectangular shape with a block face of 0.6–0.4 × 0.3–0.4 mm and a depth of 50–100 μm using a trimming diamond knife.
(g) Replace the trimming knife with a diamond knife. Set the temperature of specimen, knife, and chamber to −120 °C and cool down the chamber (**Note 9**).
(h) Carefully align the knife edge and the block face until the knife and the block are almost touching each other.
(i) Set the antistatic line and section thickness to the proper range for your specimen (**Note 10**).
(j) Start automatic sectioning until the sections appear on the knife. Use an eyelash probe to adjust and "guided" away the ribbons from the knife edge.
(k) Turn off the antistatic line.
(l) Make a mixture of 2% methylcellulose: 2.3 M sucrose with a ratio of 1:1 as a section pickup solution. Dip a 2-mm-diameter loop in the solution and lift quickly to form a large drop let on the loop (**Note 11**).
(m) Transfer the loop with the droplet to the cryo-chamber. The droplet is brought to a position near, but not yet interacting with the sections.
(n) Press the droplet gently on top of the ribbon of the sections once the edge of the droplet in the loop starts freezing (white rim).
(o) Remove the droplet from the cryo-chamber, thaw in the room temperature and place the drop with section-side down onto a formvar-/carbon-coated EM grid (**Note 12**).
(p) Put the grid with the pickup solution on a glass slide in a humid box and store at 4 °C for further staining.

8.5.2.8 Methods for Staining

(1) Positive staining for sections of plastic-embedded specimen.

(a) Prepare 3% uranyl acetate in 70% methanol and SATO lead stain solution.
(b) Gently stick up to ten specimen grids to a grid stick and insert into the staining pipettes.
(c) Suck 3% uranyl acetate into the staining pipettes and make sure all of the grids were immersed into the solution and stain for 7 min at 4 °C (**Note 13**).

(d) Rinse the stick with the pipettes pipe in and out with 70, 50, 25% methanol and distilled water, respectively, at room temperature.
(e) Dry the grids in the hood for 3–5 min.
(f) Suck about 1.25 mL SATO solution into the staining pipettes and stain for 3 min at room temperature.
(g) Rinse the stick quickly with the pipettes pipe in and out with distilled water for 5 min at room temperature.
(h) Dry the grids in the hood for 3–5 min and store the grids in a grid box for TEM analysis.

(2) Negative staining for sections of Tokuyasu cryo-sectioning.

(a) Prepare 2% methyl cellulose-4% uranyl acetate staining solution.
(b) Attach a long piece of parafilm to a bench surface.
(c) Put drops of distilled water on the parafilm.
(d) Put the grids section-side down on the drops of water and incubate for 15 min.
(e) Put drops of the staining solution on the Parafilm.
(f) Float the grids on the drops with section-side down. Transfer the grids over two drops to wash off water from the above step. Incubate the grid onto a larger drop of staining solution for 10 min.
(g) Insert a 3.5–4-mm-diameter wire loop underneath the grid and lift the grid from the drop of staining solution.
(h) Tilt the loop with grid to an angle of 45°–60° and pull the loop on a filter paper to remove excess liquid from the loop (**Note 14**).
(i) Leave the grid on the loop to dry at room temperature.
(j) When the grids are dry out, the methyl cellulose will form a thin film on the grid. Cut the film around the periphery of the grids with a pointed forceps and lift the grids away.
(k) Store the grids in a grid box for TEM analysis.

8.5.2.9 Methods for Data Acquisition by TEM

A transmission electron microscope, for example, FEI Tecnai T120, should be properly aligned and adjusted for optimal imaging. In this section, we will describe how to start the electron microscope and optimize imaging conditions. We will also describe the data acquisition method of electron tomography.

(1) Methods for image acquisition.

(a) Check the vacuum system.
(b) Add LN_2 to the Dewar flask to cool down the TEM anticontaminator.
(c) Turn on the high tension and illumination system.
(d) Load the specimen grid on the specimen holder and insert into the TEM.
(e) Wait until the vacuum is ready, open the vacuum system, and adjust eucentric height by minimizing X–Y movement when tilting the holder.

(f) Align the optical system and adjust the beam tilt X, beam tilt Y, beam shift, and rotation center to optimal conditions.
(g) Switch to a low magnification (e.g., 200–500×), find the specimen, and move to the center of the viewing.
(h) Insert the objective lens aperture and center it under diffraction mode.
(i) Find a region of interest and do a fine eucentric adjustment.
(j) Open the CCD and adjust to the proper light intensity.
(k) Switch to a high magnification for final data collection.
(l) Open the live FFT and adjust to a proper under focus value to obtain a good contrast for specimen screening.
(m) Acquire the image and store to your folder.

(2) Methods for data acquisition of electron tomography.

(a) Adjust electron microscopy system as above.
(b) Find a region of interest and set a proper magnification and beam intensity.
(c) Run the tilt-series acquisition software (For example, SerialEM, UCSFtomo, or the software provided by the microscope manufacturers).
(d) Within the software, move the region of interest to the center of the CCD. Set the proper eucentric height and defocus value.
(e) Set the tilt range (typically −60°–60°), increment (typically 0.5°–2°) and start a single tilt-series data acquisition session.
(f) Store the tilt-series images after the single tilt-series data acquisition finished.

(3) Three-dimensional reconstruction of electron tomography.

A number of software packages have been developed for the three-dimensional reconstruction of electron tomography. Among them, IMOD is one of the most popular tomography software [44]. Readers can follow the instructions of IMOD to calculate tomograms form the tilted image series. IMOD also provides a 3D graphic interface to manipulate tomograms for structural analysis and model building [41].

8.5.3 Notes

(1) Use cells in optimal physiological conditions. Work as quickly as possible and carefully to avoid any air bubble and air space. Smallest specimen volume and shallowest carrier are to accelerate the heat transfer and to get high yields of well-frozen cells.
(2) Use a syringe to suck off the substitution solution and pour the precooled acetone to the tube immediately.
(3) Use a precooled syringe to suck off the substitution solution and pour the precooled acetone to the tube with specimen immediately. Then screw the tube and close the lid immediately.
(4) Make sure to put the sapphire disk with cells upside, otherwise the cells will move away with the sapphire disk.

(5) Air bubbles should be avoided carefully. If there are air bubble, remove them with a toothpick.
(6) If possible, a "whole body perfusion" is performed for fixation.
(7) The tissues and the cell pellets infiltrated with gelatin must cut at 4 °C, because the gelatin may melt at room temperature. You can use 2.3 M sucrose as lubricating agent when cutting the tubes.
(8) At −80 °C, the specimen block will be much softer that thick sections (500 nm) can be removed without damaging the block.
(9) At −120 °C, the specimen block will be solid enough for thin sectioning.
(10) The section will adhere to the knife or fly away if the antistatic power is not proper, usually half or one third of the power is used.
(11) Lifting the wire loop fast out of the pickup solution will result in a large drop. The size of the drop is important since a larger drop will take longer to freeze and make the section picking up much easier.
(12) For immunolabeling, nickel grids should be used. Because copper grids are easily turned black due to corrosion by buffer.
(13) Wrap the staining pipette with aluminum foil to avoid light exposure.
(14) Pulling the loop faster will result in a thick film and pulling the loop slowly will make the film much thinner.

Acknowledgements We thank National Centre for Protein Science Shanghai (Electron Microscopy system) for instrumental support and technical assistance. This work is supported by the Strategic Priority Research Program of the Chinese Academy of Sciences (No. XDB08020102) and the National Natural Science Foundation of China (No. 31470735 and 31670747).

References

1. Ilangovan A, Kay CWM, Roier S, El Mkami H, Salvadori E, Zechner EL, Zanetti G, Waksman G (2017) Cryo-EM structure of a relaxase reveals the molecular basis of DNA unwinding during bacterial conjugation. Cell 169(4):708–721 e712. doi:10.1016/j.cell.2017.04.010
2. Zhang X, Jin L, Fang Q, Hui WH, Zhou ZH (2010) 3.3 Å cryo-EM structure of a nonenveloped virus reveals a priming mechanism for cell entry. Cell 141(3):472–482. doi:10.1016/j.cell.2010.03.041
3. Hayat MA (2000) Principles and techniques of electron microscopy: biological applications, 4th edn. Cambridge University Press, Cambridge
4. Gilkey JC, Staehelin LA (1986) Advances in ultrarapid freezing for the preservation of cellular ultrastructure. J Electron Microsc 3:177–210
5. McDonald K, Morphew MK (1993) Improved preservation of ultrastructure in difficult-to-fix organisms by high pressure freezing and freeze substitution: I. *Drosophila melanogaster* and *Strongylocentrotus purpuratus* embryos. Microsc Res Tech 24(6):465–473. doi:10.1002/jemt.1070240603
6. Stack JH, Horazdovsky B, Emr SD (1995) Receptor-mediated protein sorting to the vacuole in yeast: roles for a protein kinase, a lipid kinase and GTP-binding proteins. Annu Rev Cell Dev Biol 11:1–33. doi:10.1146/annurev.cb.11.110195.000245

7. Steinbrecht RA (1993) Freeze-substitution for morphological and immunocytochemical studies in insects. Microsc Res Tech 24(6):488–504. doi:10.1002/jemt.1070240605
8. Studer D, Michel M, Muller M (1989) High pressure freezing comes of age. Scan Microsc 3 (Suppl):253–268; discussion 268–259
9. Kellenberger E, Johansen R, Maeder M, Bohrmann B, Stauffer E, Villiger W (1992) Artefacts and morphological changes during chemical fixation. J Microsc 168(Pt 2):181–201
10. Weibull C, Christiansson A (1986) Extraction of proteins and membrane lipids during low temperature embedding of biological material for electron microscopy. J Microsc 142(Pt 1):79–86
11. Dubochet J, Adrian M, Chang JJ, Homo JC, Lepault J, McDowall AW, Schultz P (1988) Cryo-electron microscopy of vitrified specimens. Q Rev Biophys 21(2):129–228
12. Dubochet J, McDowall A (1981) Vitrification of pure water for electron microscopy. J Microsc 124(3):3–4
13. Dobro MJ, Melanson LA, Jensen GJ, McDowall AW (2010) Plunge freezing for electron cryomicroscopy. Methods Enzymol 481:63–82. doi:10.1016/S0076-6879(10)81003-1
14. Galway ME, Heckman JW Jr, Hyde GJ, Fowke LC (1995) Advances in high-pressure and plunge-freeze fixation. Methods Cell Biol 49:3–19
15. Nitta K, Kaneko Y (2004) Simple plunge freezing applied to plant tissues for capturing the ultrastructure close to the living state. J Electron Microsc 53(6):677–680
16. Richter T, Biel SS, Sattler M, Wenck H, Wittern KP, Wiesendanger R, Wepf R (2007) Pros and cons: cryo-electron microscopic evaluation of block faces versus cryo-sections from frozen-hydrated skin specimens prepared by different techniques. J Microsc 225(Pt 2): 201–207. doi:10.1111/j.1365-2818.2007.01732.x
17. McDonald K (1999) High-pressure freezing for preservation of high resolution fine structure and antigenicity for immunolabeling. Methods Mol Biol 117:77–97. doi:10.1385/1-59259-201-5:77
18. van Heel M, Gowen B, Matadeen R, Orlova EV, Finn R, Pape T, Cohen D, Stark H, Schmidt R, Schatz M, Patwardhan A (2000) Single-particle electron cryo-microscopy: towards atomic resolution. Q Rev Biophys 33(4):307–369
19. Jonic S, Sorzano CO, Boisset N (2008) Comparison of single-particle analysis and electron tomography approaches: an overview. J Microsc 232(3):562–579. doi:10.1111/j.1365-2818.2008.02119.x
20. Xie Q, Spear JM, Noble AJ, Sousa DR, Meyer NL, Davulcu O, Zhang F, Linhardt RJ, Stagg SM, Chapman MS (2017) The 2.8 Å electron microscopy structure of adeno-associated virus-DJ bound by a heparinoid pentasaccharide. Mol Ther Methods Clin Dev 5:1–12. doi:10.1016/j.omtm.2017.02.004
21. Bartesaghi A, Matthies D, Banerjee S, Merk A, Subramaniam S (2014) Structure of beta-galactosidase at 3.2-Å resolution obtained by cryo-electron microscopy. Proc Natl Acad Sci U S A 111(32):11709–11714. doi:10.1073/pnas.1402809111
22. Stark H, Chari A (2016) Sample preparation of biological macromolecular assemblies for the determination of high-resolution structures by cryo-electron microscopy. Microscopy 65 (1):23–34. doi:10.1093/jmicro/dfv367
23. Calabrese AN, Watkinson TG, Henderson PJ, Radford SE, Ashcroft AE (2015) Amphipols outperform dodecylmaltoside micelles in stabilizing membrane protein structure in the gas phase. Anal Chem 87(2):1118–1126. doi:10.1021/ac5037022
24. Feinstein HE, Tifrea D, Sun G, Popot JL, de la Maza LM, Cocco MJ (2014) Long-term stability of a vaccine formulated with the amphipol-trapped major outer membrane protein from *Chlamydia trachomatis*. J Membr Biol 247(9–10):1053–1065. doi:10.1007/s00232-014-9693-5
25. Tifrea DF, Sun G, Pal S, Zardeneta G, Cocco MJ, Popot JL, de la Maza LM (2011) Amphipols stabilize the *Chlamydia* major outer membrane protein and enhance its protective ability as a vaccine. Vaccine 29(28):4623–4631. doi:10.1016/j.vaccine.2011.04.065
26. Borch J, Hamann T (2009) The nanodisc: a novel tool for membrane protein studies. Biol Chem 390(8):805–814. doi:10.1515/BC.2009.091

27. Dominik PK, Borowska MT, Dalmas O, Kim SS, Perozo E, Keenan RJ, Kossiakoff AA (2016) Conformational chaperones for structural studies of membrane proteins using antibody phage display with nanodiscs. Structure 24(2):300–309. doi:10.1016/j.str.2015.11.014
28. Popot JL (2010) Amphipols, nanodiscs, and fluorinated surfactants: three nonconventional approaches to studying membrane proteins in aqueous solutions. Annu Rev Biochem 79: 737–775. doi:10.1146/annurev.biochem.052208.114057
29. Glaeser RM (2013) Invited review article: methods for imaging weak-phase objects in electron microscopy. Rev Sci Instrum 84(11):111101. doi:10.1063/1.4830355
30. Tang G, Peng L, Baldwin PR, Mann DS, Jiang W, Rees I, Ludtke SJ (2007) EMAN2: an extensible image processing suite for electron microscopy. J Struct Biol 157(1):38–46. doi:10.1016/j.jsb.2006.05.009
31. Grigorieff N (2007) FREALIGN: high-resolution refinement of single particle structures. J Struct Biol 157(1):117–125. doi:10.1016/j.jsb.2006.05.004
32. Scheres SH (2012) RELION: implementation of a Bayesian approach to cryo-EM structure determination. J Struct Biol 180(3):519–530. doi:10.1016/j.jsb.2012.09.006
33. Hohn M, Tang G, Goodyear G, Baldwin PR, Huang Z, Penczek PA, Yang C, Glaeser RM, Adams PD, Ludtke SJ (2007) SPARX, a new environment for Cryo-EM image processing. J Struct Biol 157(1):47–55. doi:10.1016/j.jsb.2006.07.003
34. Henderson R, Chen S, Chen JZ, Grigorieff N, Passmore LA, Ciccarelli L, Rubinstein JL, Crowther RA, Stewart PL, Rosenthal PB (2011) Tilt-pair analysis of images from a range of different specimens in single-particle electron cryomicroscopy. J Mol Biol 413(5):1028–1046. doi:10.1016/j.jmb.2011.09.008
35. McEwen BF, Renken C, Marko M, Mannella C (2008) Chapter 6: Principles and practice in electron tomography. Methods Cell Biol 89:129–168. doi:10.1016/S0091-679X(08)00606-7
36. De Rosier DJ, Klug A (1968) Reconstruction of three dimensional structures from electron micrograph. Nature 217:130–134
37. McEwen BF, Marko M (2001) The emergence of electron tomography as an important tool for investigating cellular ultrastructure. J Histochem Cytochem: Off J Histochem Soc 49 (5):553–564. doi:10.1177/002215540104900502
38. Liu B, Yu H, Verbridge SS, Sun L, Wang G (2014) Dictionary-learning-based reconstruction method for electron tomography. Scanning 36(4):377–383. doi:10.1002/sca.21121
39. Radermacher M (1988) Three-dimensional reconstruction of single particles from random and nonrandom tilt series. J Electron Microsc Tech 9(4):359–394. doi:10.1002/jemt.1060090405
40. Lucic V, Forster F, Baumeister W (2005) Structural studies by electron tomography: from cells to molecules. Annu Rev Biochem 74:833–865. doi:10.1146/annurev.biochem.73.011303.074112
41. He W, He Y (2014) Electron tomography for organelles, cells, and tissues. Methods Mol Biol 1117:445–483. doi:10.1007/978-1-62703-776-1_20
42. Mastronarde DN (1997) Dual-axis tomography: an approach with alignment methods that preserve resolution. J Struct Biol 120(3):343–352. doi:10.1006/jsbi.1997.3919
43. Haberfehlner G, Serra R, Cooper D, Barraud S, Bleuet P (2014) 3D spatial resolution improvement by dual-axis electron tomography: application to tri-gate transistors. Ultramicroscopy 136:144–153. doi:10.1016/j.ultramic.2013.09.006
44. Kremer JR, Mastronarde DN, McIntosh JR (1996) Computer visualization of three-dimensional image data using IMOD. J Struct Biol 116(1):71–76. doi:10.1006/jsbi.1996.0013
45. Blancard C, Salin B (2017) Plunge freezing: a tool for the ultrastructural and immunolocalization studies of suspension cells in transmission electron microscopy. J Visualized Exp: JoVE (123). doi:10.3791/54874
46. Allison DP, Daw CS, Rorvik MC (1987) The construction and operation of a simple inexpensive slam freezing device for electron microscopy. J Microsc 147(Pt 1):103–108
47. Dubochet J (2007) The physics of rapid cooling and its implications for cryoimmobilization of cells. Methods Cell Biol 79:7–21. doi:10.1016/S0091-679X(06)79001-X
48. Escaig J (1982) New instruments which facilitate rapid freezing at 83 K and 6 K. J Microsc 126:221–229

49. Meryman HT (2007) Cryopreservation of living cells: principles and practice. Transfusion 47 (5):935–945. doi:10.1111/j.1537-2995.2007.01212.x
50. Shimoni E, Muller M (1998) On optimizing high-pressure freezing: from heat transfer theory to a new microbiopsy device. J Microsc 192(Pt 3):236–247
51. Studer D, Michel M, Wohlwend M, Hunziker EB, Buschmann MD (1995) Vitrification of articular cartilage by high-pressure freezing. J Microsc 179(Pt 3):321–332
52. Adrian M, Dubochet J, Lepault J, McDowall AW (1984) Cryo-electron microscopy of viruses. Nature 308(5954):32–36
53. Dahl R, Staehelin LA (1989) High-pressure freezing for the preservation of biological structure: theory and practice. J Electron Microsc Tech 13(3):165–174. doi:10.1002/jemt.1060130305
54. Steinbrecht RA, Zierold K (1987) Cryotechniques in biological electron microscopy. Springer, Berlin
55. Moor H, Bellin G, Sandri C, Akert K (1980) The influence of high pressure freezing on mammalian nerve tissue. Cell Tissue Res 209(2):201–216
56. Riehle U, Hoechli M (1973) The theory and technique of high pressure freezing. Freeze etching techniques and applications. Société Française de Microscopie Electronique, Paris
57. Studer D, Graber W, Al-Amoudi A, Eggli P (2001) A new approach for cryofixation by high-pressure freezing. J Microsc 203(Pt 3):285–294
58. Vanhecke D, Graber W, Studer D (2008) Close-to-native ultrastructural preservation by high pressure freezing. Methods Cell Biol 88:151–164. doi:10.1016/S0091-679X(08)00409-3
59. Miiller M, Moor H (1984) Cryofixation of thick specimens by high pressure freezing. Sci Biol Specimen Prep
60. Humbel BM, Schwarz H (1989) Freeze-substitution for immunochemistry. Immuno-gold labeling in cell biology. CRC Press, Boca Raton
61. Vanharreveld A, Crowell J (1964) Electron microscopy after rapid freezing on a metal surface and substitution fixation. Anat Rec 149:381–385
62. Hippe-Sanwald S (1993) Impact of freeze substitution on biological electron microscopy. Microsc Res Tech 24(5):400–422. doi:10.1002/jemt.1070240506
63. Nicolas MT, Bassot JM (1993) Freeze substitution after fast-freeze fixation in preparation for immunocytochemistry. Microsc Res Tech 24(6):474–487. doi:10.1002/jemt.1070240604
64. White DL, Andrews SB, Faller JW, Barrnett RJ (1976) The chemical nature of osmium tetroxide fixation and staining of membranes by x-ray photoelectron spectroscopy. Biochem Biophys Acta 436(3):577–592
65. Humbel BM, Marti T, Müller M (1983) Improved structural preservation by combining freeze substitution and low temperature embedding. Beitr Elektronenmikroskop Direktabb Oberfl 16:585–594
66. Walther P, Ziegler A (2002) Freeze substitution of high-pressure frozen samples: the visibility of biological membranes is improved when the substitution medium contains water. J Microsc 208(Pt 1):3–10
67. Morphew MK, McIntosh JR (2003) The use of filter membranes for high-pressure freezing of cell monolayers. J Microsc 212(Pt 1):21–25
68. Jimenez N, Vocking K, van Donselaar EG, Humbel BM, Post JA, Verkleij AJ (2009) Tannic acid-mediated osmium impregnation after freeze-substitution: a strategy to enhance membrane contrast for electron tomography. J Struct Biol 166(1):103–106. doi:10.1016/j.jsb.2008.12.009
69. McDonald KL, Webb RI (2011) Freeze substitution in 3 hours or less. J Microsc 243(3): 227–233. doi:10.1111/j.1365-2818.2011.03526.x
70. Hawes P, Netherton CL, Mueller M, Wileman T, Monaghan P (2007) Rapid freeze-substitution preserves membranes in high-pressure frozen tissue culture cells. J Microsc 226(Pt 2):182–189. doi:10.1111/j.1365-2818.2007.01767.x
71. Tokuyasu KT (1973) A technique for ultracryotomy of cell suspensions and tissues. J Cell Biol 57(2):551–565

72. Griffiths G, McDowall A, Back R, Dubochet J (1984) On the preparation of cryosections for immunocytochemistry. J Ultrastruct Res 89(1):65–78
73. Tokuyasu KT (1978) A study of positive staining of ultrathin frozen sections. J Ultrastruct Res 63(3):287–307
74. Tokuyasu KT (1989) Use of poly(vinylpyrrolidone) and poly(vinyl alcohol) for cryoultramicrotomy. Histochem J 21(3):163–171
75. Rangell LK, Keller GA (2000) Application of microwave technology to the processing and immunolabeling of plastic-embedded and cryosections. J Histochem Cytochem: Off J Histochem Soc 48(8):1153–1159. doi:10.1177/002215540004800812
76. Takizawa T, Anderson CL, Robinson JM (2003) A new method to enhance contrast of ultrathin cryosections for immunoelectron microscopy. J Histochem Cytochem: Off J Histochem Soc 51(1):31–39. doi:10.1177/002215540305100105
77. Painter RG, Tokuyasu KT, Singer SJ (1973) Immunoferrin localization of intracellular antigens: the use of ultracryotomy to obtain ultrathin sections suitable for direct immunoferritin staining. Proc Natl Acad Sci U S A 70:1649–1653
78. Moor H (1987) Theory and practice of high-pressure freezing. Cryotechniques in biological electron microscopy. Springer, Berlin
79. McDonald KL, Sharp DJ, Rickoll W (2000) Preparing thin sections of Drosophila for examination in the transmission electron microscope. In: Drosophila: a laboratory manual. CSHL Press, Cold Spring Harbor
80. McDonald K, Muller-Reichert T (2002) Cryomethods for thin section electron microscopy. Methods Enzymol 351:96–123
81. McDonald K (2007) Cryopreparation methods for electron microscopy of selected model systems. Methods Cell Biol 79:23–56. doi:10.1016/S0091-679X(06)79002-1
82. McDonald K, Schwarz H, Muller-Reichert T, Webb R, Buser C, Morphew M (2010) "Tips and tricks" for high-pressure freezing of model systems. Methods Cell Biol 96:671–693. doi:10.1016/S0091-679X(10)96028-7

Chapter 9
Solid-State Nuclear Magnetic Resonance Spectroscopy of Membrane Proteins

Shenlin Wang, Xiaojun Xu and Yufei Yang

9.1 Introduction

Membrane proteins are critical components of biological membranes. They carry out diverse cellular functions as membrane receptors that trigger environmental stimuli, transporters that move cargo across the cell membranes, membrane enzymes to catalyze the biological reaction on membranes, and adhesion proteins to mediate cell communications among others. Many membrane proteins, e.g., the proteins of the GPCR family, are directly implicated in human diseases, which make them important drug targets. It has been estimated that about 60% of marketed drugs target on membrane proteins.

The structures of membrane proteins are key to understanding their physiological roles and can provide valuable clues for fragment-based drug design. In recent decades, the high-resolution structures of many challenging membrane proteins have been solved due to advancements in several biophysical techniques, including X-ray crystallography [1–7], nuclear magnetic resonance [8], and cryo-electron microscopy [9, 10]. Of these, solid-state NMR (ssNMR) is a flexible technique for the analysis of membrane-associated proteins in all kinds of environments [11–14], and in particular, has unique advantages in investigating membrane proteins in native-like membrane environments [15–17] without the need to form crystallines or dissolve membrane proteins in detergent solution. Impressive examples include virus channel proteins [18, 19], cell envelope proteins [20], adhesion proteins [14], microbial multi-transmembrane proteins [15], and human GPCRs [16]. In addition,

S. Wang (✉) · X. Xu · Y. Yang
College of Chemistry and Molecular Engineering, Peking University, Beijing 100871, China
e-mail: wangshenlin@pku.edu.cn

S. Wang · X. Xu · Y. Yang
Beijing NMR Centre, Peking University, Beijing 100871, China

H. Wang and G. Li (eds.), *Membrane Biophysics*,
https://doi.org/10.1007/978-981-10-6823-2_9

ssNMR is a powerful tool to depict the dynamics of membrane proteins and to visualize minor conformational states related to the molecular function [21]. For example, ssNMR has been applied to characterize the Meta II state of bovine rhodopsins [21]. Furthermore, several exciting examples have demonstrated that ssNMR is a promising technique in studying membrane proteins in heterogeneous cellular membranes [22] or in whole cells [23], providing a new opportunity for in vivo characterization of membrane proteins.

Historically, two major methods of ssNMR have been applied in membrane protein characterizations. Most early ssNMR works on single transmembrane proteins were carried out with oriented sample ssNMR (OS-ssNMR). In this method, measurements of ^{15}N chemical shift anisotropy (CSA) and ^{15}N–^{1}H dipolar couplings derive orientation of membrane proteins with respect to the lipid bilayer normal. In contrast, magic-angle-spinning ssNMR (MAS-ssNMR) uses fast spinning of a rotor with an axis inclined at 54.7° with respect to the magnetic field and is capable of obtaining high-resolution spectra of membrane proteins. With MAS-ssNMR, most of the experiments are designed to correlate the isotropic chemical shifts of investigated nuclei. As numerous distance restraints can be collected, MAS-ssNMR is powerful for full three-dimensional structure determination. Both techniques are powerful in characterization of membrane proteins in lipid environments, but each has limitations. In this chapter, we introduce the basic principles of both OS-ssNMR and MAS-ssNMR and highlight the historically important examples of membrane proteins characterized by ssNMR.

9.2 Principle

9.2.1 Oriented Sample Solid-State NMR (OS-ssNMR)

In NMR, the chemical shift of a nucleus is orientation dependent, which is termed chemical shift anisotropy (CSA). The rapid motions of the molecules in solution average out the CSA, showing only the isotropic chemical shifts, whereas in solid, chemical shifts in all possible directions are present, giving a "powder pattern." Another central orientation-dependent interaction in NMR is inter-spin dipolar coupling, which depends on the inter-spin distances and the angle of the spin–spin vector with respect to the magnetic field. Like CSA, dipolar couplings are averaged to zero in solution, but they remain anisotropic in solids. These anisotropic interactions significantly broaden the spectral lines but contain valuable information of molecular structures. Taking advantage of the flat structure of lipid bilayers, the orientation of membrane-associated proteins with respect to the lipid bilayer normal can be extracted from both the CSA and the dipolar coupling data, if the direction of the macroscopic lipid bilayers can be controlled. This leads to the experimental design of OS-ssNMR [24–26].

9.2.1.1 Sample Preparation

Controlling macroscopic lipid bilayers reconstituted with membrane proteins oriented in the same direction is mandatory to OS-ssNMR study (Fig. 9.1) [27]. This can be achieved by mechanically aligning lipid bilayers on glass plates. To prepare OS-ssNMR samples on glass plates, uniformly ^{15}N-labeled or site-specific ^{15}N-labeled protein are co-dissolved with the lipids in an organic solvent, i.e., MeOH or $CHCl_3$ are mostly used cases, to form a homogeneous solution. The mixed solution directly dropped onto the glass slides, followed by removal of organic solvents under a nitrogen or argon gas stream. Multiple pieces of such glass slides with the dried lipids are then packed on top of each other, followed by the processes of evaporation and rehydration. During evaporation, the lipids will be spontaneously arranged as several thousand of flatted bilayers, with mosaic spread proteins, aligned mechanically between each pair of glass surface. After rehydration, the fluid lipid membranes are formed, and the proteins can either remain within the membrane or emerge on the membrane surface depending on their intrinsic properties [27]. Another approach to the preparation of OS-ssNMR

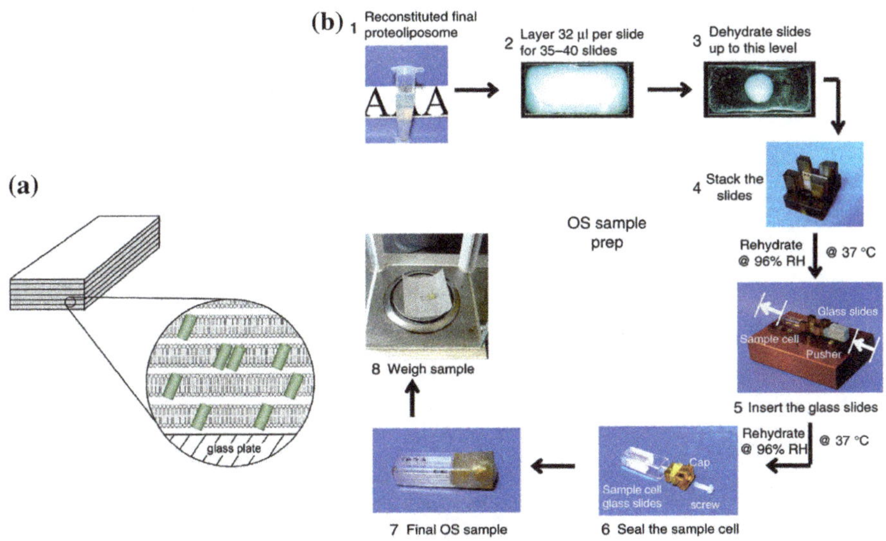

Fig. 9.1 Schematic presentation of macroscopically oriented sample and sample preparation procedures. **a** Between every each pair of glass plates, multilamellar bilayers (gray) with inserted peptides are uniformly oriented (green). Reprinted from Ref. [27], Copyright 2007, with permission from Elsevier. **b** Oriented sample preparation procedures (1) Preparation of reconstituted proteoliposome. (2) Spread the proteoliposome solution onto the slides. (3) Dehydrate slides under room temperature. (4) Stack slides and cover the top with one slide. The number of slides used depends on the size of the probe coil. (5) Rehydrate the stack to 96% relative humidity, and insert into a sample cell or wrap it with parafilm and plastic film. (6–7) Seal the sample cell. (8) Weigh the sample for hydration control. Reprinted by permission from Macmillan Publishers Ltd: Ref. [29], Copyright 2013

samples is to reconstitute membrane proteins in bicelles. Owing to their large size in solution, the bicelles are oriented intrinsically to an angle with respect to the magnetic field, owing to their large size in solution. In some cases, a small amount of paramagnetic metal ion-chelated lipids were used together to form the paramagnetic bicelles, in which the paramagnetic ions facilitate the alignment of the bicelles [28].

9.2.1.2 Orientation Dependence of CSA and Dipolar Coupling

In macroscopically oriented samples, the ^{15}N sites of backbone amides are commonly used to determine the peptide orientation. The CSA tensors of amide-^{15}N (σ_{11}, σ_{22}, σ_{33}) in the molecular frame are defined by N–C=O as *x*–*y* plane and *z* axis perpendicular to the plane (Fig. 9.2) [30]. The chemical shift can be then deduced according to:

$$\omega_0 = -\gamma B_0\left(1 + \overline{\sigma_{33}(\tau)}\right) \quad (9.1)$$

where τ denotes the orientation of the molecule, and $\sigma_{33}(\tau)$ describes the σ_{33} element in laboratory frame, which is dependent on the molecular orientation and the chemical shift principal values. Specifically, the chemical shift of ^{15}N is dependent on the angle between σ_{33}-element of the ^{15}N CSA and the external

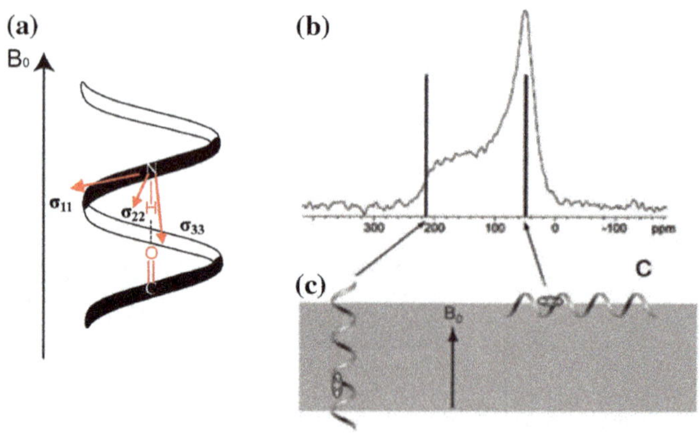

Fig. 9.2 Illustration of orientation-dependent CSA of amide-^{15}N. **a** CSA of amide-^{15}N in molecular frame defined by N–C=O as *x*–*y* plane and *z* axis perpendicular to the plane. **b** "Power pattern" spectrum, beneath which chemical shifts in all possible orientation are presented. **c** Topology of membrane proteins in the membranes can be characterized by the ^{15}N chemical shift of the amide group. ^{15}N chemical shifts at 200 ppm and 45 ppm indicate inserted state and parallel state of the peptides, respectively. Gray box shows the membranes macroscopically oriented with the lipid bilayer normal parallel to the magnetic field B_0. Reproduced from Ref. [32] by permission of John Wiley & Sons Ltd

magnetic field B_0. The σ_{33}-vector is $\sim 17°$ away from the $^{15}N–^{1}H$ bond [31], which is parallel to the helical peptide axis. The peptide orientation can thus be roughly derived by the amide-^{15}N chemical shift. As shown in Fig. 9.2b, the amide-^{15}N chemical shift around 200 ppm indicates the vertically inserted state of the peptides, and the chemical shift in the lower field around 45 ppm suggests that the peptides align parallel to the membrane surface. Although different alignments can be easily distinguished from the chemical shift of one single amide-^{15}N, it is insufficient to determine the precise orientation. The line width of ^{15}N can also report the peptide's mobility; a sharp line indicates that the peptide undergoes rapid rotation, whereas a broad line suggested a rigid segment.

Dipolar coupling is a through space interactions between nuclei in magnetic field. The strength of dipolar coupling depends on the gyromagnetic ratios of the interacting nuclei, the inter-nuclear distance r, and the orientation θ of the inter-nuclear vector with respect to the magnetic field B_0 (Fig. 9.3). The general inter-nuclear dipolar coupling is given by:

$$D = \left| 2f\hbar\gamma_1\gamma_2 \frac{\mu_0}{4\pi} \frac{1}{r_{12}^3} \frac{3\langle\cos^2\theta\rangle - 1}{2} \right| \tag{9.2}$$

where $\hbar = 1.054969 \times 10^{-34}$ Js is the value of Planck's constant, and $\mu_0/4\pi = 10^{-7}$ Hm^{-1} is magnetic field constant. γ_1 and γ_2 denote the gyromagnetic ratios of two spins, and the factor f in the case of homonuclei and for the distinct nuclei is 1.5 and 1.0, respectively. The angular brackets describe the motional averaging of the entire protein within the bilayer.

Since the dipolar coupling is also dependent on the orientation, measurements on $^{1}H–^{15}N$ dipolar coupling can yield protein orientation in OS-ssNMR sample. Like CSA, data from a single site can produce rough estimation of protein orientation, but precise orientation requires information from multiple sites.

Fig. 9.3 Illustration of inter-nuclei dipolar coupling within the peptide oriented in the membrane. r denotes the distance between a pair of spins, and θ describes the orientation between the inter-nuclear vector and the external magnetic field B_0 (N, in case the membrane normal is parallel to the magnetic field). Reprinted from Ref. [33], Copyright 2005, with permission from Elsevier

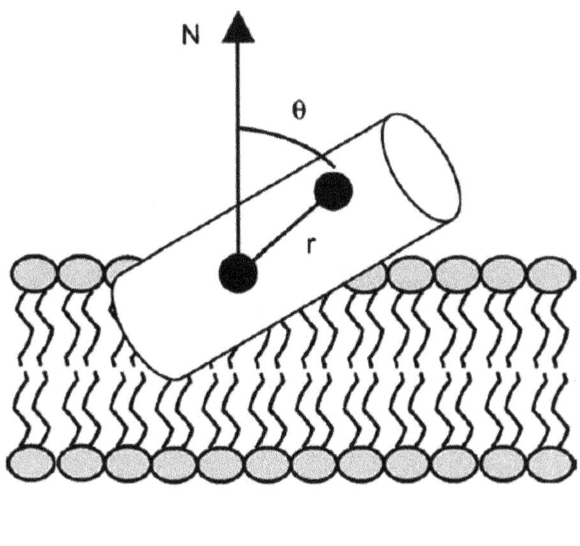

9.2.1.3 PISEMA

The classical PISEMA experiment was introduced by the Opella group in 1994 [34] and has been extensively used to precisely determination the topology and dynamics membrane-associated proteins. Using ^{15}N-labeled proteins, it correlates ^{1}H–^{15}N dipolar couplings with ^{15}N-chemical shifts to produce a characteristic wheel-like spectral pattern, termed a PISA-wheel [35]. Figure 9.4 displays simulated PISA-wheel spectra for ideal α-helical and β-strand proteins in various orientations with respect to the external magnetic field B_0. For α-helical structures, the PISA-wheel displays a circle pattern, which can be understood as a projection of the peptide helical wheel, whereas a twisted PISA-wheel can be observed for β-strands. Predicted from the theoretical simulations (Fig. 9.4), the helical proteins tilted away from the magnetic field B_0 possess good spectral dispersion, while the β-strands exhibit well-dispersed resonances when they are aligned parallel on the membrane surface. Protein orientation and conformation in lipid bilayers can be thus extracted by comparing the experimentally detected spectral patterns (by chemical shifts and dipolar couplings) with those simulated for all possible orientations. With known three-dimensional structure, the single ^{15}N-labeled membrane proteins are often designed to have ^{15}N sites around the helical wheel. It is also possible to use uniformly ^{15}N-labeled proteins without having resonance assignments, in which the center position of PISA-wheel was selected to analyze the topology.

As an extension of PISA-wheels, a plot of dipolar waves, which shows the periodic wavelike variation of ^{1}H–^{15}N dipolar couplings against primary sequence, is particularly powerful to identify non-ideal helical-type (i.e., π-helix), and to reveal the kinks in the helix [20, 37, 38]. Combination of PISA-wheels and dipolar waves is sufficient to characterize the three-dimensional conformation of membrane-bound peptides. For example, the plot of dipolar wave has been used to study the three-dimensional structure of coat protein in filamentous bacteriophages, revealing the orientation of coat protein with respect to the virus lipid bilayers and two kinks in its α-helical structure (Fig. 9.5) [20]. In contrast to the analysis of helical structure, it is more complicated to determine the β-strand structure, because the resulting twisted PISA-wheels are not easily recognizable.

9.2.2 MAS-ssNMR

MAS-ssNMR uses fast spinning of a rotor, which has a spinning axis inclined at 54.7° with respect to the magnetic field. In MAS-ssNMR, the rapid spinning along magic angle averages out CSA and dipolar coupling, showing only the isotropic chemical shift and largely suppressed the broadening effects caused by dipolar coupling. It results in much narrower spectral line width compared to power pattern spectra, dramatically improving the spectral resolution. The spectral patterns of MAS-ssNMR are similar to those of solution NMR, collecting only the isotropic chemical shifts. As for membrane protein studies, many multidimensional

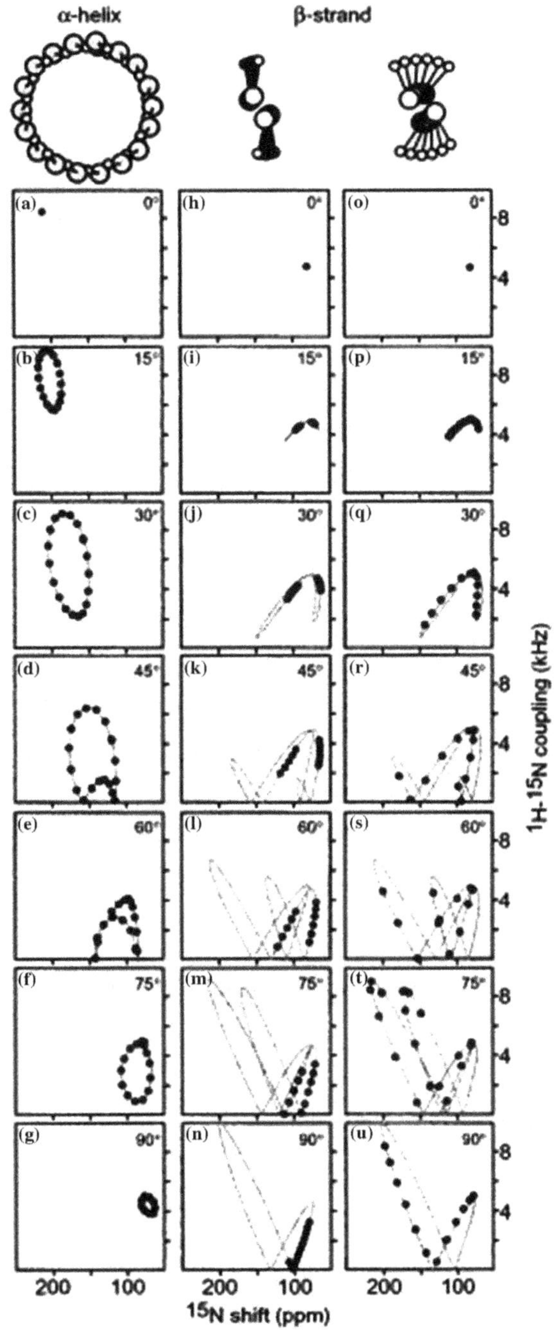

Fig. 9.4 Simulated PISA-wheel spectra for ideal α-helical and β-strand structures in different orientations with respect to the magnetic field B_0. Reprinted from Ref. [36], Copyright 2001, with permission from Elsevier

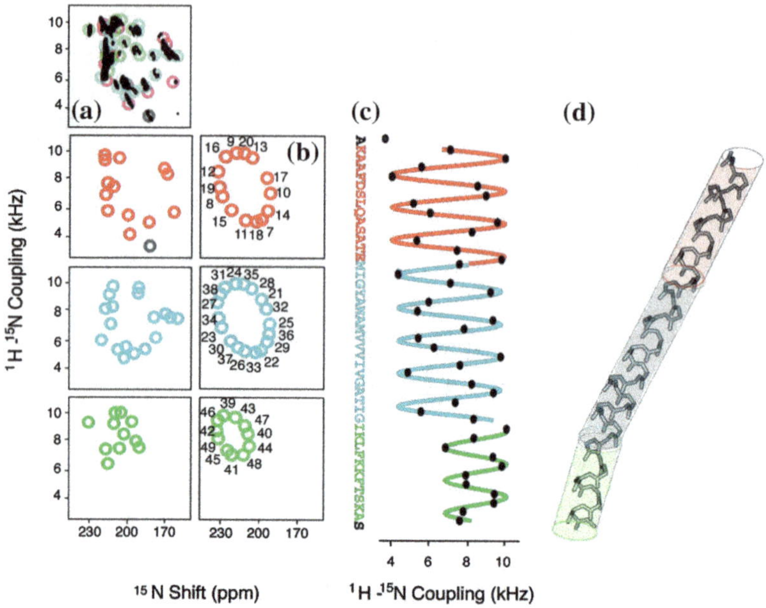

Fig. 9.5 **a** PISEMA spectrum of uniformly and selectively ^{15}N-labeled Y21M coat protein in filamentous bacteriophages particles. **b** Ideal PISA-wheels with respect to the corresponding experimental data. **c** ^{1}H–^{15}N dipolar coupling plotted against the protein sequence. **d** Three-dimensional structure of Y21M coat protein derived from PISEMA and dipolar wave data (PDB code: 1NH4). Reproduced from Ref. [20], The National Academy of Sciences of the USA, 2003

MAS-ssNMR experiments have been developed to perform chemical shift assignments. With recoupling techniques, experimental designs to obtain distance restraints for structure determination have also been well established. In recent years, MAS-ssNMR has been used extensively to investigate lipids-embedded membrane proteins with complex structures, i.e., seven-transmembrane proteins [15, 16].

9.2.2.1 Sequential Assignments

The spectroscopic studies of MAS-ssNMR on membrane proteins commonly begin from 2D ^{15}N–^{13}Cα (Fig. 9.6) and 2D ^{15}N–^{13}CO spectra. Each cross peak on ^{15}N–^{13}Cαspectra represents Cα and amide nitrogen from the same amino acids, whereas cross peaks on ^{15}N–^{13}CO spectra shows the correlation between amide nitrogen with carbonyl carbons from the proceeding amino acid residue. To obtain site-specific assignments that link the cross peaks to the amino acid residues of the primary sequence, a series of multidimensional correlation spectra are needed. Commonly, the minimal sets of three three-dimensional ssNMR spectra must be

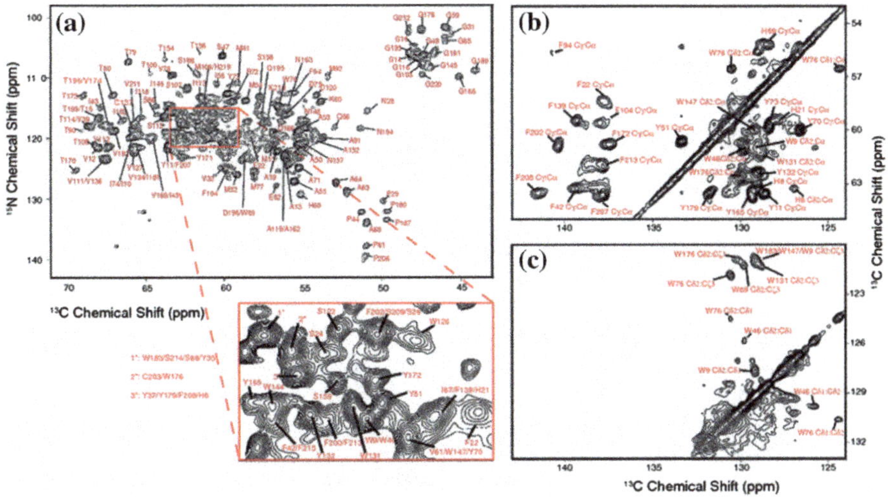

Fig. 9.6 An example of 2D ^{15}N–^{13}Cα spectrum (**a**) and part of 2D ^{13}C–^{13}C correlation spectrum (**b–c**) collected on *Anabaena* sensory rhodopsin. The site-specific assignments are highlighted in red. Reprinted from Ref. [43], with kind permission from Springer Science+Business Media

collected, including CANCO, NCACX, and NCOCX [39]. CANCO defines spin systems by linking the Cα and amide nitrogen of residue *i* and carbonyl of residue *i*−1 (the proceeding residue), NCACX links intra-residue correlations between amide nitrogen, Cα, and all other carbons from the same residue, and NCOCX connects amide nitrogen of residue *i*, and carbonyl and other carbons of residue *i*−1. The sequential backbone walk is capable of matching NCACX and NCOCX from different residues, thus linking the spin systems (Fig. 9.7). The types of amino acid residues can be derived from the chemical shifts of side chains, which show characteristic patterns for different kinds of amino acid residues. Combination of sequential walk and amino acid type identification could correlate the cross peaks to the position in the primary sequence of the proteins. As for large membrane proteins, particularly for the carbonyl resonances, some of the signals of different residues may overlap, making it difficult for both the identification of spin systems and the sequential links. Many types of three-dimensional spectra, including NCACB, N(CO)CACB, CAN(CO)CA [40], and four-dimensional spectroscopy has been advanced to overcome this problem [41, 42].

Alternatively, versatile isotope labeling approaches have been advanced to facilitate resonance assignments and to collect distance restraints for structure determination [13, 44–46]. In those approaches, metabolic precursors with different labeling patterns (i.e., 2-^{13}C and 1,3-$^{13}C_2$ glycerol are the most used cases) were supplemented in the cell growth media for protein productions, where the labeling patterns are recorded in the amino acids by biosynthesis. It results in alternate labeling patterns for most of the amino acids and simplifies the NMR spectra. Other benefits of the glycerol-labeled samples are enhancement of the spectral resolutions,

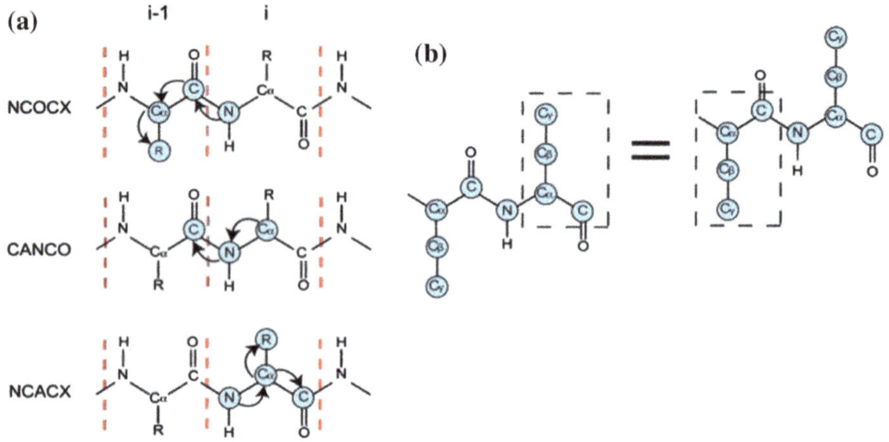

Fig. 9.7 Schematic presentation of backbone assignments strategies. **a** CANCO defines the spin system by correlating chemical shifts of Cα and N of residue *i*, and CO of residue $i - 1$ (the preceding residue). The NCOCX correlates chemical shifts of N of residue *i*, and CO and CX of residue $i - 1$. The NCACX correlates N, Cα and CX from the same residue. **b** Sequential links between two spin systems can be made by matching shifts of backbone and side chain atoms shown in dashed boxes. Reprinted from Ref. [39], with kind permission from Springer Science +Business Media

facilitation of the assignments of aromatic side chains, and aid in collecting distance restraints for structure determination.

More recently, under ultrafast MAS conditions, proton-detected experiments have also been developed to obtain multidimensional spectra [47]. Ultrafast MAS improves spectral resolution significantly, as proton line width is much narrowed due to largely suppression of ^{1}H–^{1}H dipolar couplings. With a combination of the perdeuterated approaches and high magnetic field, the spectral quality of ^{1}H-detected spectral of membrane proteins are as good as those obtained by solution NMR spectroscopy [47]. Corresponding three- and four-dimensional experiments for sequential assignments and structural determination have been well established for crystalized soluble proteins [48], some of which are combined with partially deuterated labeling approaches. Those protocols are expected to be transferable to study large membrane proteins.

9.2.2.2 Secondary Structure Analysis

The isotropic chemical shift data of amino acids in proteins contain a wealth of information about the secondary structure. Established in 1994 by the Sykes group, a simple and efficient protocol called chemical shift index (CSI) [49] was established to derive the secondary structures of investigated proteins by comparing the Cα, Cβ, and CO shifts of amino acids with those in the random coils. As for α-helical segments, the chemical shifts of Cα are higher than the random coil of the

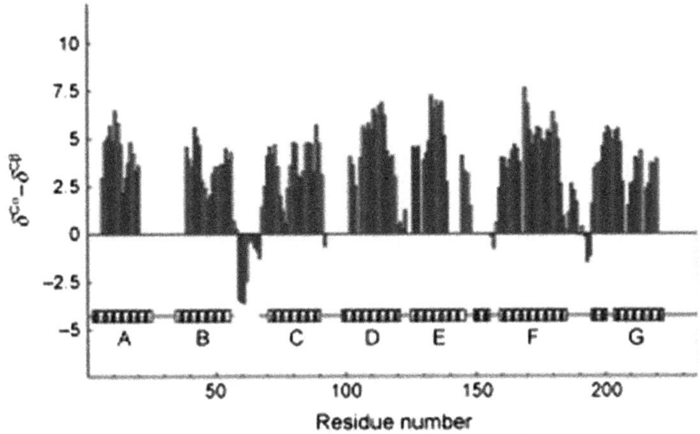

Fig. 9.8 An example of CSI analysis on secondary structure of *Anabaena* sensory rhodopsin, a seven-transmembrane protein. Site-specific CSI data are plotted as a function of residue number. The secondary structure derived from the X-ray data (PDB 1XIO) is shown at the bottom. Reproduced from Ref. [51], by permission of John Wiley & Sons Ltd

same type of amino acids and the Cβ chemical shifts are lower, whereas the rules for resides on the β-sheet are reversed. Although the CSI protocols were developed originally for solution NMR studies on soluble proteins, they are also amenable for solid-state NMR studies on membrane proteins. Figure 9.8 shows an example of CSI analysis on a seven-transmembrane protein in lipid environments. Program TALOS+ is another easy-to-use package for predicting the secondary structure and is capable of making predictions of the torsion angles of proteins [50]. Unlike the chemical shift index, which simply uses the chemical shifts, TALOS+ consider the primary sequence, the known crystal structures of protein with similar primarily sequence, and the chemical shift data. It is worthy of note that torsion angles predicted by TALOS+ program can be used as restraints for structural determination by ssNMR, leading to significant improvement on the quality of the structures.

9.2.2.3 Full Three-Dimensional Structure Determination of Membrane Proteins by MAS-ssNMR

Unlike other biophysical techniques that record the images of molecules, such as X-ray and cryo-electron microscopy, NMR techniques collect a vast number of inter-nucleus distances on the target proteins and use those restraints to reconstruct the protein structures. In MAS-ssNMR, the inter-atomic distances are mainly obtained by dipolar recoupling techniques, which is capable of measuring inter-atomic distances on isotope-labeled proteins. Proton-driven spin diffusion (PDSD) [52] or dipolar-assisted rotational resonance (DARR) [53, 54] experiments have been used extensively to probe the carbon–carbon distances within membrane proteins. Both spectra are usually performed with combination of alternate ^{13}C labeling schemes to probe long-range distances, because alternate labeling schemes are less sensitive to dipolar truncation effects, which limits long-range magnetic transfer. Other magnetization transfer methods have also been developed to probe long-range distance restraints or to measure accurate carbon–carbon distances in uniformly ^{13}C-labeled proteins, including frequency-selective recoupling techniques, such as rotational resonance (R2) [55] and homogeneously broadened rotational resonance (HBR2) [56]. The latter has been used to study *Anabaena* sensory rhodopsin and provided a number of important long-range inter-helical restraints for structure determination. Furthermore, the proton-assisted polarization techniques, including proton-assisted recoupling (PAR) [57, 58] and proton-assisted insensitive nuclei cross polarization (PAIN) [59] for homo-nuclear and hetero-nuclear distance measurements, respectively, are a new generation of recoupling techniques that have particular advantages in probing long-range distances, up to 7 Å, and are efficient at fast spinning rates. In addition, three-dimensional experiments based on the PAIN and the PAR have been developed to facilitate the resonance assignments of insoluble protein aggregates [60]. In addition, experiments based on rotational-echo double resonance (REDOR) [61] or transferred-echo double resonance (TEDOR) [62] are widely used to probe hetero-nucleus distances. These methods were initially used to determine the conformations of uniformly ^{13}C, ^{15}N-labeled peptides [63] and have also been used to probe protein interfaces [64].

Paramagnetic relaxation effects (PRE) have been used to probe even longer distance, up to 15 Å. Chemical modification on reactive Cys side chains allows to attaching a tag with stable free radical, i.e., the nitroxide radical MTSL, on the protein surface. The unpaired electrons thus enhance the transverse relaxation rates of nearby nucleus, causing line-broadening effects on the corresponding NMR signals. PRE effect is proportional to the sixth power of the electron–nucleus distance, making it possible to measure the distances between the unpaired electron and the affected nucleus. In the case of *Anabaena* sensory rhodopsin, PRE effects have been used to probe the inter-monomer interface of a trimer and to facilitate in improvement on the structure precisions [15].

To construct the membrane protein structures based on MAS-ssNMR spectroscopy, all experimental detected inter-nucleus distances and torsion angle restraints are subject to simulated annealing using molecular dynamic programs for structure calculation. In some cases, solution NMR, OS-ssNMR and EPR data can be combined with MAS-ssNMR data to produce high-resolution structures [17] or improve the precision [65]. Considering the dynamic nature of proteins, an ensemble of structures is usually obtained, starting from a number of random unfold conformations and annealing to form a family of folded structures. On MAS-ssNMR spectra, many cross peaks cannot be assigned unambiguously using only the chemical shift data, leading to ambiguous assignment of a vast majority of cross peaks. Therefore, structure calculation based on MAS-ssNMR data usually begins with spectrally unambiguous restraints to derive a low-resolution ensemble, which can be used to disambiguate the remaining peaks. These newly assigned

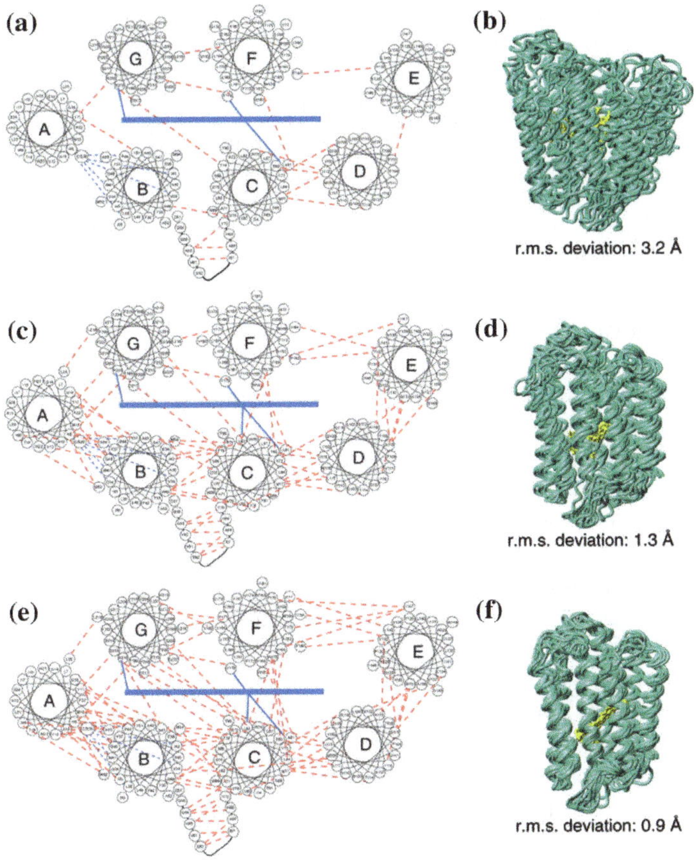

Fig. 9.9 An example of improvement of ssNMR structure precision with more unambiguous restraints in structure calculation, *Anabaena* sensory rhodopsin as an example. **a**, **c**, **e** Helical wheel representations of ASR monomer viewed from the periplasmic side. Unambiguous inter-helical restraints identified from MAS-ssNMR spectra are depicted by dashed red lines connecting pairs of correlated residues. Blue bar represents connection between retinal cofactors and proteins. **b**, **d**, **f** Ensembles of structures derived from the unambiguous restraints r.m.s. deviations are calculated for backbone atoms of α-helices. Reprinted by permission from Macmillan Publishers Ltd: Ref. [15], Copyright 2013

peaks are added to the restraints list for a new round of calculation. These processes will be repeated until the structure reaches high resolution (low root-mean-square deviation of the structure ensemble) and the majority of the cross peaks are unambiguously assigned. Figure 9.9 shows the example of structure determination on *Anabaena* sensory rhodopsin [15], which experiences progressively improvement on structure resolution as a result of obtaining more unambiguous restraints. Automatic or semi-automatic structure calculation protocols that use ambiguous restraints as starting points have also been developed and implicated in ARIA program. In this approach, all restraints are treated as ambiguous restraints and are used in the first round of structural calculation. At the second round, the initial low-resolution ensemble remove a portion of the ambiguity of the cross peaks. At the following cycles, the precision of the structure ensemble would be improved, and more cross peaks are expected to be disambiguated. The quality of ssNMR structures can be assessed by a number of programs, such as PROCHECK [66], WHATIF [67, 68], and MolProbity [69], to evaluate the compactness of the protein, Ramachandro plot of backbone torsions, and close packings.

9.2.2.4 Membrane Protein Topology

The transmembrane protein has a topological component, which describes its orientation, insertion depth in the lipid bilayer, solvent exposed segments, or polar cavities inside the protein. In addition to the PISA-wheel patterns that have been discussed in OS-ssNMR section, several protocols of MAS-ssNMR have also been developed to study membrane protein topology. Hydrogen-deuterium exchange offers the simplest way to investigate the solvent-accessible surface of membrane proteins. Incubating lipid-embedded membrane proteins in D_2O-based buffer allows the amide protons of the solvent exposed residues to be replaced by deuterons, leading to suppression of those signals on 2D ^{15}N–$^{13}C\alpha$ spectra or 3D spectra. Therefore, comparison spectral patterns of membrane protein before and after incubation in D_2O provides the information of the solvent accessible surface of membrane proteins. This approach has been used in study the solvent-exposed residues for *Anabaena* sensory rhodopsin [70] in DMPC/DMPA lipids, DAGK enzyme reconstituted in *E. coli* lipids [71], and human aquaporins in PC/PS lipids [72].

The membrane protein topology can also be investigated by ^{13}C–^{31}P REDOR experiments. The ^{13}C–^{31}P REDOR experiments estimate the distance between residues and phosphate groups of membrane, suggesting the depth of protein penetration. Using selective ^{13}C-labeled proteins, lipid insertion depth of different constructs of virus membrane fusion peptides have been characterized, which links their topologies to the physiological functions [73]. Solvent PRE effects (PRE) are another easy-to-use approach to study the penetration depth of membrane protein in lipid bilayers. Metal ions, i.e., Mn^{2+}, Cu^{2+}, and Gd^{3+}, bind to the negative-charged phosphate groups on the membrane surface and cannot penetrate in the lipid bilayers. The unpaired electrons of those metal ions would induce severe line broadening on NMR signals of the residues located on the membrane surface, whereas the residues inside the lipid bilayers would not be affected. Thus, monitoring the changes of the spectral patterns by additions of metal ions could probe the solvent exposed surface of membrane proteins. Indeed, this approach has been used in studying the insertion depth of the fragment of bacteriorhodopsin [74], the topology of M2 proton channel [75] and asymmetry of peptide penetration in lipid bilayers [76].

9.3 Application of ssNMR in Membrane Proteins

9.3.1 The M2 Proton Channel of Influenza A

The M2 proton channel protein of influenza A is critical in conducting protons across the viral membrane and in acidification of the virus interior [77, 78]. The M2 proton channel has 97 residues, composing an N-terminal domain, a transmembrane domain, and a C-terminal helical domain. It forms tetramers in native membranes. As a proven pharmacological target, the structure and the mechanism of the M2 protein channel interactions with various drugs and inhibitors have attracted considerable attention. OS-ssNMR has been used to determine the backbone structures of a truncated construct of the M2 proton channel [79–82]. PISEMA experiments determined the orientation of the transmembrane helix with respect to the bilayer normal, revealed a helical kink at position of Gly34, and indicated that the C-terminal helices are aligned nearly parallel to the membrane surface (Fig. 9.10). His37 is the key residue to proton selectivity. OS-ssNMR structure of the M2 proton channel showed that four His37 of the tetramer face the core of the helical bundle and arrange into a histidine tetrad stabilized by inter-helical H-bonds formed by adjacent histidines (Fig. 9.10c). This results in a dimer of dimers arrangement. Furthermore, the presence of this unique structure has been confirmed for the full-length M2 protein in native *E. coli* membranes [83]. Other key residues involved in the functions of the proton activities are Val27 and Trp41, both of which have a role as the gates of the channel; Val27 locates at the entrance on the extracellular side and Trp41 positions at exit at the cytoplasmic side (Fig. 9.10d). MAS-ssNMR has also been used to characterize the structure and proton-conducting mechanism of the M2 proton channel [18, 19, 84, 85]. Investigation of the pH dependence of His37 protonation states, rotameric conformation, and side chain dynamics [19], revealed that under neutral pH, His37 experiences neutral imidazole, whereas at a low pH of 4.5, the histidines form double-protonated states, presenting the open conducting state. The dynamics of histidines side chain was studied by measuring the dipolar couplings, suggesting that at acidic pH, the His ring reorientation is directly involved in proton transport. These results support the shuttle model of proton conductance.

A number of anti-influenza drugs, including amantadine (Amt), rimantadine (Rmt), and copper salts, are known to block the proton transport of the M2 proton channel [87]. The ^{13}C–^{2}D REDOR distance measurements between perdeuterated Amt and the ^{13}C-labeled protein indicated a primary Amt binding site in the core of the helix bundle, located between Val27 and His37 (Fig. 9.11a) [18], whereas at high drug/protein ratios, amantadine binds nonspecifically on the surface of the protein. Line shape analysis of ^{2}D spectra suggested that Amt experience rapid rotation in the binding pocket. Rmt was found to bind to the protein at a similar site to Amt, while most of the protein experiences significant structural rearrangements upon Rmt binding, supporting the allosteric mechanism of inhibition [88]. Finally, unlike Amt, Cu^{2+} ion binds to the M2 proton channel at a position between His37

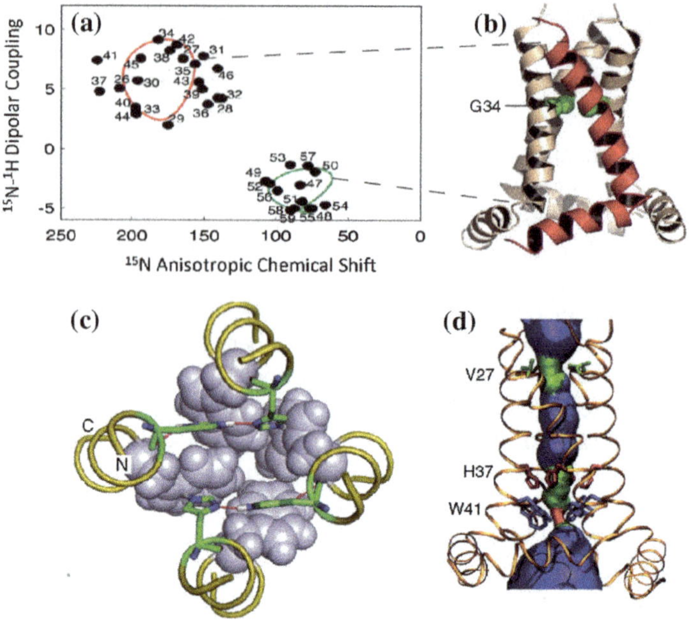

Fig. 9.10 Solid-state NMR structure of the M2 proton channel. **a** PISEMA spectrum of the M2 tetramer. **b** Tetrameric structure of the M2 proton channel. One monomer within the tetramer is shown in red. The helical kink is shown at position Gly34. The PISEMA patterns in (**a**) that derive the orientation of the helices are linked to the corresponding helices with dashed lines. **c** Packing of His37 inside the helical bundle, viewed from the cytoplasmic side. **d** Pore constriction sites at Val27 and Trp41. From [86]. Reprinted with permission from AAAS

and Trp41 (Fig. 9.11b), which was identified using the PRE effects induced by Cu^{2+} ions [89]. Although all three kinds of drugs can block the functions of the M2 proton channel, the mechanisms by which drug inhibits the channel activity are different from each other.

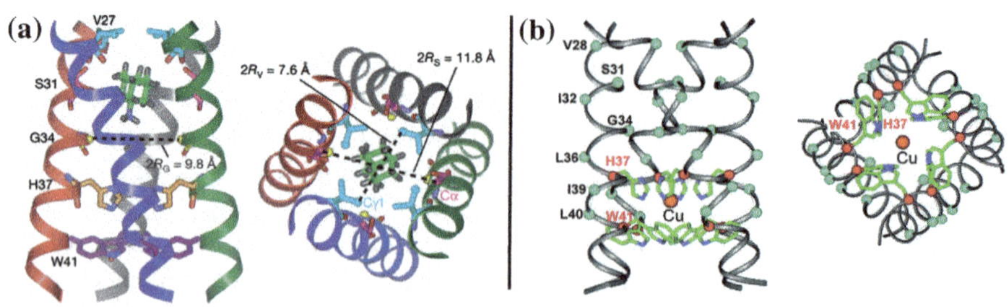

Fig. 9.11 Side view and top view of the ssNMR structures of Amt-bound M2 (**a**) and Cu^{2+}-bound of M2 (**b**). The figure **a** was reprinted by permission from Macmillan Publishers Ltd: Ref. [86], Copyright 2010. The figure **b** was reprinted with the permission from Ref. [89], Copyright 2012 American Chemical Society

9.3.2 Human Phospholamban

Human phospholamban (PLN) is an integral membrane protein that regulates the activity of Ca^{2+} cardiac sarcoplasmic reticulum ATPase (SERCA). In membrane, PLN can have two different aggregation states, monomer and pentamers, that experience variable physiological roles. Monomeric PLN can inhibit the activity of SERCA by physiological binding, whereas the pentameric form of PLN stores active monomers. It has also been argued that PLN pentamers are a Ca^{2+} channel [90].

Both the monomeric and the pentameric structures of PLN have been determined by combining solution NMR restraints and ssNMR restraints (Fig. 9.12) [17, 91–93]. To obtain the functional monomeric form of PLN, a construct of AFA-PLN with three residues mutated to Phe and Ala was engineered [94]. PISEMA data collected on monomeric PLN in lipid bilayers revealed two helical segments arranged in an L shape. One of those is oriented at a tilt angle nearly parallel to the lipid bilayer normal, showing an α-helical structure across the

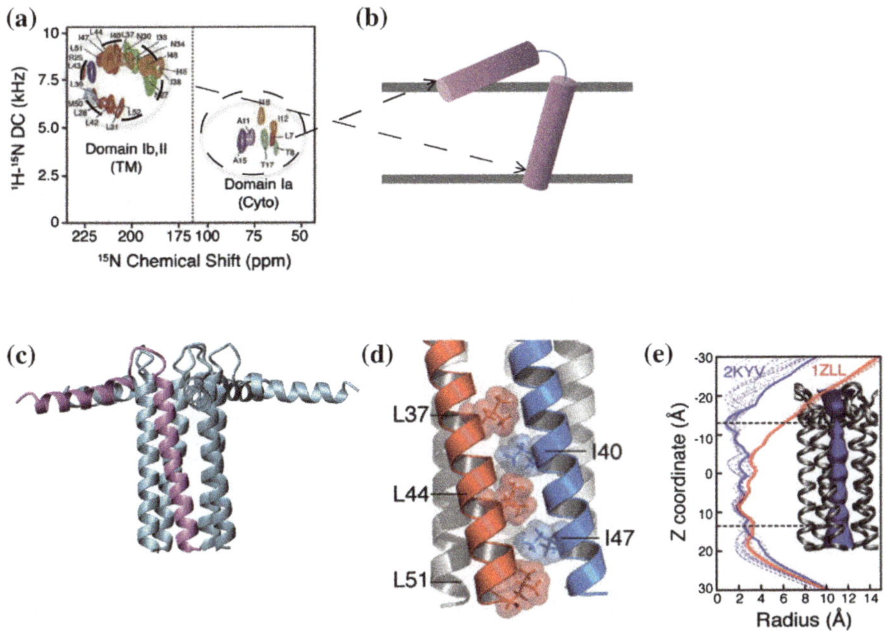

Fig. 9.12 Pentameric structure of PLN. **a** PISEMA spectrum of PLN. **b** Monomer structure within pentamer with two α-helices represented by cylinders. The PISEMA patterns in (**a**) that derive the orientation of the helices are linked to the corresponding helices with dashed lines. **c** Pentameric structure of PLN obtained using solution and solid-state NMR restraints (PDB ID: 2KYV). **d** Inter-helical side chain packing within the helical bundle of pentameric PLN, illustrating the hydrophobic surface facing the core of the protein. **e** Width of the central pore calculated. Reproduced from Ref. [17], The National Academy of Sciences of the USA, 2011

membrane, while the other spans on the membrane surface. Like the monomeric PLN, the pentamer structure of wild-type PLN has been determined using a hybrid approach, combining OS-ssNMR, MAS-ssNMR, and solution NMR [17]. Similar to the monomeric PLN, an L-shape topology was derived from the OS-ssNMR data for each monomer within the pentamers. To determine the packing between various adjacent monomers within the pentamers, the inter-monomer restraints were obtained from the proton-driven spin-diffusion spectra of MAS-SSNMR and from NOE restraints by solution NMR. Further, the detergent-protein NOEs by solution NMR were collected to study the conformation and orientation of the cytoplasmic helices. Taken together, a pinwheel model of wild-type PLN was built (Fig. 9.12c), which was also validated by RDC data from solution NMR, inter-electron distance restraints from EPR data and the Gd^{3+} caused solvent PREs [93]. This pinwheel structure identified a pentameric helical bundle with the diameter of the center of the pore only 3 Å. This size is too small to allow the Ca^{2+} ion to pass through, thus ruling out the Ca^{2+} channel activities of PLN. In addition, the core-facing residues are hydrophobic residues, which are unfavorable for Ca^{2+} binding, confirming that PLN is not a Ca^{2+} channel. Thus, the structure of PLN supported that pentameric PLN functions as an inactive state that stores the active monomeric PLN.

9.3.3 Potassium Channels

The potassium channels selectively conduct potassium ions across the membranes. They commonly have two transmembrane helices and form homo-tetramers in lipid bilayers. A conserved TVGYG motif, located in the extracellular part of the protein, constitutes the selectivity filter for potassium binding. The activities of potassium channels can be modulated by several factors, including K^+ concentration, pH, and channel blockers. MAS-ssNMR has been used to extensively investigate the inactivation mechanisms of potassium channel [95–97], using a chimeric potassium channel protein KcsA-Kv1.3 as model system. Resonance assignments allow unambiguously identification of the residues on selective filters. The addition of unlabeled channel blocker, a 38-residue long scorpion toxin kaliotoxin (KTX), to the ^{13}C-labeled KcsA-Kv1.3 induced shifting of those resonances and several other resonances corresponding to residues on the protein surface (Fig. 9.13), indicating that KTX binds at the surface of KcsA-Kv1.3 [96]. Chemical shift variations were also mapped on the spectra of ^{13}C-labeled KTX after binding with unlabeled KcsA-Kv1.3, suggesting that charged residues on the surface of KTX are the key to block the channel activity [96]. The mechanism of another channel blocker, a tetraphenylporphyrin derivative (porphyrin), was investigated in a similar approach. In contrast to KTX, porphyrin was found to lie in the middle of the pore of the selective filter region, resulting in a collapse of activated conformation, and showing a different inhibition mechanism with respect to the KTX.

The pH and K^+-dependent structure and inactivation mechanisms of the potassium channel were studied by comparison of the spectral patterns of KcsA-Kv1.3 in

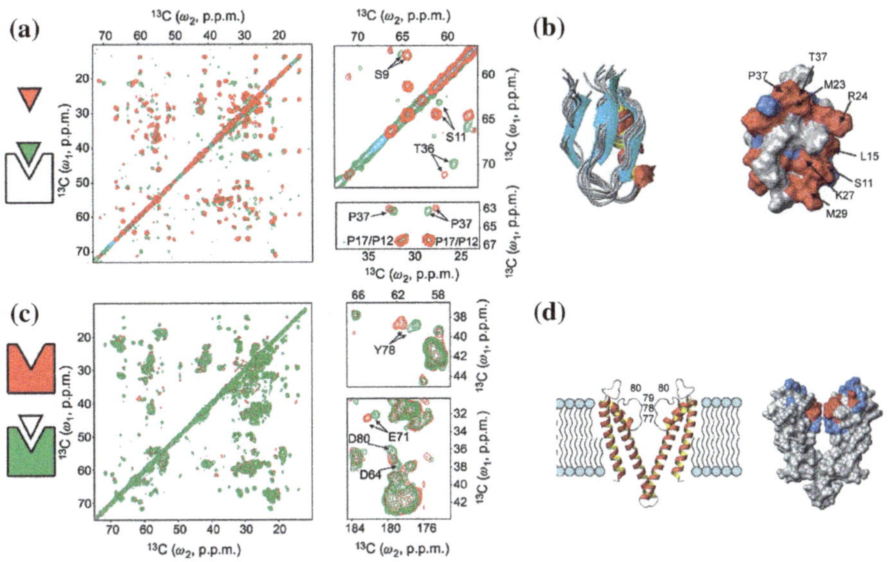

Fig. 9.13 Interaction studies between KscA-Kv1.3 and KTX. **a** A comparison of 2D ^{13}C–^{13}C correlation spectra of ^{13}C-labeled KTX (red) and in complex with unlabeled KcsA-Kv1.3 (green). KTX is shown as triangles at the left of the panel. **b** Structure ensemble of KTX and the residues experiencing significant chemical shift perturbation upon complex formation are shown in red. **c** A comparison of 2D ^{13}C–^{13}C correlation spectra of ^{13}C-labeled KcsA-Kv1.3 (red) and in complex with unlabeled KTX (green). **d** Cartoon representation of KcsA–Kv1.3 in a lipid bilayer. Reprinted by permission from Macmillan Publishers Ltd: Ref. [96], Copyright 2006

different buffer conditions [98]. At a low K^+ concentration, decrease in pH induce significant conformation changes of KcsA-Kv1.3. Sizable chemical shift variation were detected for residues located at pore-domain, indicating a bend in the TM2 helix and the collapsed conformation of selectivity filter. In contrast, at high K^+ concentration, the spectral patterns between pH 7.5 and 4.0 did not experience significant different, suggesting a closed conductive conformation formed in both pH conditions. The protonation states of the carboxyl side chains of residues E71, E118, and E120, the critical residues for K^+ binding, were also monitored using chemical shift data. At a low K^+ concentration, the carboxyl side chains of all these residues were protonated at acidic pH, whereas in the presence of K^+, they are deprotonated regardless of pH. These residues are critical in stabilizing the selectivity filter, thus playing an important role in both activation and inactivation gating mechanisms.

9.3.4 GPCRs

GPCRs are human seven-transmembrane receptors that respond to extracellular stimuli and transduce signals into the cell. They are implicated in many diseases, thus making them the most important drug targets. ssNMR has been successfully applied to determine the structure of the chemokine CXCR1 receptor in its lipid environments [16, 99], which is the only successful GPCR structure by ssNMR. OS-ssNMR is limited for membrane proteins with multiple helices, because the spectral crowding makes it difficult to distinguish the signals from different helices. To overcome this limitation, an ssNMR approach, known as rotational alignment ssNMR (RA-ssNMR) [100], is designed to combine the merits of OS-ssNMR and MAS-ssNMR. The experiments were conducted at a temperature above the lipid phase transition temperature and under MAS conditions. In such case, rapid diffusion of membrane proteins in lipid bilayers partially averages the anisotropic dipolar interactions, allowing detection of orientation restraints, and under MAS conditions, the isotropic chemical shifts and averaged ^{15}N–^{1}H and ^{13}C–^{1}H dipolar couplings can be measured. In CXCR1, seven-TM helices and kink structure in some of the helices were identified by dipolar wave patterns obtained by RA-ssNMR (Fig. 9.14). A combination of a template from known GPCR structure and the site-specific dipolar couplings allows to obtaining the backbone 3D structure of CXCR1.

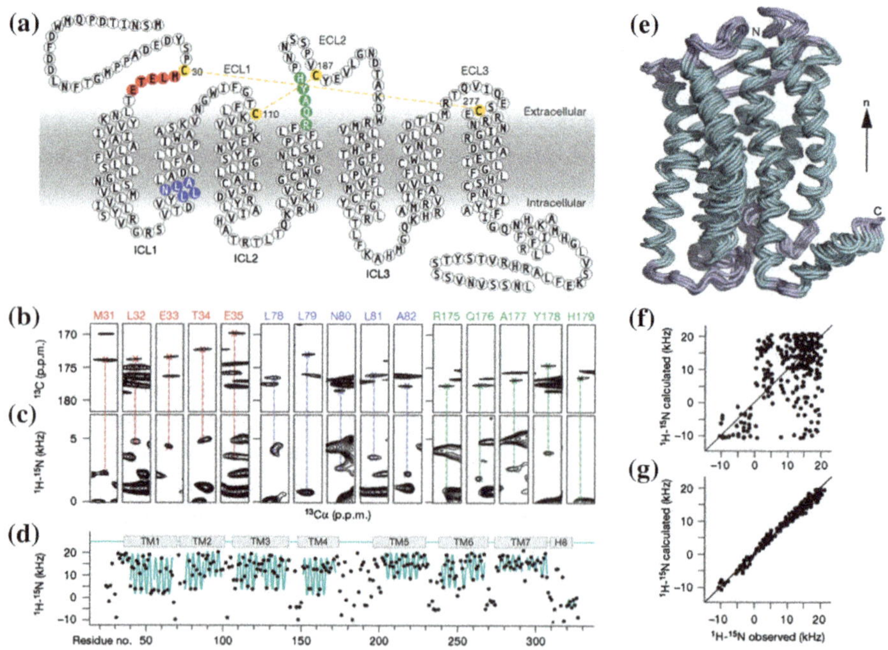

Fig. 9.14 ssNMR structure of CXCR1. **a** Topology of CXCR1. Disulfide bonds are shown in dashed lines. **b**, **c** Strip plots from 3D NCACX experiments to highlight the representative isotropic chemical shift assignments, and the corresponding ^{1}H–^{15}N dipolar coupling measured on the same residue. **d** ^{1}H–^{15}N dipolar couplings as a function of residue number. **e** Structure ensemble of CXCR1, determined by ssNMR data using CXCR4 crystal structure as a template. **f**–**g** Experimental versus back-calculated dipolar couplings before (**f**) and after (**g**) refinement against the experimental data. Reprinted by permission from Macmillan Publishers Ltd: Ref. [16], Copyright 2012

In addition to the 3D structure determinations, residue-type specific ^{13}C-labeled bovinerhodopsins were used to characterize the structure of the protein's active state [21, 101–104]. Visual rhodopsin binds 11-*cis* retinal and undergoes structure rearrangement upon illumination. To characterize the active state, the bovinerhodopsin was expressed in human cell lines, in which ^{13}C-isotope-enriched amino acids were supplied to produce residue-type specific ^{13}C-labeled samples. The expressed proteins were dissolved in detergents, and frozen after illumination, resulting in trapping of the Meta II state of the proteins. Comparison of the spectral patterns of 2D DARR of the dark-adapted state with those of the light-activated state identified the structural rearrangement upon formation of the Meta II state. Upon illumination, the C20 methyl group of retinal rotates about 90°, and the β-ionone ring of the retinal shifts closer to helix 5, which replaces the second extracellular loop. The characteristic ionic lock structure, formed by Glu134 and R135 of helix 3 and Glu247 of helix 6, was disturbed upon formation of the Meta II state, whereas an inter-helical contact between R135 of helix 3 and M257 of helix 6 was formed in the Meta II state.

9.3.5 *Anabaena* Sensory Rhodopsin

Microbial rhodopsins are photosensitive proteins, discovered in eubacteria, archaea, and unicellular eukaryotic organisms. Microbial rhodopsin typically adopts a topology of seven-transmembrane helices and covalently binds an all-*trans*-retinal at the lysyl residue of the last helix. Most of the microbial rhodopsins functions as light-activated ion pumps or sensory protein that triggers light-activated signal transduction cascades. Solid-state NMR has been used extensively to characterize the structure and dynamics of several microbial rhodopsins, including bacteriorhodopsin from *H. salinarum* [105, 106], green proteorhodopsin [42, 107, 108], Sensory rhodopsin II from *N. pharaonis* [109, 110], and *Anabaena* sensory rhodopsin (ASR) [15, 51, 70, 111].

Of these, high-resolution structure of ASR has been determined by MAS-ssNMR spectroscopy. It has been suggested that ASR binds a soluble protein, called ASRT, at the cytoplasmic side. Under illumination, ASR changes its conformation and release the ASRT, which, in turn, associates with the promoter region of DNA and regulates the expression of a number of light-harvesting proteins and proteins related to the circadian clock [112]. To characterize the structure and mechanisms of light-activated structural changes, the expressed ASR was reconstituted in DMPC/DMPA liposomes. Nearly complete resonance assignments of ASR were obtained using 3D MAS-ssNMR spectroscopy [51]. Backbone chemical shifts of ASR identified seven-transmembrane helices and a β-hairpin connecting helix B and helix C. A vast number of carbon–carbon distance restraints were collected on two complementary alternate ^{13}C-labeled samples, produced using 2-^{13}C glycerol and 1,3-$^{13}C_2$ glycerol as the carbon sources [15]. Combination of distance restraints and torsion angle restraints from TALOS+ yield a structure

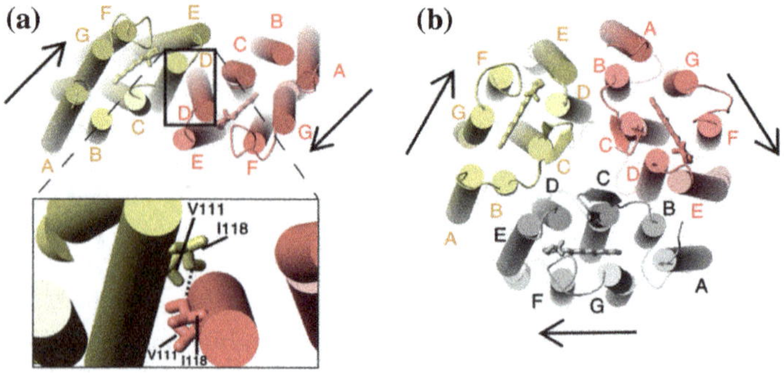

Fig. 9.15 Comparison of ASR structure in crystals (PDB code 1XIO) (**a**) and in lipids (**b**), viewed from the periplasmic side. The α-helices are shown in cylinders. Monomers are colored differently. Retinals are represented by sticks, and black arrows indicate the orientation of retinal. The enlarged box highlights the side chain packing on the inter-monomer interface of the crystal structure of ASR. Reprinted by permission from Macmillan Publishers Ltd: Ref. [15], Copyright 2013

ensemble of monomeric ASR with high resolution. ASR forms trimers in lipid environments. PRE data were used to identify inter-monomer interface and to determine the trimer structure. It showed that helix A and B from one monomer face helix D and E from adjacent monomer. However, early work showed that ASR adopted dimers in crystal, with a different inter-monomer face as observed by ssNMR (Fig. 9.15). This comparison showed that the lipid–protein interactions may be critical to facilitate the formation of ASR trimers, and that the crystal conditions promote the dissociation of trimers.

H/D exchange experiments were used to study the light-activated conformational changes of ASR [70]. In dark, most of the residues on transmembrane helices are not exchangeable with D_2O, consisting with the topology of transmembrane protein. In contrast, the spectral patterns of ASR spectra changed dramatically after H/D exchange under illumination conditions. In particular, cytoplasmic half of helix G and helix C, which located close to the retinal, are the most affected. It suggested that ASR undergoes significant light-activated structural rearrangements, potentially forming a hydrophilic cavity on the cytoplasmic side.

9.3.6 *In Vivo Characterization of Membrane Proteins*

Structure determination of membrane protein in living cells or intact cellular compartments is still challenging. In the past decade, several ssNMR research groups have developed strategies to study the membrane proteins in whole cells or in cellular environment. In common, *in cell* ssNMR characterizes membrane proteins that are overexpressed in the expression hosts, i.e., *E. coli.* Either whole cells or membrane components containing the target membrane proteins are packed into

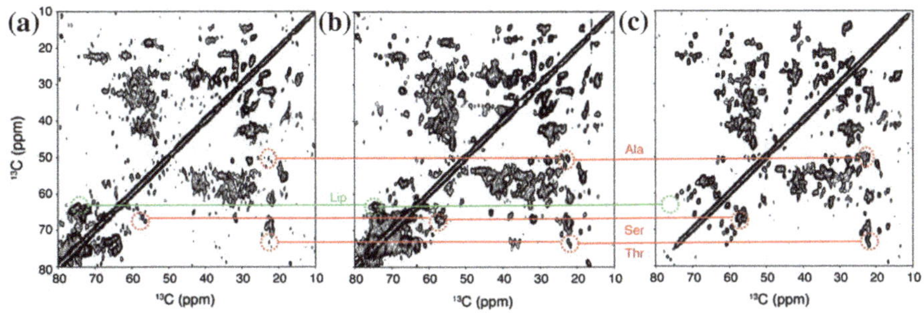

Fig. 9.16 2D ^{13}C–^{13}C correlation spectra of whole cells (**a**), cell envelopes (**b**) and purified ^{13}C,^{15}N-labeled PagL reconstituted in proteoliposomes. The characteristic cross peaks of PagL and endogenous lipids are highlighted. Reproduced from Ref. [23], The National Academy of Sciences of the USA, 2012

rotors for ssNMR characterization, without any further purification. Comparison of spectra of membrane proteins collected in native cell membranes with those in model systems, i.e., purified membrane proteins reconstituted in liposomes, provides wealthy information of the structure of membrane protein in their native environment.

Using this approach, the Bladus group investigated the bacterial membrane protein lipid A deacylase of Pseudomonas aeruginosa (PagL) in the *E. coli* outer membrane (Fig. 9.16) [23]. The characteristic cross peaks of PagL can be identified on the ssNMR spectra of whole cells and cell envelope, where the PagL expressed, and those signals compared well with those from the purified proteins reconstituted in proteoliposomes, suggesting that the conformation of PagL in model lipid systems is similar to that in native membranes. Other signals from lipopolysaccharides and peptidoglycan have also been identified. The structure of *Anabaena* sensory rhodopsin in *E. coli* membrane has also been validated in a similar strategy [22]. Using sucrose gradient, the inner membrane of *E. coli* with expressed ASR was separated from outer membrane, allowing in situ ssNMR spectroscopic characterization on ASR. Comparison of spectral patterns of ASR in *E. coli* membrane and that in DMPC/DMPA liposomes validated the structure of ASR, which is previously determined in liposomes, and confirmed trimeric topology of ASR in native membranes.

9.4 Methods

Preparation of milligram quantities of ^{15}N,^{13}C isotope enriched, either sparsely or uniformly labeled, membrane proteins are mandatory in ssNMR characterization. In OS-ssNMR, ^{15}N-labeled samples are needed to collect the site-specific ^{15}N anisotropic chemical shits for illustrating the orientation of membrane-associated proteins, whereas MAS-ssNMR uses $^{15}N,^{13}C$ isotope labeling to achieve chemical shift correlations in multidimensional spectra for resonance assignments and measurement on distance restraints. As for ultrafast MAS, proton-detected experiments in common requires membrane proteins to be perdeuterated to improve the spectral resolution, while the amide protons of deuterated samples are introduced through back exchange.

There are several ways to achieve isotope enrichment of protein, including solid-phase peptide synthesis, cell-free expression, and heterologous expression. Of those, heterologous expression, commonly using *E. coli* or *P. pastoris* as expression host, is the most economic and efficient way to produce $^{15}N,^{13}C$-labeled membrane proteins. Uniform ^{15}N and ^{13}C labeling can be achieved by supplement of ^{15}N-labeled amino salts, e.g., $^{15}N\text{–}NH_4Cl$ or $^{15}N\text{–}(NH_4)_2SO_4$, and ^{13}C-labeled carbohydrates, e.g., $^{13}C\text{–}C_6$-glucose, as sole nitrogen and carbon source, respectively, in expression medium. Amino acids are thus synthesized as isotope-labeled residues, resulting in uniformly $^{15}N,^{13}C$-labeled proteins. Many sparsely labeling protocols have also been developed for heterologous expression systems to facilitate the resonance assignments and to collect the distance restraints for structure illustration. The most used are alternate labeling strategies, in which 2-^{13}C-glycerol or 1,3-$^{13}C_2$-glycerol replaces $^{13}C\text{–}C_6$-glucose in the expression medium. In the following paragraph, we will introduce an alternate ^{13}C labeling protocol developed for *P. pastoris* expression system, using *Leptosphaeria* rhodopsin (LR) as an example.

9.4.1 Protocols

Materials and Reagents

(1) *P. pastoris* strain SMD1168H cells transformed with pPICZαA vector containing gene of LR with C-terminal 6xHis tag.
(2) 10 × YNB stock (3.4% Yeast Nitrogen Base with Ammonium Sulfate without amino acids).
(3) All-*trans*-retinal in Ethanol (10 mM).
(4) Buffered minimal glucose (BMD), pH 6 (100 mM K_2HPO_4, 2% NH_4Cl, 0.34% YNB, 0.008% Biotin, 0.5% $^{13}C_6$-Glucose).
(5) Buffered minimal methanol (BMM), pH 6.0 (100 mM K_2HPO_4, 2% NH_4Cl, 0.34% YNB, 0.008% Biotin, 0.12% Methanol, 0.37% 2-^{13}C-glycerol).
(6) Buffer A, pH 6.5 (7 mM NaH_2PO_4, 7 mM EDTA, 7 mM DTT, 1 mM PMSF).
(7) Solubilization buffer, pH 7.5 (20 mM KH_2PO_4, 1 mM PMSF, 1% Triton X-100).

(8) Washing buffer, pH 7.5 (50 mM KH_2PO_4, 400 mM NaCl, 0–50 mM of imidazole, 0.25% Triton X-100).
(9) Elution buffer, pH 7.5 (50 mM KH_2PO_4, 400 mM NaCl, 250 mM imidazole, 0.25% Triton X-100).
(10) Reconstitution buffer, pH 7.5 (50 mM KH_2PO_4, 100 mM NaCl, 0.25% Triton X-100).

Procedures

(1) *P. pastoris* SMD1168H cells were incubated into 2-ml BMD medium, grown at 30 °C, 210 rpm to reach the OD_{600} of 7.0.
(2) 1 ml of the overnight culture was added into 25 ml BMD medium in a 250 ml flask, grown at 30 °C, 210 rpm to OD_{600} higher than 7.0.
(3) The culture was centrifuged at 1500 × *g* for 10 min at 4 °C and resuspended in 800-ml BMM medium, in which the protein expression was induced by methanol.
(4) The cells were grown at 30 °C, 210 rpm for 24 h, followed by addition of 10 mM all-*trans*-retinal solution to a final concentration of 2.5 μM. The all-*trans*-retinal was used to regenerate the opsin.
(5) The cells were harvested by centrifugation at 1500 × *g* for 10 min after 40 h shaking at 30 °C, 210 rpm. The color of cell pellet was red or dark pink.
(6) The cells were resuspended in buffer A and incubated with 1 mg/ml Lyticase (Sigma) for 3 h to digest the cell walls.
(7) The cells were broken by high-pressure homogenizer, and the membrane fraction was collected by centrifugation at 50,000 × *g* for 30 min. The supernatant was discarded.
(8) The membrane fraction was solubilized in 100 ml solubilization for 12 h.
(9) The membrane solution was centrifuged at 60,000 × *g* for 30 min, and the insoluble fragments were discarded.
(10) The solubilized solution was subject to purification by Ni-NTA agarose. After binging with the proteins, the resins were washed several times with washing buffer to elute the impurities.
(11) The LR proteins were then eluted with elution buffer. The concentration of LR was determined by the absorbance of rhodopsin-bound retinal, extinction coefficient of 48,000 M^{-1} cm^{-1}.
(12) Concentrate the LR protein to ~1 mg/ml and change the buffer conditions to reconstitution buffer.
(13) To reconstitute LR protein into lipid environments, the LR protein solution was mixed with liposome solution (DMPA/DMPC = 1:9) at a protein–liposome ratio of 2.5:1 and stirred for 1 h.
(14) The bio-beads SMII (biorad) were added to the mixture to a final ratio of 0.6 g/ml. The mixture was incubated 24 h to remove the detergents.
(15) The bio-beads were washed exhaustively to collect the liposome solution. To collect pellet, the liposome solution was ultracentrifuged at 900,000 × *g* for 3 h. The obtained pellet was ready for ssNMR studies.

9.5 Conclusions and Perspectives

Solid-state NMR has been emerging rapidly as a powerful technique to characterize membrane proteins. Solid-state NMR has been applied in a variety of membrane proteins, ranging from virus channel proteins to human GPCRs. Improvement of spectral quality and advancements of NMR spectroscopy enabled it to determine the structures for membrane proteins in high resolution, and characterize the proteins' dynamic and intermediate states, providing insight into the mechanisms of membrane proteins. More recently, in vivo or in situ ssNMR open a new venue to study membrane protein in native membranes. However, there is still technical challenging. Low sensitivity is the intrinsic weakness of these NMR, which results in requirements of large amount of samples and long instrumental time. It poses the challenge to study proteins with short lifetime or proteins with low expression yield. Methods for dynamic nuclear polarization (DNP) have been developed for biology systems to overcome this bottleneck. DNP has potential to largely enhance the signals and applied to characterize the site-specific structure of membrane proteins and the low populated state of the protein conformation [113–115]. Additionally, ultrafast MAS condition significantly improves proton line width, which provides more flexibility to experimental design as a consequence of allows to designing more complicated experiments [116]. Thus, these developments would facilitate ssNMR to contribute more challenging membrane protein systems.

Acknowledgements This work was supported by the National Key Research and Development Program of the Ministry of Science and Technology of the People's Republic of China (No. 2016YFA0501203), the National Natural Science Foundation of China (No. 31470727) and the Beijing National Laboratory for Molecular Sciences.

References

1. Palczewski K, Kumasaka T, Hori T, Behnke CA, Motoshima H, Fox BA, Le Trong I, Teller DC, Okada T, Stenkamp RE, Yamamoto M, Miyano M (2000) Crystal structure of rhodopsin: A G protein-coupled receptor. Science 289:739–745. doi:10.1126/science.289.5480.739
2. Cherezov V, Rosenbaum DM, Hanson MA, Rasmussen SGF, Thian FS, Kobilka TS, Choi H-J, Kuhn P, Weis WI, Kobilka BK, Stevens RC (2007) High-resolution crystal structure of an engineered human β_2-adrenergic G protein-coupled receptor. Science 318:1258–1265. doi:10.1126/science.1150577
3. Rasmussen SGF, Choi H-J, Rosenbaum DM, Kobilka TS, Thian FS, Edwards PC, Burghammer M, Ratnala VRP, Sanishvili R, Fischetti RF, Schertler GFX, Weis WI, Kobilka BK (2007) Crystal structure of the human β_2 adrenergic G-protein-coupled receptor. Nature 450:383–387. doi:10.1038/nature06325
4. Jaakola V-P, Griffith MT, Hanson MA, Cherezov V, Chien EYT, Lane JR, Ijzerman AP, Stevens RC (2008) The 2.6 angstrom crystal structure of a human A(2A) adenosine receptor bound to an antagonist. Science 322:1211–1217. doi:10.1126/science.1164772
5. Kopfer DA, Song C, Gruene T, Sheldrick GM, Zachariae U, de Groot BL (2014) Ion permeation in K^+ channels occurs by direct Coulomb knock-on. Science 346:352–355. doi:10.1126/science.1254840

6. Staus DP, Strachan RT, Manglik A, Pani B, Kahsai AW, Kim TH, Wingler LM, Ahn S, Chatterjee A, Masoudi A, Kruse AC, Pardon E, Steyaert J, Weis WI, Prosser RS, Kobilka BK, Costa T, Lefkowitz RJ (2016) Allosteric nanobodies reveal the dynamic range and diverse mechanisms of G-protein-coupled receptor activation. Nature 535:448–452. doi:10.1038/nature18636
7. Yang T, Liu Q, Kloss B, Bruni R, Kalathur RC, Guo YZ, Kloppmann E, Rost B, Colecraft HM, Hendrickson WA (2014) Structure and selectivity in bestrophin ion channels. Science 346:355–359. doi:10.1126/science.1259723
8. Kim H, Howell SC, Van Horn WD, Jeon Y, Sanders CR (2009) Recent advances in the application of solution NMR spectroscopy to multi-span integral membrane proteins. Prog Nucl Magn Reson Spectrosc 55:335–360. doi:10.1016/j.pnmrs.2009.07.002
9. Li N, Wu J, Ding D, Cheng J, Gao N, Chen L (2017) Structure of a pancreatic ATP-sensitive potassium channel. Cell 168:101–110. doi:10.1016/j.cell.2016.12.028
10. Sun J, MacKinnon R (2017) Cryo-EM structure of a KCNQ1/CaM complex reveals insights into congenital long QT syndrome. Cell 169:1042–1050. doi:10.1016/j.cell.2017.05.019
11. Gardiennet C, Schuetz AK, Hunkeler A, Kunert B, Terradot L, Boeckmann A, Meier BH (2012) A sedimented sample of a 59 kDa dodecameric helicase yields high-resolution solid-state NMR spectra. Angew Chem Int Ed 51:7855–7858. doi:10.1002/anie.201200779
12. Fragai M, Luchinat C, Martelli T, Ravera E, Sagi I, Solomonov I, Udi Y (2014) SSNMR of biosilica-entrapped enzymes permits an easy assessment of preservation of native conformation in atomic detail. Chem Commun 50:421–423. doi:10.1039/c3cc46896h
13. Castellani F, Van Rossum B, Diehl A, Schubert M, Rehbein K, Oschkinat H (2002) Structure of a protein determined by solid-state magic-angle-spinning NMR spectroscopy. Nature 420:98–102. doi:10.1038/nature01070
14. Shahid SA, Bardiaux B, Franks WT, Krabben L, Habeck M, van Rossum BJ, Linke D (2012) Membrane-protein structure determination by solid-state NMR spectroscopy of microcrystals. Nat Methods 9:1212–1217. doi:10.1038/nmeth.2248
15. Wang S, Munro RA, Shi L, Kawamura I, Okitsu T, Wada A, Kim S, Jung K, Brown L, Ladizhansky V (2013) Solid-state NMR spectroscopy structure determination of a lipid-embedded heptahelical membrane protein. Nat Methods 10:1007–1010. doi:10.1038/nmeth.2635
16. Park S, Das BB, Casagrande F, Tian Y, Nothnagel HJ, Chu M, Kiefer H, Maier K, De Angelis AA, Marassi FM, Opella SJ (2012) Structure of the chemokine receptor CXCR1 in phospholipid bilayers. Nature 491:779–783. doi:10.1038/nature11580
17. Verardi R, Shi L, Traaseth NJ, Walsh N, Veglia G (2011) Structural topology of phospholamban pentamer in lipid bilayers by a hybrid solution and solid-state NMR method. Proc Natl Acad Sci U S A 108:9101–9106. doi:10.1073/pnas.1016535108
18. Cady SD, Schmidt-Rohr K, Wang J, Soto CS, DeGrado WF, Hong M (2010) Structure of the amantadine binding site of influenza M2 proton channels in lipid bilayers. Nature 463:689–692. doi:10.1038/nature08722
19. Hu F, Luo W, Hong M (2010) Mechanisms of proton conduction and gating in influenza M2 proton channels from solid-state NMR. Science 330:505–508. doi:10.1126/science.1191714
20. Zeri AC, Mesleh MF, Nevzorov AA, Opella SJ (2003) Structure of the coat protein in fd filamentous bacteriophage particles determined by solid-state NMR spectroscopy. Proc Natl Acad Sci U S A 100:6458–6463. doi:10.1073/pnas.1132059100
21. Ahuja S, Hornak V, Yan ECY, Syrett N, Goncalves JA, Hirshfeld A, Ziliox M, Sakmar TP, Sheves M, Reeves PJ, Smith SO, Eilers M (2009) Helix movement is coupled to displacement of the second extracellular loop in rhodopsin activation. Nat Struct Mol Biol 16:168–175. doi:10.1038/nsmb.1549
22. Ward ME, Wang S, Munro R, Ritz E, Hung I, Gor'kov PL, Jiang Y, Liang H, Brown LS, Ladizhansky V (2015) In situ structural studies of Anabaena sensory rhodopsin in the *E. coli* membrane. Biophys J 108:1683–1696. doi:10.1016/j.bpj.2015.02.018

23. Renault M, Tommassen-van Boxtel R, Bos MP, Post JA, Tommassen J, Baldus M (2012) Cellular solid-state nuclear magnetic resonance spectroscopy. Proc Natl Acad Sci U S A 109:4863–4868. doi:10.1073/pnas.1116478109
24. Opella SJ (2013) Structure determination of membrane proteins by nuclear magnetic resonance spectroscopy. In: Cooks RG, Pemberton JE (eds) Annual review of analytical chemistry. Annual Reviews, Palo Alto, pp 305–328
25. Murray DT, Das N, Cross TA (2013) Solid state NMR strategy for characterizing native membrane protein structures. Acc Chem Res 46:2172–2181. doi:10.1021/ar3003442
26. Opella SJ, Marassi FM (2004) Structure determination of membrane proteins by NMR spectroscopy. Chem Rev 104:3587–3606. doi:10.1021/cr0304121
27. Durr UHN, Waskell L, Ramamoorthy A (2007) The cytochromes P450 and b(5) and their reductases—promising targets for structural studies by advanced solid-state NMR spectroscopy. BBA-Biomembranes 1768:3235–3259. doi:10.1016/j.bbamem.2007.08.007
28. Tang WX, Knox RW, Nevzorov AA (2012) A spectroscopic assignment technique for membrane proteins reconstituted in magnetically aligned bicelles. J Biomol NMR 54: 307–316. doi:10.1007/s10858-012-9673-y
29. Das N, Murray DT, Cross TA (2013) Lipid bilayer preparations of membrane proteins for oriented and magic-angle spinning solid-state NMR samples. Nat Protoc 8:2256–2270. doi:10.1038/nprot.2013.129
30. Bechinger B, Zasloff M, Opella SJ (1998) Structure and dynamics of the antibiotic peptide PGLa in membranes by solution and solid-state nuclear magnetic resonance spectroscopy. Biophys J 74:981–987
31. Hartzell CJ, Whitfield M, Oas TG, Drobny GP (1987) Determination of the ^{15}N and ^{13}C chemical shift tensors of L-^{13}C-alanyl-L-^{15}N alanine from the dipole-coupled powder patterns. J Am Chem Soc 109:5966–5969. doi:10.1021/ja00254a012
32. Strandberg E, Ulrich AS (2004) NMR methods for studying membrane-active antimicrobial peptides. Concepts Magn Reson 23A:89–120. doi:10.1002/cmr.a.20024
33. Ulrich AS (2005) Solid state ^{19}F NMR methods for studying biomembranes. Prog Nucl Magn Reson Spectrosc 46:1–21. doi:10.1016/j.pnmrs.2004.11.001
34. Wu CH, Ramamoorthy A, Opella SJ (1994) High-resolution heteronuclear dipolar solid-state NMR-spectroscopy. J Magn Reson A 109:270–272. doi:10.1006/jmra.1994.1169
35. Marassi FM, Opella SJ (2000) A solid-state NMR index of helical membrane protein structure and topology. J Magn Reson 144:150–155. doi:10.1006/jmre.2000.2035
36. Marassi FM (2001) A simple approach to membrane protein secondary structure and topology based on NMR spectroscopy. Biophys J 80:994–1003
37. Zeri AC, Mesleh MF, Nevzorov AA, Opella SJ (2007) Structure of the coat protein in fd filamentous bacteriophage particles determined by solid-state NMR spectroscopy. Proc Natl Acad Sci U S A 104:2024. doi:10.1073/pnas.0610764104
38. Mesleh MF, Opella SJ (2003) Dipolar waves as NMR maps of helices in proteins. J Magn Reson 163:288–299. doi:10.1016/s1090-7807(03)00119-8
39. Shi L, Ladizhansky V (2012) Magic angle spinning solid-state NMR experiments for structural characterization of proteins. In: Clifton NJ (ed) Methods in molecular biology, pp 153–165
40. Schuetz A, Wasmer C, Habenstein B, Verel R, Greenwald J, Riek R, Boeckmann A, Meier BH (2010) Protocols for the sequential solid-state NMR spectroscopic assignment of a uniformly labeled 25 kDa protein: HET-s(1-227). ChemBioChem 11:1543–1551. doi:10.1002/cbic.201000124
41. Li Y, Berthold DA, Gennis RB, Rienstra CM (2008) Chemical shift assignment of the transmembrane helices of DsbB, a 20-kDa integral membrane enzyme, by 3D magic-angle spinning NMR spectroscopy. Protein Sci 17:199–204. doi:10.1110/ps.073225008
42. Shi L, Lake EMR, Ahmed MAM, Brown LS, Ladizhansky V (2009) Solid-state NMR study of proteorhodopsin in the lipid environment: secondary structure and dynamics. BBA Biomembr 1788:2563–2574. doi:10.1016/j.bbamem.2009.09.011

43. Wang SL, Shi LC, Okitsu T, Wada A, Brown LS, Ladizhansky V (2013) Solid-state NMR ^{13}C and ^{15}N resonance assignments of a seven-transmembrane helical protein Anabaena Sensory Rhodopsin. Biomol NMR Assign 7:253–256. doi:10.1007/s12104-012-9421-y
44. Hong M, Jakes K (1999) Selective and extensive ^{13}C labeling of a membrane protein for solid-state NMR investigations. J Biomol NMR 14:71–74. doi:10.1023/a:1008334930603
45. Loquet A, Giller K, Becker S, Lange A (2010) Supramolecular interactions probed by ^{13}C–^{13}C solid-state NMR spectroscopy. J Am Chem Soc 132:15164–15166. doi:10.1021/ja107460j
46. Loquet A, Lv G, Giller K, Becker S, Lange A (2011) ^{13}C Spin dilution for simplified and complete solid-state NMR resonance assignment of insoluble biological assemblies. J Am Chem Soc 133:4722–4725. doi:10.1021/ja200066s
47. Linser R, Bardiaux B, Andreas LB, Hyberts SG, Morris VK, Pintacuda G, Sunde M, Kwan AH, Wagner G (2014) Solid-state NMR structure determination from diagonal-compensated, sparsely nonuniform-sampled 4D proton-proton restraints. J Am Chem Soc 136:11002–11010. doi:10.1021/ja504603g
48. Agarwal V, Penzel S, Szekely K, Cadalbert R, Testori E, Oss A, Past J, Samoson A, Ernst M, Boeckmann A, Meier BH (2014) De novo 3D structure determination from sub-milligram protein samples by solid-state 100 kHz MAS NMR spectroscopy. Angew Chem Int Ed 53:12253–12256. doi:10.1002/anie.201405730
49. Wishart DS, Sykes BD (1994) The ^{13}C chemical-shift index—a simple method for the identification of protein secondary structure using ^{13}C chemical-shift data. J Biomol NMR 4:171–180
50. Shen Y, Delaglio F, Cornilescu G, Bax A (2009) TALOS plus: a hybrid method for predicting protein backbone torsion angles from NMR chemical shifts. J Biomol NMR 44:213–223. doi:10.1007/s10858-009-9333-z
51. Shi LC, Kawamura I, Jung KH, Brown LS, Ladizhansky V (2011) Conformation of a seven-helical transmembrane photosensor in the lipid environment. Angew Chem Int Ed 50:1302–1305. doi:10.1002/anie.201004422
52. Szeverenyi NM, Sullivan MJ, Maciel GE (1982) Observation of spin exchange by two-dimensional fourier-transform ^{13}C cross polarization-magic-angle spinning. J Magn Reson 47:462–475. doi:10.1016/0022-2364(82)90213-x
53. Morcombe CR, Gaponenko V, Byrd RA, Zilm KW (2004) Diluting abundant spins by isotope edited radio frequency field assisted diffusion. J Am Chem Soc 126:7196–7197. doi:10.1021/ja047919t
54. Takegoshi K, Nakamura S, Terao T (2001) ^{13}C–^{1}H dipolar-assisted rotational resonance in magic-angle spinning NMR. Chem Phys Lett 344:631–637. doi:10.1016/s0009-2614(01)00791-6
55. Raleigh DP, Levitt MH, Griffin RG (1988) Rotational resonance in solid-state NMR. Chem Phys Lett 146:71–76. doi:10.1016/0009-2614(88)85051-6
56. Peng X, Libich D, Janik R, Harauz G, Ladizhansky V (2008) Dipolar chemical shift correlation spectroscopy for homonuclear carbon distance measurements in proteins in the solid state: application to structure determination and refinement. J Am Chem Soc 130:359–369. doi:10.1021/ja076658v
57. De Paepe G, Lewandowski JR, Loquet A, Bockmann A, Griffin RG (2008) Proton assisted recoupling and protein structure determination. J Chem Phys 129:1–21. doi:10.1063/1.3036928
58. Lewandowski JR, De Paepe G, Eddy MT, Struppe J, Maas W, Griffin RG (2009) Proton assisted recoupling at high spinning frequencies. J Phys Chem B 113:9062–9069. doi:10.1021/jp810280t
59. Lewandowski JR, De Paepe G, Griffin RG (2007) Proton assisted insensitive nuclei cross polarization. J Am Chem Soc 129:728–729. doi:10.1021/ja0650394
60. Donovan KJ, Silvers R, Linse S, Griffin RG (2017) 3D MAS NMR experiment utilizing through-space ^{15}N-^{15}N correlations. J Am Chem Soc 139:6518–6521. doi:10.1021/jacs.7b01159

61. Gullion T, Schaefer J (1989) Rotational-echo double-resonance NMR. J Magn Reson 81:196–200. doi:10.1016/0022-2364(89)90280-1
62. Hing AW, Vega S, Schaefer J (1992) Transferred-echo double-resonance NMR. J Magn Reson 96:205–209. doi:10.1016/0022-2364(92)90305-q
63. Jaroniec CP, MacPhee CE, Astrof NS, Dobson CM, Griffin RG (2002) Molecular conformation of a peptide fragment of transthyretin in an amyloid fibril. Proc Natl Acad Sci U S A 99:16748–16753. doi:10.1073/pnas.252625999
64. Nieuwkoop AJ, Rienstra CM (2010) Supramolecular protein structure determination by site-specific long-range intermolecular solid state NMR spectroscopy. J Am Chem Soc 132:7570–7571. doi:10.1021/ja100992y
65. Milikisiyants S, Wang S, Munro RA, Donohue M, Ward ME, Bolton D, Brown LS, Smirnova TI, Ladizhansky V, Smirnov AI (2017) Oligomeric structure of Anabaena sensory rhodopsin in a lipid bilayer environment by combining solid-state NMR and long-range DEER constraints. J Mol Biol 429:1903–1920
66. Laskowski RA, Macarthur MW, Moss DS, Thornton JM (1993) PROCHECK—a program to check the stereochemical quality of protein structures. J Appl Crystallogr 26:283–291. doi:10.1107/s0021889892009944
67. Vriend G (1990) WHAT IF—a molecular modeling and drug design program. J Mol Graph 8:52–56. doi:10.1016/0263-7855(90)80070-v
68. Rodriguez R, Chinea G, Lopez N, Pons T, Vriend G (1998) Homology modeling, model and software evaluation: three related resources. Bioinformatics 14:523–528. doi:10.1093/bioinformatics/14.6.523
69. Chen VB, Arendall WB, Headd JJ, Keedy DA, Immormino RM, Kapral GJ, Murray LW, Richardson JS, Richardson DC (2010) MolProbity: all-atom structure validation for macromolecular crystallography. Acta Crystallogr Sect D Biol Crystallogr 66:12–21. doi:10.1107/s0907444909042073
70. Wang SL, Shi LC, Kawamura I, Brown LS, Ladizhansky V (2011) Site-specific solid-state NMR detection of hydrogen-deuterium exchange reveals conformational changes in a 7-helical transmembrane protein. Biophys J 101:L23–L25. doi:10.1016/j.bpj.2011.06.035
71. Chen YK, Zhang ZF, Tang XQ, Li JP, Glaubitz C, Yang J (2014) Conformation and topology of diacylglycerol kinase in *E. coli* membranes revealed by solid-state NMR spectroscopy. Angew Chem Int Ed 53:5624–5628. doi:10.1002/anie.201311203
72. Wang SL, Ing C, Emami S, Jiang YJ, Liang HJ, Pomes R, Brown LS, Ladizhansky V (2016) Structure and dynamics of extracellular loops in human Aquaporin-1 from solid-state NMR and molecular dynamics. J Phys Chem B 120:9887–9902. doi:10.1021/acs.jpcb.6b06731
73. Qiang W, Sun Y, Weliky DP (2009) A strong correlation between fusogenicity and membrane insertion depth of the HIV fusion peptide. Proc Natl Acad Sci U S A 106:15314–15319. doi:10.1073/pnas.0907360106
74. Tuzi S, Hasegawa J, Kawaminami R, Naito A, Saito H (2001) Regio-selective detection of dynamic structure of transmembrane alpha-helices as revealed from ^{13}C NMR spectra of 3-^{13}C Ala-labeled bacteriorhodopsin in the presence of Mn^{2+} ion. Biophys J 81:425–434
75. Li C, Yi M, Hu J, Zhou HX, Cross TA (2008) Solid-state NMR and MD simulations of the antiviral drug amantadine solubilized in DMPC bilayers. Biophys J 94:1295–1302. doi:10.1529/biophysj.107.112482
76. Su Y, Mani R, Hong M (2008) Asymmetric insertion of membrane proteins in lipid bilayers by solid-state NMR paramagnetic relaxation enhancement: a cell-penetrating peptide example. J Am Chem Soc 130:8856–8864. doi:10.1021/ja802383t
77. Grambas S, Bennett MS, Hay AJ (1992) Influence of amantadine resistance mutations on the pH regulatory function of the M2-protein of influenza-A viruses. Virology 191:541–549. doi:10.1016/0042-6822(92)90229-i
78. Lamb RA, Zebedee SL, Richardson CD (1985) Influenza virus-M2 protein is an integral membrane-protein expressed on the infected-cell surface. Cell 40:627–633. doi:10.1016/0092-8674(85)90211-9

79. Kovacs FA, Denny JK, Song Z, Quine JR, Cross TA (2000) Helix tilt of the M2 transmembrane peptide from influenza A virus: an intrinsic property. J Mol Biol 295:117–125. doi:10.1006/jmbi.1999.3322
80. Song ZY, Kovacs FA, Wang J, Denny JK, Shekar SC, Quine JR, Cross TA (2000) Transmembrane domain of M2 protein from influenza A virus studied by solid-state ^{15}N polarization inversion spin exchange at magic angle NMR. Biophys J 79:767–775
81. Sharma M, Yi M, Dong H, Qin H, Peterson E, Busath DD, Zhou H-X, Cross TA (2010) Insight into the mechanism of the influenza a proton channel from a structure in a lipid bilayer. Science 330:509–512. doi:10.1126/science.1191750
82. Wang JF, Kim S, Kovacs F, Cross TA (2001) Structure of the transmembrane region of the M2 protein H^+ channel. Protein Sci 10:2241–2250. doi:10.1110/ps.17901
83. Miao Y, Qin H, Fu R, Sharma M, Can TV, Hung I, Luca S, Gor'kov PL, Brey WW, Cross TA (2012) M2 Proton channel structural validation from full-length protein samples in synthetic bilayers and *E. coli* membranes. Angew Chem Int Ed 51:8383–8386. doi:10.1002/anie.201204666
84. Andreas LB, Eddy MT, Chou JJ, Griffin RG (2012) Magic-angle-spinning NMR of the drug resistant S31N M2 proton transporter from influenza A. J Am Chem Soc 134:7215–7218. doi:10.1021/ja3003606
85. Hu F, Schmidt-Rohr K, Hong M (2012) NMR detection of pH-dependent histidine-water proton exchange reveals the conduction mechanism of a transmembrane proton channel. J Am Chem Soc 134:3703–3713. doi:10.1021/ja2081185
86. Sharma M, Yi MG, Dong H, Qin HJ, Peterson E, Busath DD, Zhou HX, Cross TA (2010) Insight into the mechanism of the Influenza A proton channel from a structure in a lipid bilayer. Science 330:509–512. doi:10.1126/science.1191750
87. Gandhi CS, Shuck K, Lear JD, Dieckmann GR, DeGrado WF, Lamb RA, Pinto LH (1999) Cu(II) inhibition of the proton translocation machinery of the influenza A virus M_2 protein. J Biol Chem 274:5474–5482. doi:10.1074/jbc.274.9.5474
88. Andreas LB, Eddy MT, Pielak RM, Chou J, Griffin RG (2010) Magic angle spinning NMR investigation of influenza A M2(18-60): support for an allosteric mechanism of inhibition. J Am Chem Soc 132:10958–10960. doi:10.1021/ja101537p
89. Su Y, Hu F, Hong M (2012) Paramagnetic Cu(II) for probing membrane protein structure and function: inhibition mechanism of the influenza M2 proton channel. J Am Chem Soc 134:8693–8702. doi:10.1021/ja3026328
90. Kovacs RJ, Nelson MT, Simmerman HKB, Jones LR (1988) Phospholamban forms Ca^{2+}-selective channels in lipid bilayers. J Biol Chem 263:18364–18368
91. Mascioni A, Karim C, Zamoon J, Thomas DD, Veglia G (2002) Solid-state NMR and rigid body molecular dynamics to determine domain orientations of monomeric phospholamban. J Am Chem Soc 124:9392–9393. doi:10.1021/ja026507m
92. Abu-Baker S, Lu JX, Chu S, Shetty KK, Gor'kov PL, Lorigan GA (2007) The structural topology of wild-type phospholamban in oriented lipid bilayers using ^{15}N solid-state NMR spectroscopy. Protein Sci 16:2345–2349. doi:10.1110/ps.072977707
93. Traaseth NJ, Verardi R, Torgersen KD, Karim CB, Thomas DD, Veglia G (2007) Spectroscopic validation of the pentameric structure of phospholamban. Proc Natl Acad Sci U S A 104:14676–14681. doi:10.1073/pnas.0701016104
94. Traaseth NJ, Shi L, Verardi R, Mullen DG, Barany G, Veglia G (2009) Structure and topology of monomeric phospholamban in lipid membranes determined by a hybrid solution and solid-state NMR approach. Proc Natl Acad Sci U S A 106:10165–10170. doi:10.1073/pnas.0904290106
95. Schneider R, Ader C, Lange A, Giller K, Hornig S, Pongs O, Becker S, Baldus M (2008) Solid-state NMR spectroscopy applied to a chimeric potassium channel in lipid bilayers. J Am Chem Soc 130:7427–7435. doi:10.1021/ja800190c
96. Lange A, Giller K, Hornig S, Martin-Eauclaire MF, Pongs O, Becker S, Baldus M (2006) Toxin-induced conformational changes in a potassium channel revealed by solid-state NMR. Nature 440:959–962. doi:10.1038/nature04649

97. Lange A, Giller K, Pongs O, Becker S, Baldus M (2006) Two-dimensional solid-state NMR applied to a chimeric potassium channel. J Recept Signal Transduction 26:379–393. doi:10.1080/10799890600932188
98. Ader C, Schneider R, Hornig S, Velisetty P, Vardanyan V, Giller K, Ohmert I, Becker S, Pongs O, Baldus M (2009) Coupling of activation and inactivation gate in a K^+ channel: potassium and ligand sensitivity. EMBO J 28:2825–2834. doi:10.1038/emboj.2009.218
99. Park SH, Casagrande F, Chu M, Maier K, Kiefer H, Opella SJ (2012) Optimization of purification and refolding of the human chemokine receptor CXCR1 improves the stability of proteoliposomes for structure determination. BBA Biomembr 1818:584–591. doi:10.1016/j.bbamem.2011.10.008
100. Opella SJ (2013) Structure determination of membrane proteins in their native phospholipid bilayer environment by rotationally aligned solid-state NMR spectroscopy. Acc Chem Res 46:2145–2153. doi:10.1021/ar400067z
101. Goncalves JA, South K, Ahuja S, Zaitseva E, Opefi CA, Eilers M, Vogel R, Reeves PJ, Smith SO (2010) Highly conserved tyrosine stabilizes the active state of rhodopsin. Proc Natl Acad Sci U S A 107:19861–19866. doi:10.1073/pnas.1009405107
102. Ahuja S, Eilers M, Hirshfeld A, Yan ECY, Ziliox M, Sakmar TP, Sheves M, Smith SO (2009) 6-*s-cis* Conformation and polar binding pocket of the retinal chromophore in the photoactivated state of rhodopsin. J Am Chem Soc 131:15160–15169. doi:10.1021/ja9034768
103. Ahuja S, Crocker E, Eilers M, Hornak V, Hirshfeld A, Ziliox M, Syrett N, Reeves PJ, Khorana HG, Sheves M, Smith SO (2009) Location of the retinal chromophore in the activated state of rhodopsin. J Biol Chem 284:10190–10201. doi:10.1074/jbc.M805725200
104. Crocker E, Eilers M, Ahuja S, Hornak V, Hirshfeld A, Sheves M, Smith SO (2006) Location of Trp265 in metarhodopsin II: implications for the activation mechanism of the visual receptor rhodopsin. J Mol Biol 357:163–172. doi:10.1016/j.jmb.2005.12.046
105. Creuzet F, McDermott A, Gebhard R, Vanderhoef K, Spijkerassink MB, Herzfeld J, Lugtenburg J, Levitt MH, Griffin RG (1991) Determination of membrane-protein structure by rotational resonance NMR—bacteriorhodopsin. Science 251:783–786. doi:10.1126/science.1990439
106. Kawamura I, Tanabe J, Ohmine M, Yamaguchi S, Tuzi S, Naito A (2009) Participation of the BC loop in the correct folding of bacteriorhodopsin as revealed by solid-state NMR. Photochem Photobiol 85:624–630. doi:10.1111/j.1751-1097.2009.00536.x
107. Shi LC, Ahmed MAM, Zhang WR, Whited G, Brown LS, Ladizhansky V (2009) Three-dimensional solid-state NMR study of a seven-helical integral membrane proton pump-structural insights. J Mol Biol 386:1078–1093. doi:10.1016/j.jmb.2009.01.011
108. Ward ME, Shi L, Lake E, Krishnamurthy S, Hutchins H, Brown LS, Ladizhansky V (2011) Proton-detected solid-state NMR reveals intramembrane polar networks in a seven-helical transmembrane protein proteorhodopsin. J Am Chem Soc 133:17434–17443. doi:10.1021/ja207137h
109. Etzkorn M, Martell S, Andronesi OC, Seidel K, Engelhard M, Baldus M (2007) Secondary structure, dynamics, and topology of a seven-helix receptor in native membranes, studied by solid-state NMR spectroscopy. Angew Chem Int Ed 46:459–462. doi:10.1002/anie.200602139
110. Gordeliy VI, Labahn J, Moukhametzianov R, Efremov R, Granzin J, Schlesinger R, Buldt G, Savopol T, Scheidig AJ, Klare JP, Engelhard M (2002) Molecular basis of transmembrane signalling by sensory rhodopsin II-transducer complex. Nature 419:484–487. doi:10.1038/nature01109
111. Wang SL, Munro RA, Kim SY, Jung KH, Brown LS, Ladizhansky V (2012) Paramagnetic relaxation enhancement reveals oligomerization interface of a membrane protein. J Am Chem Soc 134:16995–16998. doi:10.1021/ja308310z
112. Jung KH (2007) The distinct signaling mechanisms of microbial sensory rhodopsins in Archaea, Eubacteria and Eukarya. Photochem Photobiol 83:63–69. doi:10.1562/2006-03-20-ir-853

113. Mak-Jurkauskas ML, Bajaj VS, Hornstein MK, Belenky M, Griffin RG, Herzfeld J (2008) Energy transformations early in the bacteriorhodopsin photocycle revealed by DNP-enhanced solid-state NMR. Proc Natl Acad Sci U S A 105:883–888. doi:10.1073/pnas.0706156105
114. Bajaj VS, Mak-Jurkauskas ML, Belenky M, Herzfeld J, Griffin RG (2009) Functional and shunt states of bacteriorhodopsin resolved by 250 GHz dynamic nuclear polarization-enhanced solid-state NMR. Proc Natl Acad Sci U S A 106:9244–9249. doi:10.1073/pnas.0900908106
115. Ni QZ, Daviso E, Can TV, Markhasin E, Jawla SK, Swager TM, Temkin RJ, Herzfeld J, Griffin RG (2013) High frequency dynamic nuclear polarization. Acc Chem Res 46:1933–1941. doi:10.1021/ar300348n
116. Reif B (2012) Ultra-high resolution in MAS solid-state NMR of perdeuterated proteins: implications for structure and dynamics. J Magn Reson 216:1–12. doi:10.1016/j.jmr.2011.12.017

Chapter 10
Mass Spectrometry of Membrane Proteins

Ling-Peng Zhan, Chao-Zi Liu and Zong-Xiu Nie

10.1 General

The advent of electrospray ionization (ESI) and matrix-assisted laser desorption/ionization (MALDI) has facilitated the analysis of biological macromolecules, especially proteins. Mass spectrometry has become an essential technique for protein identification in the post-genome era, and proteomic analysis is already a routine job in most biological laboratories. Mass spectrometry has many advantages in analyzing proteins, such as low sample consumption, high throughput, high sensitivity and high speed. In the past decades, technological improvements in mass spectrometry have made it an irreplaceable tool not only for mass analysis of biomolecules but also for structural characterization of proteins. Structural information of proteins in different levels, from the primary structure to quaternary assembly, can be obtained from mass analysis of proteins combined with specific processing, as shown in Fig. 10.1. The identification of individual membrane proteins can be achieved from their peptide mass fingerprints, which are obtained via bottom-up or top-down proteomic or from the mass of intact molecule detected by ESI and MALDI. With the hydrogen/deuterium exchange (HDX) and oxidative labeling techniques, the secondary structure or higher-order structure of the protein in native states or definitive conditions has been probed according to the mass changing in some structural domains. Native mass spectrometry (native MS), which has been widely used in analyzing higher-order structures of soluble proteins and protein complexes for many years, also enables native structure of membrane proteins to be assessed in the gas phase. Evidence shows that protein structure is

L.-P. Zhan · C.-Z. Liu · Z.-X. Nie (✉)
Key Laboratory of Analytical Chemistry for Living Biosystems, Institute of Chemistry Chinese Academy of Sciences, Beijing 100190, China
e-mail: znie@iccas.ac.cn

L.-P. Zhan · C.-Z. Liu · Z.-X. Nie
University of Chinese Academy of Sciences, Beijing 100190, China

H. Wang and G. Li (eds.), *Membrane Biophysics*,
https://doi.org/10.1007/978-981-10-6823-2_10

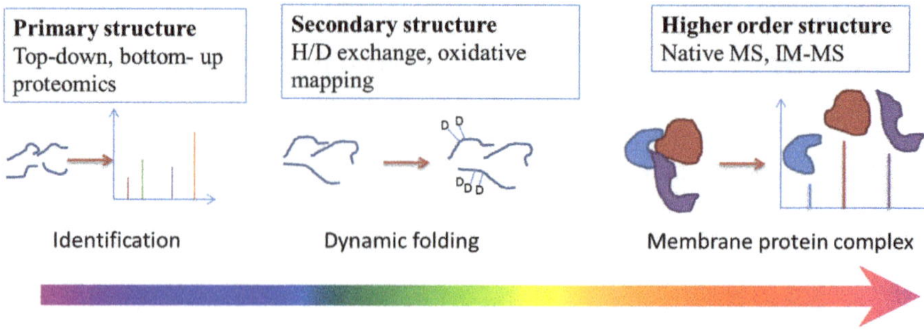

Fig. 10.1 Mass spectrometry analyses of membrane proteins in several levels

somewhat preserved when it is transferred from non-denaturing solution to gas phase via ESI. The secondary structure features of proteins, such as helix, sheet, and salt bridge, have also been demonstrated to be retained in the gas phase by the infrared multiphoton dissociation (IRMPD) [1] and electron capture dissociation (ECD) [2]. Ion mobility mass spectrometry (IM-MS), which separate ions in the gas phase, can provide information on mass, charge, size, and shape of the protein ions. IM-MS has been widely applied in characterizing protein–protein (ligand) interactions [3, 4], protein conformation dynamics [5], and oligomer analyses [6]. By releasing intact membrane protein complexes from their detergent micelle, the higher-order structures of membrane proteins have been successfully deciphered by the combination of native MS and ion mobility. This strategy has been further applied in investigating the folding states and conformational changes of membrane proteins upon small molecule binding.

Membrane proteins (MPs) play essential roles in biological activities, such as transporting, signaling, and catalyzing. More than half of the drug targets are membrane proteins. Thus, the identification, as well as structure-function analysis of membrane proteins, is of great importance. However, their intrinsic hydrophobic character, dynamic properties, and low abundance bring big challenges to membrane protein analyses. An approach combining the aforementioned MS-based strategies will accelerate the analyses of membrane proteins at different levels [7, 8].

10.2 Principle

To get a thorough understanding of membrane protein structures at different levels, several MS-based approaches are involved. Bottom-up and top-down proteomics are two leading methods to determine the primary structures of membrane proteins. Intact masses of individual membrane proteins can also be measured by direct analysis of intact denatured proteins by ESI and MALDI, which can identify the proteoforms of membrane proteins with different post-translational modifications (PTM). Labeling methods have been developed to probe the secondary structures or conformational dynamics of membrane proteins, such as hydrogen-exchange (H/D) mass spectrometry and fast photochemical oxidation. These approaches can reveal the secondary structure features in natural lipid membranes or detergent micelle according to the extent of hydrogen-exchange or oxidative labeling. Native MS has become an important method for analyzing protein-ligand/protein interaction and protein topology. We will give a brief introduction of these mass spectrometry techniques.

10.2.1 Introduction to Mass Spectrometry

Mass is an intrinsic character of the molecule, which can be used to identify molecules. The mass spectrometer is an instrument for measuring the mass of ions, which mainly consists of three parts, ionization source, mass analyzer, and detector, as shown in Fig. 10.2a. The key step of MS analysis is ionization. A lot of ionization methods have been developed to satisfy the needs of ionizing various compounds. Thanks to the innovation of "soft" ionization, biological molecules can be determined as intact molecules. ESI and MALDI are two main soft ionization sources in biological mass spectrometry, which can ionize large molecules without breaking any chemical (covalent) bonds inside the molecules. ESI ionizes molecules by applying a strong electric field to the solution containing analytes, which pass through a capillary with a weak flux. The electric field causes charge

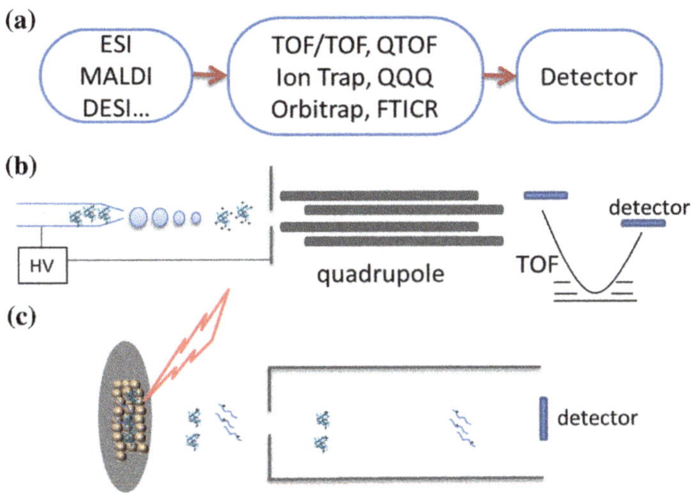

Fig. 10.2 Biological mass spectrometers. **a** Schematic view of biological mass spectrometers, which mainly consist of the ionization source, mass analyzer, and detector. **b** ESI-QTOF, ions are yielded via electrospray and transferred to the quadrupole, subsequently analyzed by the mass analyzer TOF. **c** MALDI-TOF, ions are produced via matrix-assisted laser desorption/ionization and fly through the high-vacuum flight tube, then detected by the detector

accumulation at the liquid surface at the end of the capillary, which breaks to form highly charged droplets. The droplets pass through a curtain of heated inert gas and a heated capillary, where the solvents evaporate, and subsequently the ions form. In positive ionization mode, $[M+H]^+$, $[M+Na]^+$, $[M+K]^+$, and $[M+NH_4]^+$ ions ($[M+nH]^{n+}$ for large biomolecule) are often obtained; while in negative ionization mode, $[M-H]^-$ ions are the dominant ones. For MALDI ionization of biomolecule, a compound named matrix is mixed with analytes, which forms analyte-doped matrix crystals. The matrix should have strong absorption at the work laser wavelength. When laser irradiates at the analyte-matrix co-crystals, it induces rapid heating of the crystals and subsequently causes localized sublimation of the matrix crystals and expansion of the matrix and intact analyte into the gas phase. In protein analysis, ESI generally produces multiply charged ions while MALDI mainly yields singly charged protein ions. Due to their excellent performances in analyzing large biomolecules, the innovators of these two ionization methods, Fenn and Tanaka, have been awarded the Nobel prizes in 2002. The methods have been described as "make the 'elephant' fly."

Two widely used biological mass spectrometers, ESI-Quadruple time-of-flight (QTOF) and MALDI-TOF, have been displayed in Fig. 10.2b, c, respectively. After formed in the source, the biomolecule ions will be transferred to mass analyzer and subsequently detected by the detector. The mass analyzers, such as time-of-flight (TOF), quadrupole mass filter (Q), ion trap (IT), and Orbitrap and Fourier transfer ion cyclotron resonance (FTICR), transfer or separate the ions based on different theories. For example, the TOF analyzer utilizes a simple physical theory. The ions are initially accelerated in the source to gain kinetic energy and then drift in a field-free flight tube. The ion's velocity is determined from the tube length and flight time, and the kinetic energy is calculated from the accelerating voltage; thus, the mass is determined from these parameters as

$$(m/z)^{1/2} = \left(\frac{\sqrt{2eV}}{L}\right)t \tag{10.1}$$

The acceleration voltage V and flight distance L are constant for a given mass spectrometer, and e is a constant parameter. The m/z can be determined from the value of flight time t, which is measured by the analyzer. The TOF analyzer is well suitable to the pulsed nature of laser desorption ionization known as MALDI-TOF. The mass analyzers can be hyphenated with each other to perform tandem mass spectrometry (MS/MS or tandem MS), which fragments the biomolecule ions to obtain structure-related product ions. The tandem MS can be divided into collision-, electron- and photon-based strategies, which bring external energies or transfer electrons to the ions to induce fragmentation. The MS/MS has been widely used to determine biomolecule structure, such as protein sequences [9, 10], post-translational modification (PTM) [11], protein complex stoichiometry [12], even the secondary structure features [1, 13], and unfolding conformation information [2, 14] of proteins.

10.2.1.1 Identification by Proteomics and Intact Mass Determination

To identify the membrane proteins, bottom-up and top-down proteomics is utilized to determine the protein sequence information. As shown in Fig. 10.3, top-down protein analysis deduces the protein sequence by fragmenting intact protein ions directly, while bottom-up proteomics identifies protein from their digested peptides. For bottom-up proteomic analysis of membrane proteins, the solubilization and digestion efficiency are crucial. Due to the hydrophobic character of MPs, the membrane-spanning segments are not readily accessible to the proteolytic enzymes when conducting in-gel digestion. However, digestion is a limited step in a bottom-up approach. To improve the digestion efficiency of membrane proteins and the recovery of hydrophobic peptides, many efforts in different aspects have been made over the past years, such as protein cleavage reagents, filter-aided sample preparation, gel-based shotgun methods, and on-membrane digestion. For example, a gel absorption-based sample preparation method is developed for shotgun analysis, in which membrane proteins are directly immobilized in gel, and the protein lysate with detergent is entrapped in a vacuum-dried polyacrylamide gel [15]. This approach avoids the limitation of protein loss and chemical modifications, thereby improving the protein digestion and peptide recovery. The bottom-up proteomic has been widely used to determining membrane proteins and is reviewed by abundant reports [7, 16–19]. Compared to "bottom-up" mass spectrometry, top-down proteomics addresses more detailed information of individual proteins or their specific modifications [9]. In the top-down approach, multiply-charged membrane proteins are fragmented directly via collision-induced dissociation or electron-activated dissociation to obtain the sequence information. Two strategies have been proposed to conduct a top-down analysis of membrane proteins. One involves purification and solubilization of intact membrane proteins using aqueous/organic solvent mixtures compatible with soft ionization mass spectrometry [10, 11, 20, 21].

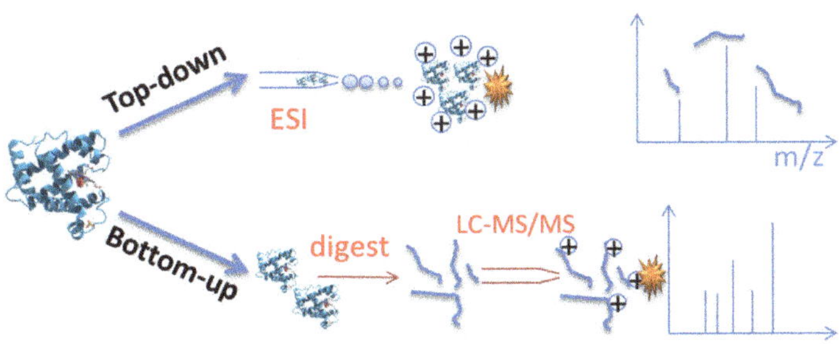

Fig. 10.3 Top-down and bottom-up proteomics. In the top-down method, the whole protein ions were fragmented directly via collision-induced dissociation or electron-based dissociation, which yielded different kinds of products. In bottom-up strategy, proteins were digested at first by trypsin or other enzymes, and the digested peptides were separated by HPLC and identified by MS/MS

Another approach utilizes the detergent to solubilize MP and sprays directly from the micelle solution. The membrane protein is released from micelle, then collides with neutral gas in the collision cell and subsequently fragments. This approach is extended from native MS of MPs and can reveal the oligomeric state and sequence information of membrane protein complex in a single experiment [22].

Another approach to determine the membrane proteoforms is measuring the mass of the intact protein without digestion. Solubilizing/denaturing agents are important factors influencing the detection efficiency by ESI-MS. However, the commonly used detergents in membrane proteins are not always compatible with mass spectrometry. The reason is that they always dominate the mass spectrum because of that they can be easily ionized in ESI. The removal of detergents becomes a key step in membrane protein proteomics. Therefore, specific organic solvent conditions or sample preparations are developed to keep the MP soluble and compatible with ESI-MS [23, 24], such as formic acid/isopropanol.

MALDI is high throughput ionization method in protein analysis. But the membrane protein's insolubility prevents the co-crystallization of proteins and matrix, which tremendously reduces the signal of MP in the mass spectrum. MALDI has some advantages over ESI, such as producing a simpler spectrum, high sensitivity, high throughput, and low sample consumption. However, the premise of a good MALDI experiment is that the sample and matrix are co-crystallized well. Thus, the key to using MALDI MS to analyze membrane proteins is to find out suitable matrix, which can assist the laser desorption and ionization of hydrophobic proteins (or peptides digested from membrane proteins). Another obstacle of membrane protein analysis by MALDI MS is the detergents, which are often used to extract membrane proteins from tissues. During MALDI analysis, the detergents will lead to signal decrease and peak broadening and make the target signal difficult to observe. Conventional MALDI matrixes are mainly hydrophilic and are not compatible with hydrophobic peptides or proteins. Hence the principle of the design of new matrix is to solve the solubility problem and, at the same time, help desorption and ionization of the target proteins/peptides. Several matrices have been introduced to improve the co-crystallization of hydrophobic proteins and matrix, which are fabricated by adding alkyl chain to conventional matrix molecules to increase their hydrophobicity (Fig. 10.4a). However, the most commonly used matrix for membrane proteins is R-cyano-4-hydroxycinnamic acid (CCA) [25, 26]. Not only the use of the matrix, but also the sample preparation can also be improved to facilitate detection of membrane proteins by MALDI-TOF. A good example is using formic acid-based formulations (sinapinic acid matrix) to analyze the intact proteins from *E. coli* K12 cells, which has got 119 singly charged ions, with most of them having not been detected by sinapinic acid matrix itself (Fig. 10.4b–d) [27]. In Fig. 10.4d, the heterogeneity characteristic of lipid acylation in a major outer membrane lipoprotein *lpp* was revealed.

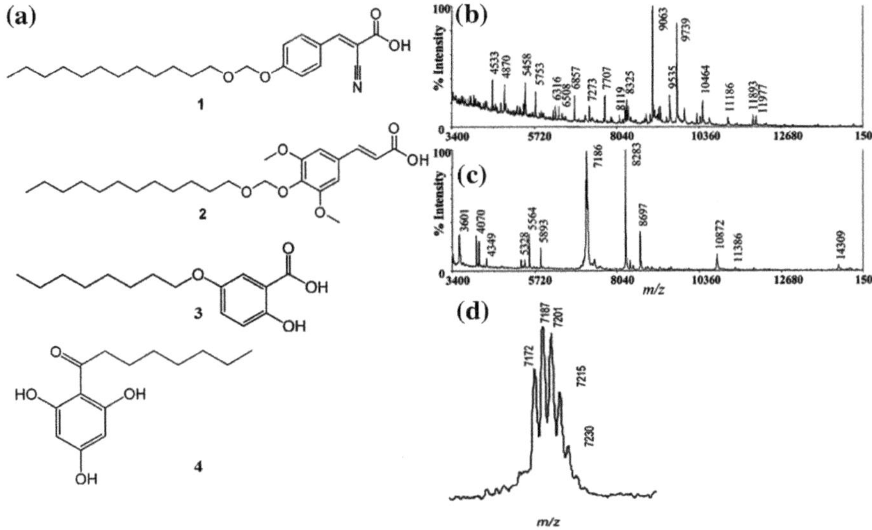

Fig. 10.4 MALDI-TOF analysis of membrane proteins. **a** Specific matrixes developed for membrane proteins, which are alkylation forms of traditional MALDI matrix molecules. They are alkylated derivatives of HCCA (1) [28], sinapinic acid (2) [28], dihydroxybenzoic acid (3) [29], trihydroxyacetophenone (4) [30, 31]; MALDI mass spectra obtained from lyophilized E. coli cells desorbed from sinapinic acid using **b** standard and **c** formic acid-based matrix conditions; **d** Expansion of B from m/z 7000–7500 revealed heterogeneity characteristic of lipid acylation in *lpp* P69776. Reprinted with the permission from Ref. [27]. Copyright 2007 American Chemical Society

10.2.2 Labeling Method

Hydrogen/deuterium exchange (HDX) mass spectrometry is a key technique for interrogation of protein conformation and dynamics. It is a hydrogen-exchange labeling procedure monitored by mass spectrometry (MS) to show structural and dynamic changes of proteins in solution. A schematic diagram of HDX-MS of peptides and membrane proteins is displayed in Fig. 10.5. As is known to all, hydrogen atoms of O–H, N–H, and S–H groups can exchange with the surrounding water, thus can be replaced by deuterium in D_2O solution. When exposing proteins or peptides to a D_2O-containing environment, the hydrogen atoms of proteins undergo exchanges with deuterium atoms of D_2O, leading to an increase of one unit mass per replace event. This is referred to as an "exchange-in" scenario, distinguishing from relatively less used "exchange-out" scenario, in which the protein is first completely deuterated before placed in H_2O, occurring D→H exchange. HDX mostly takes place at the N–H groups of protein backbone, as well as at side chains. Based on the fact that every residue of a protein possesses an amide N–H group (except for proline and the first amino acid in the chain), HDX can probe features affecting the entire protein. The key to the application of HDX for structural studies

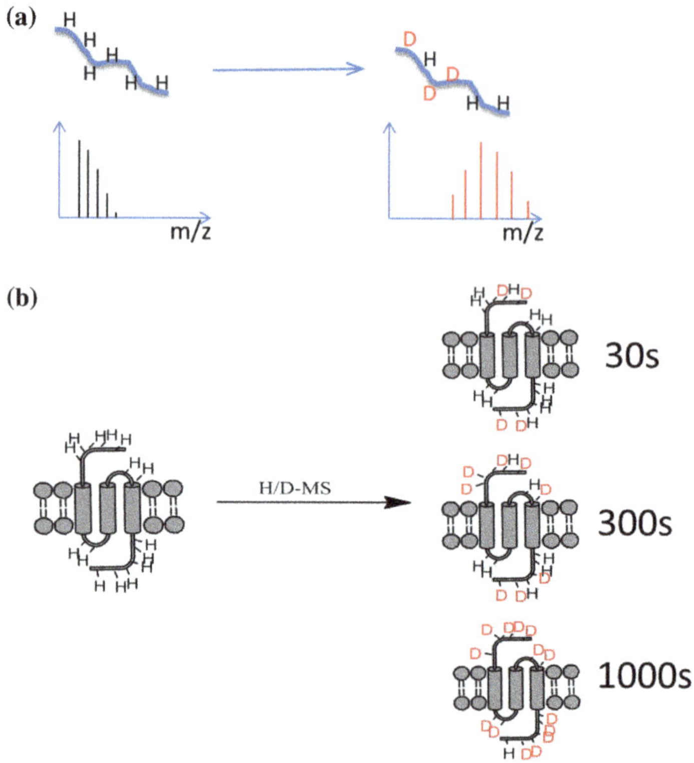

Fig. 10.5 Hydrogen-Deuterium exchange mass spectrometry for **a** peptides and **b** membrane proteins. The reaction time can be tuned to monitor the dynamic conformation in defined condition, such as light exposure

is that the rate of isotope exchange differs, depending on the conformational properties of proteins. For completely solvent-exposed amides without hydrogen bond, the exchange goes fastest. While in ordered regions, a multitude of N–H… O = C hydrogen bonds exist and the solvent access to amino sites is limited. Both factors protect the affected amide, leading to a steep decrease in the exchange rate. However, the protected amides can also undergo HDX at measurable rates owing to the conformational fluctuations of protein. In this case, H/D exchange can only occur during the short-lived transition to an "open" conformation. For membrane proteins solubilized in the detergent or lipid, the HDX always occurs on the non-transmembrane segments, which are accessible to solvents.

Most HDX/MS studies employed continuous labeling strategy. In this approach, a native protein is exposed to D_2O, and the deuterium incorporation can be monitored by MS as a function of exposure time. The time points typically vary from minutes to hours (sometimes days), while the mass of protein gradually increases along with the time, as shown in Fig. 10.6. The normalized ESI spectra can be converted to a figure that plotted with reactive HDX level versus time, which can

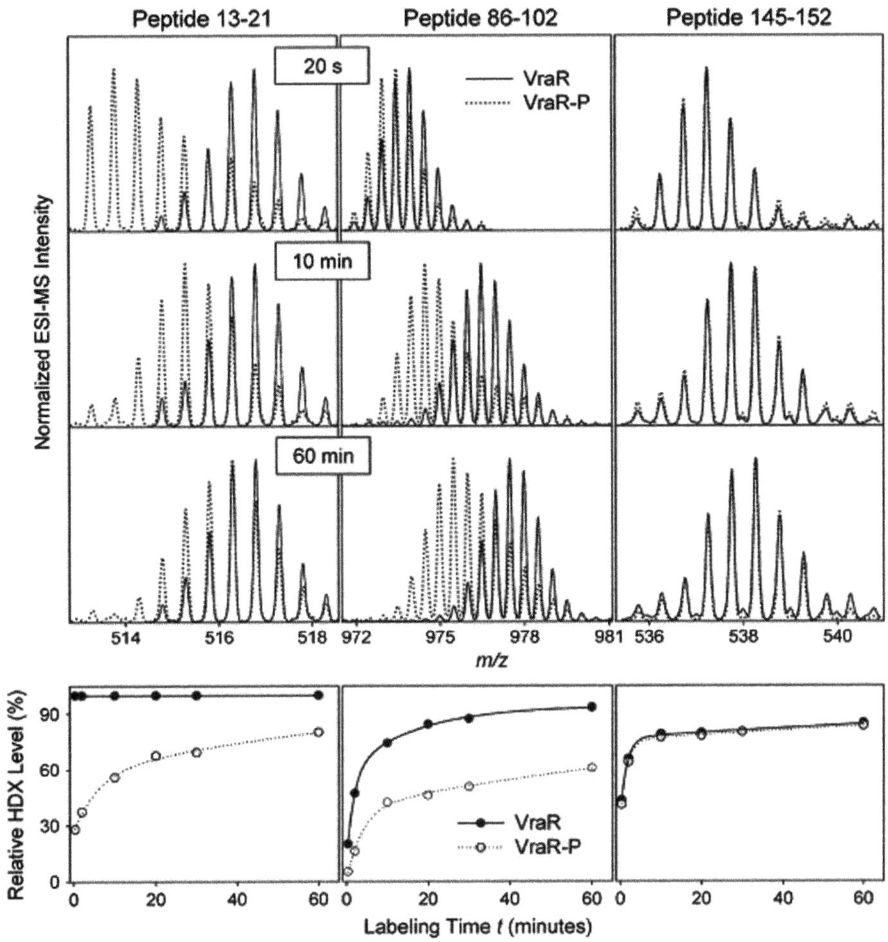

Fig. 10.6 HDX mass spectrometry of three different regions of VraR for deuteration times of 20 s, 10 min, and 60 min. Continuous lines represent VraR, and dotted lines represent VraR-P. Reprinted from Ref. [32]. Copyright 2009, with permission from Elsevier

reflect the conformation dynamic of a specific region in the protein. The graphs at the bottom of Fig. 10.6 are examples of conformation dynamic analyses, which investigate the phosphorylation-induced activation of the vancomycin-resistance-associated regulator (VraR). Regions that stay longer on open conformations undergo faster exchange than those that are more tightly folded.

When applying HDX-MS technique to membrane proteins, one should be careful to handle the detergents used for mimicking natural membrane environments to solubilize integral membrane proteins, because these detergents may be hard to separate from proteins in condition compatible with HDX-MS. Furthermore, the choice of detergents will probably affect the protein's confirmation as well as reduces the percentage of sequence coverage.

Alternative strategies based on a chemical modification to mapping membrane proteins' secondary or higher-order structural features have also been developed, such as fast photochemical oxidation of proteins (FPOP), cross-linking, and chemical labeling [33]. The labeling approaches rely on the principle that reactive sites are located on the surfaces of membrane proteins, which are correlated with protein conformation, conformational dynamics, and solvent accessibility. Similarly, covalent labeling causes a mass change of peptides or proteins of interest, resulting in peak shifts in mass spectra. Unlike labile H/D exchange, the permanent covalent modifications fix protein conformation and facilitate conformation analysis in certain aspects [34]. For instance, covalent labeling methods can be applied to membrane proteins within their natural lipid environment or even inside living cells. A representative workflow of covalent labeling analysis of membrane proteins is displayed in Fig. 10.7a. Membrane proteins firstly react with labeling reagents in solution. After a certain period of time, the reaction is quenched. The labeling proteins are subsequently digested and then analyzed by LC-MS/MS. The labeling proteins can also be analyzed via top-down approach, i.e., fragment intact labeled protein ions without proteolysis in the gas phase [35]. An example of laser-induced oxidative labeling of a model membrane protein bacteriorhodopsin (BR) is shown in Fig. 10.7b and c. The X-ray structure was displayed in Fig. 10.7b, with methionine (Met) shown in red. The solvent-accessible Met would appear as +16 Da adducts in laser-inducing oxidation condition. Figure 10.7c shows that the peptide T8 was extensively oxidized as the mass increase 16 Da after laser-induced oxidative labeling. However, it was found that nearly no oxidation occurred in the peptide T7 by comparing the spectra before and after the reaction. The results revealed that the methionine in the solvent-accessible domain (T8) was oxidized, while the residues in the transmembrane domain (T7) were not oxidized. The laser-induced labeling method FPOP is a chemical footprinting method where solvent-accessible amino acid residues of proteins are covalently labeled through oxidation with hydroxyl radicals, which are produced by photolysis of hydrogen peroxide. The data provide information about solvent accessibility and protein conformation at the peptide or even the amino acid level. FPOP must ensure that only the native conformation is labeled, like other chemical footprinting techniques. Although oxidation via hydroxyl radical induces protein unfolding on a time scale of milliseconds or longer, FPOP is designed to limit hydroxyl radical exposure to 1 μs or less by employing a pulsed laser for initiation of the radicals and a radical scavenger to limit their lifetimes.

In addition, oxidative species can also be generated on the flying using ESI-MS with high-needle voltages and oxygen as a reactive nebulizer. Before, there are several ways to generate radicals like oxidative Fenton chemistry using Fe(II)-EDTA, which uses Fe(III) (S)-1-(p-Bromoacetamido-benzyl) ethylene diamine tetra-acetic acid as a thiol-reactive reagent and incorporates a Fe(III) EDTA moiety into a purified bait protein. The bromoacetyl group allows the reagent to attach covalently to thiol groups. When using ascorbate and peroxide to activate it, the Fe (III) EDTA group cleaves peptide bonds, acting as an artificial protease via

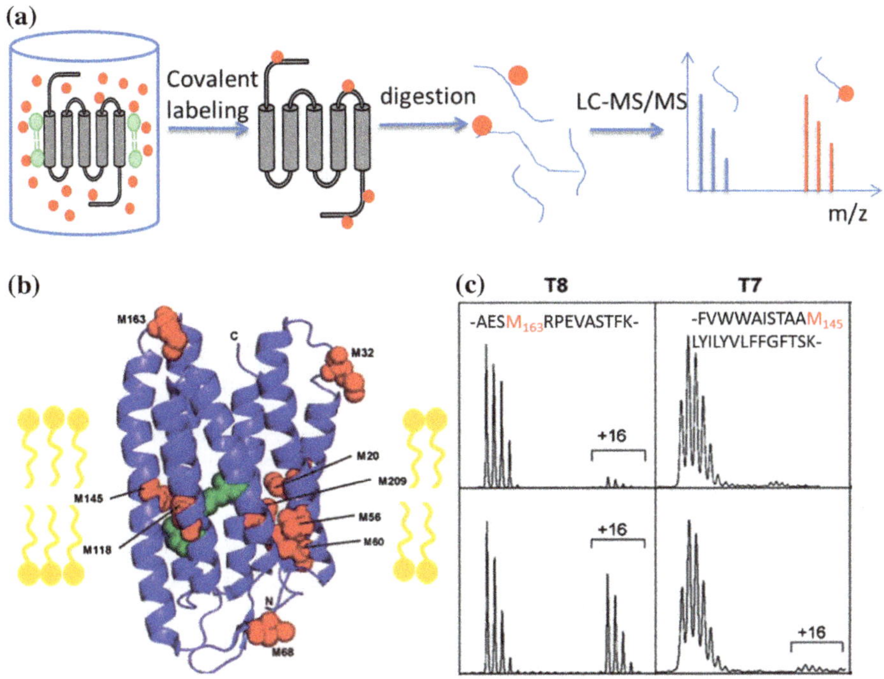

Fig. 10.7 Workflow and examples of covalent labeling method for membrane proteins. **a** A representative covalent labeling approach for membrane proteins. An example of laser-induced oxidative labeling of bacteriorhodopsin (BR) is shown in (**b** and **c**). **b** X-ray structure of BR (pdb code 1XJI). Met residues are red, and the retinal chromophore is green. **c** ESI mass spectrum of BR tryptic peptides T8 (left column) and T7 (right column) before (upper) and after (under) laser-induced oxidative labeling. The peptide sequence is displayed, and Met residues are highlighted in red. M163 was heavily oxidized while M145 was slightly oxidized. Reprinted with the permission from Ref. [36]. Copyright 2009 American Chemical Society

oxidative Fenton chemistry. However, the utilization of ionizing radiation for creating the active radicals from water in radiolysis process is the most commonly used approach, which can be achieved by high-density X-Ray (X-Ray radiolysis footprinting MS, i.e., XF-MS) [37]. Although a different mechanism is used to generate radicals, the principles of XF are essentially the same as other oxidative labeling approaches. To generate reactive hydroxyl radicals in situ, a high brilliance flux of white-light photons of different energies in ranges from 5 to 30 keV range is usually applied. The reaction leading to the generation of reactive hydroxyl radicals in non-oxygenated solutions is $2H_2O + e^- \rightarrow H_2O^+ + H_2O^* + e^-_{aq}$. Some covalent labeling methods are summarized in Table 10.1. The strategy of labeling method in membrane proteins can provide not only information of the secondary structures, but also the membrane protein-ligand interaction, such as cross-linking and the native structure features in living cells, such as genetically targetable peroxidase enzyme (APEX) [38].

Table 10.1 Several labeling approaches to investigate the membrane protein structure

Membrane proteins	Labeling methods	Labeling residues	Application	Ref.
Human erythrocyte membrane band 3 protein	Chloramine T	Methionine	Detergent exposure cause structural changes	[39]
Membrane proteomic	Fenton chemistry	Met and aromatic residues, but also for Asp, Leu, Val, and Ser	Structural transitions of outer membrane in living cells	[40]
Nicotinic acetylcholine receptor	Hydrophobic photoreactive probe with UV light	Multiple labeling agents	Switching events in response to changes in transmembrane potential	[41]
Rhodopsin	Sulfosuccinimidyl acetate, acetylation of lys	Lys	Interaction of rhodopsin and transducing-derived short peptide	[42]
Membrane-attached Fenna-Matthews-Olson (FMO) antenna protein	Carboxyl-group modification	Carboxyl group on glutamic acid (E), aspartic acid (D)	The FMO protein interacts with the transmembrane reaction center complex	[43]
Tyrosine Kinase Her4	Carboxyl-group footprinting	Carboxyl group on glutamic acid (E), aspartic acid (D)	Dimerization interface and phosphorylation-induced conformational changes	[44]
Inner mitochondrial membrane proteome	Genetically targetable peroxidase enzyme (APEX)	Solvent-exposed tyrosine	Architecture and connectome of the IMM proteome in cell	[38]

10.2.3 Native Mass Spectrometry of Membrane Proteins

Native mass spectrometry is referred to mass spectrometry analysis of biological molecules or complexes in their physiological conditions, which are sprayed from non-denaturing solution. It is considered that these ions in the gas phase are near the native folded structure in solution. Despite the debate that whether native MS retains protein solution structure in the gas phase remains [1, 45], it has been widely used in protein structure characterization and structure-function analysis. Native MS of biomolecule complex can preserve non-covalent interactions in the gas phase, and it has succeeded in obtaining subunit stoichiometry, protein topology, and protein-ligand-binding thermodynamic, therefore deciphering some complex biological phenomena. Due to the inherent insolubility of membrane proteins in buffers, which are compatible with ESI and the ready dissociation of subunit interaction, native MS of membrane proteins had remained big challenges, until the first report on the detection of naked membrane protein ions at 2008 [46]. Membrane proteins always solubilize in the detergent micelle, which mimics the lipid bilayer of cells. However, the vast excess of detergent aggregates in solution would lead to the suppression of protein ionization and dominate in mass spectra. The approach to conduct native mass spectrometry of membrane proteins reported by Robinson is shown in Fig. 10.8a. The micelle containing membrane proteins or complex was ionized via nano-ESI or ESI. Then, the charged micelle ions collided with neutral gas in the collision cell. Finally, the charged membrane proteins were released and detected after stripping the detergent cover in the collision cell. A common mass spectrum of membrane protein micelle is displayed in the bottom of Fig. 10.8b, which was dominant by micelle complex ions. After a collision with neutral gas, the proteins appeared, as shown in the upper spectrum of Fig. 10.8b. The nonionic detergent n-dodecyl-b-D-maltoside (DDM) was used to maintain solubility of $BtuC_2D_2$. The solution containing DDM above critical micelle concentration (CMC) and $\sim$5 μM membrane proteins were introduced from a gold-coated nanoflow capillary to the QTOF via nano-ESI. After collisional activation in the electrospray source and collision cell without precursor selection, membrane proteins $BtuC_2D_2$ were released from the protein-detergent assembly and detected as intact protein complex. While under the same collision voltage conditions, all soluble protein complexes dissociated. The result implied that the release of detergent aggregates protects protein complex from dissociation [47].

As protein ions produced in native mass spectrometry are always in low charge states, the mass-to-charge (*m/z*) of protein complex ions will be higher than that in denaturing MS. The mass spectrometer used in native MS should be suitable for transferring and detecting high-*m/z* ions. The most popular mass analyzer in native mass spectrometry is quadrupole-time-of-flight (QTOF), because the TOF analyzer has no mass limit theoretically. But for native MS of the protein complex, the frequency of the transfer electrode should be reduced, and the pressure of transmission region should be raised to improve the transmission of high-m/z protein ions. In the early days, such QTOFs were not commercial and should be modified

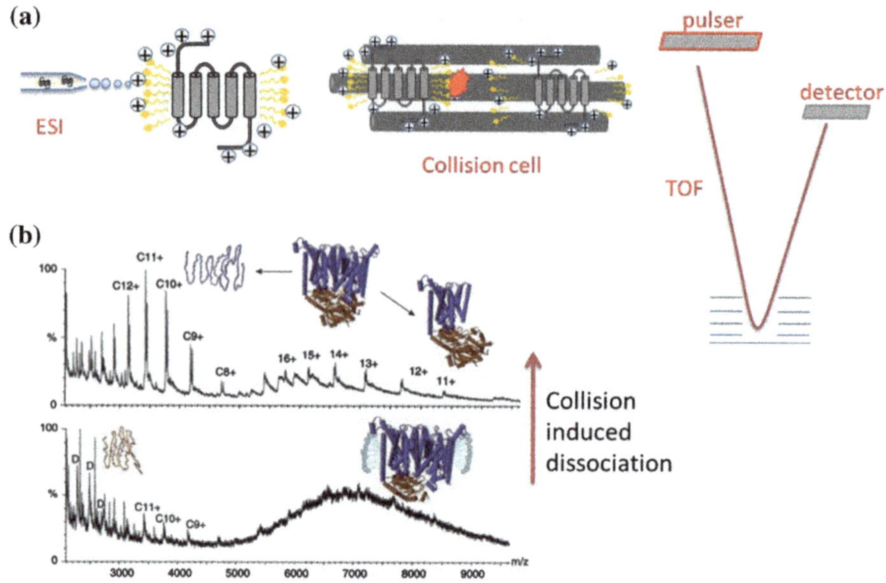

Fig. 10.8 Native mass spectrometry of membrane proteins. **a** Schematic workflow of native MS of membrane proteins. The membrane protein micelle sprayed from nano-ESI was collided with gas in the collision cell, the detergent was stripped, and the protein released from the micelle was detected by the TOF mass analyzer. **b** The native mass spectrum of membrane proteins $BtuC_2D_2$. From [46]. Reprinted with permission from AAAS

[48], which limited the development of native MS of the protein complex. The modified QTOF utilized a custom-built radio frequency (RF) generator to reduce the frequency of the quadrupole, to allow mass selection above 32000 *m/z* and enhance transmission of the high-mass protein complex. The pressure in source and vacuum system at various stages during the flight path was increased. Nowadays, the commercial QTOF and FT-based mass spectrometers have such options for large molecule analysis. The Waters Synapt G2 and Thermo Orbitrap EMR (extend mass range) are good examples of instruments for native MS analysis of proteins/protein complexes. The modified Orbitrap for high-resolution measurement of membrane protein complex has been reported firstly in 2016 [49, 50], and soluble protein complex has been studied since 2012 [51]. The high-resolution mass spectra of membrane protein complex will expose unexpected proteoforms and enable covalent or non-covalent binding substrates to be determined [52]. Due to the development over past five years, FTICR has also been applied to native MS research of membrane proteins. The performance of various mass analyzers, including QTOF, Orbitrap, and FTICR, for native MS analysis of membrane protein has been compared by seven-transmembrane helical protein bacteriorhodopsin [53]. The study predicts that FTICR platform will become a routine instrument for native MS analysis of membrane proteins, which have distinct advantages in terms of levels of native desolvation, reducing background noise, and increasing S/N levels.

Additionally, ion mobility (IM) coupled with QTOF has enabled measurement of the topology of the protein complex, which facilitates the structure-function study. Ion mobility is an instrument that separates ions based on their mass, charge, size, and shapes, which determine their mobility or ability to move through a certain medium in the gas phase. As shown in Fig. 10.9a, two kinds of protein or peptide ions with identical sequence or same *m/z* transmit through the ion mobility cell where they collide with neutral gas molecule under the influence of a certain electric field. These two kinds of ions are separated at definitive drift time because of their different sizes and shapes. The drift time can be converted to collision cross section (CCS), which correlates with the ion conformation. A commercial traveling voltage wave ion mobility mass spectrometry (TWIMS) is displayed in Fig. 10.9b. The ion mobility cell is installed after quadrupole and before TOF analyzer. There are other commercial IM-MS including trapped-ion-mobility mass spectrometry (TIMS) supplied from Bruker and drift-tube IM-MS 6560 from Agilent. Taking advantage of the ability of IM-MS to determine protein topology, researchers have found that protein conformation can be retained somewhat after transferring from solution to

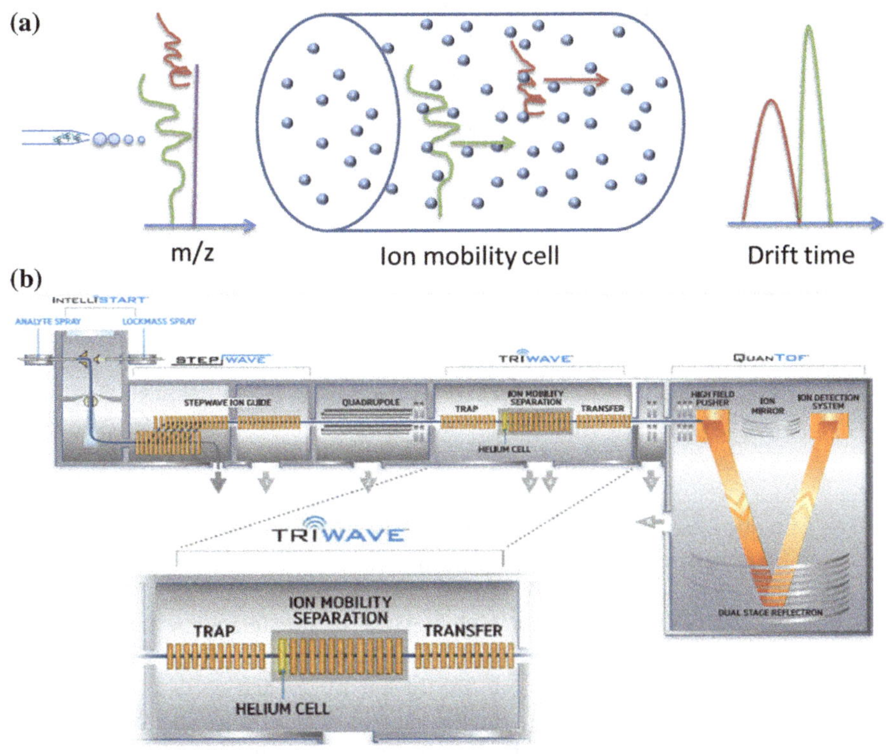

Fig. 10.9 Schematic of ion mobility mass spectrometry. **a** Principles of ion mobility mass spectrometry. **b** Schematic representation of a commercial TWIMS instrument, SynaptG2 (Waters, UK)

gas phase [54], which is a longstanding debate in native mass spectrometry [1, 45]. For membrane proteins liberated from micelle upon collisional activation, protein unfolding or dissociation may happen under the harsh activation [55]. Unexpectedly, IM-MS study of potassium channel KirBac3.1 showed that the CCSs measured are in accordance with the native states determined by X-ray crystallography [56], which implied that the native structure of membrane protein was preserved in the gas phase. Moreover, IM-MS analyses of membrane proteins have demonstrated that detergent releasing prolongs the lifetime of membrane protein in the gas phase [57] and stabilizes membrane protein complex [47].

To obtain good results of the native mass spectrum of membrane proteins, several conditions should be considered, such as solubilization platform, detergent concentration, collision voltage, and source/collision gas pressure [58–60]. Choosing the suitable detergent or solubilization platform to solubilize membrane proteins is the first step. Different membrane protein complexes need specific detergents to preserve their native structure from collisional activation, which is used to rid the detergent cover from membrane protein complexes. The most commonly used detergent in native MS of membrane proteins is n-dodecyl-β-D-maltoside (DDM) [46, 59]. However, it requires moderate-to-high levels of collisional energy to strip the detergent cover, i.e., 150–200 V, which may result in the loss of structure integrity. For example, one study used DDM as the solubilization detergent, but only the ion channel Mscl monomer was detected, indicating that the activation condition was too harsh to maintain the native oligomeric state of Mscl [61]. But when conducted native MS of Mscl in Triton X-100, the protein pentamer ions appear in the spectrum at very mild declustering condition, i.e., $\sim$50 V.

It has been found that the optimal molar ratio of detergent: protein is $\leq$100:1 while maintaining the detergent concentration above the critical micelle concentration (CMC). Another critical parameter is activation coefficient, which is derived from cone and collision voltages.

$$\text{Activation coefficient} = \text{Average charge state} \times \left[\frac{\text{cone voltage}}{\text{maximum cone voltage}} + \frac{\text{collision voltage}}{\text{maximum collision voltage}}\right] \tag{10.2}$$

The activation energy required to release proteins from micelle is proportional to the mass of protein complex, and this parameter is higher for membrane protein than that for soluble complexes.

Although detergents have been widely used in membrane protein structural biology, they have some drawbacks. For example, detergents promote unfolding of membrane proteins and fail to mimic the lateral forces and curvature of cellular membranes [60]. Amphipathic polymers, bicelle, and nanodisc have been introduced into the native MS analysis of membrane proteins and show the better protective effect on protein structure. For *E. coli* diacylglycerol kinase (DgkA), a trimeric cytoplasmic membrane protein, native MS in the presence of DDM showed

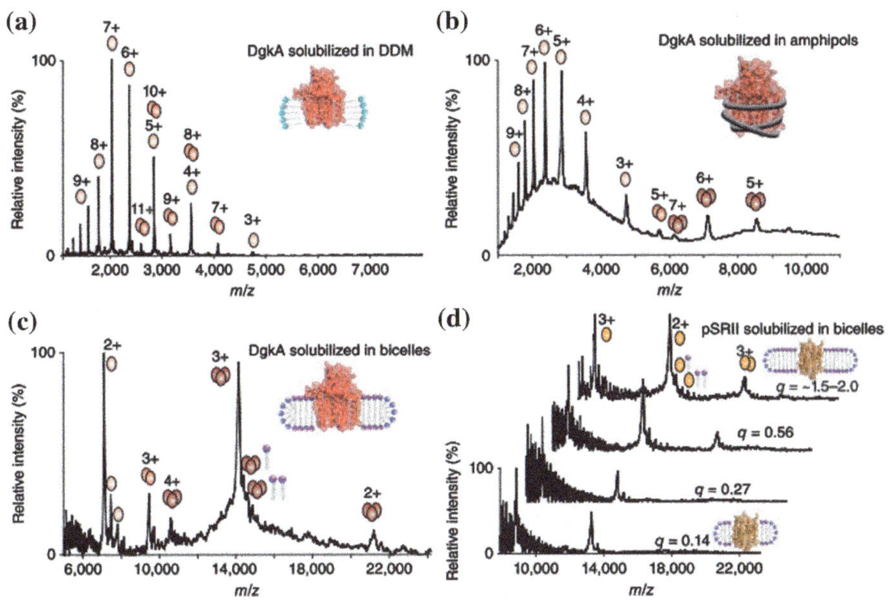

Fig. 10.10 Comparison of Native MS of membrane proteins in **a** detergent, **b** amphipols, and **c**, **d** bicelles. Reprinted by permission from Macmillan Publishers Ltd: Ref. [60], Copyright 2013

no trimeric ions (Fig. 10.10a). While native MS analysis of DgkA solubilized in amphipols (Fig. 10.10b) and bicelles (Fig. 10.10c) showed that the trimers appeared in the spectrum and bicelles were better than amphipols. Bicelles are bilayered discoidal lipid-detergent assemblies, which mimic cell membranes and have been an excellent medium for structure determination of membrane proteins by NMR spectroscopy [62]. The proteins liberated from bicelle exhibit in low charge states. It is suggested that lipids in bicelle shielded the complex from charging during electrospray ionization and protein maintained a compact native-like state in the bilayer before activation. Moreover, native MS of bicelle complex in the presence of lipids showed that lipids remained binding to the membrane proteins after collisional activation (Fig. 10.10c and d).

Amphipathic polymers (amphinols) [63, 64] are designed to bind non-covalently to membrane proteins and used in native MS of membrane proteins, which show better property than DDM in stabilizing membrane protein structure in the gas phase [64]. The spectrum in Fig. 10.10b also shows the trimeric state of DgkA, which implies that amphinols maintain the membrane protein complex near their native structure [60]. By comparing the ESI-IM-MS result of membrane proteins in DDM with that in amphinols (A8-35), one study indicated that solubilizing membrane proteins in amphinols resulted in similar charge state distribution as less gas-phase unfolding states [64]. The low charge state ions that produced by A8-35-solubilized membrane proteins exhibit CCS comparable with those calculated from high-resolution data. The study highlights the alternative solubilization of amphinols for native MS analysis of membrane proteins.

Nanodiscs are cell membrane-like lipid bilayers and have been developed as vesicles for membrane proteins in native MS analyses. Nanodiscs consist of the lipid bilayer and engineered amphipathic membrane scaffold proteins (MSP), where lipid bilayer is encircled by MSP. For DgkA solubilized in nanodiscs, which reconstituted with membrane scaffold protein 1D1 and 1-palmitoyl-2-oleoyl-sn-glycero-3-phosphocholine (POPC), the trimer can be liberated intact from nanodisc complex [60]. The energy required to release membrane proteins from nanodiscs is higher than that from DDM. But when the activation energy is low, membrane protein assemblies with a large number of lipids bound can be obtained in several defined subsets. Using the high-resolution Orbitrap modified for high-mass ion and the Bayesian deconvolution program UniDec, individual lipid molecules bound to membrane proteins were counted in the complex mass spectra [49]. By comparing the lipid distributions measured by native MS and results of molecular dynamics, it revealed that the distributions corresponding to distinct lipid shells varied, according to the type of protein-lipid interactions. These results highlight that nanodiscs offer the potential for native MS to study the interactions of membrane proteins and lipid environment.

10.3 Applications of Mass Spectrometry in Membrane Proteins

10.3.1 Redox-Linked Conformational Transition of Enzyme Monitored by HDX-MS

Here, we briefly illustrate an application of hydrogen-deuterium exchange MS on membrane protein research using DDM as detergent. In 2006, Busenlehner et al. [65] demonstrated that the transitions between catalytic intermediates in cytochrome c oxidase (CcO) (a redox-driven proton pump) are orchestrated with opening and closing of specific proton pathways, which provides an alternating access for protons to the two sides of the membrane. CcO is an integral membrane protein catalyzing the four-electron reduction of O_2 to H_2O. The crystal structure is displayed with HDX data undergo redox-dependent conformational changes in Fig. 10.11g. An additional uptake of four protons is also required in the reduction, and the protons are taken from the cytosolic or negative side of the membrane by pumping across the membrane. The location of residues involved in accepting protons, and the organizing water molecules are essential for determining the mechanism of redox-coupled proton translocation. In general, CcO has two proton uptake pathways, described as K pathway and D pathway. Meanwhile, it has four redox states that are sufficiently stable for analysis, including the oxidized state (O), the four-electron reduced state (R), the "peroxy" intermediate (Pm), and the "ferryl" intermediate (F), as shown in Fig. 10.11h. During protons transportation and O_2 reduction, the intermediate redox states cycle was turned as O→R→A→P_m→F→O. The K and D pathways peptide on four intermediate states of CcO were selected for deuterium incorporation. Briefly, catalytic intermediate states of CcO were dissolved with DDM and then pulsed with deuterium oxide (D_2O) at pH 7.0 for varying times (15 s to 6 h). After acid quenching and proteolysis, the resulting peptides were separated by HPLC, with an additional wash step after each injection to remove detergent and undigested proteins. Deuterium incorporation was obtained as a function of time by MS. The kinetics of H/D exchange monitored differences in structure and dynamics among discrete intermediate states.

Deuterium incorporation for residues related either to K pathway or to D pathway is shown in Fig. 10.11a–f, where different colors represent four intermediate states from which the residues originate (O: red, R: black, Pm: blue, and F:

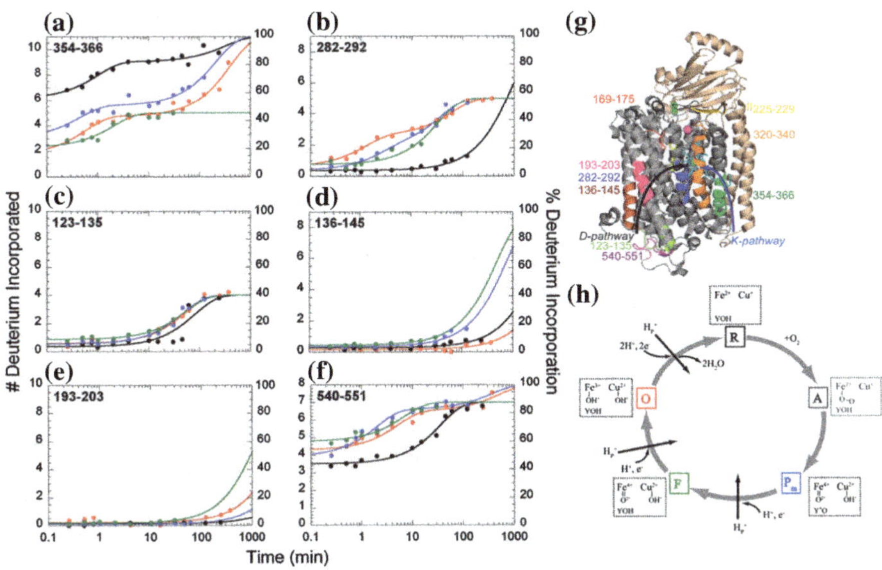

Fig. 10.11 Amide H/D exchange mass spectrometry of cytochrome c oxidase (CcO). Kinetic profiles of selected K and D pathway peptides for the four intermediate states of CcO are shown from **a** to **f**. Deuterium incorporation into selected peptides for each intermediate state (O; red), (R; black), (P_m; blue), and (F; green) are displayed as well. **a** Peptide 354–366 of subunit I contains K pathway residues. **b** Peptide 282–292 contains the D pathway residue E286, and the Y288-H284 covalent cross-link. Y288 is in the K pathway, and H284 is a CuB ligand. **c** Peptide 123–135 contains D132 at the entrance of the D pathway. **d** Peptide 136–145 contains the D pathway residue N139. **e** Peptide 193–203 contains the D pathway residue S201. **f** Peptide 540–551 of subunit I resides in a loop at the cytosolic entrance to the D pathway. **g** Peptides that undergo redox-dependent conformational changes. **h** The four redox states of CcO: the oxidized state (O), the four-electron reduced state (R), the "peroxy" intermediate (P_m), and the "ferryl" intermediate (F). Reprinted from Ref. [65]. Copyright (2006) National Academy of Sciences, USA

green). Generally, peptide 354–366 (a) is related to K pathway, while peptide 123–135 (c), peptide 136–145 (d), peptide 193–203 (e), and peptide 540–551 (f) are related to D pathway, and peptide 282–292 (b) is related to both D and K pathway. The total number (left axis) or percentage (right axis) of deuterium incorporated into the peptide backbone as a function of time is fit to exponential equation

$$D = N - \sum_{i=1}^{N} \exp(-k_i t) \tag{10.3}$$

where N represents the number of amide protons that exchange at rate constant k_i. Much information can be obtained from the figures, for example, K pathway opens to solvent during O→R transition. In the O state, peptide 354–366, which embodies most of the K pathway, was in a protected conformation as shown by the low percentage of amide hydrogen exchange in the fast phase (18% deuterium incorporation within 15 s) (Fig. 10.11a). However, in the R state, the deuterium incorporation of peptide 354–366 increased (63% within 15 s), revealing that the amide hydrogen of this residue exposed to base-catalyzed exchange after full reduction (R state). As a result, by monitoring amide HDX kinetics along protein backbone, the distinct redox-linked conformational transitions in the catalytic cycle of CcO were detected.

HDX-MS analysis has also been applied to other kinds of membrane proteins, such as receptors [66], enzymes [67], and ion channels [68].

10.3.2 Membrane Protein Conformational Dynamic Monitoring by IM-MS

IM-MS has been applied to measure the size and shape of protein assembly [3] and has provided information on the stoichiometry, topology, and cross section of the protein complex. It enables characterization of the conformational changes upon partner binding, such as lipids, drugs, and nucleotides. A previous study using IM-MS to examine the gas-phase structure of two tetrameric membrane protein complexes has suggested that the native-like structure can be maintained under the harsh activation condition required to release protein complexes from micelle [56]. A good example for IM-MS to monitor membrane protein conformational changes is the study on structural changes of an ion channel during its gating [61]. The pentamer structure models with a different number of wild type (WT) and cysteine point-mutated subunits (G22C) of the mechanosensitive channel of large conductance (MscL) are shown in Fig. 10.12c. By using this kind of heteropentameric MscLs, the opening of MscL can be controlled. As the number of G22C increased, the channels opened further. From previous studies on this ion channel, the pore diameter is ~3.5 Å in its closed state while it becomes as large as ~30 Å in the fully open state. IM-MS can detect several coexisting states with a change as small as 3%. Based on this system, the covalent binding of a positively charged

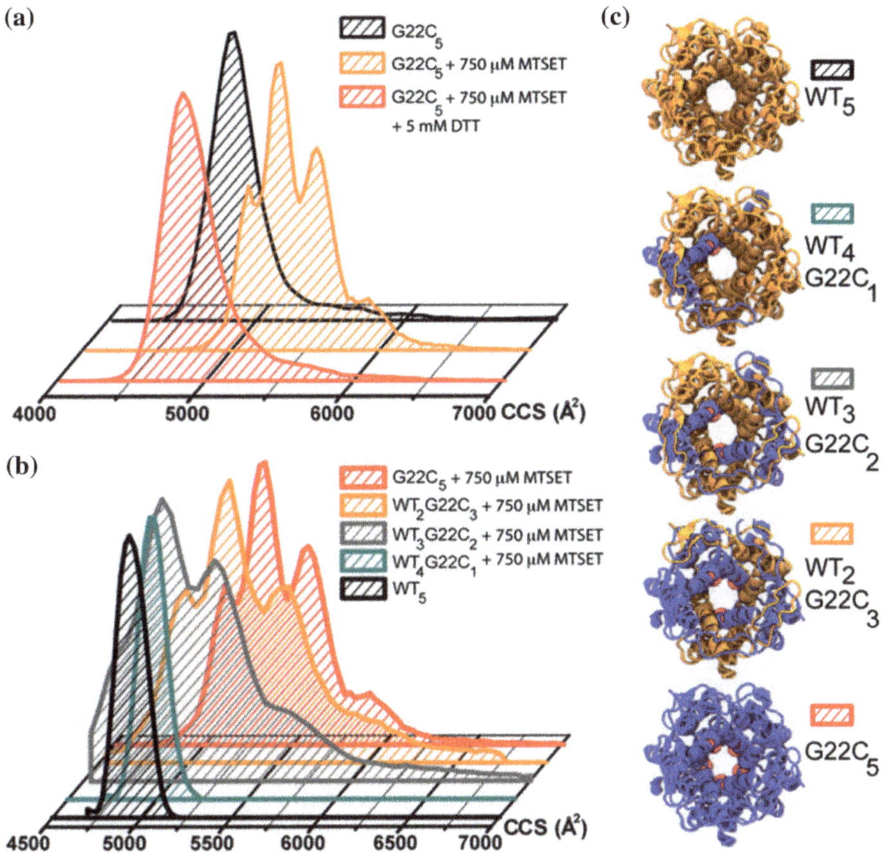

Fig. 10.12 Opening of MscL monitored by IM-MS. **a** Charge-induced opening of MscL is reversible. **b** Increasing the number of MTSET-labeled cysteines in the pore lining allows MscL to occupy more extended conformations. **c** Top view of homo- and heteropentameric MscL with different numbers of WT (yellow) and G22C (blue) subunits. MTSET binds to the cysteine at position 22 of MscL monomer, which is highlighted in red. Reprinted from Ref. [61]. Copyright (2014) National Academy of Sciences, USA

cysteine-specific channel activator [2-(trimethyl ammonium) ethyl] methanethiosulfonate bromide (MTSET) was used to test whether the MscL gating could be reversibly detected by IM-MS. In the presence of MTSET, the CCS of G22C5 (16+ charge state) increased and showed four distinct peaks (Fig. 10.12a), suggesting various sub-open conformations during the ion channel's gating. When adding the reducing agent DTT, MTSET was cleaved off the protein, and the CCS of protein returned to its original values. Moreover, MTSET did not cause a structural change in CCS of WT_5 alone. For tracking the gating pathway from closed to open state, a different number of chargeable G22C subunits were analyzed in the presence of 750 μM MTSET (Fig. 10.12b). The increase of CCSs was in accordance with the increasing number of G23C subunits. These results were in good agreement with the patch-clamp single channel measurements and highlighted the important application of native IM-MS for detecting conformational changes of membrane proteins upon activation.

The conformational changes of membrane protein induced by binding other small molecule have also been investigated by ion mobility mass spectrometry in detail, such as lipid, drug, and nucleotide [69]. For example, the effects of nucleotide binding on both ATP hydrolysis and proton translocation of V-Type ATPases have been revealed by native IM-MS [70].

10.3.3 Membrane Protein-Ligand-Binding Study

Since membrane proteins function as ion channels, enzymes, and receptors, identifying the binding partner and determining the strength of the interactions are important for understanding the protein functionality. Though a diversity of methods have been developed to study protein-ligand-binding events, the analyses are generally ensemble measurements, which contain the contribution from free and bound species, such as isothermal titration calorimetry (ITC) and surface plasmon resonance (SPR). Another challenging situation in these approaches to discern is multi-ligand-binding event. Mass spectrometry has been applied widely to the study of protein-ligand binding [71, 72], such as saccharide-protein [73], lipid-protein, and drug-protein binding, even the binding strength of protein-ligand. Utilizing native MS to determine the binding thermodynamics has been reported recently. It has many advantages over other approaches such as low sample consumption, high throughput, and the ability to detect individual ligand-binding events. To determine the binding thermodynamics of membrane protein-lipid through van't Hoff analysis, an instrument which can alter and control the temperature of analyte solution (T_{sample}) and air surrounding the mass spectrometer source chamber (T_{air}) is constructed (Fig. 10.13a). The T_{sample} equilibrium can be established at $\pm$ 0.3 °C within $\sim$40 s after moving the nano-ESI stage into the source chamber. The T_{sample} can be set to a desired point by monitoring the T_{air} conveniently with a calibration curve between T_{air} and T_{sample}. The procedure to record a native mass spectrum of membrane-ligand binding is shown in Fig. 10.13b. First, slide the nano-ESI capillary containing sample into the source chamber, then adjust T_{sample} to a set temperature and incubate online for a given time to reach binding equilibrium. Finally, acquire native mass spectrum in an optimized condition. After that, process and deconvolute the native MS data and convert the protein (P) and protein-ligand (PL) species to a molar fraction for a given ligand titration. For simplicity, one ligand-binding event is considered:

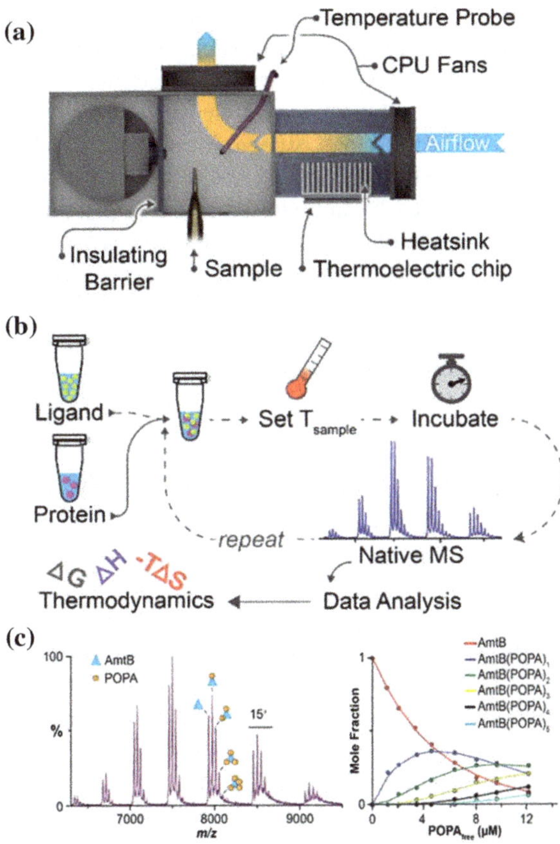

Fig. 10.13 Native MS determines the binding thermodynamics. **a** Instruments; the heat sink and thermoelectric chip along with the direction of airflow (yellow and blue arrows) through the use of two central processing unit (CPU) fans, and the temperature probe (T-type thermocouple) to monitor T_{air} to get a desired T_{sample}. **b** The workflow to measure the thermodynamic parameters. **c** A representative native mass spectrum of membrane protein titrated with lipid. Up to five binding events were detected as shown on the left plots of molar fraction for AmtB and AmtB $(\mathrm{POPA})_{1-5}$ are shown on the right. Reprinted with the permission from Ref. [74]. Copyright 2016 American Chemical Society

$$
\begin{aligned}
P + L \leftrightarrow PL \\
K_A = \frac{[PL]}{[P][L]}
\end{aligned}
\tag{10.4}
$$

K_A is apparent equilibrium association constant.

$$
InK_A = \frac{-\Delta G}{RT} = \frac{T\Delta S - \Delta H}{RT} = -\frac{\Delta H}{R} \cdot \frac{1}{T} + \frac{\Delta S}{R} \tag{10.5}
$$

Through van't Hoff analysis (Eq. 10.5), the enthalpy change (ΔH), entropy change (ΔS), and Gibbs free energy (ΔG) can be determined from the plot of the natural logarithm of K_A as a function of reciprocal of temperature. While for membrane protein-lipid-binding analysis, self-association of lipids must be taken into account [74], because a fraction of lipids are free to bind and others can't bind. The multiple binding events and lipid self-association model processing can be found in reference [74]. Using this protein-binding thermodynamic measurement, ammonia channel (AmtB) was selected to investigate the thermodynamics for lipids interacting with membrane proteins and to understand the molecular basis for their recognition. A representative mass spectrum of AmtB incubated with lipid POPA, which contains 1-palmitoyl-2-oleoyl (PO, 16:0–18:1) tails and headgroup phosphatidic acid (PA), is shown in Fig. 10.13c. Up to five binding events were detected, and the equilibrium association constant (K_A) for each lipid-binding event was determined by applying lipid-binding models. The results revealed that five different lipids bind in different pathways, which are driven by hydrophobic and van der Waals interactions. Enthalpy-entropy compensation for binding lipids with variable chain length was found. The thermodynamic signatures of binding become more hydrophobic in nature with a longer chain length that is consistent with the physiochemical properties of lipid.

10.3.4 Membrane Proteins-Drug Binding

More than half of the drug targets are membrane proteins. Therefore, membrane protein-drug binding analysis is important to evaluate the drug efficiency and provides insights into drug development. Native MS has provided direct evidence for the binding of HIV protease inhibitors (PIs) to membrane protein ZMPSTE24 [75]. ZMPSTE24 is a human integral metalloprotease on the membrane, which catalyzes prelamin A to lamin A. Some HIV PIs have been found to block ZMPSTE24 function. However, direct evidence of HIV PIs binding to ZMPSTE24 at the molecular level remains lacking, until the report of native MS analysis on this system. First, high-mass resolution modified Orbitrap confirmed the presence of zinc binding, which is essential for binding and cleavage of prelamin A peptides. After incubated with equimolar amounts of a 26-mer farnesylated synthetic peptide corresponding to the C-terminus of prelamin A, the catalytic cleavage of prelamin peptide by ZMPSTE24 was monitored in real time by mass spectrometry. Then, HIV PIs were added to an equimolar solution of ZMPSTE24. The results indicated that the catalysis of prelamin A was blocked in the presence of HIV PIs. The drug-bound enzyme conjugate has been assigned in the recorded high-resolution mass spectra. A competitive drug-binding assay was performed, and the results were shown in Fig. 10.14a. The intensities of ZMPSTE24-drug conjugates showed similar trends to the intensity profiles observed when drugs were added separately. Based on the peak height of drug-bound enzyme ions and the results of the competitive binding assay, which were performed under the same procedures and solution concentrations, the order of affinity of drugs to enzyme had been established as lopinavir > ritonavir > amprenavir > darunavir. Finally, the effects of HIV protease inhibitor drugs binding to ZMPSTE24 were evaluated by IM-MS, as shown in Fig. 10.14b. The transition from compact folded state to the first intermediate state was highlighted with white arrows in Fig. 10.14b. The onset of the intermediate states occurred at activation voltages $\sim$80, $\sim$85, and $\sim$ 95 V for apo, ritonavir, and lopinavir bound state, respectively. From these results, it was

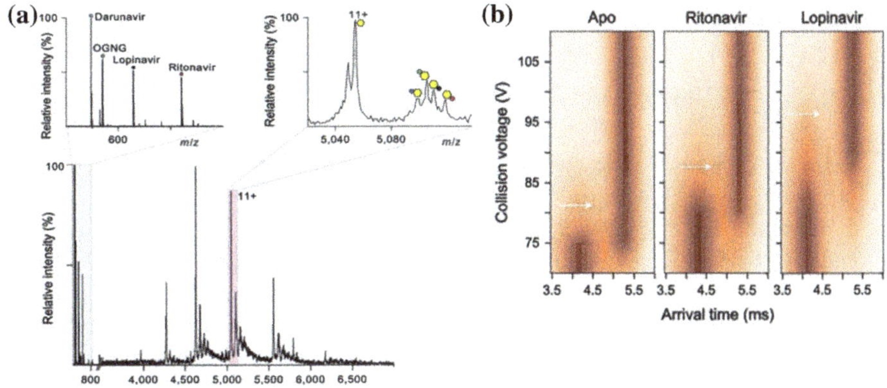

Fig. 10.14 Native mass spectrometry of membrane protein-drug conjugate. **a** Competitive binding of HIV PI drugs to ZMPSTE24 recorded by Orbitrap. Interaction between ZMPSTE24 and a 1:1:1 solution of three HIV drugs lopinavir, ritonavir, and darunavir is shown on the left. Expansion of the mass spectrum for m/z 5020-5135 reveals binding of CHS, OGNG, lopinavir, and ritonavir to the 11+ charge state of ZMPSTE24 in the right. 11+ charge states are labeled: yellow hexagon, zinc-bound ZMPSTE24; dark blue circle, ZMPSTE24-CHS; green circle, ZMPSTE24-OGNG; black circle ZMPSTE24-lopinavir; red circle, ZMPSTE24-ritonavir. **b** IM-MS of free ZMPSTE24 and drug-bound states. Comparison of the collision-induced unfolding from 70 to 110 V of the 11+ charge state of Zn^{2+}-bound ZMPSTE24 with ritonavir or lopinavir. The transition from compact folded state to the first intermediate state is highlighted (white arrows). This transition occurs at a higher voltage for lopinavir-bound ZMPSTE24 than ritonavir-bound ZMPSTE24. Reprinted by permission from Macmillan Publishers Ltd: Ref. [75], Copyright 2016

concluded that both ritonavir and lopinavir stabilize ZMPSTE24. However, the effect is greater for lopinavir, because the activation voltage was higher of lopinavir than that of ritonavir. This study highlighted the combination of multiple mass spectrometers (Orbitrap and IM-QTOF) to solve important biological issues.

10.4 Protocol

A protocol for native MS analysis of membrane proteins has been published yet [59], where readers can find more detailed description of the experiment, especially the sample preparation. Here we summarize some notes and reprint the procedures of that protocol.

10.4.1 Procedures

10.4.1.1 Preparation of the Purified Membrane Protein Complex

1. Membrane protein preparation for native MS is achieved by either using a buffer-exchange device (option A) or by gel filtration (option B) on a purified membrane protein complex, as shown in Fig. 10.15. Regardless of the option chosen, the final concentration of the membrane protein complex should be, at the minimum, in the high nanomolar range. Typically, one can begin with option A and use option B if a membrane protein requires an additional purification step.
2. Adjust several MS parameters to the following suggested starting conditions for membrane protein complexes: collision gas pressure, 0.3 MPa; backing pressure, 6.0 μbar; cone voltage, 180 V; collision energy, 195 V; capillary voltage, 1.5–1.9 kV; collision gas, Argon.

10.4.1.2 Nanoflow ES and Optimizing MS Parameters

1. Load a gold-coated capillary into the capillary holder and tighten it into position.
2. By using a microscope, trim the tip of the capillary with tweezers.

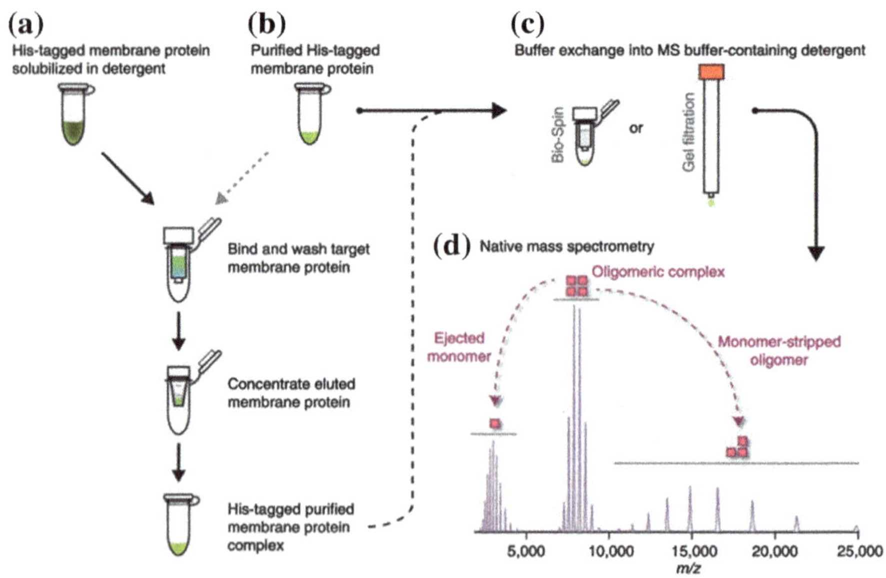

Fig. 10.15 Overview of a typical membrane protein purification and preparation, as well as analysis by mass spectrometry of intact complex. Reprinted by permission from Macmillan Publishers Ltd: Ref. [59], Copyright 2013

3. Load the capillary with 1–2 μl of prepared membrane protein complex solution with a gel-loading tip.
4. Place the capillary onto the stage of the electrospray source and apply nanoflow pressure until the loaded solution forms a drop at the capillary tip, and then remove the nanoflow pressure.
5. Slide the stage into position. As a starting point, adjust the capillary position to be level with the height of the source inlet, and then move the capillary tip to be roughly at a 2-cm distance by following an imaginary line at a 45° angle from the normal of the source inlet.
6. Start data acquisition and monitor the ion current. Perform fine capillary adjustments to achieve a stable ion current.
7. Optimization of the mass spectrometer parameters will depend on the outcome of the mass spectrum. The parameters, which can be tuned to obtain better results, are activation energy (cone and collision energy voltage), collision gas pressure, source pressure, and collision gas.

10.4.1.3 Lipid-Binding Studies

1. Prepare several titrations of lipid stock, for example, in the range of 250 ng μl^{-1}–100 μg ml^{-1}, in Mem MS buffer.
2. Thoroughly mix equal volumes of the membrane protein preparation with titrated lipid stock and incubate it at the desired temperature for a minimum of 10 min.
3. Analyze the equilibrated membrane protein and lipid preparations.
4. Calibrate the instrument/spectra using CsI solution and analyze the spectra.

10.4.2 Notes

1. Detergent used in native MS should be screened to obtain the best results for particular membrane protein system as mentioned above. A general protocol for screening detergent is introduced in ref [59]. Nonionic detergent DDM is the most popular solubilization to transfer membrane protein complex intact to the mass spectrometer. The solubilization platforms compatible with mass spectrometry have been summarized in Table 10.2, which not only contain detergent, but also include amphipol (A8-35), nanodisc, and bicelle.
2. The mass spectrometer used in this protocol is QTOF, which is modified for transmitting high-mass molecular ions. Other mass spectrometers can also be used in this study, such as FTICR and Orbitrap, which are high-resolution MS analyzers. If the membrane protein or complex is high in mass, one should consider whether the instruments have the ability to transfer and detect high-m/z

Table 10.2 Solubilization platform compatible with MS of intact membrane protein complexes. The list of nonionic detergent is adapted from Ref. [59]. Reprinted by permission from Macmillan Publishers Ltd: Ref. [59], Copyright 2013. The alternative solubilization methods are from Ref. [60]

Detergent	Abbreviation	CMC(%)
Glucosides n-Octyl-β-d–glucopyranoside n-Nonyl-β-d-glucopyranoside	OG NG	0.53 0.2
Thioglucosides n-Octyl-β-d-thioglucopyranoside	OTG	0.28
Maltosides n-Decyl-β-d–maltopyranoside n-Dodecyl-β-d–maltopyranoside n-Undecyl-β-d–maltopyranoside n-Tridecyl-β-d–maltopyranoside Cymal-5 Cymal-6	DM DDM UDM TDM Cy5 Cy6	0.087 0.0087 0.029 0.0017 0.12 0.028
Thiomaltosides n-Dodecyl-β-d-thiomaltopyranoside	DDTM	0.0026
Alkyl glycosides Octyl glucose neopentyl glycol	OGNG	0.058
Polyoxyethylene glycols Triton X-100 C8E4 C8E5 C10E5 C12E8 C12E9 Anapoe-58 (Brij-58)	TX100 C_8E_4 C_8E_5 $C_{10}E_5$ $C_{12}E_8$ $C_{12}E_9$ $C_{16}E_{20}$	0.015 0.25 0.25 0.031 0.0048 0.003 0.00045
Amphipol A8-35 [63]	A8-35	Protein/A8-35 1:5 (w/w)
Nanodisc [56] Membranescaffold protein 1D1 (MSP1D1), lipid DMPC	nanodisc	
Bicelle [60] 1,2-dimyristoyl-sn-glycero-3-phosphocholine (DMPC)–1,2-diheptanoyl-sn-glycero-3-phosphocholine (DHPC-c7)	DMPC-DHPC	Long: short chain lipids (q) = 0.27

ions. For small membrane proteins, low resolution and mass range mass spectrometer can also be used, such as ion trap and triple quadruple.

3. Some softwares used for native MS data analysis of protein complex have been published. The pioneers in the native MS fields have offered the link to download the software (http://robinsonweb.chem.ox.ac.uk/software.aspx).

Acknowledgements This work was supported by the National Natural Science Foundation of China (No. 21625504).

References

1. Seo J, Hoffmann W, Warnke S, Bowers MT, Pagel K, von Helden G (2016) Retention of native protein structures in the absence of solvent: a coupled ion mobility and spectroscopic study. Angew Chem Int Edit 55(45):14173–14176
2. Breuker K, Bruschweiler S, Tollinger M (2011) Electrostatic stabilization of a native protein structure in the gas phase. Angew Chem 50(4):873–877
3. Uetrecht C, Rose RJ, van Duijn E, Lorenzen K, Heck AJ (2010) Ion mobility mass spectrometry of proteins and protein assemblies. Chem Soc Rev 39(5):1633–1655
4. Young LM, Saunders JC, Mahood RA, Revill CH, Foster RJ, Tu L-H, Raleigh DP, Radford SE, Ashcroft AE (2014) Screening and classifying small-molecule inhibitors of amyloid formation using ion mobility spectrometry–mass spectrometry. Nat Chem 7(1): 73–81
5. Wyttenbach T, Pierson NA, Clemmer DE, Bowers MT (2014) Ion mobility analysis of molecular dynamics. Annu Rev Phys Chem 65:175–196
6. Bleiholder C, Dupuis NF, Wyttenbach T, Bowers MT (2010) Ion mobility–mass spectrometry reveals a conformational conversion from random assembly to β-sheet in amyloid fibril formation. Nat Chem 3(2):172–177
7. Savas JN, Stein BD, Wu CC, Yates JR (2011) Mass spectrometry accelerates membrane protein analysis. Trends Biochem Sci 36(7):388–396
8. Mehmood S, Allison TM, Robinson CV (2015) Mass spectrometry of protein complexes: from origins to applications. Annu Rev Phys Chem 66(1):453–474
9. Whitelegge J, Halgand F, Souda P, Zabrouskov V (2006) Top-down mass spectrometry of integral membrane proteins. Expert Rev Proteomics 3(6):585–596
10. Carroll J, Altman MC, Fearnley IM, Walker JE (2007) Identification of membrane proteins by tandem mass spectrometry of protein ions. Proc Natl Acad Sci U S A 104(36):14330–14335
11. Ryan CM, Souda P, Bassilian S, Ujwal R, Zhang J, Abramson J, Ping P, Durazo A, Bowie JU, Hasan SS, Baniulis D, Cramer WA, Faull KF, Whitelegge JP (2010) Post-translational modifications of integral membrane proteins resolved by top-down fourier transform mass spectrometry with collisionally activated dissociation. Mol Cell Proteomics 9(5):791–803
12. Gupta K, Donlan JAC, Hopper JTS, Uzdavinys P, Landreh M, Struwe WB, Drew D, Baldwin AJ, Stansfeld PJ, Robinson CV (2017) The role of interfacial lipids in stabilizing membrane protein oligomers. Nature 541(7637):421–426
13. Morrison LJ, Brodbelt JS (2016) 193 nm ultraviolet photodissociation mass spectrometry of tetrameric protein complexes provides insight into quaternary and secondary protein topology. J Am Chem Soc 138(34):10849–10859
14. Breuker K, McLafferty FW (2008) Stepwise evolution of protein native structure with electrospray into the gas phase, 10(−12)–10(2) s. Proc Natl Acad Sci U S A 105(47):18145–18152
15. Zhou J, Xiong J, Li J, Huang S, Zhang H, He Q, Lin Y, Chen P, Wang X, Liang S (2010) Gel absorption-based sample preparation for the analysis of membrane proteome by mass spectrometry. Anal Biochem 404(2):204–210
16. Wu CC, MacCoss MJ, Howell KE, Yates JR 3rd (2003) A method for the comprehensive proteomic analysis of membrane proteins. Nat Biotechnol 21(5):532–538
17. Wu CC, Yates JR (2003) The application of mass spectrometry to membrane proteomics. Nat Biotechnol 21(3):262–267
18. Zhang X (2015) Less is more: membrane protein digestion beyond urea-trypsin solution for next-level proteomics. Mol Cell Proteomics 14(9):2441–2453
19. Speers AE, Wu CC (2007) Proteomics of integral membrane proteins-theory and application. Chem Rev 107(8):3687–3714

20. Catherman AD, Li M, Tran JC, Durbin KR, Compton PD, Early BP, Thomas PM, Kelleher NL (2013) Top down proteomics of human membrane proteins from enriched mitochondrial fractions. Anal Chem 85(3):1880–1888
21. Skinner OS, Catherman AD, Early BP, Thomas PM, Compton PD, Kelleher NL (2014) Fragmentation of integral membrane proteins in the gas phase. Anal Chem 86(9):4627–4634
22. Konijnenberg A, Bannwarth L, Yilmaz D, Kocer A, Venien-Bryan C, Sobott F (2015) Top-down mass spectrometry of intact membrane protein complexes reveals oligomeric state and sequence information in a single experiment. Protein Sci Publ Protein Soc 24(8):1292–1300
23. Carroll J, Fearnley IM, Walker JE (2006) Definition of the mitochondrial proteome by measurement of molecular masses of membrane proteins. Proc Natl Acad Sci U S A 103 (44):16170–16175
24. Whitelegge JP, Gundersen CB, Faull KF (1998) Electrospray-ionization mass spectrometry of intact intrinsic membrane proteins. Protein Sci 7(6):1423–1430
25. Alves ID, Sachon E, Bolbach G, Millstine L, Lavielle S, Sagan S (2007) Analysis of an intact G-protein coupled receptor by MALDI-TOF mass spectrometry: molecular heterogeneity of the tachykinin NK-1 receptor. Anal Chem 79(6):2189–2198
26. Cadene M, Chait BT (2000) A robust, detergent-friendly method for mass spectrometric analysis of integral membrane proteins. Anal Chem 72(22):5655–5658
27. Loo RRO, Loo JA (2007) Matrix-assisted laser desorption/ionization-mass spectrometry of hydrophobic proteins in mixtures using formic acid, perfluorooctanoic acid, and sorbitol. Anal Chem 79(3):1115–1125
28. Norris JL, Porter NA, Caprioli RM (2005) Combination detergent/MALDI matrix: functional cleavable detergents for mass spectrometry. Anal Chem 77(15):5036–5040
29. Fukuyama Y, Tanimura R, Maeda K, Watanabe M, Kawabata S, Iwamoto S, Izumi S, Tanaka K (2012) Alkylated dihydroxybenzoic acid as a MALDI matrix additive for hydrophobic peptide analysis. Anal Chem 84(9):4237–4243
30. Fukuyama Y, Nakajima C, Furuichi K, Taniguchi K, Kawabata S, Izumi S, Tanaka K (2013) Alkylated trihydroxyacetophenone as a MALDI matrix for hydrophobic peptides. Anal Chem 85(20):9444–9448
31. Fukuyama Y, Nakajima C, Izumi S, Tanaka K (2016) Membrane protein analyses using alkylated trihydroxyacetophenone (ATHAP) as a MALDI matrix. Anal Chem 88(3): 1688–1695
32. Liu Y-H, Belcheva A, Konermann L, Golemi-Kotra D (2009) Phosphorylation-induced activation of the response regulator VraR from staphylococcus aureus: insights from hydrogen exchange mass spectrometry. J Mol Biol 391(1):149–163
33. Gupta S, Bavro VN, D'Mello R, Tucker SJ, Vénien-Bryan C, Chance MR (2010) Conformational changes during the gating of a potassium channel revealed by structural mass spectrometry. Structure 18(7):839–846
34. Pan Y, Konermann L (2010) Membrane protein structural insights from chemical labeling and mass spectrometry. Analyst 135(6):1191–1200
35. Konermann L, Pan J, Liu Y-H (2011) Hydrogen exchange mass spectrometry for studying protein structure and dynamics. Chem Soc Rev 40(3):1224–1234
36. Pan Y, Stocks BB, Brown L, Konermann L (2009) Structural characterization of an integral membrane protein in its natural lipid environment by oxidative methionine labeling and mass spectrometry. Anal Chem 81(1):28–35
37. Bavro VN, Gupta S, Ralston C (2015) Oxidative footprinting in the study of structure and function of membrane proteins: current state and perspectives. Biochem Soc Trans 43(5): 983–994
38. Lee SY, Kang MG, Shin S, Kwak C, Kwon T, Seo JK, Kim JS, Rhee HW (2017) Architecture mapping of the inner mitochondrial membrane proteome by chemical tools in live cells. J Am Chem Soc 139(10):3651–3662

39. Li C, Takazaki S, Jin X, Kang D, Abe Y, Hamasaki N (2006) Identification of oxidized methionine sites in erythrocyte membrane protein by liquid chromatography/electrospray ionization mass spectrometry peptide mapping. Biochemistry 45(39):12117–12124
40. Zhu Y, Guo TN, Park JE, Li X, Meng W, Datta A, Bern M, Lim SK, Sze SK (2009) Elucidating in vivo structural dynamics in integral membrane protein by hydroxyl radical footprinting. Mol Cell Proteomics 8(8):1999–2010
41. Leite JF, Blanton MP, Shahgholi M, Dougherty DA, Lester HA (2003) Conformation-dependent hydrophobic photolabeling of the nicotinic receptor: electrophysiology-coordinated photochemistry and mass spectrometry. Proc Natl Acad Sci 100(22):13054–13059
42. Wang X, Kim S-H, Ablonczy Z, Crouch RK, Knapp DR (2004) Probing rhodopsin–transducin interactions by surface modification and mass spectrometry. Biochemistry 43(35): 11153–11162
43. Wen J, Zhang H, Gross ML, Blankenship RE (2009) Membrane orientation of the FMO antenna protein from Chlorobaculum tepidum as determined by mass spectrometry-based footprinting. Proc Natl Acad Sci 106(15):6134–6139
44. Zhang H, Shen W, Rempel D, Monsey J, Vidavsky I, Gross ML, Bose R (2011) Carboxyl-group footprinting maps the dimerization interface and phosphorylation-induced conformational changes of a membrane-associated tyrosine kinase. Mol Cell Proteomics 10(6):1–16
45. Hopper JT, Rawlings A, Afonso JP, Channing D, Layfield R, Oldham NJ (2012) Evidence for the preservation of native inter- and intra-molecular hydrogen bonds in the desolvated FK-binding protein. FK506 complex produced by electrospray ionization. J Am Soc Mass Spectrom 23(10):1757–1767
46. Barrera NP, Di Bartolo N, Booth PJ, Robinson CV (2008) Micelles protect membrane complexes from solution to vacuum. Science 321(5886):243–246
47. Reading E, Liko I, Allison TM, Benesch JL, Laganowsky A, Robinson CV (2015) The role of the detergent micelle in preserving the structure of membrane proteins in the gas phase. Angew Chem 54(15):4577–4581
48. Sobott F, Hernandez H, McCammon MG, Tito MA, Robinson CV (2002) A tandem mass spectrometer for improved transmission and analysis of large macromolecular assemblies. Anal Chem 74(6):1402–1407
49. Marty MT, Hoi KK, Gault J, Robinson CV (2016) Probing the lipid annular belt by gas-phase dissociation of membrane proteins in nanodiscs. Angew Chem Int Edit 55(2):550–554
50. Gault J, Donlan JAC, Liko I, Hopper JTS, Gupta K, Housden NG, Struwe WB, Marty MT, Mize T, Bechara C, Zhu Y, Wu BL, Kleanthous C, Belov M, Damoc E, Makarov A, Robinson CV (2016) High-resolution mass spectrometry of small molecules bound to membrane proteins. Nat Methods 13(4):333–336
51. Rose RJ, Damoc E, Denisov E, Makarov A, Heck AJR (2012) High-sensitivity orbitrap mass analysis of intact macromolecular assemblies. Nat Methods 9(11):1084–1088
52. Hoi KK, Robinson CV, Marty MT (2016) Unraveling the composition and behavior of heterogeneous lipid nanodiscs by mass spectrometry. Anal Chem 88(12):6199–6204
53. Campuzano IDG, Li HL, Bagal D, Lippens JL, Svitel J, Kurzeja RJM, Xu H, Schnier PD, Loo JA (2016) Native MS analysis of bacteriorhodopsin and an empty nanodisc by orthogonal acceleration time-of-flight, orbitrap and ion cyclotron resonance. Anal Chem 88(24):12427–12436
54. Ruotolo BT, Giles K, Campuzano I, Sandercock AM, Bateman RH, Robinson CV (2005) Evidence for macromolecular protein rings in the absence of bulk water. Science 310(5754): 1658–1661
55. Ruotolo BT, Benesch JL, Sandercock AM, Hyung SJ, Robinson CV (2008) Ion mobility-mass spectrometry analysis of large protein complexes. Nat Protoc 3(7):1139–1152
56. Wang SC, Politis A, Di Bartolo N, Bavro VN, Tucker SJ, Booth PJ, Barrera NP, Robinson CV (2010) Ion mobility mass spectrometry of two tetrameric membrane protein complexes reveals compact structures and differences in stability and packing. J Am Chem Soc 132(44):15468–15470

57. Borysik AJ, Hewitt DJ, Robinson CV (2013) Detergent release prolongs the lifetime of native-like membrane protein conformations in the gas-phase. J Am Chem Soc 135(16):6078–6083
58. Barrera NP, Isaacson SC, Zhou M, Bavro VN, Welch A, Schaedler TA, Seeger MA, Miguel RN, Korkhov VM, van Veen HW, Venter H, Walmsley AR, Tate CG, Robinson CV (2009) Mass spectrometry of membrane transporters reveals subunit stoichiometry and interactions. Nat Methods 6(8):585–587
59. Laganowsky A, Reading E, Hopper JT, Robinson CV (2013) Mass spectrometry of intact membrane protein complexes. Nat Protoc 8(4):639–651
60. Hopper JT, Yu YT, Li D, Raymond A, Bostock M, Liko I, Mikhailov V, Laganowsky A, Benesch JL, Caffrey M, Nietlispach D, Robinson CV (2013) Detergent-free mass spectrometry of membrane protein complexes. Nat Methods 10(12):1206–1208
61. Konijnenberg A, Yilmaz D, Ingolfsson HI, Dimitrova A, Marrink SJ, Li ZL, Venien-Bryan C, Sobott F, Kocer A (2014) Global structural changes of an ion channel during its gating are followed by ion mobility mass spectrometry. Proc Natl Acad Sci U S A 111(48):17170–17175
62. De Angelis AA, Nevzorov AA, Park SH, Howell SC, Mrse AA, Opella SJ (2004) High-resolution NMR spectroscopy of membrane proteins in aligned bicelles. J Am Chem Soc 126(47):15340–15341
63. Leney AC, McMorran LM, Radford SE, Ashcroft AE (2012) Amphipathic polymers enable the study of functional membrane proteins in the gas phase. Anal Chem 84(22):9841–9847
64. Calabrese AN, Watkinson TG, Henderson PJ, Radford SE, Ashcroft AE (2015) Amphipols outperform dodecylmaltoside micelles in stabilizing membrane protein structure in the gas phase. Anal Chem 87(2):1118–1126
65. Busenlehner LS, Salomonsson L, Brzezinski P, Armstrong RN (2006) Mapping protein dynamics in catalytic intermediates of the redox-driven proton pump cytochrome c oxidase. Proc Natl Acad Sci 103(42):15398–15403
66. Zhang X, Chien EYT, Chalmers MJ, Pascal BD, Gatchalian J, Stevens RC, Griffin PR (2010) Dynamics of the beta(2)-Adrenergic G-Protein coupled receptor revealed by hydrogen-deuterium exchange. Anal Chem 82(3):1100–1108
67. Vahidi S, Bi Y, Dunn SD, Konermann L (2016) Load-dependent destabilization of the γ-rotor shaft in FOF1ATP synthase revealed by hydrogen/deuterium-exchange mass spectrometry. Proc Natl Acad Sci 113(9):2412–2417
68. Pan Y, Brown L, Konermann L (2011) Hydrogen exchange mass spectrometry of bacteriorhodopsin reveals light-induced changes in the structural dynamics of a biomolecular machine. J Am Chem Soc 133(50):20237–20244
69. Marcoux J, Wang SC, Politis A, Reading E, Ma J, Biggin PC, Zhou M, Tao HC, Zhang QH, Chang G, Morgner N, Robinson CV (2013) Mass spectrometry reveals synergistic effects of nucleotides, lipids, and drugs binding to a multidrug resistance efflux pump. Proc Natl Acad Sci U S A 110(24):9704–9709
70. Zhou M, Morgner N, Barrera NP, Politis A, Isaacson SC, Matak-Vinkovic D, Murata T, Bernal RA, Stock D, Robinson CV (2011) Mass spectrometry of intact V-Type ATPases reveals bound lipids and the effects of nucleotide binding. Science 334(6054):380–385
71. Deng L, Broom A, Kitova EN, Richards MR, Zheng RB, Shoemaker GK, Meiering EM, Klassen JS (2012) Kinetic stability of the streptavidin-biotin interaction enhanced in the gas phase. J Am Chem Soc 134(40):16586–16596
72. Hyung SJ, Robinson CV, Ruotolo BT (2009) Gas-phase unfolding and disassembly reveals stability differences in ligand-bound multiprotein complexes. Chem Biol 16(4):382–390
73. El-Hawiet A, Kitova EN, Klassen JS (2013) Quantifying protein interactions with isomeric carbohydrate ligands using a catch and release electrospray ionization-mass spectrometry assay. Anal Chem 85(16):7637–7644

74. Cong X, Liu Y, Liu W, Liang X, Russell DH, Laganowsky A (2016) Determining membrane protein-lipid binding thermodynamics using native mass spectrometry. J Am Chem Soc 138(13): 4346–4349
75. Mehmood S, Marcoux J, Gault J, Quigley A, Michaelis S, Young SG, Carpenter EP, Robinson CV (2016) Mass spectrometry captures off-target drug binding and provides mechanistic insights into the human metalloprotease ZMPSTE24. Nature chemistry 8(12): 1152–1158

Chapter 11
Infrared Spectroscopy for Studying Plasma Membranes

Lie Wu and Xiue Jiang

11.1 Introduction

Vibrational spectroscopy is one of the most promising label-free biophysical techniques for probing plasma membranes since it provides a "fingerprint" of vibrational modes based on which molecular orientation and structure can be elucidated. Variations in peak frequency, bandwidth, and intensity are sensitive to structural transitions of both lipid and protein components, and the vibrations of individual groups provide structural information on highly localized regions of the membrane. Due to the low energy that the radiation uses, infrared (IR) spectroscopy essentially belongs to a non-perturbing technique. External probe is not required since the absorptions of lipids and proteins can reflect their genuine environments. Initially, the major challenge in study of plasma membranes was the strong absorption of liquid water over much of the IR spectrum, which limited the analysis of samples in their natural state. Infrared reflection absorption spectroscopy (IRRAS) is an early approach for solid/liquid interface study in aqueous environment. The incident IR beam passes the solution and is reflected at the supported solid surface on which the membrane has been formed, which necessitates the use of high concentrations, low light-path length cells (less than 50 μm), or deuterium oxide as solvent. Attenuated total reflection Fourier transform infrared spectroscopy (ATR-FTIR) is another approach for IR study in aqueous solution. In ATR configuration, the incident IR beam is reflected on the internal reflection element (IRE) surface and generates an evanescent wave that penetrates several micrometers into the solution. Only molecules located in the first (few) micrometer(s) over the IRE surface are interacting with the IR light, thus the contribution from the solution is largely eliminated. Though the number of reflections can be raised by optical

L. Wu · X. Jiang (✉)
State Key Laboratory of Electroanalytical Chemistry, Changchun Institute of Applied Chemistry, Chinese Academy of Sciences, 130022 Changchun, China
e-mail: jiangxiue@ciac.ac.cn

H. Wang and G. Li (eds.), *Membrane Biophysics*,
https://doi.org/10.1007/978-981-10-6823-2_11

design to allow for an appropriate signal-to-noise ratio and meaningful spectrum, stacked multilayer membrane is typically required hampering the analysis of orientation and structure due to the heterogeneity of sample. Besides optical difficulties, a major drawback is the fact that vibrational contributions from solvent increase concomitantly with sample signal.

Surface-enhanced infrared absorption spectroscopy (SEIRAS), uses nanostructured metal film as enhancing substrate, resulting in an enhanced signal of 10–1000 fold for adsorbed molecules [1]. The enhancement factor can even reach 10^4–10^5 when applying tailored plasmonic nano-array substrates [2, 3]. Thus, SEIRAS has ultra-sensitivity that can record IR spectrum of monolayer sample and detect minute absorption change (as small as 10^{-5}) [4–6]. In ATR-SEIRAS, the rapidly decayed enhancement with increased distance from the metal surface ($\sim$10 nm, vide infra) perfectly eliminates the contribution of bulk solution, achieving selective detection of signals from the adsorbate even when immersed in water [7–9]. Such property endows SEIRAS with the ability to investigate membrane in physiological condition. Besides, SEIRAS is compatible with electrochemical modulation since the metal film can serve as an electrode. The impact of membrane potential on plasma membrane is investigated which is nearly impossible in conventional IR technique. Thus, ATR-SEIRAS is an in situ label-free, and exquisitely surface-sensitive tool that provides structure/function information of plasma membrane on monolayer level. Although the investigation of "real" membrane on molecular level is still in its infancy, ATR-SEIRAS can be applied to model systems that represent segments of organism under close-to-physiological conditions. In this chapter, we focus on the emerging ATR-SEIRAS and discuss the theory and principles, the application in plasma membrane studies, and detailed protocols.

11.2 Theory and Principle

11.2.1 Infrared Spectroscopy

11.2.1.1 General Principles

Covalent bonds in molecules are not rigid sticks but behave more like springs. At room temperature, molecules are always in motion, as their bonds stretch, bend, and twist. These complex vibrations can be broken down mathematically into individual vibrational modes; a few of which are illustrated in Fig. 11.1. Each individual vibration has its own inherent frequency (ν). If a molecule is exposed to electromagnetic radiation that matches the frequency of one of its vibrational modes, it will, in most cases, absorb energy from the radiation and jump to a higher vibrational energy state—which means that the amplitude of the vibration will increase, but the vibrational frequency will remain same. The difference in energy between the two vibrational states is equal to the energy associated with the wavelength of

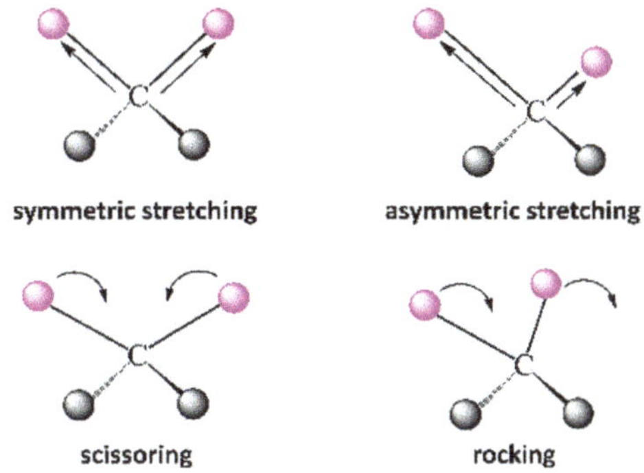

Fig. 11.1 Typical vibrational modes

radiation that was absorbed. Thus, in IR radiation, the frequencies at which there are absorptions of IR radiation ("peaks" or "signals") can be correlated directly to chemical bonds.

Molecular vibrational modes can be simply described by the use of harmonic oscillator model where the atoms are treated as point masses connected by massless springs. The frequency (v) of a two-atomic oscillator is given by

$$v = \frac{\sqrt{k/m_{\mathrm{r}}}}{2\pi}, \tag{11.1}$$

where k is the force constant between the two atoms, and m_{r} is the reduced mass ($1/m_{\mathrm{r}} = 1/m_1 + 1/m_2$). According to the formula, two factors govern the vibration frequency. Firstly, when the force constant increases, which means that the electron density in the bond between the two atoms increases, the frequency rises. Any inter- or intramolecular factor that alters the electron density in a specific bond will affect the vibrational spectrum. Secondly, the larger the masses of vibrating atoms are, the slower the vibrational frequency is. Chemical bonds in different chemical environments will absorb varying intensities and at varying frequencies. Thus, IR spectroscopy provides a "fingerprint" of molecular structure and chemical environment.

11.2.1.2 The IR Absorption

As an optical method, the extent of the IR absorption is displayed as the absorbance A (in OD = optical density) by the use of the Lambert–Beer law:

$$A = -\log\left(\frac{I}{I_{\mathrm{Ref}}}\right) = \varepsilon \cdot c \cdot d. \tag{11.2}$$

Here, I_{Ref} and I are the intensities of the IR radiation before and after passing through the analyte. For a given vibrational mode, the resulting IR absorbance is on one hand dependent on the measurement conditions reflected by the analyte concentration c and the optical path length d, and on the other hand, it is associated with the molar absorption coefficient ε including the quantum mechanical probability of the transition between the initial and final vibrational state, which is proportional to the square of the transition dipole moment. Thus, only the vibrational mode whose transition dipole moment is nonzero is IR-active. This consideration holds for all three Cartesian coordinates ($q = x, y, z$). Therefore, the absorbance A of unpolarized light of randomly oriented molecules arises from the sum of the transition dipole moments along all three components, while polarized light on an oriented sample allows to address the individual component of the transition dipole moment.

11.2.1.3 Information Derived from the IR Absorption Spectrum

Chemical structure. The chemical structure of a molecule is the dominating factor that determines the strengths of the vibrating bonds and the masses of the vibrating atoms. It often provides a fingerprint by which molecules can be identified. Thus, IR spectrum is valuable and informative for elucidation of molecular structure.
Hydrogen bonding. IR spectroscopy is one of the few methods that directly report on the strength of hydrogen bonds. Generally, hydrogen bonding lowers the frequency of stretching vibrations due to the weakening of the restoring force, while increases the frequency of bending vibrations due to the production of an additional restoring force.
Conformational freedom. For a given vibrational mode, the band position of each conformer is usually slightly different, leading to a heterogeneous band broadening. Structures with more conformational freedom will exhibit broader bands than rigid structures. Thus, bandwidth can be taken as a measure of conformational freedom.
Bond length and strength. Force constants of the vibrating bonds are determined by a number of physicochemical parameters, among which bond length and bond strength are main dominating factors. Vibrational frequencies of stretching vibrations are a direct function of the force constants of the vibrating bonds and thus provide information of bond length and bond strength, which are very helpful for the interpretation of detailed structure changes.
Bond angle and conformation. Individual vibration is often coupled with nearby vibrations, and such coupling depends greatly on the molecular geometry. Vibrational coupling therefore often provides insight into the three-dimensional structure of molecules due to the frequency–geometry correlations.
Electric field. An electric field applied from external or produced by local environment can change the electron density distribution of a given molecule.

11.2.2 Surface-Enhanced Infrared Absorption Spectroscopy

In 1980, Hartstein et al. reported an adsorption enhancement of organic compounds adsorbed to the Ag and Au island films [10], which was defined as surface-enhanced infrared absorption (SEIRA) effect. Since then, a number of SEIRA spectra have been observed on various metals and started the research of SEIRAS. SEIRAS, taking advantages of enhancement effect of metal film, is capable of in situ monitoring dipole vibrations of adsorbed molecules on monolayer level and providing information of orientational and conformational changes of adsorbed molecules. These inherent characters of SEIRAS make it a powerful tool for interfacial analysis and draw more and more attention in various areas of scientific research.

11.2.2.1 Enhancement Mechanisms for SEIRA

Similar with surface-enhanced Raman scattering (SERS), thin films composed of metal islands play a crucial role in the enhancement of SEIRAS. Though the detailed enhancement mechanism of SEIRA is still unclear, new effects and models emerge continuously. So far, it is generally accepted that the observed total enhancement is provided by at least two different mechanisms: the electromagnetic (EM) mechanism and chemical mechanism [1, 11, 12].

Electromagnetic mechanism—plasmon resonance. The electric field of incident IR photon polarizes the metal islands through the excitation of collective electron resonances or localized plasmon modes, and the dipole (*p*) induced in the metal island generates a local EM field which is stronger than the incident photon field around the island (Fig. 11.2). Adsorbed molecules can interact with this enhanced EM field and cause transitions between vibrational states. The induced local enhanced EM field is essentially polarized along the surface normal at every point on the surface of the island, and its intensity decays sharply with the sixth power of the distance to the metal surface (d). This mechanism is quite convenient for explaining the relatively short-ranged enhancement (near field effect) and the surface selection rule. Due to the sharply decayed intensity of EM field, effective

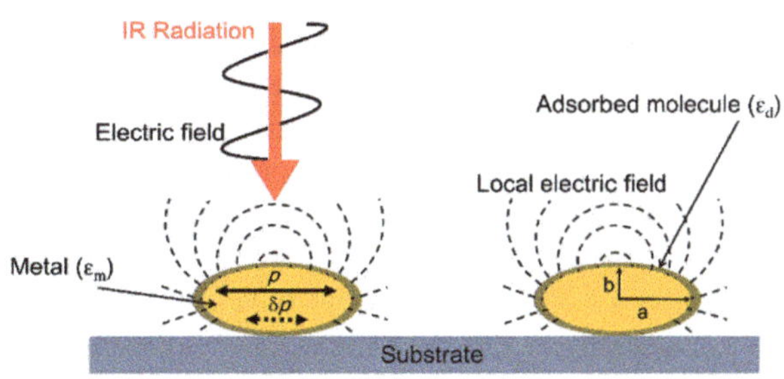

Fig. 11.2 Schematic representation of the electromagnetic mechanism of SEIRA on metal island film

enhancement distance is estimated to be 8–10 nm [13]. Thus, the contribution from species far from that distance is barely observed. Since the vector of local enhanced EM field is perpendicular to the surface, vibrational modes having dipole moment derivative components perpendicular to the surface are preferentially enhanced. This surface selection rule makes SEIRAS capable of determining molecular orientation.

Electromagnetic mechanism—perturbation of optical properties of the metal film. The oscillating dipoles in the adsorbed molecules can induce dipoles (δp) in the metal island, resulting in the change of the dielectric function of the metal. Since the perturbation should be largest at frequencies of the molecular vibrations, the spectrum of the metal island film should be nearly identical to the spectrum of the molecule. Both the absorptivity and volume fraction of metals are much larger than those of the surface molecules (nearly monolayer). Thus, the change in optical properties of the island film will dominate the spectrum and causes an effective enhancement of the IR absorption.

Chemical Mechanism—donor-acceptor interaction and charge transfer. Molecules chemisorbed on metal surface show a larger enhancement than physical stacked ones. Chemisorption will change the absorption coefficient of the adsorbate. Donor–acceptor interaction between the adsorbate and the metal changes the vibrational polarizability of the molecules. Thus, charge transfer between molecular orbitals of adsorbate and metal surface enlarges the absorption coefficient of the adsorbate by “intensity borrowing” from charge transfer.

Chemical Mechanism—uniform orientation. Chemisorption aligns the surface molecules with a preferential orientation in respect to the metal surface. Due to the surface selection rule of SEIRAS, vibrational modes perpendicularly oriented to the metal surface give rise to three times higher IR absorption than that of randomly oriented situation.

11.2.2.2 Geometry and Optical Configuration

The most widely used optical arrangement of SEIRAS is internal reflection (Kretschmann attenuated total reflection) configuration (Fig. 11.3a), also known as ATR-SEIRAS. High-refractive index IR-transparent materials such as Si, Ge, and ZnSe are suitable to serve as supporting substrates for ATR-SEIRAS. Due to its high chemical stability, Si is the most commonly used reflection element. In ATR-SEIRAS configuration, the incident IR beam irradiates the lateral side of the prism and is total reflected at the substrate surface, and then is directed to the detector. The evanescent wave generated during total reflection can reach at the surface of the metal film and couple to the enhancing field, resulting in enhanced vibrational signal of sample molecules. Due to the distance-dependent decay of evanescent wave and enhancing field, such optical configuration nearly perfectly eliminates the contribution from bulk solution and is specifically sensitive to species adsorbed on the metal surface. Furthermore, under such internal reflection

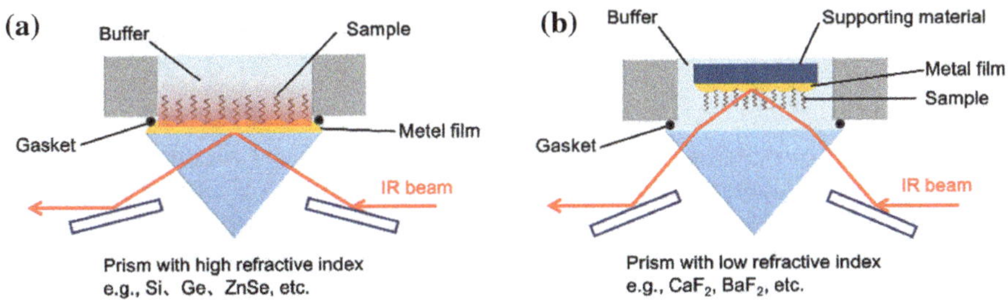

Fig. 11.3 Schematic representation of the attenuated total reflection mode (**a**) and external reflection mode (**b**) of SEIRAS

configuration, the probing IR beam is protected from atmospheric absorption and the enhancing substrate is accessible to manipulation during measurements. Thus, ATR-SEIRAS is quite suitable for biological related studies. Low-refractive index IR-transparent materials, such as CaF_2 and BaF_2, are suitable for external reflection mode (Fig. 11.3b). Enhancing substrate is prepared by depositing the metal film on the supporting material or roughening the bulk supporting material. The IR beam is directed through the substrate, passes the solution, and is reflected at the metal surface. Under external reflection configuration, the probing IR beam passes through the solution layer between IR substrate and supporting material twice. Thus, the sensitivity of surface-adsorbed species is relatively low. But this configuration is superior in detecting solution phase species, and suitable for electrocatalytic studies. To alleviate the interference from bulk solution, the solution layer should be designed as thin as possible.

11.3 Application of IR Spectroscopy in Plasma Membrane Studies

Owing to the non-perturbing and "molecular fingerprint" features, IR spectroscopy has been widely applied in plasma membrane-related studies. IR spectroscopy not only serves as a characterization tool to determine physicochemical parameters of membranes, such as phase transition, hydration state, membrane fluidity, but also provides detailed information of molecular conformation and orientation. In this part, we will first briefly discuss the application of traditional IR techniques (IRRAS, ATR-FTIR) in characterization of membranes (Sect. 11.3.1), then focus on the emergence and development of ATR-SEIRAS through giving representative examples to illustrate the potential of ATR-SEIRAS in plasma membrane-related studies (Sects. 11.3.2 and 11.3.3).

11.3.1 Physicochemical Characterization of Membrane

Lipid bilayer membrane is the basis of plasma membrane and serves as simplified model in plasma membrane-related studies. Characterization of physicochemical properties of lipid membrane model and regulation factors is the foundation of plasma membrane studies. Also, the impact of bioactive substrates on the membrane can be evaluated by the changes of the physicochemical properties. In this section, we briefly summarize the application of IR techniques in physicochemical characterization of membranes.

Structural and orientational characterization of membrane. Polarization modulation infrared reflection absorption spectroscopy (PM-IRRAS) is a powerful tool to monitor reorientation, conformation, and hydration of lipid molecules in membranes supported at the metal surface. In PM-IRRAS, the integrated intensities of the IR absorption bands are related to the orientation of the vector of transition dipole moment with respect to the vector of the electric field of the photon, which is perpendicular to the metal surface. Therefore, the integrated intensities are used to calculate the angle (θ) between the direction of the transition dipole and the surface normal using the following equation:

$$\cos^2\theta = \frac{1}{3}\left(\frac{A_{\mathrm{E}}}{A_{\mathrm{random}}}\right) \tag{11.3}$$

where A_{E} and A_{random} are the integrated intensities of the bands in the membrane at the metal surface and in a hypothetical bilayer of randomly distributed molecules, respectively. Since the vectors of transition dipole ν (CH_2) are perpendicular to each other and are perpendicular to the line of a fully extended *all-trans* hydrocarbon chain, the chain tilt angle θ_{chain} can be calculated from θ_{s} (CH_2) and θ_{as} (CH_2) through the following formula:

$$\cos^2\theta_{\mathrm{s}} + \cos^2\theta_{\mathrm{as}} + \cos^2\theta_{\mathrm{chain}} = 1 \tag{11.4}$$

Jacek Lipkowski et al. systematically studied the potential-driven conformational and orientational changes of molecules in biomimetic membrane of single lipid component, mixed lipid components, and lipid membrane protein mixed components supported at Au(111) electrode and revealed the effect of different components on the structure and conformation of membrane [14–20]. The conformational and orientational changes of lipid molecules were elucidated by constructing the potential-dependent change of title angle (θ) of individual transition dipole vector. The change in order of chains was indicated by the change in band frequency and bandwidth of CH_2 vibration, and the change in hydration state was suggested by the peaks shift of ν (C = O) and ν (PO_2^-) vibrations. Furthermore, they studied the effect of perfluorinated compounds interaction and protein binding on structural and orientational changes of membranes [21, 22]. Besides, ATR-FTIR is also employed to monitor the kinetics of bilayer formation or vesicle adsorption by recording integrated area of the CH_2 symmetric stretching band [23, 24].

Fluidity/ordering and phase transition characterization of membrane. The peak frequency of ν (CH_2) is quite sensitive to its environment, and its peak width is related to its conformational freedom and thus the fluidity/ordering of membranes can be evaluated by the peak frequency and width of CH_2 vibration. The higher frequency and broader band of ν (CH_2) indicate the more disorder of acyl chain of membrane, and vice versa. The phase transition temperature (T_m) is determined by the temperature-dependent frequency change of ν (CH_2), and the first derivative of the curve is usually calculated to improve the precision of the T_m value [25]. Combination with specific peak assignment, fluidity/ordering, phase transition characterization, the influence of bioactive substrates, and external stimuli on membranes can be investigated, such as proteins [25, 26], nanoparticles [27], drug compounds [28, 29], shear force [30], ions [31, 32], and amino acids [33]. For more information on applying IR tools in phase transition and phase organization studies of membranes, please refer to the recently published review [34].

Structure and dynamics of water in membranes. The plasma membrane provides a boundary between a living cell and its surroundings, where water is the major constituent. The structural and dynamic properties of the water at the membrane interface play a significant role in structure and function of membranes. Water molecules exhibit a complex broad ν (OH) band in the range of 3000–3800 cm^{-1} and a δ (OH) band in the range of 1600–1700 cm^{-1}. The shape and position of water absorption in IR are greatly influenced by intermolecular hydrogen banding. The frequency of ν (OH) band shifts to lower wave number with the increasing strength of the hydrogen bond, while the δ (OH) band exhibits a reverse tendency. And the broad ν (OH) band of water is resulted from the overlapping of water molecules with different hydrogen banding state. Since the δ (OH) band is weaker due to weaker transition dipole moment and has weaker frequency dependence on the hydrogen bonding strength than the ν (OH) band, it is less frequently explored. Besides, the peak positions of ν (C = O) and ν (PO_2^-) in lipids also shift to lower wave numbers when membrane hydration degree increases, which is caused by the more hydrogen bond formation with water molecules. Deep analysis of the ν (OH) band and lipid-specific band can reveal the change of water population species with membranes. For instance, Binder studied the order and H-bond characteristics of the hydration water by varying either the temperature or the relative humidity of the ambient atmosphere of membrane [35]; Bridelli et al. reported the water sorption kinetics and the changes of the characteristic absorption bands of lipid with increasing hydration level and revealed that dehydration–rehydration cycles irreversibly modified the lipid packing and water distribution [36]; Disalvo et al. studied the relationship of water structure and acyl chains ordering of membrane and found that changes in the type of water populations are concomitant with the blue shift of ν (CH_2) to higher values, and the increase in isolated populations of CH_2 is congruent with the formation of highly ordered water cluster bonded by hydrogen [37]. These results suggest the formation of confined water pockets in between the lipid acyl chains. They also found that the structures of confined water in membranes are changed when membranes are in different phase states and are in agreement with the exposure of carbonyl groups [38]. For studies of dynamics of membrane interfacial water, ultrafast multi-dimensional vibrational spectroscopy (such as two-dimensional infrared (2D-IR) vibrational echo spectroscopy) is applied. It is demonstrated that phosphate vibration [39] and the *v* (OH) of water in D_2O are capable of revealing dynamics of membrane interfacial water [40]. Recently, Kubarych et al. synthesized a vibrational labeled cholesterol derivative and successfully employed this probe to measure the dynamics of interfacial water [41].

Structural dynamics of membrane. The structural dynamics of the inside membrane is another important physical properties of the membrane. Taking nonpolar and water insoluble tungsten hexacarbonyl ($W(CO)_6$) as the probe of vibrational dynamics, Fayer et al. reported the ultrafast structural dynamics of the interior alkane regions of membranes and investigated the influence of cholesterol concentration, the degree of headgroup hydration on the interior membrane dynamics [42]. Further, the effect of vesicle sizes and the differences between lipid vesicles and planar lipid bilayers were reported [43, 44]. They found that spectral diffusion caused by the structural dynamics of the membrane in the time scale of ~200 fs to ~200 ps was varied as a function of the number of water molecules hydrating the head groups and as a function of cholesterol content for a fixed hydration level. As the cholesterol concentration increased at a certain concentration (between 10 and 15% cholesterol for lipid vesicles, and 25 and 30% for planar bilayers), the dynamics abruptly slowed down. Moreover, the interior structural dynamics became faster as the size of the vesicles decreased, and the structural dynamics was faster in the curved vesicle bilayers than in the planar bilayers.

Structural and orientational analysis of membrane proteins. Biological plasma membranes are composite structures that are composed of lipids and proteins. Studying the binding, inserting, and incorporating of proteins to membranes and their mutual effects are the foundation of deeply understanding the cellular function of plasma membrane. IR spectroscopy is a powerful tool to analyze the secondary structure of proteins [45, 46]. IRRAS and ATR-FTIR are two suitable techniques to study the structure and organization of membrane proteins and membrane-bound peptides in biologically relevant membranes. A special issue entitled "FTIR in membrane proteins and peptide studies," which contains 11 review articles discussing the application of IR spectroscopy in plasma membrane-related proteins, is available in *Biochimica et Biophysica Acta (BBA)–Biomembranes*. For more information, please refer to these articles. Besides, the factors regulating the interaction between membranes and proteins have also been investigated. For instance, Garcia-Alvarez et al. applied ATR-FTIR to compare the structure, orientation, and lipid–protein interactions of a native palmitoylated surfactant protein C with those of a non-palmitoylated recombinant form in air-exposed multilayer membrane environments and found that palmitoylation did not affect the secondary structure of the protein but modulated lipid–protein interactions at the headgroup region of phospholipid layers, resulting in a more tilted transmembrane orientation [47]; Chen et al. studied the effect of lipid composition on the membrane orientation of G protein-coupled receptor through ATR-FTIR and revealed that phosphatidylinositol 4′, 5′-bisphosphate (PIP_2) had a specific interaction with pleckstrin homology domain and played an important role in optimal function of proteins [48]; and recently, Auger et al. and Mäntele et al. reported the effect of ions in regulating the interactions between proteins and membranes [49, 50]. Furthermore, it should be mentioned that Kötting et al. successfully studied the extraction of a membrane-bound protein (Rab GTPases) from the model membrane by a soluble protein (GDP dissociation inhibitor, GDI) by applying multiple reflection ATR-FTIR and calculated the EC_{50} values [51]. This work demonstrated the great potential of ATR-FTIR in studies of protein–protein interaction in a membrane environment.

11.3.2 Structural and Functional Studies of Membrane Proteins

Membrane proteins exert a variety of fundamental functions such as electron/proton transfer, enzymatic transformation, and information/substrate transduction across the membrane. Any change of structures and functions of membrane proteins are crucial to living cells. Thus, the structural and functional studies of membrane proteins are of great significance. The application of IR spectroscopy in studies of membrane protein has already been discussed [52]. We focus on demonstrating the potential of ATR-SEIRAS to investigate membrane proteins on monolayer level through discussing some recent studies.

11.3.2.1 Probe the Impact of Membrane Potential on the Regulatory Mechanism of a Membrane Photoreceptor-Sensory Rhodopsin II

Sensory Rhodopsin II (SR II) is the primary light sensor for the photophobic response of halobacteria. The cofactor of SR II (chromophore retinal) is excited by blue-green light and generates a conformational strain to the protein, which relaxes back to the original state through several structural intermediate states (K, M, and O). Such cyclic structural change propagates the signal to its downstream transducer proteins and eventually leads to flagella motion to make the cell escaping from light. Membrane potential is ubiquitously created and employed in every living cell and is evidenced to influence membrane proteins. However, due to the experimental difficulties of elaborating the impact of membrane potential on the molecular level, such essential component is scarcely considered for the functional study of membrane proteins. The metal film of SEIRAS enhancing substrate can serve as work electrode, enabling the electrochemical modulation compatibility. Through application of the electrode potential to supply a voltage drop across the protein monolayer, membrane potential can be established in vitro. In such novel approach, the impact of the membrane potential on the photoreaction of monolayer SR II was investigated.

The recombinant SR II from *N. pharaonis* was immobilized through its C-terminus His-tag on a Ni-NTA modified enhancing substrate, resulting in an orientation with the extracellular surface facing the bulk aqueous solution. After formation of the oriented protein monolayer, SR II was reconstituted into halobacterial lipids, which is its native functional environment. The light-induced difference spectra, difference spectra between the ground (dark) state, and the photo-stationary state (constant illumination with green light) were recorded by ATR-SEIRAS and demonstrated the function of obtained reconstituted protein-lipid layer. Figure 11.4 shows the light-induced difference spectra of SR II monolayer reconstituted in halobacterial lipids (A) or solubilized in detergent (B). Negative bands correspond to the resting state which was taken as a reference, whereas positive bands refer to the photo-stationary state which was established under continuous illumination. Difference peaks can be mainly categorized into three spectral regions: bands of carboxylic acid and various amide side chains (1690–1770 cm^{-1}), amide I bands (1620–1680 cm^{-1}), and cofactor retinal plus amide II bands (1500–1600 cm^{-1}). The negative band at 1544 cm^{-1} assigned to the ethylenic C = C stretching vibration of the *all-trans* retinal is the most prominent difference of spectra in Fig. 11.4a, b. The disappearance of this band (negative peak) accompanied by a positive band at 1763 cm^{-1} indicates that the M-intermediate is predominating. The peak at 1763 cm^{-1} was assigned to the primary proton acceptor aspartate 75 and is an indicator of proton transfer. The absence of ethylenic C = C vibration of retinal in Fig. 11.4a suggests that the

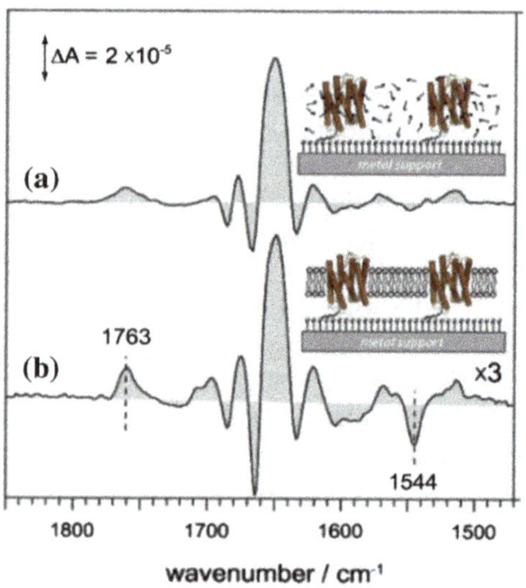

Fig. 11.4 Light-induced SEIRA difference spectra of a monolayer of SR II from *N. pharaoniswith* reconstituted in halobacterial polar lipids (**a**), and solubilized in detergent (**b**). Reprinted from Ref. [53], Copyright 2013, with permission from Elsevier

predominating state is O-intermediate for lipid-reconstituted proteins. Moreover, in the amide I region, differences between the two spectra are also noted. These findings reveal considerable changes in the protein structure and photoreaction cycle induced by lipid reconstitution. It should be noted that this is the first time the effect of lipid on the structure and function of SR II is observed, owing to the unique advantage of ATR-SEIRAS.

Figure 11.5a shows the light-induced difference SEIRA spectra of SR II monolayer under the various electrode potential. Generally, this series of difference spectra can be divided into two parts with the potential of −0.4 V as demarcation. At E > −0.4 V, the observation of positive band of C = O vibrational mode of aspartate 75 at 1757–1765 cm^{-1}, the missing of ethylenic C = C vibration mode of retinal at 1548 cm^{-1}, and the appearance of small positive band at 1536 cm^{-1} suggest that the photo-stationary state is dominated by the contribution of the O-intermediate. The overall difference spectrum is almost invariant in this potential range, except for the intensity of the C = O stretching mode of aspartate 75. The relative intensity of the aspartate 75 band reaches the maximum at + 0.2 V and decreases with the external potential shifting to more negative values. The decrease in band intensity suggests less protonated aspartate 75, which indicates a

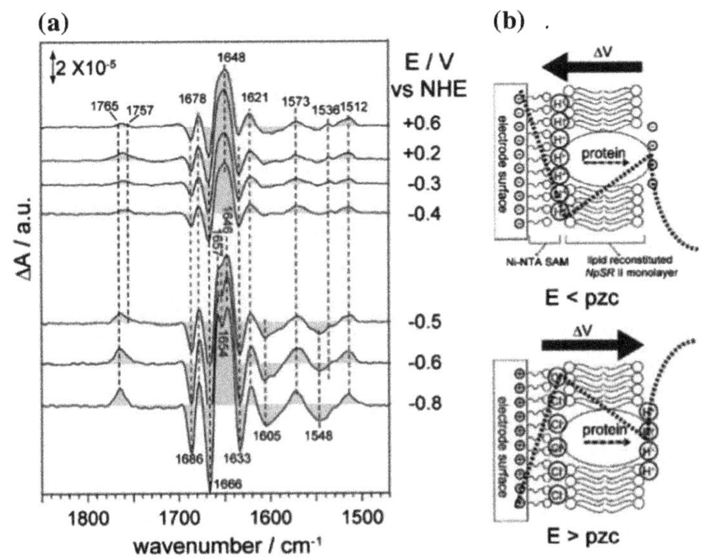

Fig. 11.5 **a** Difference spectra of SR II from *N. pharaonis* were recorded at potentials from +0.6 V to −0.8 V (from top to bottom). Applied voltages are given versus the normal hydrogen electrode. **b** Model of the protein monolayer at the solid–liquid interface immersed in the electrochemical double layer. The local potential profile is denoted by dashed line. The bold arrow indicates the direction of the net electric fields created by the potential difference between the electrode surface and the bulk electrolyte. The gray arrow in the protein represents internal proton translocation during photoreaction. Adapted with the permission from Ref. [4]. Copyright 2010 American Chemical Society

suppression of proton transfer from retinal Schiff base to aspartate 75 in the O-intermediate state. This observation is explained in terms of an interaction between the involved proton and the directional applied electric field (Fig. 11.5b). Since the C-terminus His-tag/Ni-NTA approach assembled SR II is oriented with the extracellular side facing the bulk aqueous solution, the direction of internal proton transfer is from the electrode surface to the bulk solution. When the electrode potential is below + 0.2 V, the electrode surface is negatively charged, which creates a dipole against the direction of proton translocation from retinal Schiff base to aspartate 75. As the potential shifts to negative, the energy barrier is sufficiently high to suppress proton translocation in the protein. Though the application of an electric field opposite to the physiological proton transfer direction leads to the halt of light-induced proton transfer reaction at the stage of aspartate 75, other structural rearrangements that accompany light-induced activity of the membrane protein are essentially unaffected. This indicates that proton transfer from Schiff base to aspartate 75 is not the cause for the structural changes in the helical protein backbone and is sensitive to the change of membrane potential. At $E < -0.4$ V, the observed bands display spectral features of the M-intermediate. The shift of the photo-stationary equilibrium from an O dominant equilibrium toward an M dominant equilibrium as changing electrode potential from positive to negative is explained in terms of local pH change raised by the compensation of excess charge from the electrode. When the applied potential is set at very negative values, the negative surface charge of electrode attracts protons in diffuse double layer to its surface, which may in turn cause a raise of local pH at the vicinity of the membrane surface and will be significantly higher than that of the bulk, favoring the adsorbed SR II to be in the M predominated state. Furthermore, it was found that the increasing of solution pH from acidic to alkaline exhibited similar effects on the light-induced SEIRA difference spectra as shift of positive to negative potential. However, proton transfer of aspartate 75 becomes insensitive to the electric field exerted by the electrode when the solution pH is much higher than 5.8. This is likely due to the protonation state of surface residues at the extracellular side, especially the proton releasing group. It is inferred that the deprotonation of proton releasing group creates a local polar environment surrounding aspartate 75 as a consequence of hydrogen-bonding rearrangements that exceed the energy of the external dipole. These results establish a molecular model for the physiological regulation of SR II photo-cycle by modulating the potential drop across the membrane which acts through the interplay between the change in local pH at the membrane surface and the external electric field. Detailed discussion and conclusion can be found in the original publications (Refs. [4, 5]).

This research represents a unique example of monitoring the effect of the membrane potential on the functionality of a membrane protein at atoms and bonds level and demonstrates the great potential of ATR-SEIRAS in studying the influence of membrane potential on membrane proteins.

11.3.2.2 Study of the Uptake of Cations by Gramicidin A

Ion channels are pore-forming membrane proteins present in the membranes of all cells which gate the flow of ions across the membrane. Their functions include establishing a resting membrane potential, shaping action potentials, and other electrical signals, which are essential to cells. Gramicidin A (gA), which specifically transports monovalent cations across membranes, is an ionic-channel membrane protein. By ATR-SEIRAS, ion uptake-induced structural and conformational changes are resolved, which could help to further understand the structure-function relationship of ionic channel at the molecular level.

Lipid-tethered bilayer lipid membranes are constructed on the nanostructured SEIRAS enhancing substrate to incorporate gA. It consists of a mixed self-assembled monolayer of synthetic lipid-thiols like (cholesterylpolyethylenoxy) thiol (CPEO3) with a "dilution" small thiols like 6-mercaptohexanol (6MH) which is used to create a sub-membrane reservoir attached to metal surface, and a lipid layer on the top of the layer formed by fused unilamellar vesicles (Fig. 11.6a). The formation of bilayer lipid membrane by vesicle spreading and fusion on the mixed monolayer was monitored by ATR-SEIRAS through taking the spectrum of the mixed monolayer as a reference, as shown in Fig. 11.6b. The positive and negative bands represent contributions of the species binding to and removing from the substrate surface, respectively. The positive bands at 2800–3050 cm^{-1} and

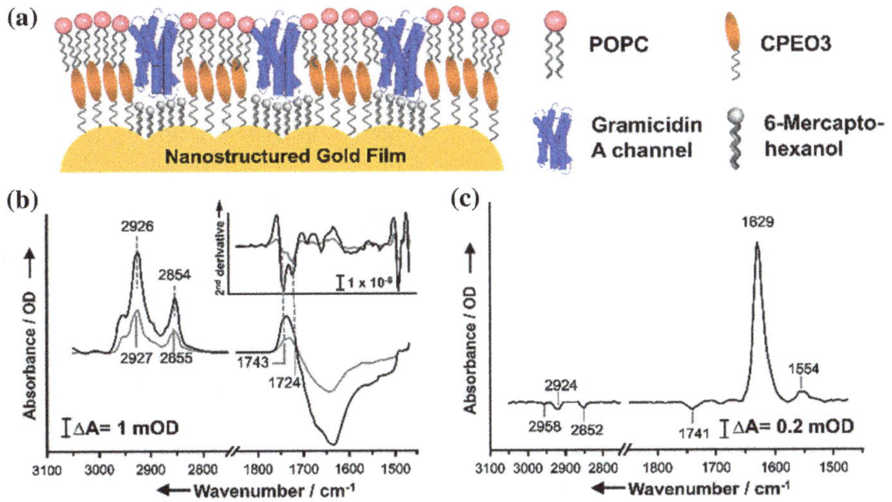

Fig. 11.6 **a** Schematic representation of the gA-incorporated bilayer membrane assembled on the nanostructured Au electrode. **b** SEIRA difference spectra obtained during the spreading of lipid vesicles, recorded 3 min (gray) and 3 h (black) after incubation, and related to the SEIRA spectrum of the mixed monolayer; the inset shows the second-derivative spectra. **c** SEIRA difference spectrum recorded during the incorporating of gA. Reproduced from Ref. [54] by permission of John Wiley & Sons Ltd

1733 cm^{-1}, assigned to C–H stretching modes and C = O stretching mode, respectively, are the characteristic peaks of lipid molecules. The broad negative band centered at 1645 cm^{-1} originates from the removal of water molecules close to the mixed thiols monolayer that have been replaced by lipid molecules. The slight red shift of ν_{as} (CH_2) and ν_s (CH_2) peaks of lipids indicates that acyl chain order of lipid membrane increases as the bilayer membrane forms. In C = O stretching region, both position and relative intensities of the bands display significant time-dependent changes. These changes, which are particularly evident in the second derivatives of the spectra (Fig. 11.6b, inset), indicate the existence of two distinct lipid components inside the bilayer membrane. A two-step formation process is proposed based on the kinetic analysis of the two components. The fast process is ascribed to the adsorption and spreading of lipid vesicles onto the monolayer, and the slow process represents the reorientation of some lipid molecules to fill the gaps in the mixed monolayer that may involve a "flip over" of lipids. Figure 11.6c shows the SEIRA spectrum of bilayer membrane after incorporation of gA. The appearance of amide I (1629 cm^{-1}) and II (1554 cm^{-1}) modes suggests the successful incorporation of gA. The band positions of amide I and II modes indicate the $\beta^{6,3}$ species of incorporated gA. The replacement of several lipid molecules that induced by incorporation of gA is also evidenced by small negative band at C = O stretching region. The incorporated gA helix adopts a perpendicular alignment with respect to the surface, which is suggested by the high amide I/amide II intensity ratio (ca. 16). The incorporation of gA into membrane is also a two-step process, which is revealed by the bi-exponential behavior of time-dependent changes in the amide I intensity. The fast phase attributed to the direct incorporation of gA into the upper lipid layer, and the slow phase may reflect the translocation of the peptide from the upper into the lower layer of the membrane.

Taking the spectrum of the Ba^{2+}-containing buffer as the reference, the uptake of cations by gA was studied by measuring SEIRA difference spectra of gA in the presence of Li^+, Na^+, K^+, Cs^+, and Tl^+. Both the cation-induced vibrational change of lipid membrane in C = O stretching region and the structural change of gA in amide I region are clearly observed (Fig. 11.7). The replacing Ba^{2+} ions at the binding site in the gA channel with the various monovalent cations lead to the reversible spectral changes of amide I. The intensities of the positive difference bands in the amide I region (1620–1629 cm^{-1}) induced by the uptake of monovalent cations are correlated well with conductive states measured by electrochemical impedance spectroscopy (EIS) for the corresponding cation-exchange experiments. This provides a spectroscopy-based approach for conductivity measurement. In addition, the amide I frequency is found to increase with the size of the cation, except for Tl^+, which has a diameter similar to that of K^+. This discrepancy is tentatively explained by the influence of ions on the interaction between gA and membranes, which is due to the higher polarizability of the Tl^+ ion.

The present study establishes an approach for spectroscopic study of interaction between ions and channel proteins and demonstrates the high potential of ATR-SEIRAS in structural and functional analysis of membrane channel proteins at a molecular level.

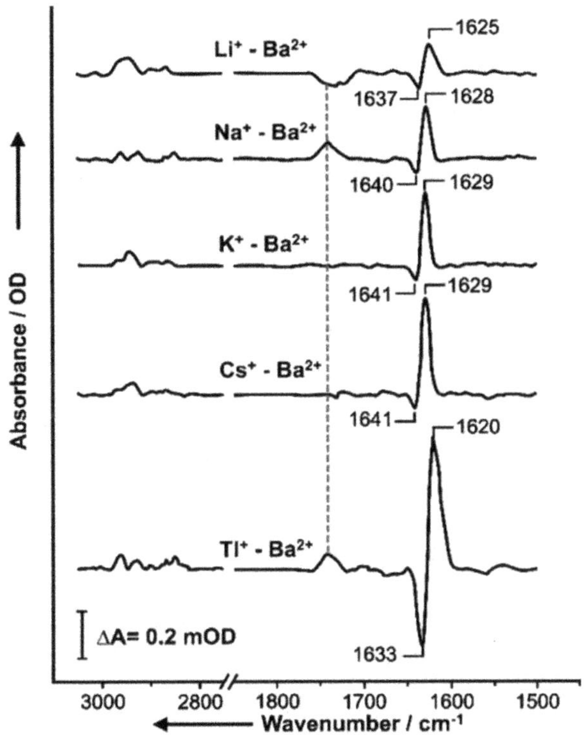

Fig. 11.7 Baseline-corrected SEIRA difference spectra of gA incorporated into bilayer membrane in the presence of different monovalent cations, taking the spectrum of the sample in the presence of Ba^{2+} as a reference. Reproduced from Ref. [54] by permission of John Wiley & Sons Ltd

11.3.3 Interfacial Interaction of Biomimetic Membrane

Plasma membrane is the natural barrier of single cell and various compartmented organelles. Many cellular functions (e.g., nutrient uptake, energy transport, signal transduction) involve substance passing through and interacting with membrane. Understanding the interfacial interaction of membrane with external substance is of significant importance. ATR-SEIRAS is an exquisitely surface-sensitive technique that is capable of providing valuable "fingerprint" information of interfacial molecules. By using simplified biomimetic membrane, ATR-SEIRAS serves as a powerful tool to study the mechanism of interfacial interactions of membranes. In this section, some recent achievements will be discussed to give an insight of ATR-SEIRAS in the study of membrane interfacial interactions.

11.3.3.1 Analyzing the Interaction Force and Regulation Mechanism of Biomimetic Membrane and Proteins

Cytochrome *c* (cyt *c*) is a small heme protein that transfers electrons between the integral membrane protein complexes of the respiratory chain in mitochondrion.

Besides transfer electron in respiratory chain, cyt *c* is believed to play a significant role in early apoptosis due to its enhanced peroxidase activity resulting from the structural transition when interacted with cardiolipin (CL). Resolving interaction mode, detailed structural changes, and factors that regulate such transitions are very important in understanding the apoptotic mechanisms. In situ label-free ATR-SEIRAS is applied to explore the interactions between cyt *c* and CL membranes in physiological conditions.

The adsorption of cyt *c* on CL membranes which were constructed by vesicle fusion onto a pre-modified hydrophobic monolayer was monitored in situ by ATR-SEIRAS (Fig. 11.8a, i). Since the spectrum of CL membrane is taken as reference, the bands in sample spectra include contributions from adsorption of protein and fluctuation of CL membrane. Under condition of neutral pH with low ionic strength (Fig. 11.8a, ii), two positive peaks at 1660 and 1550 cm^{-1} corresponding to the amide I and II bands indicate the adsorption of cyt *c*. However, some weak negative peaks are resolved at 3010, 2960, 2917, and 2852 cm^{-1} overlapped with the respective positive bands in C–H vibrational region. These negative peaks should be derived from structural changes in the CL membrane since the surface bound species are expected to produce positive absorptions. Besides, a weak negative band at 1095 cm^{-1} in ν_s (PO_2^-) region accompanied by a very small positive peak at 1088 cm^{-1} that assigned to the hydrogen-bonded ν_s (PO_2^-) is observed, which suggests the formation of a hydrogen-bond network

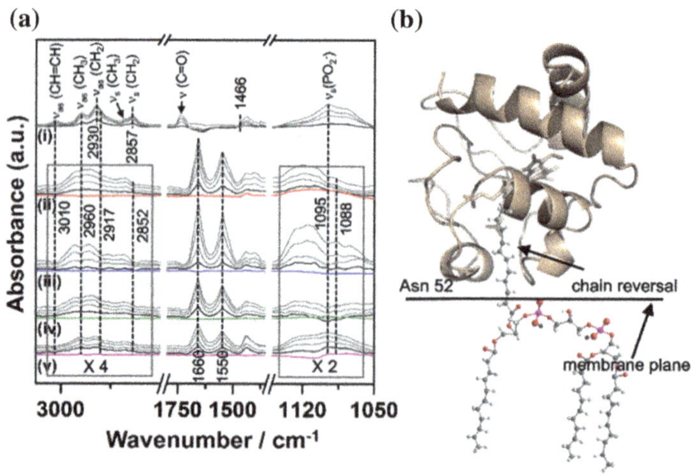

Fig. 11.8 **a** The SEIRA spectra of CL vesicles fused on the pre-modified surface (i) and subsequent adsorption of cyt *c* in the presence of (ii) 10 mM phosphate solution at pH 7.0, (iii) 10 mM phosphate solution at pH 5.0, (iv) 150 mM NaCl solution at pH 5.0, and (v) 150 mM phosphate solution at pH 7.0 (from bottom to top, 10 s, 150 s, 13.5 min, and 90.5 min), respectively. Colored lines are the corresponding spectra of pure CL membrane in different additives for 90 min. **b** Illustration to elucidate the possibility of insertion of a CL acyl chain into the hydrophobic pocket of cyt *c*. Adapted with the permission from Ref. [55]. Copyright 2016 American Chemical Society

around the PO_2^- group of CL after protein adsorption. This finding is further supported by the more efficient absorption of proteins and greatly enhanced couplet peaks in ν_s (PO_2^-) region under acidic condition with low ionic strength (Fig. 11.8a, iii) due to the enhanced hydrogen bonding by increasing the protonation degree of CL membrane. Therefore, the couplet indicates the presence of hydrogen bonding between CL and cyt *c*. The enhanced hydrogen bonding facilitates structural changes of the underlying CL membranes as more obvious negative peaks in C–H vibrational region are observed in acidic condition. In the environment of acidic pH with high ionic strength (Fig. 11.8a iv), the adsorption of cyt *c* is largely weakened due to the screening of electrostatic interaction. While negative peaks in C–H vibrational region can still be resolved, suggesting a definitive contribution from hydrogen bonding to this phenomenon. In the presence of high concentration of phosphate (Fig. 11.8a, v), which is applied to screen the hydrogen bonding between the phosphate group of CL and protein, such negative peaks in C–H vibrational region and couplet in ν_s (PO_2^-) region are not visible. This result indicates that the hydrogen bond between phosphate group of CL and protein is responsible for the structural changes in the underlying CL membranes. The negative peaks in C–H vibrational region are ascribed to the extension of acyl chain of CL into the hydrophobic channel of cyt *c* (Fig. 11.8b), which would cause the acyl chains move away from the substrate surface and resulting in negative peaks. Hydrogen bonding may play an important role in such extension once this process happens.

Through deep analysis of the adsorption kinetics of cyt *c* on CL membrane under different binding forces, a heterogeneous binding mode is proposed. It is found that the kinetic data can be only fitted well to the bi-exponential equation when the adsorption is dominated by one kind of driving forces (e.g., electrostatic interaction, hydrogen bonding, or hydrophobic interaction), illustrating the heterogeneous nature of the CL-bound cyt *c* ensembles. However, in the condition of acidic pH with low ionic strength where electrostatic interaction and hydrogen bonding both contribute to the adsorption, the kinetic data are fitted well with the mono-exponential expression, suggesting the homogeneous nature of CL-bound cyt *c* under such condition. The proposed heterogeneous binding mode is also supported by the desorbing test of CL-bound cyt *c*. High ionic strength treatment can desorb nearly half of cyt *c* that pre-adsorbed on CL membrane at neutral pH, while CL-bound cyt *c* that pre-adsorbed at acidic pH is insensitive to ionic strength. Thus, it is believed that the synergistic interactions of hydrogen bonding and electrostatic interaction may promote the homogeneous orientation of CL-bound cyt *c*.

Since cyt *c* is redox active, potential-induced SEIRA difference spectra are recorded to investigate the structural properties of CL membrane adsorbed cyt *c* by different binding forces (Fig. 11.9a). Based on the position shift and intensity changes of bands assigned to different components of secondary structure, the structural changes of CL-bound cyt *c* induced by different interaction forces are revealed and the binding mode of cyt *c* on CL membrane driven by different interaction force is proposed (Fig. 11.9b). The competition and/or the synergy of

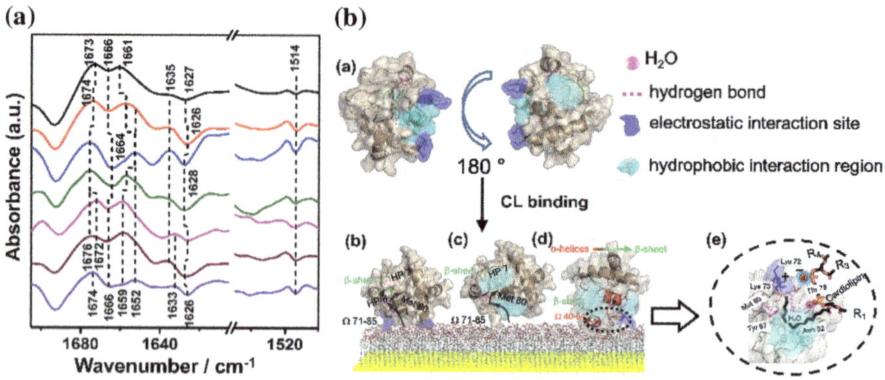

Fig. 11.9 **a** Potential-induced difference spectra of cyt *c*. **b** Cartoons to elucidate the interaction mode of cyt *c* with CL membrane. (*a*) Structure of cyt *c* and the possible binding sites with CL; (*b*) The pure electrostatic interaction mode; (*c*) The hydrophobic interaction mode; (*d*) The hydrogen bonding and electrostatic synergic interaction mode; (*e*) The insertion of acyl chain into the hydrophobic channel. Adapted with the permission from Ref. [55]. Copyright 2016 American Chemical Society

electrostatic, hydrophobic interactions, and hydrogen bonding result in the adsorption of cyt *c* to CL membranes in heterogeneous or homogeneous modes. The strong electrostatic interactions between the electrostatic interaction site (blue region) and CL membranes induce the structural transformation from α-helices to loop/random coil. Such unfolding of partial α-helices may further impose changes in the microenvironment of β-sheets via intra-protein hydrogen-bond network. Hydrophobic interactions mediate another binding mode, which only involves the hydrophobic channel between the two nonpolar polypeptide strands (cyan region), leading to a change in the microenvironment of β-turn type III and β-sheets. The third interaction mode results from the synergy of electrostatic interactions with hydrogen bonding, under which the adsorption orientation of protein is homogeneous. Such synergic interaction is the driving force for structural transition between α-helices and β-sheets. The hydrogen bonding between CL and cyt *c* may modify the hydrogen-bond network around asparagine 52, leading to the adjustments of the heme pocket which may allow for possible extension of CL acyl chains to the hydrophobic channel. In the absence of electrostatic interactions, the enhanced hydrogen bonding only causes the disturbance of the β-turn type III microenvironment. For detailed discussions and conclusions, please refer to the original publication (Ref. [55]).

This work represents a unique case of exploring ATR-SEIRAS in investigating the complex interfacial interactions between proteins and membranes. ATR-SEIRAS enables the simultaneous probing of structural changes of proteins and underlying membranes, which is essential for mutual interaction.

11.3.3.2 Probing the Molecular Mechanism of Interaction Between Biomimetic Membrane and Nanomaterials

With the boom of nanotechnology, nanomaterials have been widely utilized in biological and biomedical fields and receive considerable attention for the potential impact of these nanomaterials and their production on bio-systems. Biomembrane is the first barrier that nanomaterial may encounter when it is presented in a biological environment. Understanding the underlying mechanism of how nanomaterial interacts with biomembrane is of great importance. It is not only for guiding future biological application of biomaterial but also for managing nano-bio effect of nanomaterial. Graphene oxide (GO) has drawn worldwide attention in various biomedical fields, and great progress has been made in the past years. Thus, probing the interaction between GO and lipid membrane is a crucial issue. However, studies on interactions between GO and membrane have obtained conflicting results and the molecular mechanism is still unclear. ATR-SEIRAS has been successfully applied to reveal the molecular mechanism of interactions between GO and zwitterionic lipid membrane by combining the advantage of extremely large extinction coefficient of water in the mid-infrared region and strict surface sensitivity of SEIRAS.

GO-induced SEIRA difference spectra of zwitterionic hybrid lipid bilayer are shown in Fig. 11.10. After addition of GO, some negative peaks gradually appear in both the ν (CH_n) and ν ($C = O$) regions. The study of molecular dynamic (MD) simulation suggested that GO could destructively extract lipid molecules from membranes [56]. Therefore, these negative peaks are ascribed to the loss of phospholipid molecules in the membrane due to the extractive effect of GO.

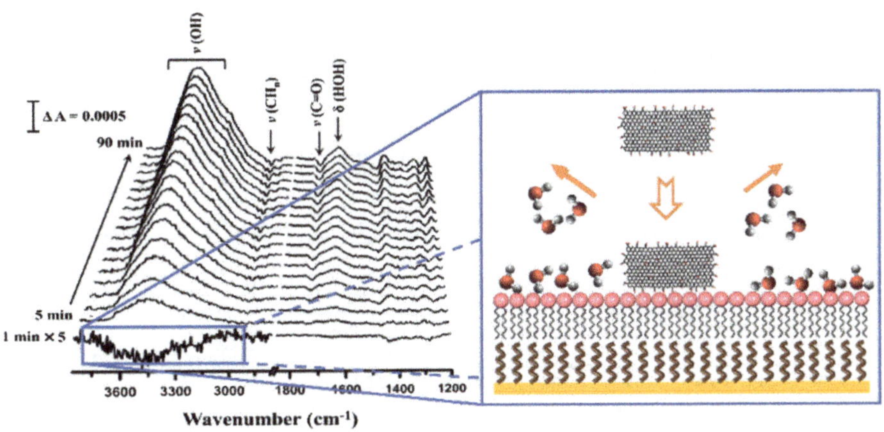

Fig. 11.10 3D plot of GO-induced SEIRA spectral changes of zwitterionic hybrid lipid bilayer in the presence of GO in water at 5 min interval. The first spectrum (1 min) obtained after addition of GO was shown with fivefold magnification in ν (OH) region. Adapted with the permission from Ref. [57]. Copyright 2015 American Chemical Society

This result provides supporting experimental evidence for the theoretical simulation studies. Furthermore, a weak but distinguishable negative peak in v (OH) region is observed immediately after addition of GO. The interfacial water along the membrane surface will be expelled and reorganized by the enriched sample (Fig. 11.10). As a benefit of ATR-SEIRAS, such process could be clearly observed in situ. Thus, the negative peak in v (OH) region in GO-induced different spectrum of membrane is ascribed to the replacement of interfacial water by adsorbed GO. This initial interfacial water replacement suggests the initial interaction between membrane and GO and is taken as an indicator of adsorption of GO. Based on the analysis of changes of initial interfacial water induced by GO in the presence of different intermolecular force shielding agents (Ca^{2+} pre-treatment to compensate the electrostatic repulsion from phosphate group, choline chloride to weaken the interaction from choline group, NaCl to screen the electrostatic interaction, phosphate to shield interaction from phosphate group), a non-electrostatic attractive force from phosphate group of lipid is identified to contribute to the adsorption of GO on membranes.

To reveal the origin of this non-electrostatic attractive force, GO-induced SEIRA difference spectra of phospholipid membrane in the presence of NaCl and phosphate is closely compared and analyzed (Fig. 11.11). The appearance of positive v_{as} (PO_2^-) band of lipids and red shift of v (C = O) band of GO suggests that the non-electrostatic attractive force is hydrogen-bonding interaction between carboxyl

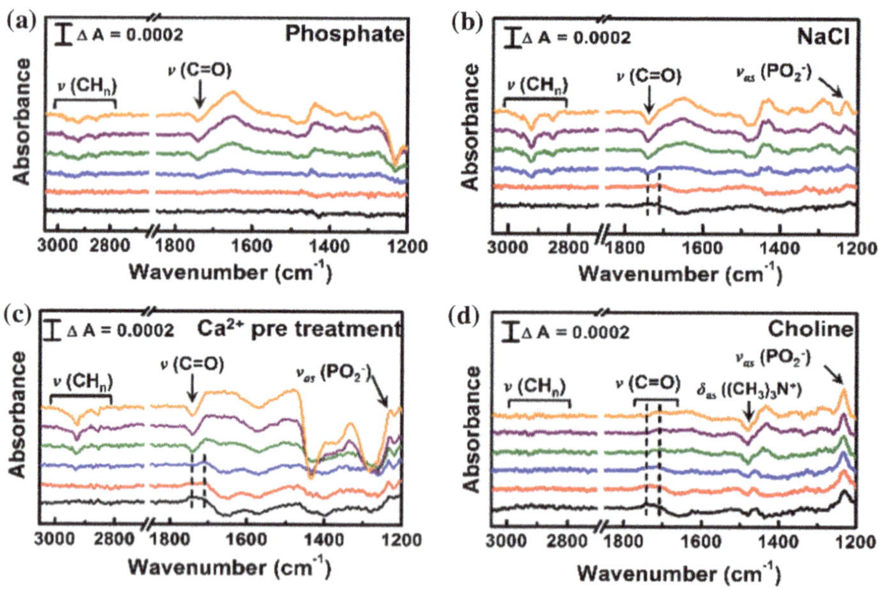

Fig. 11.11 GO-induced SEIRA difference spectra of lipid membrane at selected time points (1, 5, 10, 30, 60, and 90 min; from bottom to top) in phosphate solution (**a**), in NaCl solution (**b**), with Ca^{2+} pre-treated membrane in water (**c**), and in choline solution (**d**). Reprinted with the permission from Ref. [57]. Copyright 2015 American Chemical Society

groups of GO and phosphate groups of lipids. This phenomenon is much more obvious when GO interacts with Ca^{2+} pre-treated membranes. Both the electrostatic attraction from choline groups and repulsion from phosphate groups are weakened in NaCl solution, while Ca^{2+} pre-treatment only compensates the negative charge of phosphate groups and thus facilitates the attraction between GO and choline groups. Considering the different screening targets under these two conditions, a long-range electrostatic attraction between GO and choline groups may facilitate the short-range hydrogen-bonding interaction between GO and phosphate groups. Since no significant position change of ν_{as} (PO_2^-) band is observed during the interaction, the hydrogen bonds between COOH groups of GO and phosphate groups of phospholipids are believed to be mediated by water molecules. Hydrogen bonding is a weak force, but a multivalent binding with lipid membrane might be enabled by the plenty of carboxyl groups on the edge of GO, which achieves effective absorption. Another important observation is the extraction of lipid molecules from membranes by GO disappears in the presence of choline (Fig. 11.11). The strong non-electrostatic absorption of choline on GO sheet confirms that hydrophobic interaction existed between choline groups of lipid membrane and GO, and the extractive effect was eliminated due to the reduced hydrophobicity of GO in the presence of choline. These results show an overall insight of interaction between GO and lipid membrane: adsorption of GO on lipid membrane and extraction of lipid molecule are a complex process resulting from the balance of at least four interaction forces: electrostatic attraction, electrostatic repulsion, hydrogen bonding, and hydrophobic interaction. For detailed discussions and conclusions, please refer to the original publication (Ref. [57]).

In addition, combined with other techniques, ATR-SEIRAS is also employed to study the mechanism of passive penetration of D-penicillamine-coated CdSe/ZnS quantum dots (QDs) through red blood cell membranes [6], the mechanism of hemolysis induced by mercaptosuccinic acid-capped CdTe QDs [58], toxicity mechanism of graphene material in red blood cells [59]. These successful cases demonstrate the great potential of applying ATR-SEIRAS in nano-biomembrane interface and suggest the great power of ATR-SEIRAS in revealing the mechanism of interfacial interactions at the molecular level.

11.4 Protocol for ATR-SEIRAS in Studying Plasma Membranes

For studying plasma membranes by ATR-SEIRAS, preparing the metal film as enhancing substrate and constructing biomimetic membrane are the two essential steps. In this section, we will describe how to deposit nanostructured gold film on silicon substrate, how to modify the gold film, and how to construct model membranes. These detailed practical procedures will provide basic knowledge for researchers in related field. All solutions should be prepared with ultrapure water and analytical grade reagents at room temperature (RT, 25 °C) and stored at 4 °C (unless indicated otherwise). Diligently follow all waste disposal regulations when disposing waste materials.

11.4.1 Treatment of Silicon Substrate and Preparation of Gold Nanostructured Film

Nano-structured gold film on silicon is the most widely used enhancing substrate for SEIRAS. Preparation of Au film with high and stable enhancement factor is the first and essential step for SEIRAS applications. Chemical (electroless) deposition is commonly employed to prepare SEIRA-active gold film. Procedures for preparing a thin Au film on silicon surface are described below.

11.4.1.1 Materials

(1) 1.0 μm Al_2O_3 polishing powder.
(2) 40% *w/v* NH_4F solution. Weight 20 g of NH_4F in pan paper and transfer to a plastic beaker. Add water to a volume of 40 mL. Dissolve the solid completely by sonication. Make up to 50 mL with water. Transfer the solution to a plastic bottle, and store at RT (see **Note 1**).
(3) Solution A (0.3 M Na_2SO_3 + 0.1 M $Na_2S_2O_3$ + 0.1 M NH_4Cl). Weight 0.268 g NH_4Cl and 1.891 g Na_2SO_3 in pan paper and transfer to a glass beaker. Add water to a volume of 20 mL and dissolve the solid. Weight 1.241 g $Na_2S_2O_3{\cdot}5H_2O$ in pan paper and transfer to the 20 mL solution. Dissolve the solid completely and make up to 50 mL with water. Store at 4 °C.
(4) Solution B (0.03 M $NaAuCl_4$). Weight 0.119 g $NaAuCl_4{\cdot}2H_2O$ in pan paper and transfer to a glass bottle. Add water to a volume of 8 mL and dissolve the solid. Make up to 10 mL with water and store at 4 °C in dark.
(5) Solution C (2.5% *v/v* HF). Measure 100 μL 40% *v/v* HF solution and dilute with 1500 μL water to a plastic bottle. Store at RT in a fume hood (see **Note 1**).
(6) 0.1 M H_2SO_4 solution (see **Note 2**).

11.4.1.2 Methods

(1) **Remove gold film**. Dissolve the Au film by immersing in a boiling solution of 1:1:1 mixture of HCl (32%), H_2O_2 (30%), and H_2O (see **Note 2**). Once the shiny color disappears, wash the silicon crystal thoroughly with water.
(2) **Surface polish**. Polish the flat surface of a prism or hemisphere of silicon crystal with 1.0 μm Al_2O_3 slurry and then wash thoroughly with water.

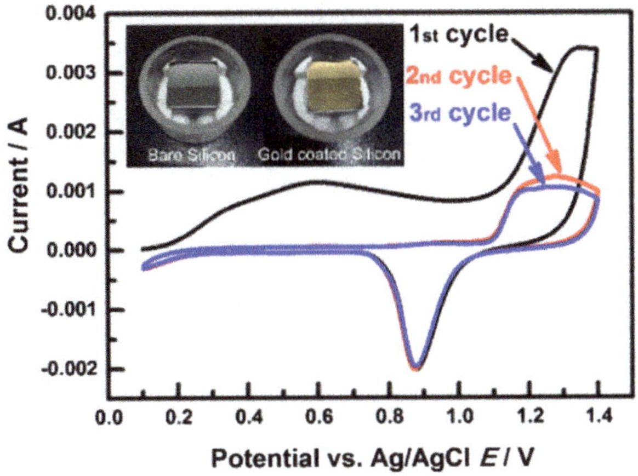

Fig. 11.12 Electrochemical cleaning of the as-prepared gold film (Insert, picture of silicon crystal before and after gold film deposition)

Ultrasonic cleaning sequential in water and ethanol is optional to completely remove the Al_2O_3 particles from the silicon surface (see **Note 3**).

(3) **NH_4F treatment**. Cover the surface with 40% *w/v* NH_4F solution for a few minutes (typically 1–3 min) to remove the oxide layer and terminate the surface with hydrogen; then thoroughly rinse with water (see **Note 1**).

(4) **Gold film deposit**. Place the silicon crystal in water bath at 60 °C for 10 min, and keep the liquid level slightly below the cleaned surface of silicon (see **Note 4**); cover the surface of silicon crystal with a freshly prepared 1:1:1 volume mixture of solution A, B, and C for 90 s, then a shiny Au film is formed (Fig. 11.12, inset); terminate the reaction by rinsing with water (see **Note 1**).

(5) **Electrochemical cleaning**. Transfer the gold film-coated silicon crystal into a 0.1 M H_2SO_4 solution (see **Note 2**); electrochemical clean the gold film by several oxidation–reduction cycles in the range between 0.1 and 1.4 V (vs. Ag/AgCl) until the cyclic voltammogram of polycrystalline gold appears to remove any remaining contaminator (Fig. 11.12); thoroughly rinse with water and completely blow off any water droplet on silicon crystal with an aurilave.

(6) **Accessory assembly**. Mount the gold-coated silicon crystal into SEIRAS accessory cell, and ready for measurement.

11.4.1.3 Notes

(1) NH_4F and HF are corrosive and skin contact must be avoided. HF would be produced if NH_4F solution is heated. NH_4F and HF solution should be handled and stored with plastic bottle under a fume hood. Any solution containing NH_4F or HF should be completely neutralized with $CaCl_2$ or $CaCO_3$ before disposal.

(2) A large amount of heat would be released when acids are mixed with water. Please always add acids into water.

(3) A clean and hydrophobic surface should be appeared after polishing. Complete removal of Al_2O_3 particle is essential to deposit a stable gold film; otherwise, the Au film would be easily flaked off after electrochemical cleaning.
(4) If the surface of the silicon crystal is with any water droplet, blow off it completely with an aurilave.

11.4.2 Preparation of Unilamellar Vesicles

Several strategies are successfully employed to construct biomimetic membrane on nanostructured Au film, including hybrid lipid bilayer, lipid-tethered bilayer lipid membrane, and protein-tethered bilayer lipid membrane (vide infra). Unilamellar vesicles are used as lipid source in all strategies, and the species and compositions of membranes can be regulated by single or mixed components vesicles. In some cases, membrane proteins are introduced into biomimetic membrane system by membrane proteins containing vesicles. The procedure for preparing unilamellar vesicles is described below.

11.4.2.1 Materials

(1) Phospholipids: e.g., phosphatidylcholine (PC) lyophilized powder.
(2) Membrane proteins: e.g., human voltage-dependent anion channel 1 (hVDAC1) in detergent micelles.
(3) Detergents: e.g., *N*-lauryl-*N*,*N*-dimethylamine-*N*-oxide (LDAO).
(4) Bio-Beads SM2.

11.4.2.2 Methods

(1) **Pure phospholipid vesicles**

  (a) Make lipid stock solution. For lyophilized powder, dissolve lipid in chloroform to make 10 mg/mL concentrated solution (stock solution), and store at −20 °C in dark (see **Note 1**).
  (b) Sample solution. Add 100 μL of the stock solution (1 mg of lipid) to a clean round-bottomed glass test tube (see **Note 2**), and dilute the concentrated stock solution with more chloroform (approx. 400 μL).
  (c) Evaporation. Blow-dry the sample using a stream of N_2 under a fume hood; a thin film of lipids should be visible at the bottom of the tube. Place the tube in a vacuum desiccator for at least 90 min to remove any residual chloroform.

(d) Hydration. Add 1 mL water or buffer solution needed (to yield a final concentration of 1 mg/mL) to hydrate the dry lipid sample for 1 h (see **Note 3**), and vortex briefly until the solution appears milky.
(e) Sonication. Place the test tube in a holder with ice/water mixture to prevent overheating; insert the tip of sonicator inside the lipid suspension and do sonication on a 50% duty cycle (3 s on followed by 3 s off) with a total running time of at least 10 min (more if liquid has not cleared) (see **Note 4**).
(f) Centrifugation. Spin the clear solution in a centrifuge to remove any titanium particle from the sonicator probe and remaining non-unilamellar vesicles (centrifuging at 12000 rpm for 30 min is sufficient); transfer the supernatant into a glass vial, store unilamellar vesicles at 4 °C, and use within 24 h.

(2) **Membrane protein-reconstituted phospholipid vesicles**

(a) Gently mix as-prepared phospholipid vesicle with LDAO to a final concentrations of 0.5 mg/mL phospholipid vesicle and 6.5 mM LDAO.
(b) Add detergent-solubilized hVDAC1 to the vesicle/detergent mixture with a final concentration of 1 mM, and incubate the mixture at RT for 30 min.
(c) Add 50 mg Bio-Beads SM2 to the vesicle/detergent/protein mixture, and incubate the mixture overnight at 4 °C to absorb the detergents.
(d) Remove the Bio-Beads SM2, store the hVDAC1 containing phospholipid vesicles at 4 °C, and use within 24 h.

11.4.2.3 Notes

(1) Use glass pipettes and containers, and work quickly to avoid oxidation of the lipids. Container caps should be sealed securely and rapped with parafilm.
(2) Leave the stock solution at room temperature for a while before opening the container cap to avoid moisture absorption.
(3) It's important to hydrate the lipid film above their transition temperature. Pre-heat the hydration solution and incubate the test tube in a water bath are recommend if room temperature is below the transition temperature of lipids.
(4) The tip of the sonicator should be carefully placed just below the surface of the liquid, and do not touch the sides of the test tube; the solution would become clear and nearly transparent if the sonication is successful.

11.4.3 Construction of Hybrid Lipid Bilayer Membranes

Hybrid lipid bilayer membranes are constructed by vesicle spreading and fusion on hydrophobic surface of a self-assembled monolayer (SAM) on Au film. Step-by-step assembly can be monitored in situ by ATR-SEIRAS (Fig. 11.13). The details of preparing a hybrid lipid bilayer membrane are described below.

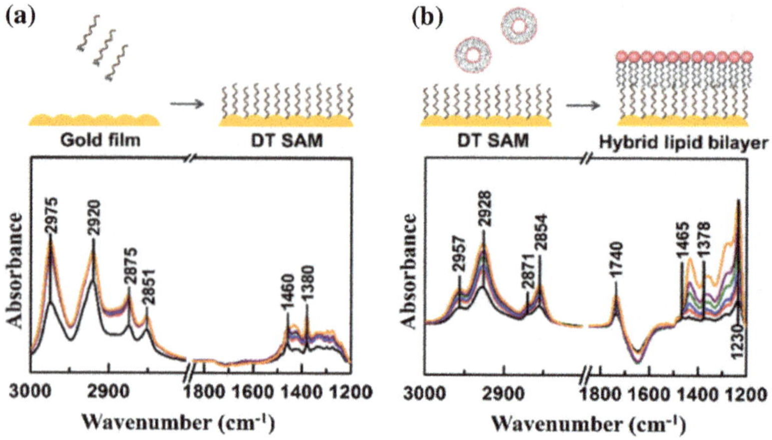

Fig. 11.13 In situ ATR-SEIRAS monitor of hybrid lipid bilayer membrane formation. **a** Formation of DT monolayer by self-assembly; **b** Formation of planar bilayer membrane

11.4.3.1 Materials

(1) 2 mM 1-dodecanethiol (DT) solution. Measure 100 μL DT and dissolve into 8250 μL ethanol to get a 50 mM stock solution, and store at 4 °C; freshly dilute 50 mM stock solution into 2 mM DT solution before use.
(2) Lipid vesicles.

11.4.3.2 Methods

(1) **Formation of hydrophobic surface.**

  (a) Add 500 μL ethanol to completely cover the gold film in SEIRAS accessory cell, and take the reference spectrum.
  (b) Add 500 μL 2 mM DT solution into the SEIRAS cell, and incubate for at least 1 h (see **Note 1**); thoroughly rinse the modified gold surface with ethanol after formation of DT monolayer. For in situ monitor the adsorption process, record the series sample spectra immediately after addition of DT solution.
  (c) Blow-dry the DT-modified gold surface with N_2 stream (see **Note 2**).

(2) **Formation of lipid bilayer**

  (a) Completely cover the DT monolayer modified surface with 500 μL water or buffer, and take the reference spectrum (see **Note 3**).
  (b) Add 500 μL 1 mg/mL lipid vesicles (to reach a final concentration of 0.5 mg/mL), and incubate for 3 h to enable the vesicle spreading and

fusion. For in situ monitor the formation of bilayer membrane, record the series sample spectra immediately after addition of lipid vesicles.
(c) Thoroughly rinse the modified surface with water or buffer after bilayer membrane formation (see **Note 4**).

11.4.3.3 Notes

(1) Incubation the gold film with DT solution overnight is recommended.
(2) Complete removal of residual ethanol is essential.
(3) Completely covering the hydrophobic surface is necessary.
(4) Exposure of bilayer membrane to air should be avoided during washing.

11.4.4 Construction of Lipid-Tethered Bilayer Membranes

Lipid-tethered bilayer membrane consists of a mixed phase-separated SAM and a lipid layer on top of the SAM formed from fused unilamellar vesicles (Fig. 11.6a). Synthetic lipid-thiols (serve as "tether-lipid") and small short chain thiols (serve as "spacer-thiol") are co-assembled to from a phase-separated surface induced by differences in the chain length and chemistry, creating a sub-membrane reservoir for incorporation of transmembrane proteins. Membrane proteins are co-immobilized during membrane formation by protein-reconstituted vesicles or incorporated by incubation with membrane protein after the membrane formation, depending on the protein studied. The details of preparing a hybrid lipid bilayer membrane are described below.

11.4.4.1 Materials

(1) 0.1 mM (cholesterylpolyethylenoxy)thiol (CPEO3) isopropanol solution.
(2) 0.1 mM 6-mercaptohexanol (6MH) isopropanol solution. Measure 50 μL 6MH and dissolve into 7258 μL isopropanol to get a 50 mM stock solution, store at 4 °C; freshly prepare 0.1 mM 6MH solution by dilution before use.
(3) Lipid vesicles: membrane protein-reconstituted or pure lipid vesicles.

11.4.4.2 Methods

(1) Mix 0.1 mM CPEO3 and 0.1 mM 6MH solution (1/1, *v*/*v*) to get a mixture of 0.05 mM CPEO3 and 0.05 mM 6MH in isopropanol (see **Note 1**).
(2) Add 500 μL isopropanol to completely cover the gold film, and completely remove the isopropanol after taking the reference spectrum.

(3) Add 500 μL CPEO3/6MH mixture and incubate overnight; thoroughly wash the modified gold surface after formation of mixed monolayer with iso-propanol. For in situ monitor the adsorption process, record the series sample spectra immediately after addition of CPEO3/6MH mixture (see **Note 2**).
(4) Rinse the mix monolayer with water or buffer, and take the reference spectrum (see **Note 3**); add 1 mg/mL lipid vesicles (membrane protein-reconstituted or pure vesicles) to reach a final concentration of 0.5 mg/mL, and incubate for 3 h to enable the bilayer formation. For in situ SEIRAS monitor, record the series sample spectra immediately after addition of lipid vesicles.
(5) Thoroughly rinse the modified surface with buffer after bilayer formation; in case to incorporate membrane protein after bilayer formation, expose the membrane to membrane protein and incubate for certain time, then thoroughly rinse the bilayer to remove un-incorporated proteins (see **Note 4**).

11.4.4.3 Notes

(1) Prepare CPEO3/6MH mixtures one day before the SAM formation and equilibrate the mixtures overnight at −20 °C is important to obtain reproducible SAMs. Fresh solutions tend to result in scattered surface concentration ratios.
(2) The ratio of CPEO3/6MH on the Au surface plays a crucial role for the quality of the bilayer membrane, as very low concentrations of CPEO3 do not induce vesicle fusion while at high concentrations, not enough space may be provided to insert integral membrane proteins. For adsorption of 1/1-CPEO3/6MH molar ratio (in solution), mixed monolayer with surface composition of 80% CPEO3 and 20% 6MH is achieved, which is believed to combine both requirements of pure lipid–lipid bilayer patches and good bilayer membrane quality.
(3) Complete the rinsing step as fast as possible, since long incubation time of the mixed SAM with buffer may result in bad bilayer membrane properties.
(4) Exposure of bilayer membrane to air should be avoided during washing.

11.4.5 Formation of Protein-Tethered Bilayer Membranes

Protein-tethered bilayer membrane, one of the models of cell membrane, is superior to structural and functional analysis of membrane proteins due to the capability of controlling the orientation and the surface concentration of target protein. This novel approach combines two common experimental concepts (Fig. 11.14). First, the membrane proteins are oriented-attached to solid support by affinity-based immobilization. Then, the oriented, surface-attached, and detergent-solubilized proteins are reconstituted into a lipid bilayer environment by in situ dialysis. The combined concept ensures both the oriented immobilization and the restoration of a physiologically natural matrix that enhances the stability of the membrane protein and preserves its functionality. The details of constructing a protein-tethered bilayer membrane are described below.

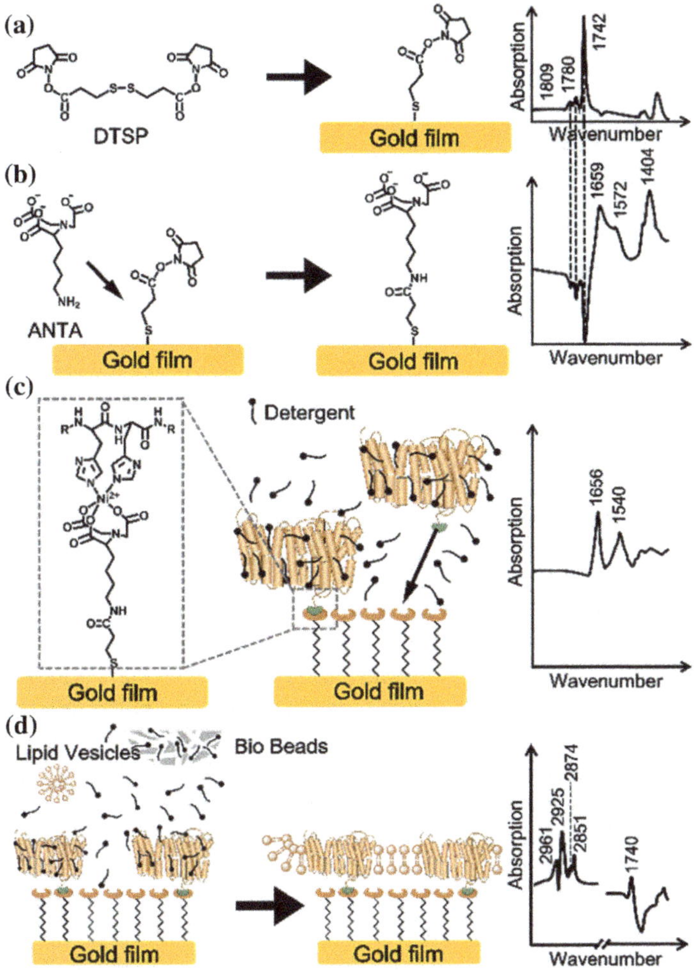

Fig. 11.14 In situ ATR-SEIRAS monitor of the construction of protein-tethered bilayer membrane. **a** Formation of TSP monolayer by self-assembly; **b** Covalent linkage of ANTA; **c** Oriented-attachment of membrane protein; **d** Reconstitution of tethered protein into lipid bilayer membrane

11.4.5.1 Materials

(1) 1 mg/mL dithiobis (succinimidylpropionate) (DTSP) in DMSO. Weight 10 mg DTSP power in pan paper and transfer to a test tube, add 1 mL dry DMSO to completely dissolve the solid; and then freshly dilute into 2 mg/mL solution before use (see **Note 1**).

(2) 0.5 M K_2CO_3 buffer.

(3) 150 mM $N_{\alpha'}$,$N_{\alpha''}$-bis(carboxymethyl)-L-lysine (ANTA). Weight 40 mg ANTA and transfer to a test tube. Add 1 mL 0.5 M K_2CO_3 buffer to dissolve the solid.
(4) 100 mM $NiSO_4$ solution. Weight 1.314 g $NiSO_4 \cdot 6H_2O$ and transfer to glass beaker. Add water to a volume of 50 mL and dissolve the solid. Store at 4 °C.
(5) Buffer solution: e.g., 50 mM phosphate buffer (pH 8.0) with 0.1% dodecylmaltoside (DDM).
(6) Detergent-solubilized His-tag fused membrane proteins: e.g., 0.1% DDM solubilized cytochrome *c* oxidase (C*c*O) in buffer.
(7) Detergent-destabilized lipid vesicles: e.g., 0.1% DDM-destabilized PC vesicles.
(8) Bio-Beads SM2.

11.4.5.2 Methods

(1) **Formation of TSP monolayer**. Add 500 μL dry DMSO to completely cover the Au film in SEIRAS accessory cell, and take the reference spectrum; then add 500 μL 2 mg/mL DTSP solution into the SEIRAS cell, and incubate for 10–30 min; rinse the modified surface after monolayer formation with DMSO, and dried with N_2. For in situ monitor the adsorption process, record the series sample spectra immediately after addition of DTSP solution (Fig. 11.14 a).
(2) **Covalent linkage of ANTA**. Completely cover the self-assembled monolayer with 500 μL 0.5 M K_2CO_3 buffer, and remove the buffer after taking the reference spectrum (see **Note 2**); add 500 μL 150 mM ANTA and incubate for 90 min to enable the formation of carboxamide linkage between ANTA and DTSP; thoroughly rinse the NTA-terminated surface with water to remove excess ANTA. For in situ monitor the cross-linking process, record the series sample spectra immediately after addition of ANTA (Fig. 11.14b).
(3) **Complexation of Ni^{2+} by NTA**. Cover the NTA-terminated surface with 500 μL water, and take the reference spectrum; add 500 μL 100 mM $NiSO_4$ solution and incubate for 30 min to enable the chelating of Ni^{2+} by NTA; thoroughly rinse Ni-NTA coated surface with water and 50 mM phosphate buffer. For in situ monitor the chelating process, record the series sample spectra immediately after addition of $NiSO_4$ solution.
(4) **Affinity adsorption of membrane protein**. Completely cover the Ni-NTA coated surface with 50 mM phosphate buffer with 0.1% DDM, and take the reference spectrum; add 0.1% DDM solubilized C*c*O to reach a final concentration of 2 μM, and incubate for 90 min to enable the affinity adsorption of C*c*O; excess C*c*O in solution is removed by thorough rinsing with buffer solution (see **Note 3**). For in situ monitor the adsorption process, record the series sample spectra immediately after addition of C*c*O solution (Fig. 11.14c).
(5) **Reconstitution of tethered protein into lipid bilayer membrane**. Expose the surface-tethered C*c*O to 1 mL of 0.1% DDM-destabilized PC vesicles (1 mg/mL) in buffer solution, and take the reference spectrum; add 50 mg macroporous Bio-Beads, and incubate for at least 1 h to completely remove the detergent and

enable the assembly of lipids around the protein and formation of protein-tethered bilayer lipid membrane (see **Note 4**); after formation of the bilayer lipid membrane, Bio-Beads and excess PC vesicles in solution are removed by thorough rinsing the surface with buffer solution (see **Note 3**). For in situ monitor the adsorption process, record the series sample spectra immediately after addition of Bio-Beads (Fig. 11.14d).

11.4.5.3 Notes

1: DTSP is sensitive to water due to hydrolysis. Leave the stock DTSP powder (stored at −20 °C) at room temperature for a while before opening the container cap to avoid moisture absorption; and always use dry DMSO as solvent; freshly prepare the DTSP solution before use.
2: Record the reference spectrum as fast as possible, since TSP monolayer undergoes hydrolysis immediately after addition of 0.5 M K_2CO_3 buffer.
3: Exposure of membrane to air should be avoided during washing process.
4: 50 mg Bio-Beads are added in two portions (25 mg each) to increase the efficiency of detergent removal (Fig. 11.14).

Acknowledgements This work was financially supported by the National Natural Science Foundation of China (No. 21675149 and 21505130), the Science and Technology Development Program of Jilin Province (No. 20150519014JH, 20170414037GH, 20170520133JH), and the Key Research Program of Frontier Sciences, CAS (QYZDY-SSW-SLH019).

References

1. Osawa M (2006) Surface-enhanced infrared absorption spectroscopy. In: Chalmers JM, Griffiths PR (eds) Handbook of vibrational spectroscopy. Wiley, Hoboken, pp 765–799
2. Adato R, Aksu S, Altug H (2015) Engineering mid-infrared nanoantennas for surface enhanced infrared absorption spectroscopy. Mater Today 18(8):436–446
3. Neubrech F, Huck C, Weber K, Pucci A, Giessen H (2017) Surface-enhanced infrared spectroscopy using resonant nanoantennas. Chem Rev 117(7):5110–5145
4. Jiang X, Engelhard M, Ataka K, Heberle J (2010) Molecular impact of the membrane potential on the regulatory mechanism of proton transfer in sensory rhodopsin II. J Am Chem Soc 132(31):10808–10815
5. Jiang X, Zaitseva E, Schmidt M, Siebert F, Engelhard M, Schlesinger R, Ataka K, Vogel R, Heberle J (2008) Resolving voltage-dependent structural changes of a membrane photoreceptor by surface-enhanced IR difference spectroscopy. Proc Natl Acad Sci U S A 105 (34):12113–12117
6. Wang T, Bai J, Jiang X, Nienhaus GU (2012) Cellular uptake of nanoparticles by membrane penetration: a study combining confocal microscopy with FTIR spectroelectrochemistry. ACS Nano 6(2):1251–1259
7. Ataka K, Yotsuyanagi T, Osawa M (1996) Potential-dependent reorientation of water molecules at an electrode/electrolyte interface studied by surface-enhanced infrared absorption spectroscopy. J Phys Chem 100(25):10664–10672

8. Osawa M, Ataka K, Yoshii K, Yotsuyanagi T (1993) Surface-enhanced infrared ATR spectroscopy for in situ studies of electrode/electrolyte interfaces. J Electron Spectrosc Relat Phenom 64–65:371–379
9. Ataka K, Heberle J (2003) Electrochemically induced surface-enhanced infrared difference absorption (SEIDA) spectroscopy of a protein monolayer. J Am Chem Soc 125(17): 4986–4987
10. Hartstein A, Kirtley JR, Tsang JC (1980) Enhancement of the infrared absorption from molecular monolayers with thin metal overlayers. Phys Rev Lett 45(3):201–204
11. Osawa M (2001) Surface-enhanced infrared absorption. In: Kawata S (ed) Near-field optics and surface plasmon polaritons, vol 81. Topics in applied physics. Springer, Berlin Heidelberg, pp 163–187
12. Aroca RF, Ross DJ, Domingo C (2004) Surface-enhanced infrared spectroscopy. Appl Spectrosc 58(11):324A–338A
13. Osawa M, Ataka K, Yoshii K, Nishikawa Y (1993) Surface-enhanced infrared spectroscopy: the origin of the absorption enhancement and band selection rule in the infrared spectra of molecules adsorbed on fine metal particles. Appl Spectrosc 47(9):1497–1502
14. Bin X, Zawisza I, Goddard JD, Lipkowski J (2005) Electrochemical and PM-IRRAS studies of the effect of the static electric field on the structure of the DMPC bilayer supported at a Au (111) electrode surface. Langmuir 21(1):330–347
15. Garcia-Araez N, Brosseau CL, Rodriguez P, Lipkowski J (2006) Layer-by-layer PMIRRAS characterization of DMPC bilayers deposited on a Au(111) electrode surface. Langmuir 22 (25):10365–10371
16. Zawisza I, Bin X, Lipkowski J (2007) Potential-driven structural changes in Langmuir-Blodgett DMPC bilayers determined by in situ spectroelectrochemical PM IRRAS. Langmuir 23(9):5180–5194
17. Brosseau CL, Leitch J, Bin X, Chen M, Roscoe SG, Lipkowski J (2008) Electrochemical and PM-IRRAS a glycolipid-containing biomimetic membrane prepared using Langmuir-Blodgett/ Langmuir-Schaefer deposition. Langmuir 24(22):13058–13067
18. Laredo T, Dutcher JR, Lipkowski J (2011) Electric field driven changes of a gramicidin containing lipid bilayer supported on a Au(111) surface. Langmuir 27(16):10072–10087
19. Matyszewska D, Bilewicz R, Su Z, Abbasi F, Leitch JJ, Lipkowski J (2016) PM-IRRAS studies of DMPC bilayers supported on Au(111) electrodes modified with hydrophilic monolayers of thioglucose. Langmuir 32(7):1791–1798
20. Bin X, Lipkowski J (2006) Electrochemical and PM-IRRAS studies of the effect of cholesterol on the properties of the headgroup region of a DMPC bilayer supported at a Au (111) electrode. J Phys Chem B 110(51):26430–26441
21. Leitch JJ, Brosseau CL, Roscoe SG, Bessonov K, Dutcher JR, Lipkowski J (2013) Electrochemical and PM-IRRAS characterization of cholera toxin binding at a model biological membrane. Langmuir 29(3):965–976
22. Matyszewska D, Leitch J, Bilewicz R, Lipkowski J (2008) Polarization modulation infrared reflection–absorption spectroscopy studies of the influence of perfluorinated compounds on the properties of a model biological membrane. Langmuir 24(14):7408–7412
23. Hernandez MR, Towns EN, Moore J, Lee H, German JB, Lebrilla CB, Parikh AN, Land DP (2012) Use of attenuated total reflectance Fourier transform infrared spectroscopy to study lactosylceramide and GD3 DMPC bilayers. Colloids Surf B 94:374–377
24. Towns EN, Parikh AN, Land DP (2015) Influence of vesicle size and aqueous solvent on intact phospholipid vesicle adsorption on oxidized gold monitored using attenuated total reflectance Fourier transform infrared spectroscopy. J Phys Chem C 119(5):2412–2418
25. Sikorska E, Dawgul M, Greber K, Iłowska E, Pogorzelska A (1838) Kamysz W (2014) Self-assembly and interactions of short antimicrobial cationic lipopeptides with membrane lipids: ITC, FTIR and molecular dynamics studies. Biochim Biophys Acta, Biomembr 10:2625–2634
26. Forstner MB, Yee CK, Parikh AN, Groves JT (2006) Lipid lateral mobility and membrane phase structure modulation by protein binding. J Am Chem Soc 128(47):15221–15227

27. Brisebois PP, Arnold AA, Chabre YM, Roy R, Marcotte I (2012) Comparative study of the interaction of fullerenol nanoparticles with eukaryotic and bacterial model membranes using solid-state NMR and FTIR spectroscopy. Eur Biophys J 41(6):535–544
28. Cieślik-Boczula K, Maniewska J, Grynkiewicz G, Szeja W, Koll A, Hendrich AB (2012) Interaction of quercetin, genistein and its derivatives with lipid bilayers—an ATR IR-spectroscopic study. Vib Spectrosc 62:64–69
29. Kuć M, Cieślik-Boczula K, Rospenk M (2016) Influence of inhalation anesthetics on the chain-melting phase transition of DPPC liposomes. Near-infrared spectroscopy studies supported by PCA analysis. Vib Spectrosc 85:55–61
30. Schwörer F, Trapp M, Ballauff M, Dahint R, Steitz R (2015) Surface-active lipid linings under shear load—a combined in-situ neutron reflectivity and ATR-FTIR study. Langmuir 31 (42):11539–11548
31. Adams EM, Casper CB, Allen HC (2016) Effect of cation enrichment on dipalmitoylphosphatidylcholine (DPPC) monolayers at the air-water interface. J Colloid Interface Sci 478:353–364
32. Šegota S, Vojta D, Pletikapić G, Baranović G (2015) Ionic strength and composition govern the elasticity of biological membranes. A study of model DMPC bilayers by force-and transmission IR spectroscopy. Chem Phys Lipids 186:17–29
33. Peng B, Ding X-Y, Sun C, Liu W, Zhang JZH, Zhao X (2016) The effect of POPC acyl chains packing by aromatic amino acid methyl esters investigated by ATR-FTIR combined with QM calculations. RSC Adv 6(51):45569–45577
34. Lewis RNAH, McElhaney RN (2013) Membrane lipid phase transitions and phase organization studied by Fourier transform infrared spectroscopy. Biochim Biophys Acta, Biomembr 1828(10):2347–2358
35. Binder H (2007) Water near lipid membranes as seen by infrared spectroscopy. Eur Biophys J 36(4–5):265–279
36. Bridelli MG, Capelletti R, Mora C (2013) Structural features and functional properties of water in model DMPC membranes: thermally stimulated depolarization currents (TSDCs) and Fourier transform infrared (FTIR) studies. J Phys D Appl Phys 46(48):485401
37. Disalvo EA, Bouchet AM (1828) Frias MA (2013) Connected and isolated CH_2 populations in acyl chains and its relation to pockets of confined water in lipid membranes as observed by FTIR spectrometry. Biochim Biophys Acta, Biomembr 8:1683–1689
38. Disalvo EA, Frias MA (2013) Water state and carbonyl distribution populations in confined regions of lipid bilayers observed by FTIR spectroscopy. Langmuir 29(23):6969–6974
39. Levinger NE, Costard R, Nibbering ETJ, Elsaesser T (2011) Ultrafast energy migration pathways in self-assembled phospholipids interacting with confined water. J Phys Chem A 115(43):11952–11959
40. Costard R, Levinger NE, Nibbering ETJ, Elsaesser T (2012) Ultrafast vibrational dynamics of water confined in phospholipid reverse micelles. J Phys Chem B 116(19):5752–5759
41. Osborne DG, Dunbar JA, Lapping JG, White AM, Kubarych KJ (2013) Site-specific measurements of lipid membrane interfacial water dynamics with multidimensional infrared spectroscopy. J Phys Chem B 117(49):15407–15414
42. Kel O, Tamimi A, Thielges MC, Fayer MD (2013) Ultrafast structural dynamics inside planar phospholipid multibilayer model cell membranes measured with 2D IR spectroscopy. J Am Chem Soc 135(30):11063–11074
43. Kel O, Tamimi A, Fayer MD (2014) Size-dependent ultrafast structural dynamics inside phospholipid vesicle bilayers measured with 2D IR vibrational echoes. Proc Natl Acad Sci U S A 111(3):918–923
44. Kel O, Tamimi A, Fayer MD (2015) The influence of cholesterol on fast dynamics inside of vesicle and planar phospholipid bilayers measured with 2D IR spectroscopy. J Phys Chem B 119(29):8852–8862
45. Barth A, Zscherp C (2003) What vibrations tell about proteins. Q Rev Biophys 35(4):369–430
46. Barth A (2007) Infrared spectroscopy of proteins. Biochim Biophys Acta, Bioenerg 1767 (9):1073–1101

47. Roldan N, Goormaghtigh E, Pérez-Gil J, Garcia-Alvarez B (2015) Palmitoylation as a key factor to modulate SP-C-lipid interactions in lung surfactant membrane multilayers. Biochim Biophys Acta, Biomembr 1848(1, Part A):184–191
48. Yang P, Homan KT, Li Y, Cruz-Rodríguez O, Tesmer JJG, Chen Z (2016) Effect of lipid composition on the membrane orientation of the G protein-coupled receptor kinase 2-Gβ1γ2 complex. Biochemistry 55(20):2841–2848
49. Potvin-Fournier K, Lefèvre T, Picard-Lafond A, Marcotte C, Dufresne C, Cantin L, Salesse C, Auger M (2016) Discriminating lipid-from protein-calcium binding to understand the interaction between recoverin and phosphatidylglycerol model membranes. Biochemistry 55 (24):3481–3491
50. Güler G, Gärtner RM, Ziegler C, Mäntele W (2016) Lipid-protein interactions in the regulated betaine symporter BetP probed by infrared spectroscopy. J Biol Chem 291(9):4295–4307
51. Gavriljuk K, Itzen A, Goody RS, Gerwert K, Kötting C (2013) Membrane extraction of Rab proteins by GDP dissociation inhibitor characterized using attenuated total reflection infrared spectroscopy. Proc Natl Acad Sci U S A 110(33):13380–13385
52. Baenziger JE, daCosta CJB (2008) Membrane protein structure and conformational change probed using Fourier transform infrared spectroscopy. In: Pebay-Peyroula E (ed) Biophysical analysis of membrane proteins. Wiley-VCH Verlag GmbH & Co, KGaA, pp 259–288
53. Ataka K, Stripp ST, Heberle J (2013) Surface-enhanced infrared absorption spectroscopy (SEIRAS) to probe monolayers of membrane proteins. Biochim Biophys Acta, Biomembr 1828(10):2283–2293
54. Kozuch J, Steinem C, Hildebrandt P, Millo D (2012) Combined electrochemistry and surface-enhanced infrared absorption spectroscopy of gramicidin A incorporated into tethered bilayer lipid membranes. Angew Chem Int Ed 51(32):8114–8117
55. Zeng L, Wu L, Liu L, Jiang X (2016) Analyzing structural properties of heterogeneous cardiolipin-bound cytochrome c and their regulation by surface-enhanced infrared absorption spectroscopy. Anal Chem 88(23):11727–11733
56. Tu YS, Lv M, Xiu P, Huynh T, Zhang M, Castelli M, Liu ZR, Huang Q, Fan CH, Fang HP, Zhou RH (2013) Destructive extraction of phospholipids from *Escherichia coli* membranes by graphene nanosheets. Nat Nanotechnol 8(8):594–601
57. Wu L, Zeng L, Jiang X (2015) Revealing the nature of interaction between graphene oxide and lipid membrane by surface-enhanced infrared absorption spectroscopy. J Am Chem Soc 137(32):10052–10055
58. Wang T, Jiang X (2015) The broken of phosphodiester bond: a key factor induced hemolysis. ACS Appl Mater Interfaces 7(1):129–136
59. Wang T, Zhu S, Jiang X (2015) Toxicity mechanism of graphene oxide and nitrogen-doped graphene quantum dots in RBCs revealed by surface-enhanced infrared absorption spectroscopy. Toxicol Res 4(4):885–894

Chapter 12 Computer Simulations to Explore Membrane Organization and Transport

Huiying Chu, Yuebin Zhang, Yan Li and Guohui Li

12.1 Introduction

Ever since the first modeling of elastic collision between rigid spheres, the molecular dynamics (MD) simulation technique has been greatly developed to achieve atomic-level information. It has become a critical component of the widely used tool set, and been applied to both material science and biological systems, such as proteins, nucleic acids and lipid membranes. MD simulations not only allow the examination of experimental findings at the atomic level, which can test new hypotheses, but also provide data that cannot be obtained from experiments, such as the pressure profile of membranes [1]. Cells are normally surrounded and protected by plasma membranes which consist of different types of lipids, proteins, and carbohydrates. In the last decade, computer simulation has opened new ways to study bilayers at the atomic level, yielding a detailed picture of the structure and dynamics of membranes and membrane proteins [2–4].

Membranes serve many critical biological functions, such as forming barriers between intracellular and extracellular environments, regulating the transport of substances [5], detecting and transmitting electric and chemical signals through protein receptors [6], mediating the communications between cells [7] and so forth. Also, membrane proteins have been found to comprise approximately one-third of the human genome [8], and over half of these are known as drug targets. Thus, the biological functions of membrane proteins have become an important focus in fundamental research. Unfortunately, even with advanced experimental techniques,

H. Chu · Y. Zhang · Y. Li · G. Li (✉)
Laboratory of Molecular Modeling and Design, State Key Laboratory of Molecular Reaction Dynamics, Dalian Institute of Chemical Physics, Chinese Academy of Science, 457 Zhongshan Road, Dalian 116023, Liaoning, China
e-mail: ghli@dicp.ac.cn

H. Wang and G. Li (eds.), *Membrane Biophysics*,
https://doi.org/10.1007/978-981-10-6823-2_12

it is still difficult to achieve sufficient details of protein structures at the molecular or even atomic level, not to mention the relationship between structural information and functionality. To solve this problem, MD simulation is now accepted as an indispensable tool to achieve structural and dynamical information not available via experiments.

In principle, all details of molecular structures and interactions can be depicted by first principles using quantum mechanics. Unfortunately, most of the problems involving membrane proteins cannot be handled by quantum mechanics for its high computational cost. Hence, MD simulation applies molecular force fields, which are mainly based on a kind of potential energy descriptions at different atomic and molecular levels, to describe the topological structures and dynamic behaviors of membrane protein molecules. Molecular force fields are usually adopted to calculate the energies of molecules by using positions of atoms, and greatly speed up calculations compared to quantum mechanics. Thus it can be used to study the systems that contain tens of thousands of atoms. A lot of studies have shown that molecular force fields could help to explain many physical problems. In addition, one of the main approximations of the additive all-atom force fields is related to the description of electrostatic properties. Additive all-atom force fields of lipid, such as AMBER [9, 10], CHARMM [11–13], OPLS-AA [14], and united-atom force field of lipid, such as GROMOS [15], treat the electrostatic interactions with fixed atomic charges. The fixed partial charge is placed at the nucleus of each atom to represent its electrostatic properties. In additive force fields, the charge is a parameter which can be tuned to represent atom polarization effects in an average way through a mean-field approximation, which responds to different environments that a molecule might experience [16]. For lipid bilayers, the polar headgroups of lipids face the high-dielectricity water environment on one side and interact with the low-dielectric hydrocarbon core on the other side [17]. The electronic polarization experienced exterior or interior of the membrane by an embedded molecule is very different [18, 19]. Roux et al. [20] have investigated the ion selectivity of several membrane-binding channels and transporters. The results indicated that although the fundamental physical properties could be described using the non-polarizable models, more detailed understanding of the conformation-driven super-selectivity depended on improvements in force field models considering the explicit polarizability.

Several advances are made in both software and hardware aspects for simulating membrane and membrane proteins in the last decade, including the easy-to-use software for setting up MD simulations, the massively parallel algorithms and the GPU accelerated computing.

In the following sections, we first introduce the technical principle of MC and MD. Then the applications of both CG and All-atom model are introduced. And finally, the protocol is presented.

12.2 Technical Principle

12.2.1 Monte Carlo Simulations and Molecular Dynamics Simulation

12.2.1.1 Monte Carlo Simulations

The Monte Carlo simulation was first applied to perform computer simulation on the molecular system, therefore it occupies a special position in the history of molecular modeling. The Monte Carlo simulations obtain the conformations of a system through random changes of the positions of atoms, and meanwhile, change the system to appropriate orientations and conformations. Monte Carlo methods are a broad class of computational algorithms that rely on repeated random sampling to obtain numerical results. Their essential idea is using randomness to solve problems that might be deterministic in principle. They are often used in physical and mathematical problems, which can be summarized into three distinct classes [21]: optimization, numerical integration and generating draws from a probability distribution. Based on the position of atoms, the system potential energy of each conformation, and the other values of properties can be calculated. Thus, Monte Carlo samples are from a 3N-dimensional space of the particles.

The classical expression for partition function Q:

$$Q = c\iint \mathrm{d}p^N \mathrm{d}r^n \exp\left[-H\left(r^N p^N\right)/K_\mathrm{B}T\right] \tag{12.1}$$

where r^N is the coordinates of all N particles, p^N is the corresponding momenta, and c is a constant of proportionality. The $H(r^N p^N)$ is the Hamiltonian of the system, which depends on the 3N positions and 3N momenta of the particles in the system. It can be written as the sum of the kinetic and potential energies of the system:

$$H\left(r^N p^N\right) = \sum_{i=1}^{N} \frac{|p_i|^2}{2m} + V\left(r^N\right) \tag{12.2}$$

From the above two equations, the canonical ensemble partition can be separated into two separate integrals, one is over the positions part, and the other is over the momenta part.

Though Monte Carlo methods diversify in different aspects, they still follow a particular pattern: (1) defining a domain of possible inputs; (2) generating inputs randomly from a probability distribution over the domain; (3) performing a deterministic computation on the inputs; and (4) aggregating the results.

Here we take a circle inscribed in a unit square for an example. Given that the area ratio of the circle to the square is π/4, the value of π can be approximated using a Monte Carlo method [22]: (1) Draw a square and inscribe a circle within it; (2) Uniformly scatter objects of uniform size over the square; (3) Counting the number of objects inside the circle and the total number of objects; (4) The ratio of the two counts is an estimate of the ratio of the two areas, which is π/4; (5) Multiply the result by 4 to estimate π. In this procedure, the domain of inputs is the square that circumscribes the circle. Random inputs are generated by scattering grains over the square, then computations on each input are conducted (test whether it falls within the circle). Finally, the results are aggregated to obtain the final output, which is the approximation of π.

There are two important points to be noted here. Firstly, if the grains are not uniformly distributed, our approximation will be poor. Secondly, there should be a large number of inputs, because the approximation is usually poor if only a few grains are randomly dropped into the whole square. Generally, the approximation improves as more grains are dropped.

12.2.1.2 Molecular Dynamics Simulation

Molecular dynamics (MD) is a computer simulation method for studying the physical movements of atoms and molecules and belongs to a type of many-body simulation. The atoms and molecules are allowed to interact for a fixed period of time, giving a view of the dynamical evolution of the system. In the most common version, the trajectories of atoms and molecules are determined by numerically solving Newton's equations of motion for a system of interacting particles, where forces between the particles and their potential energies are calculated using interatomic potentials or molecular mechanics force fields. The method was originally used in the field of theoretical physics in the late 1950s [23, 24], and now it is widely applied in various fields, such as chemical physics, materials science and the modeling of biomolecules. For example, MD is frequently used to refine three-dimensional structures of proteins and other macromolecules based on experimental constraints from X-ray crystallography or NMR spectroscopy. In biophysics and structural biology, the method is used to study the motions of biological macromolecules such as proteins and nucleic acids, which is useful for interpreting the results of certain biophysical experiments and modeling interactions between molecules.

In principle, MD can be used for ab initio prediction of protein structure by simulating folding of the polypeptide chain from random coil. The trajectory is obtained by solving the different equations embodied in Newton's second law ($F = ma$):

$$\frac{\mathrm{d}^2 x}{\mathrm{d}t^2} = \frac{F_i}{m_i} \tag{12.3}$$

The equation describes the motion of a particle of mass m_i along one coordinate (x_i) with F_{xi} being the force on the particle in the corresponding direction, and can be written as:

$$F_i = -\nabla U(r_i, \ldots, r_N) \tag{12.4}$$

where U $(r_i, \ldots, r_N)$ is the potential energy function of N particles which contain the bonded and non-bonded interactions. The bonded interactions describe the interactions of the covalently bound atoms in proteins and lipid molecules, and the non-bonded interactions can be decomposed into four pieces: Coulomb energy between two atoms, Polarization interaction between atoms, Dispersion (van der Waals) potential, and Short-range repulsion. The Coulomb energy can be calculated as:

$$U_{\text{Coul}} = \frac{q_i q_j}{4\pi\varepsilon_0 r_{ij}} \tag{12.5}$$

And the van der Waals potential is calculated based on the Lennard-Jones potential:

$$U_{\text{LJ}} = 4\varepsilon\left[(\sigma/r)^{12} - (\sigma/r)^{6}\right] \tag{12.6}$$

Here ε is the depth of potential at the minimum ($r = 2^{1/6}\sigma$), and the potential vanishes at $r = \sigma$.

Implementation of the Coulomb and LJ terms is straight forward, but a calculation of the induced polarization requires the iteration of the polarization equations, which increases the computational cost by several-fold.

Integration Algorithms. Given the position and velocities of N particles at time t, a straight forward integration of Newton's equation of motion yields the following at $t + \Delta t$

$$v_i(t + \Delta t) = v_i(t) + \frac{F_i(t)}{m_i}\Delta t \tag{12.7}$$

$$r_i(t + \Delta t) = r_i(t) + v_i(t)\Delta t + \frac{F_i(t)}{2m_i}\Delta t^2 \tag{12.8}$$

In the popular Verlet algorithm, one eliminates velocities by adding the time-reversed position at $t - \Delta t$:

$$r_i(t - \Delta t) = r_i(t) - v_i(t)\Delta t + \frac{F_i(t)}{2m_i}\Delta t^2 \tag{12.9}$$

While r_i $(t + \Delta t)$ can be given as:

$$r_i(t + \Delta t) = 2r_i(t) - r_i(t - \Delta t) + \frac{F_i(t)}{m_i}\Delta t^2 \tag{12.10}$$

This is especially useful in situations where one is interested only in the positions of the atoms. If required, velocities can be calculated from

$$v_i(t) = \frac{1}{2\Delta t}[r_i(t+\Delta t) - r_i(t-\Delta t)] \tag{12.11}$$

The Verlet algorithm has several drawbacks: (1) positions are obtained by adding a small quantity to large ones, which may lead to a loss of precision; (2) velocity at time t is available only at the next time step $t + \Delta t$; (3) it is not self-starting, i.e., at t_0, there is no position at $t - \Delta t$. These drawbacks can be avoided in the leap-frog algorithm, where the positions and velocities are calculated at different times separated by $\Delta t/2$:

$$v_i(t+\Delta t/2) = v_i(t-\Delta t/2) + \frac{F_i(t)}{m_i}\Delta t \tag{12.12}$$

$$r_i(t+\Delta t) = r_i(t) + v_i(t+\Delta t/2)\Delta t \tag{12.13}$$

The initial coordinate can be taken from the Protein Data Bank. After energy minimization, the coordinates give the $t = 0$ time atom positions. The initial velocities are sampled from a Maxwell-Boltzmann distribution:

$$P(v_{ix}) = \left(\frac{m_i}{2\pi kT}\right)\exp\left[-m_i v_{ix}^2/2kT\right] \tag{12.14}$$

Boundaries and Ensembles. In MD simulations, however, the system size is so small that one should consider the boundary effects. Using vacuum is not realistic for bulk simulations because a vacuum creates an ordering of surface waters, which could influence the dynamics of a biomolecule separated by a few layers of water from the surface. The most common solution is to use periodic boundary conditions, that is, the simulation box is replicated in all directions just like in a crystal. The cube and rectangular prism are the obvious choices for a box shape, though other shapes are also possible. Application of the periodic boundary conditions results in an infinite system which, in turn, raises the question of accurate calculation of the long-range Coulomb interactions. This problem has been resolved using Ewald's sum, where the long-range part is separately evaluated in the reciprocal Fourier space.

MD simulations are typically performed in the NVE ensemble, where all three quantities (number of atoms, volume, and energy) are constant. Due to truncation errors, keeping the energy constant in long MD simulations can be problematic. To avoid this problem, the alternative NVT and NPT ensembles are employed. The temperature of the system is obtained from the average kinetic energy:

$$\langle K\rangle = \frac{3}{2}NkT \tag{12.15}$$

Thus, an obvious way to keep the temperature constant at T_0 is to scale the velocities as:

$$v_i(t) \to \lambda v_i(t),\ \lambda = \sqrt{T_0/T(t)} \tag{12.16}$$

Because the kinetic energy has considerable fluctuations, this is a rather crude method. A better method, which achieves the same result more smoothly, is the Berendsen thermostat, where the atoms are weakly coupled to an external heat bath with the desired temperature T_0:

$$m_i \frac{\mathrm{d}^2}{\mathrm{d}t^2} r_i = F_i + m_i \gamma_i \left[\frac{T_0}{T(t)} - 1 \right] \frac{\mathrm{d}r_i}{\mathrm{d}t} \tag{12.17}$$

If $T(t) > T_0$, the coefficient of the coupling term is negative, which invokes a viscous force slowing the velocity, and vice versa for $T(t) < T_0$. Similarly, in the NPT ensemble, the pressure can be kept constant by simply scaling the volume. Again, this is very crude, and a better method is to weakly couple the pressure difference to atoms using a similar force as above (Langevin piston), which will maintain the pressure at the desired value of ~1 atm.

12.2.2 Additive Force Field

Computational treatment of molecular dynamics is based on inter-atomic forces, which can be derived by solving the Schrödinger's equation. And the related solving approaches are categorized as quantum mechanical method. However, high calculation costs limit the use of those approaches to relatively simple systems. In 1930, Andrews [25] first proposed the basic conception of molecular force fields, a bead-spring model was applied to describe the bond length and bond angle, and compute the interactions of non-bonded atoms by using van der Waals interaction expressions. Then Lifson and Warshel described consistent force field (CFF) called empirical function force field in the 1960s, which could be attributed to the modern molecular force field [26]. MD simulation, on the other hand, builds an empirical function to model the potential energy of the system. The function can be constructed via estimating the intermolecular interaction energies from isolated monomer wave functions (namely the perturbative method) or via the energy differences between isolated monomers and corresponding dimers (namely the super-molecular calculations) [27]. For different atoms or atoms in different environments, parameter sets are introduced as the variables in the potential energy function. Due to the diversity of interatomic interactions in biological systems, as well as the complex electrostatic environments, it is challenging to build a uniform set of parameters that can model the motions of atoms in different situations.

The classical way to estimate the interatomic interactions includes treating atoms as rigid spheres with fixed charges located on the nucleus. Electron distribution among bonded atoms based on electronegativity can be empirically illustrated as a partial charge, either positive or negative, on each atom. Thus, each atom responds to the surrounding electrostatic environment in an average way (mean-field approximation). From this simple treatment, interatomic electrostatic energies can be simply estimated via the Coulomb's law, and the total electrostatic energy of the system is the summation of those pairwise energies. Then the complete potential energy function can be estimated via a summation of bonded energy terms (including bond lengths, bond angles, dihedral angles, and may also include improper dihedrals and other empirical correction terms), van der Waals interaction term (usually described by the Lennard-Jones potential), and the electrostatic energy. According to the targeting system, different force fields have been built. To accurately model the properties of small organic or inorganic molecules such as metals, crystals, polymers and nanoparticles in materials, force fields such as CFF [28–31], MM3 [32], MMFF94 [33–35], UFF [36], and DREIDING [37] are implemented. For the dynamics of macromolecules, force fields like AMBER [38, 39], CHARMM [40–47], GROMOS [48–59], and OPLS [60, 61] have also been built.

Currently, three force fields AMBER, OPLS, and CHARMM are also used for the modeling of the ionic liquids. In addition, the MM series force fields and CFF are suitable for the system of organic compounds. In the 1980s, molecular force fields such as AMBER, CHARMM, OPLS and GROMOS produce a positive impact on the research of life science and promote the development of the molecular force fields targeting life science.

12.2.2.1 Assisted Model Building with Energy Refinement (AMBER) Force Field

AMBER force field is one of the earliest molecular force field used for the research of biological macromolecules and covers the simulations of proteins, DNA, monosaccharide, and polysaccharide. In this force field, –CH2– and –CH3 are regarded as united atom and used to treat hydrogen bonding interactions. The simulation results show that the AMBER force field can obtain reasonable molecular geometry, conformation energy, vibration frequency and solvation free energy. The parameters of the AMBER force field are obtained as follows: (1) the parameters of equilibrium bonds length and angles are from the experimental data of microwave, neutron scattering and molecular mechanics calculations; (2) the distorted constants are built by microwave, NMR, and molecular mechanics calculations; (3) the non-bonded parameters are obtained through the unit cell calculations; and (4) the parameters of atomic charges are given by the calculations of local charge model and ab initio quantum mechanics. For non-bonded interactions within neighboring four atoms in the AMBER force field, the electrostatic interactions reduce to 1/1.2 of other atoms, while the van der Waals interactions reduce to 1/2 of other atoms. The bond stretching and angle bending energies in the AMBER force field are calculated using the harmonic oscillator model, dihedral angle torsion energy is described by Fourier series form, Lennard-Jones potential is chosen to represent the van der Waals force, and the Coulomb formula is applied to estimate the electrostatic interactions. The functional form of AMBER force field is shown as follows:

$$\varepsilon_{ij} = \frac{4\varepsilon_{ii}\varepsilon_{jj}}{\left(\varepsilon_{ii}^{1/2} + \varepsilon_{jj}^{1/2}\right)^2} \tag{12.18}$$

$$U = \sum_{\text{bonds}} K_r(\gamma - \gamma_{\text{eq}})^2 + \sum_{\text{angle}} K_\theta(\theta - \theta_{\text{eq}})^2 + \sum_{\text{dihedral}} \frac{1}{2} U_n[1 + \cos(n\varphi - \gamma)] + \sum_{i<j}\left[\frac{A_{ij}}{R_{ij}^{12}} - \frac{B_{ij}}{R_{ij}^{6}} + \frac{q_i q_j}{\varepsilon R_{ij}}\right] + \sum_{\text{H-bonds}}\left[\frac{C_{ij}}{R_{ij}^{12}} - \frac{D_{ij}}{R_{ij}^{10}}\right] \tag{12.19}$$

where r, θ, φ are the bond length, angle, and dihedral angle, respectively. The fourth term represents the sum of the van der Waals and the electrostatic interactions, and the fifth term is the hydrogen bonding interactions.

12.2.2.2 Optimized Potentials for Liquid Simulations (OPLS) Force Field

The OPLS force field includes united-atom model (OPLS-UA) and all-atom model (OPLS-AA), and it is suitable for the simulations of organic molecules and peptides [62]. The bond stretching and bending parameters of OPLS force field are obtained based on the modifications of the AMBER force field. This force field is committed to calculate conformation energies of gas-phase organic molecules, solvation free energies of pure organic liquids and other thermodynamic properties. The OPLS force field is represented as follows:

$$U(R) = \sum_{\text{bonds}} K_b(b - b_0)^2 + \sum_{\text{angle}} K_\theta(\theta - \theta_0)^2 + \sum_{\text{dihedral}} \frac{k_\varphi}{2}[1 + \cos(n\varphi - \varphi_0)] + \sum_{\text{nonbond}}\left\{4\varepsilon_{ij}\left[\left(\frac{\sigma_{ij}}{r_{ij}}\right)^{12} - \left(\frac{\sigma_{ij}}{r_{ij}}\right)^{6}\right] + \frac{q_i q_j}{r_{ij}}\right\} \tag{12.20}$$

12.2.2.3 Chemistry at Harvard Molecular Mechanics (CHARMM) Force Field

The CHARMM force field is developed by Harvard University, and the force field parameters are not only from the experimental results but also involve many results of quantum chemical calculations. This force field is mostly used to study multi-molecular systems including small organic molecules, solutions, polymers, biochemical molecules etc. [63]. It can also be used to perform energy minimization, molecular dynamics (MD) and Monte Carlo (MC) simulations. The form of CHARMM force fields is as follows:

$$U = \sum k_b(r - r_0)^2 + \sum k_\theta(\theta - \theta_0)^2 + \sum\left[|k_\varphi| - k_\varphi \cos(n\varphi)\right] + \sum k_\aleph(\aleph - \aleph_0)^2 + \sum_{i,j} \frac{q_i q_j}{4\pi\varepsilon_0 r_{ij}} + \sum_{i,j}\left(\frac{A_{ij}}{r_{ij}^{12}} - \frac{B_{ij}}{r_{ij}^{6}}\right)\mathrm{sw}\left(r_{ij}^2 r_{\mathrm{on}}^2, r_{\mathrm{off}}^2\right) \tag{12.21}$$

In the CHARMM force, hydrogen bonding interaction energies are computed by the expression form as follow:

$$E = \left(\frac{A}{r_{\mathrm{AD}}^6} - \frac{A}{r_{\mathrm{AD}}^9}\right)\cos^m(\varphi_{\mathrm{A-H-D}})\cos^n(\varphi_{\mathrm{AA-H-D}})\mathrm{sw}\left(r_{\mathrm{AD}}^2, r_{\mathrm{on}}^2, r_{\mathrm{off}}^2\right) \times \mathrm{sw}\left[\cos^2(\varphi_{\mathrm{A-H-D}}), \cos^2(\varphi_{\mathrm{on}}), \cos^2(\varphi_{\mathrm{off}})\right] \tag{12.22}$$

where sw is defined as a switching function, and it is used to control the range of the hydrogen bonding interaction. The subscripts on and off indicate the start and termination point to calculate the bond lengths and angle values relating to hydrogen bonds in this function.

Force fields in themselves are not correct forms. If the performance of one force field is better than another one, it should be desirable. According to selected different simulation unit, the force field can be divided into all-atom models such as OPLS-AA and united-atom models such as OPLS-UA model.

12.2.2.4 Polarizable Force Field

As a well-established technique, additive force field has its intrinsic limitations. For systems involving a frequent and large change of electrostatic environment, such as the passage of small molecules or ions through lipid membrane bilayers, or the binding of substrate to the hydrophobic interior of an enzyme in water solution, the electron distribution change of those molecules can hardly be reflected by the fixed charge model. Thus, extended descriptions of the electrostatic interactions have been proposed to add polarization effects into the force field.

In general, several theoretical models have been developed to treat the polarization explicitly during the MD simulations: (1) The fluctuating charge/charge equilibration model; (2) Drude oscillator model, which is used in the CHARMM Drude FF [64]; (3) Induced dipole model, which has been implemented in the development of AMOEBA force field. Basic concepts of these models as well as their strengths and weaknesses will be described below.

Fluctuating Charge Model. The Fluctuating charges (FQ) model [65] treats the charges on the atoms as dynamical variables, and the topology can vary during the MD simulations. FQ model is based on the principle of electronegativity equalization: charges can redistribute among atoms until instantaneous electronegativities are equalized, though the overall charge on the whole molecule is maintained [66]. The charge distribution can be derived from Taylor series expansion of the energy required to create a charge (q_a) on an atom (a) to the second order:

$$U_{\mathrm{ele}} = E_{a0} + \chi_a q_a + \frac{1}{2} J_{aa} q_a^2 \tag{12.23}$$

Here E_{a0} is the electrostatic energy with zero charge being created ($q_a = 0$). χ_a is the "Mulliken electronegativity" [67] and J_{aa} is the "absolute hardness" [68].

Considering a system of multiple atoms, the charges are also placed on centres of the atoms, and the electrostatic interactions between atoms must also be counted and expressed by the column's law. Thus, the total electrostatic energy of a system containing N atoms can be expressed as:

$$U_{\mathrm{ele}} = \sum_{a=1}^{N} (E_{a0} + \chi_a q_a + \frac{1}{2} J_{aa} q_a^2) + \sum_{a=1}^{N} \sum_{b>a} J_{ab}(r_{ab}) q_a q_b \tag{12.24}$$

Here, the Coulomb potential $J_{aa}(r_{ab})$ [69] between unit charges on atoms a and b separated by a distance r_{ab}, can be written as:

$$J_{ab}(r_{ab}) = \frac{\frac{1}{2}(J_{aa} + J_{bb})}{\sqrt{1 + \frac{1}{4}(J_{aa} + J_{bb})^2 r_{ab}^2}} \tag{12.25}$$

$J_{ab}(r_{ab})$ becomes equal to $r_{ab} - 1$ with a large distance ($r_{ab} > 2.5$ Å), making this component equal to that of a traditional non-polarizable force field.

Practically the extended Lagrangian method [70] can be applied with the charge on each atom being treated as dynamic particles and a "fictitious" mass being assigned to these particles, while the positions of atoms are propagated based on Newton's equations of motion. The force on each charge is equal to the deviation of its own electronegativity to the averaged one. The Lagrangian strategy is also used in other polarizable models due to the requirement to perform self-consistent field calculations.

Compared with other models, one significant advantage of the FQ model is that the number of interactions being calculated is not increased, however, there are also some drawbacks. FQ model may cause non-physical charge distribution among atoms with large separations. Thus, FQ model leads to a non-physical charge distribution between infinitely separated atoms. This error exists in dealing with large polymers, making the polarization increase fast along the polymer chain [71]. To solve this problem, some variations of the FQ model, including atom-atom charge transfer (AACT) [72] and bond-charge increment (BCI) [73, 74] have been developed by restricting charge distributions between directly bonded atoms. However, these approaches cannot reproduce out-of-plane polarization in planar systems, such as benzene [75]. Besides, over polarization may also be resulted from an intrinsic reduction of polarizability in the condensed phase. As an example, extreme charges on polar atoms was obtained in condensed-phase simulations using electrostatic parameters derived from gas-phase experimental results [71]. To compensate for the over-polarization effect, hardness and electronegativity can be scaled or treated as a function of atom charge [76]. On the basis of the electronegativity equalization method, a modified electronegativity equalization method (MEEM) was developed by Yang et al. [77, 78], and the method was further developed to the atom-bond electronegativity equalization method (ABEEM), which allows more accurate estimation of electron distribution and electrostatic energy of large molecules.

CHEQ model has been successfully performed on the investigation of proteins, ion solvation, etc. [76]. Patel and co-workers have developed a polarizable force field for dimirystoylphosphatidylcholine (DMPC) and dipalmitolphosphatidylcholine (DPPC) based on the charge equilibration (CHEQ) force field approach [8, 79]. The CHEQ force field has been applied to the studies of bilayers and monolayers of lipids, as well as membrane-bounded protein channels, such as gramicidin A [80]. Taking the water permeation for example [79], the simulations using the polarizable force field showed higher permeation than the results with non-polarizable models. It was suggested that fixed-charge force field could not produce the expected dielectric property of the nonpolar hydrocarbon region, and water molecules in membrane interior had large dipole moments similar to the waters in the bulk [81].

Drude Oscillator Model. In the Drude oscillator model [82], a polarizable point dipole is introduced to each atom by connecting to a Drude particle with a harmonic spring, which is a direct extension of additive force field. A "core" charge and an opposite "shell" charge are assigned to the parent atom and the Drude particle [83] to maintain the normal atom charge state, that is to simulate the induced polarization via its displacement under the influence of an electric field.

Thus, the dipole moment of this two-particle system in presence of an electric field (E) can be expressed as:

$$\mu_i = q_i d_i = \frac{q_i^2 E}{k} \tag{12.26}$$

Here, the induced dipole moment μ_i is dependent on the charge (q_i) of Drude particle and the spring distance d_i, which is controlled by the spring force constant k. Both q_i and k are adjustable parameters in this model. In MD simulation, initial positions of Drude particles can be achieved by energy minimization, with positions of atoms being fixed. Then those Drude particles will be involved in the simulation to dynamically get the corresponding dipole moments. Contributions to the total electrostatic energy from these induced dipoles can be separated into three parts: the interaction with static fields (charges, dipoles, etc.), induced dipole-induced dipole interaction, and polarization energy:

$$U_{\mathrm{ind}} = U_{\mathrm{stat}} + U_{\mu\mu} + U_{\mathrm{pol}} \tag{12.27}$$

Here U_{stat} and $U_{\mu\mu}$ can be calculated by Coulomb's law, and the polarization energy is equal to the spring potential, which is:

$$U_{\mathrm{pol}} = \frac{1}{2}\sum_{i=1}^{N} k_i d_i^2 \tag{12.28}$$

Different from additive FFs, electrostatic interactions between bonded atoms are included to obtain the correct molecular polarization response. In the Drude particle model, electrostatic interactions can be treated similarly to charge-charge interactions. However, adding Drude particles greatly increases the computational cost, thus in fact only heavy atoms are attached to Drude particles.

Drude oscillator based on polarizable force field has also been developed [84, 85], which includes a board classes of molecules such as proteins [86], carbohydrates [87–89], and DNA [90, 91]. And the parameters of RNA is close to completion [92]. Furthermore, Drude force fields of DPPC [17], cholesterol [93], and sphingomyelin [93] have been established recently. In all the simulations using Drude force field, the description of the membrane dipole potential has been improved as a result of the inclusion of atomic polarizabilities.

Induced Dipole Model. When accounting for higher-order contributions approximately via a modification of additive force field, the idea of explicit treatment of first-order induction was introduced [94]. The induced dipole model is implemented in FFs such as AMBER ff02 [95], AMOEBA (Atomic Multipole Optimized Energetics for Biomolecular Applications) [96] etc. AMOEBA [96] was developed by Ren and Ponder. The electrostatic energy in AMOEBA includes contributions from both permanent and induced multipoles. Permanent electrostatic interactions are computed with higher order moments where

$$M_i = \left[q_i, d_{ix}, d_{iy}, d_{iz}, Q_{ixx}, Q_{ixy}, Q_{ixz}, Q_{iyx}, Q_{iyy}, Q_{iyz}, Q_{izx}, Q_{izy}, Q_{izz}\right]^T \tag{12.29}$$

is a multipole composed of charge, q_i, dipoles, $d_{i\alpha}$, and quadrupoles, $Q_{i\alpha\beta}$. The interaction energy between two multipole sites is

$$U_{\mathrm{emp}}^{\mathrm{perm}} = \begin{bmatrix} q_i \\ d_{ix} \\ d_{iy} \\ d_{iz} \\ Q_{ixx} \\ \vdots \end{bmatrix}^T \begin{bmatrix} 1 & \frac{\partial}{\partial x_j} & \frac{\partial}{\partial y_j} & \frac{\partial}{\partial z_j} & \cdots \\ \frac{\partial}{\partial x_i} & \frac{\partial^2}{\partial x_i \partial x_j} & \frac{\partial^2}{\partial x_i \partial y_j} & \frac{\partial^2}{\partial x_i \partial x_j} & \cdots \\ \frac{\partial}{\partial y_i} & \frac{\partial^2}{\partial y_i \partial x_j} & \frac{\partial^2}{\partial y_i \partial y_j} & \frac{\partial^2}{\partial y_i \partial y_j} & \cdots \\ \frac{\partial}{\partial z_i} & \frac{\partial^2}{\partial z_i \partial x_j} & \frac{\partial^2}{\partial z_i \partial y_j} & \frac{\partial^2}{\partial z_i \partial z_j} & \cdots \\ \vdots & \vdots & \vdots & \vdots & \ddots \end{bmatrix} \frac{1}{R_{ij}} \begin{bmatrix} q_j \\ d_{jx} \\ d_{jy} \\ d_{jz} \\ Q_{jxx} \\ \vdots \end{bmatrix} \tag{12.30}$$

Since induced dipoles are introduced to represent polarization, the charge on each atom can be directly derived from experimental values in gas phase or high-level QM calculations. This is even more straightforward than the approach in additive force fields, in which partial charges are assigned to atoms to represent the polarization effects. However, this approach suffers from an important issue: polarization catastrophe. Thole [97] developed a series of approaches to solve this problem by mimicking smeared charge distributions between atoms of short distances using a set of fitting functions. In this way, the dipole field tensor, T_{ij}, is modified so that it is not approximated with $r_{ij} - 3$ with a small atom separation. Outlined by Thole [97], if two dipoles are close to each other, the induced dipole calculated from the equation will be unphysically amplified. Thus, damping methods are important when dealing with dipole-dipole interactions with short distances.

At a very close distance, when the electron clouds overlap, the multipole approximation becomes inadequate. In 2015, the penetration effects were introduced into AMOEBA force field [98], and the method proposed by Piquemal et al. [99] was revisited. The charge of an atom is divided into a core and an electron cloud, and therefore the total electrostatic energy between two atoms can be calculated as three components, core-core, core-electron, and electron-electron interactions. The electrostatic energy can be written as

$$E_{qq}(r) = \left[\begin{array}{c} Z_1Z_2 - Z_1(Z_2 - q_2)(1 - \exp(-\alpha_2 r)) \\ -Z_2(Z_1 - q_1)(1 - \exp(-\alpha_1 r)) \\ +(Z_1 - q_1)(Z_2 - q_2)(1 - \exp(-\beta_1 r))(1 - \exp(-\beta_2 r)) \end{array} \right] \Big/ r \tag{12.31}$$

where γ is the distance between two atoms, Z is the positive core charge, q is the net charge of the atom, $(Z - q)$ can be considered as the electron cloud, and the α and β are two parameters controlling the magnitude of the damping of the electron cloud when the atom is interacting with the core and with electrons from other atoms. The α is intuitively set to the same as the number of valence electrons. When the distance between two atoms increases, Eq. 12.9 will reduce to additive Coulomb law. Thus, in the medium and long distances, the electrostatic energy is still calculated via multipole expansion accurately, and the penetration diminishes rapidly with distance, It is worth to remark that the penetration is only significant when distance is shorter than the sum of atomic van der Waals radii. The results of this method show the polarization response using perturbation theory rather than a variational approach to achieve the SCF condition, and produce an improvement in computational efficiency.

12.2.2.5 MARTINI Coarse-Grained (CG) Model

Although there are various coarse-grained (CG) approaches available, MARTINI model, developed by the groups of Marrink and Tieleman [100], is actually one of the most successful and broadly utilized CG force fields. The MARTINI model was initially developed to study the self-assembly and fusogenicity of small lipid vesicles in 2004 and later extended to investigate the interactions between membrane proteins and their lipid environments. Currently, the MARTINI force field provides parameters for a variety of biomolecules and materials, including the majority of lipid molecules, cholesterol, all native amino acids, carbohydrates, nucleotides, fullerene, polymers, and surfactants.

As a CG model, MARTINI adopts a four-to-one coarse-grained mapping scheme to reduce the resolution of the representation of a system. Four heavy atoms from all-atom models are represented by a single CG bead to discard degrees of freedom (DoF) of the system by assuming that the dynamic behavior of a given system is less strongly associated with those DoF. The ring-like molecules (e.g. benzene, cholesterol, and several of the aromatic amino acids) are mapped with higher resolution (up to two-to-one). The Martini model averages atomic properties to chemical entities and neglects individual atoms. A total of four main types of sites: polar (P), non-polar (N), apolar (C), and charged (Q) are defined to account for the interactions of a system. The parameterizations of non-bonded interactions of the chemical building blocks are extensively calibrated against thermodynamic data such as oil/water partitioning coefficients using a Lennard-Jones (LJ) 12-6 potential. In addition to the LJ interaction, charged groups (type Q) bear a charge $\pm e$ and interact via a Coulombic energy function. Coulombic interactions are screened implicitly with a relative dielectric constant $\varepsilon_{\mathrm{rel}} = 15$ to account for the reduced set of partial charges and resulting dipoles that occur in an atomistic force field.

Bonded interactions are described by a standard set of potential energy functions that are common in classical force fields, including harmonic bond, angle potentials, and multimodal dihedral potentials. Proper dihedrals are primarily used to impose secondary structure on the peptide backbone. Improper dihedrals are mainly used to prevent out-of-plane distortions of planar groups. LJ interactions between nearest neighbors are excluded. The detailed parameterization process of MARTINI force filed can be found in Ref. [100].

Building a membrane protein system of interest using the Martini force field can be fulfilled by CHARMM-GUI Martini Make [101]. CHARMM-GUI [102] is a web-based graphical user interface to generate various molecular simulation systems and input files for major MD engines (e.g. CHARMM, NAMD, GROMACS, AMBER, and OpenMM programs) to facilitate and standardize the usage of common and advanced simulation techniques. By taking advantages of the frameworks in all-atom CHARMM-GUI modules, Wonpil Im and co workers [103] recently have provided a convenient interface to build complex bilayers, micelles, vesicles, and more, with proteins embedded, which supports the force field including martini, martini with polarizable water, dry martini, and ElNeDyn (an elastic network model for proteins).

12.3 Applications of Computer Simulations

12.3.1 Coarse-Grained Molecular Dynamic Simulation Case Study

12.3.1.1 Binding Sites Between Cholesterol in β_2-Adrenergic Receptor

In the human genome, the integral membrane proteins represent a larger portion. Among them are G-protein coupled proteins (GPCR), which own the seven transmembrane domains and have over one thousand members, comprising the largest membrane protein family [104]. GPCRs primarily participate in the transduction of signals across the plasma membrane through their response to diverse extracellular environment, such as light, peptides, small molecules, protons, etc. Therefore, GPCRs are the major targets for the development of novel drug candidates in all clinical areas [105].

According to the sequence alignment, the GPCRs have been divided into five classes [104, 106]. β_2-adrenergicreceptors (β_2AR) belong to the class A receptors, which can be further divided into groups associated with ligand specificities, such as the opsin, amine, peptide, cannabinoid, and olfactory receptors [107]. GPCRs obviously take part in many physiological processes, which contain the neurotransmission, cellular metabolism, secretion, cellular differentiation, growth, inflammatory and immune response [108]. The adrenergic receptor modifications are associated with various diseases, such as asthma, hypertension, and heart failure [109]. β_2AR is one of the best-characterized GPCRs, and expressed in pulmonary and cardiac myocyte tissue [110, 111]. β_2AR can modulate the signal in the erythrocytes during the malarial infection [108, 112].

The cellar membrane can partly functionally module a lot of membrane proteins [113–115], and the functional modulation is associated with the physical or chemical interactions between the phospholipids, sphingolipids, and cholesterol etc. [116]. Cholesterol is an essential component of eukaryotic membranes and plays a critical role in membrane organization, dynamics, and functions. The equilibrium state of the proteins is sensitive to the presence and amount of the cholesterol [116]. Increasing the amount of cholesterol in the membrane moves the equilibrium to the inactive conformation of the proteins. Some works have found that cholesterol can modulate the physiological function of GPCRs, and it is high associated with the kinetic, energetic and mechanical stability of the β_2AR [108]. Moreover, a study has shown that cholesterol seems to be helpful in crystallizing β_2AR [107]. In 2007, Cherezov et al. published an X-ray crystallography model of human β_2AR [107], and the model showed that cholesterol bound to the surface formed by α-helices H1, H2, H3, H4 and H8. Compared to rhodopsin, the ligand-binding pocket was formed by structurally conserved and divergent helices, which was also found to be present in most class A GPCRs. The observation of this complex structure formed by β_2AR and cholesterol suggested a possible interaction between them. In addtion, Zocher et al. [116] found that cholesterol considerably increased the strength of interactions stabilizing structural segments of β_2AR, and the interactions increased the stability of all the structure segments of β_2AR except for the structure core segment and the binding of cholesterol. From the results, they speculated that the structural properties of the GPCRs in the presence of cholesterol might cause GPCRs to respond differently to environmental changes. Therefore, the amount of cholesterols that bind to GPCRs is important in the research.

To show the dynamics, functions and interaction energies of the binding cholesterol to β_2AR, a series of microsecond (μs) level coarse-grained (CG) molecular dynamics (MD) including 8 μs CG MD simulations on β_2AR embedded in DOPC, 1:1 DOPC/cholesterol, 3:1 DOPC/cholesterol, 6:1 DOPC/cholesterol mixture membrane were conducted, and β_2AR is modeled via the Martini models combined with the elastic network [117], which conserve the tertiary and quaternary structures more faithfully without sacrificing realistic dynamics of a protein. The results of the simulations validated the cholesterol binding site with the crystal structure in β_2AR and the interaction energy of the β_2AR in different scale of DOPC/cholesterol mixture membrane.

12.3.1.2 Methods

Simulation Systems. Four systems of β_2AR monomer, which embedded in the four different scales DOPC/cholesterol were constructed for the CG MD simulations. The model of β_2AR monomer was designed to reproduce the shape, surface polarity and dynamics of the β_2AR monomer as reported by the 3D4S crystal structure without the binding of cholesterol [118]. In the crystal structure of β_2AR, the intracellular loop, which was located between the Helix 5 and Helix 6 and connected them, was lost, and the loop deletion was left [118]. The ELNEDIN term was used in the whole simulation progress, and the elastic network was used as the structure scaffold to describe and maintain the overall shape of β_2AR. The ELMEDIN models are comparable to the atomistic protein models, and they can build good results of structure and dynamic properties of proteins, including the collective motions [117]. In the simulations, the ElNeDyn term is selected. The topology options set the elastic bond force constant to 500 kJ mol^{-1} nm^{-2} (-ef 500) and 200 kJ mol^{-1} nm^{-2} (-ef 200), and an upper bond length cut-off 0.9 nm.

The different scales of the mixed complexes of β_2AR and DOPC/cholesterol were performed CG MD simulations to validate the cholesterol binding site with the crystal structure in β_2AR and the interaction energy of the β_2AR in different scales of DOPC/cholesterol mixture membrane [119]. The system β_2AR monomer in the DOPC/cholesterol mixture membrane contained one β_2AR monomer embedded in DOPC lipid bilayer which DOPC/cholesterol scales are 1:0 DOPC, 1:1 DOPC/cholesterol, 3:1 DOPC/cholesterol, 6:1 DOPC/cholesterol. The β_2AR monomer embedded in DOPC/cholesterol mixture membrane was built by the CHARMM-GUI, and the total amount of DOPC and cholesterol is 512. The cholesterol molecules were randomly dispersed in the DOPC lipid bilayer. Then

Table 12.1 The detail information of the above simulation systems

Systems	DOPC number	Cholesterol number	Elastic bond force constant (kJ mol^{-1} nm^{-2})	Time (μs)
β_2AR_DOPC_cholestrol_11_200	169	226	200	1
β_2AR_DOPC_cholestrol_11_500	169	226	500	1
β_2AR_DOPC_cholestrol_31_200	264	95	200	1
β_2AR_DOPC_cholestrol_31_500	264	95	500	1
β_2AR_DOPC_cholestrol_61_200	306	53	200	1
β_2AR_DOPC_cholestrol_61_500	306	53	500	1
β_2AR_DOPC_200	349	0	200	1
β_2AR_DOPC_500	349	0	500	1

β_2AR inserted into the mixture membrane using the GROMACS. The details information of the above simulation systems are listed in Table 12.1.

CG MD Simulations. CG MD simulations were carried out with the GROMACS 4.5.3. The martini_2.1 force field was employed for all CG MD simulations, and the force field parameters of martini_2.2_lipids and martini_2.0_cholesterol were applied to the DOPC/cholesterol mixture bilayer [120]. During the simulations, all bonds were constrained using the LINCS algorithm, and the integration time step was set to 20 fs. The particle mesh Ewald (PME) method was employed to treat long-range electrostatic interactions, and a cut-off value of 12 Å was used for non-bonded interactions.

Prior to the MD runs, all systems were minimized to remove the conflicting contacts. Then, the systems were heated to 300 K within 1 ns. Each system was equilibrated for a further 1 μs with the constraint only imposed on the protein. The NPT simulation was performed and periodic boundary condition.

Spatial Distribution of Cholesterol around β_2AR. Spatial distribution function (SDF) was used to reveal potential cholesterol-binding sites on the β_2AR surface, and the SDF of the cholesterol molecules around β_2AR was calculated as the 3D spatial distribution function of cholesterol model [121]. The SDF was calculated using the last 0.35 μs MD trajectory of β_2AR in different scales of DOPC/cholesterol mixture membrane through the *g* spatial module in the Gromacs package. In general, the SDF reflects the average 3D density distribution of cholesterol CG models. Therefore, the peaks of SDF imply the locations where cholesterol molecules reside with a higher probability as previous paper described.

12.3.1.3 Spatial Distribution of Cholesterol Molecules Around β_2AR in Different Scales of DOPC/Cholesterol Mixture Membrane

The binding sites of cholesterol around β_2AR (PDB code 2RH1) [107] are shown in Fig. 12.1. The SDF of cholesterol is displayed as isosurfaces around the surface of 2RH1. Three cholesterol molecules are binding to 2RH1: one binds to the surface of

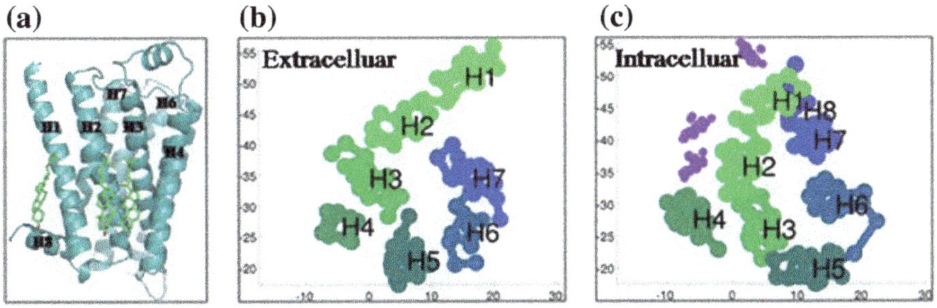

Fig. 12.1 Spatial distribution of cholesterol around the crystal structure of β_2AR (PDB code 2RH1). **a** The crystal structure of β_2AR. β_2AR is shown in cartoon and colored in teal, and the cholesterol molecules are shown in sticks and colored in green. **b**, **c** Two-dimension projections of SDFs on the upper and lower membrane planes. The purple sites are the cholesterol binding sites of β_2AR (PDB code 2RH1)

helix H1 and H8, and the other two bind to the surface which is constructed by helix H1, H2, H3, H4.

The SDF also expresses in the average spatial distribution of β_2AR in DOPC/cholesterol membrane, which is based on the upper and lower planes. The membrane is divided into extracellular part and intracellular part. The analysis clearly shows the high cholesterol density site is in the surface of the helix of β_2AR.

From Fig. 12.2, there are several higher cholesterol distributions in the surface of β_2AR in the 1:1 DOPC/cholesterol mixture membrane. In this process, the elastic

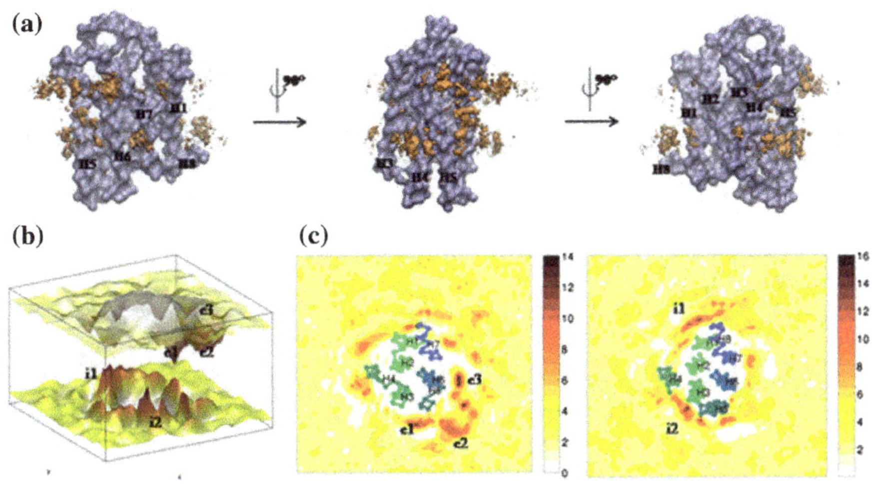

Fig. 12.2 The spatial distribution of cholesterol around β_2AR in the 1:1 DOPC/cholesterol membrane, and the elastic bond force constant of β_2AR is set to 200 kJ mol^{-1} nm^{-2}. **a** The SDF of cholesterol is shown in isosurface, and the average structure though the simulation is fitted to the isosurface. **b** The surface plot of the average SDF of cholesterol beads. **c** The 2D projections of the SDF around the helix of β_2AR

bond force constant of β_2AR is set to 200 kJ mol^{-1} nm^{-2} (β_2AR_DOPC_cholestrol_11_200). And all cholesterol located near the helix of β_2AR. In the intracellular part, the first peak is located on the surface of H1 and H8, which is consistent with the crystal structure of β_2AR (Fig. 12.1c). Another peak exists near the surface constructed of H3, H4 and H5. In the extracellular part, almost all the peaks distribute in the surface of H4, H5 and H6.

From Fig. 12.3, there are more cholesterol distribution peaks in the surface of β_2AR. In this process, the elastic bond force constant of β_2AR is set to 500 kJ mol^{-1} nm^{-2} (β_2AR_DOPC_cholestrol_11_500). In the intracellular part, the first peak is located on the surface of H1 and H8, which is consistent with the crystal structure of β_2AR and the results of upon simulation. The elastic bond force is set to 200 kJ mol^{-1} nm^{-2} (Fig. 12.2c). The second peak exists near the surface constructed of H2, H3 and H4. Almost press closes to H4. In the crystal structure of 2RH1, there is one cholesterol molecule hold in the similar position. Also in the intracellular part, there are three small peaks. These three peaks are near the H3, H6 and H7. And in the extracellular part, most of the peaks distribute in the surface of H1, H5, H6 and H7.

To the β_2AR in the 3:1 and 6:1 DOPC/cholesterol mixture membrane, the results are shown in Table 12.2.

A potential cholesterol-binding site detected by the CG MD is located in the interface of H1 and H8, which is detected by all the simulations, and it is consistent with the crystal structure 2RH1. The results demonstrate that MD simulations could

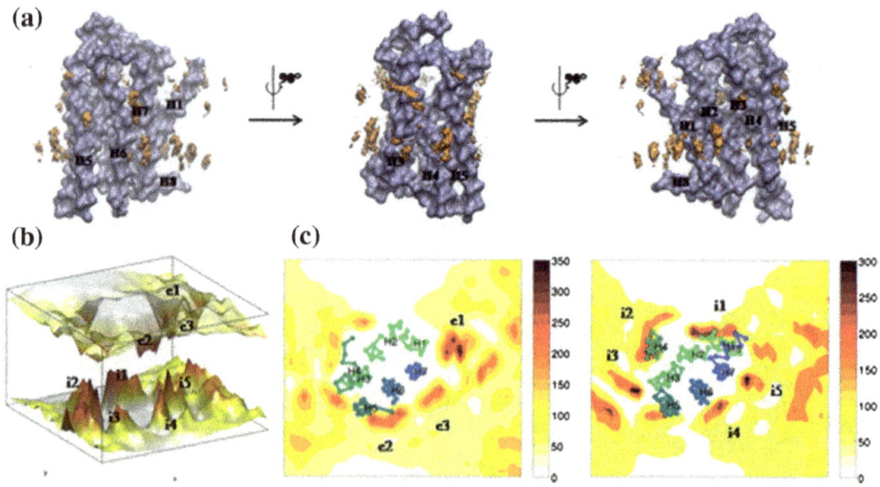

Fig. 12.3 The spatial distribution of cholesterol around β_2AR in the 1:1 DOPC/cholesterol membrane, and the elastic bond force constant of β_2AR is set to 500 kJ mol^{-1} nm^{-2}. **a** The SDF of cholesterol is shown in isosurface, and the average structure though the simulation is fitted to the isosurface. **b** The surface plot of the average SDF of cholesterol beads. **c** The 2D projections of the SDF around the helix of β_2AR

Table 12.2 Interaction energies of the eight helixes in different conditions (KJ/mol)

	β_2AR_DOPC_ chol_11_200	β_2AR_DOPC_ chol_31_200	β_2AR_DOPC_ chol_61_200
H1-else	66.10	45.54	65.85
H2-else	113.83	151.49	153.23
H3-else	7.87	33.43	129.02
H4-else	−40.89	−17.24	2.09
H5-else	67.41	5.88	172.75
H6-else	37.15	35.98	156.16
H7-else	78.72	44.55	92.39
H8-else	18.48	11.67	41.11

be employed to reproduce the binding model of cholesterols to β_2AR. The cholesterol can mediate the dimeric structure, which has been reported in 2007. However, skepticism still exists: whether it is a physiologically relevant form or just a crystal packing artifact because the two cholesterol molecules that mediate the dimer are located on the crystal packing interface. Another potential cholesterol-binding site in the intracellular surface is also detected in most simulations except the systems of β_2AR_DOPC_cholestrol_11_200 and β_2AR_DOPC_cholestrol_61_500, which locates the interface of H2, H3, and H4.

The interaction energies of the eight helixes, which the elastic bond force constant of β_2AR is set to 200 kJ mol^{-1} nm^{-2}, are shown in Table 12.2. It can be seen that when the density of cholesterol increases, the interaction energies between H3, H4 with other helixes become weaken. The results suggest that the cholesterol increasing affect the conformation of H3 and H4 β_2AR obviously.

12.3.2 *All-Atom Molecular Dynamic Simulation Case Study*

12.3.2.1 Calcium Facilitated Chloride Permeation in Bestrophin

Calcium-activated chloride channels (CaCCs) perform a variety of physiological roles in regulating photo-transduction, olfactory transduction, vasculartone, epithelial electrolyte secretion and neuronal and cardiac excitability [122]. Despite their broad distribution and important functions [123], the molecular identify of CaCCs remains cloudy. Significant progress has been made in recent years to identify the family members of CaCCS. Three groups of proteins (TMEM16, LRRC8 and bestrophins) have been regarded as CaCCs so far [124]. However, only bestrophin was demonstrated to have a chloride conducting pore, while the formation of anion channels by TMEM16 and LRRC8 was just indirectly evidenced.

Human Bestrophin 1 (hBest1) is highly expressed at the basolateral surface of retinal pigment epithelial (RPE) cells to regulate retinal homeostasis [125]. Mutations in hBest1 cause multiple retinal degeneration disorders, typically the autosomal dominant vitelliform macular dystrophy (Best disease) [126]. The chloride channel activity of hBest1 is stimulated by the intracellular calcium with a K_d of 150 nM [127]. Although there is evidence indicating that the activation is directly regulated by the binding of Ca^{2+} at the cytosolic region of the protein, it is still unclear how Ca^{2+} participates in gating the channel.

Recent available X-ray structures of chicken BEST1 (Best1cryst) [128] and the bacterial homolog KpBest1 [129] open up a new avenue in understanding the mechanisms of calcium facilitated chloride permeation and selectivity of the bestrophin family. The chicken BEST1 shares 74% sequence identity with hBest1 and the protein assembles in a form of symmetrical homo-pentamer around a central axis. A single ~95 Å long, continuous ion pore located along the central axis of the protein forms the anion permeation pathway with a narrow necked fined by the conserved hydrophobic residues Ile76, Phe80 and Phe84 of each subunit. Mutations in the neck region significantly influence the channel property. Especially, the I76E mutation in hBEST1 flips the ion selectivity to Na^+ and the mutations of F80E and F84E impair the Cl^- permeability [129]. Below the neck, the pore opens a large inner cavity with a maximum radius of 10 Å and ~45 Å long at the cytosolic region, in which Ca^{2+} might be accommodated. At the bottom of the channel's cytosolic region, there is an aperture surrounded by Val205 (Ile 205 in hBes1). Replacing Ile205 by Threonine in hBest1 significantly decreased the chloride conductance [130], suggesting the important role of the aperture to contribute the anion selectivity.

Another prominent feature of the X-ray structure of Best1cryst is that each subunit has a strong Ca^{2+} binding cavity comprised by the acidic cluster (Glu300, Asp301, Asp302, Asp303 and Asp304). The coordination of Ca^{2+} in Best1cryst is similar to those observed in the EF hand domains [131] and the Ca^{2+} bowl' of the BK potassium channel [132]. The Ca^{2+} clasps formed by the acidic cluster resemble a pentagonal geometry and locate at the midsection of the channel, near the membrane–cytosol interface. Mutations around the Ca^{2+} clasp in hBEST1 impair the interactions between the transmembrane domains and the cytosolic domains [133], resulting in a dysfunctional channel.

Although approximately 200 distinct mutations in bestrophins have been identified to cause the retinal degenerative diseases [128] and most of the mutations lead to a dysfunction of the chloride channel, the molecular mechanism of Ca^{2+} dependent chloride channel activity of bestrophin is still not fully understood. Here, in order to gain a molecular insight of calcium facilitated chloride permeation along the channel of Best1cryst, all-atom MD simulations are utilized to compare the chloride permeation property of Best1cryst in the presence of Ca^{2+} and Na^+, respectively. The main purpose of this section is to illustrate how MD simulations could be employed to investigate the ion transporting process at the atomic level.

12.3.2.2 System Setup

The MD simulation systems were prepared using the recent available X-ray structure of chicken BEST1 (PDBid:4RDQ) [128]. The assembling of Best1cryst into the bilayer was employed using the CHARMM-GUI web server [134]. The co-crystallized Fab fragments were deleted and the Best1cryst was merged into a heterogeneous bilayer composed of 400 POPE/POPG lipids with a mixture ratio of 3:1 to mimic the experimental liposome condition [128]. The five Ca^{2+} ions coordinated by the acidic cluster in the Ca^{2+} clasps were retained during the system preparation. Then the systems were solvated with 41,959 TIP3P water molecules and the charges of the systems were balanced to neutral using 0.1 M $CaCl_2$ and 0.2 M NaCl, respectively. The systems containing ~207032 atoms were placed into an orthogonal box of 115 × 115 × 150 Å^3. All MD simulations were performed using Gromacs [135] 5.0.4 package with CHARMM 36 force field [136] under NPT condition. The leap-frog integrator [137] was used with an integration time-step of 2 fs. The calculation of electrostatic interactions was performed using the Particle-Mesh Ewald algorithm [138] with a cut-off of 1.2 nm. The same cut-off value was chosen for treating the van der Waals interactions. The semi-isotropic pressure coupling was employed using the Parrinello-Rahmanbarostat [139] to control the pressure at 1 bar with a coupling constant of 5 ps when production run was performed. The Nose-Hoover thermostat [140] was employed to couple the temperature of the systems around 303.15 K with a time constant of 1 ps.

After 50 ns MD simulations under NPT ensemble, the calculations of the PMFs along the reaction coordinate of chloride permeation of Best1cryst were performed using umbrella sampling technique. The initial conformations for the umbrella sampling simulations were obtained from the last frames of two 50 ns independent standard MD simulations with the ion concentrations of 0.1 M $CaCl_2$ and 0.2 M NaCl, respectively. The z distance between Cl^- and the Best1cryst's center of mass (COM) has been divided into 180 uniformly spaced bins with a length of 0.5 Å, which covers a distance of 90 Å. In the simulations of each window, the chloride anion was subjected to a harmonic potential with a spring constant of 6000 kJ/mol/nm^2, which is implemented using the PLUMED free energy calculation library [141]. A cylinder constraint was also applied if Cl^- shifted away larger than 8 Å from the COM of Best1cryst in the x–y plane. A 2 ns umbrella sampling MD simulation of each bin was conducted and the last 1.8 ns trajectories were used for the weighted histogram analysis. Then, the 1D potential of mean force (PMF) for the chloride permeation was estimated using the WHAM package [142] with a convergence tolerance of 10^{-6}.

12.3.2.3 Free Energy Profiles of Chloride Permeation in the Presence of Ca^{2+} and Na^+

The potential of mean force (PMF) profiles of chloride permeation in the presence of $Ca^{2+}$and Na^+ are compared to understand the mechanisms of Ca^{2+} facilitating

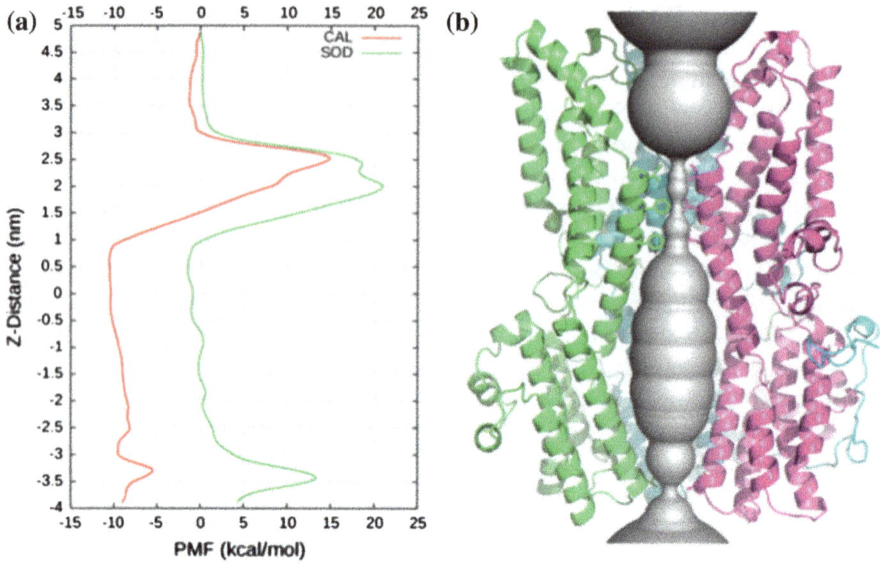

Fig. 12.4 **a** Potential of mean force (PMF) profiles for Cl^- permeating along the channel in the presence of Ca^{2+} (red) and Na^+ (green), respectively. **b** Ion pore along the Best1cryst

chloride permeation in bestrophins. As shown in Fig. 12.4a, both of the PMF profiles have two distinct free energy maxima, corresponding to the Cl^- anion permeating through the neck and the aperture. However, when Ca^{2+} is present, the free energy barriers are considerably reduced compared to those in the presence of Na^+, especially for Cl^- passing through the hydrophobic gate (z from 1.0 to 3.0 nm). Two peaks have been identified on the PMF profiles at the neck region, which correspond to the locations of the Cl^- anion at the pores defined by I76 (peak1) and F80 (peak2). The free energy barrier is lowered by 3 kcal/mol at peak1 (z = 2.5 nm) when Ca^{2+} is presence, whereas the free energy barrier is dramatically reduced about 12 kcal/mol by Ca^{2+} when Cl^- crosses the peak2 (z = 2.0 nm).

As Cl^- permeates further into the inner cavity from peak2, the PMF profile demonstrates a strong downhill character with the free energy difference is about 20 kcal/mol. This process corresponds to the Cl^- anion transiting from a partially dehydrated configuration to the fully hydrated state once entering the inner cavity. During the permeation of Cl^- in the inner cavity, the anion nearly faces no free energy barriers until it approaches the aperture defined by V205. The free energy barrier is about 5 kcal/mol when the Cl^- passes through the aperture in the presence of Ca^{2+}, while it changes to $\sim$13 kcal/mol when Na^+ is present.

12.3.2.4 The Energetic Barriers Raised by the Dehydration of Chloride

In Fig. 12.5, the average water coordination numbers along the anion permeation pathway are depicted to understand the causes of the large energetic barriers for Cl^- permeating. By comparing the water coordination profiles and the PMFs, it can be seen that the free energy maxima are directly raised by the partially dehydrated state of Cl^- (Fig. 12.4). The lower the coordination number, the higher the energetic compensation for Cl^- permeation, indicating that the hydration states of chloride are strongly correlated with the permeation energetic barriers. In bulk water, the average water coordination number of chloride is 8 when CHARMM force field [143] is used, and the absolute free energy of hydration of a chloride ion is −77.2 kcal/mol [144]. As shown in Fig. 12.5, Cl^- exhibits the lowest coordination number around 3.2 when passing through the hydrophobic gate (z = 1.5–3 nm), which explains the high free energy barrier at the neck region on the PMF profiles. In addition, the water coordination profile of Na^+ at the neck shows smaller coordination numbers than that of Ca^{2+}, also explaining the higher free energy barrier when Na^+ is present. Again, at the aperture, Cl^- exhibits a coordination number of 4 in the presence of Na^+ while Cl^- shows a larger average coordination number by one when Ca^{2+} is present, leading to a free energy height of ~7 kcal/mol for Cl^- permeating the aperture.

When Cl^- enters the inner cavity, the Cl^- ion is recovered to the fully hydrated state with the average coordination numbers around 7.5, which is similar to the anion-water interactions at the extracellular region (z = 3.5–5 nm). The occurrence of sudden jumps on the water coordination profiles from z = −3 to 0 nm indicate the interactions between Cl^- and the cations in the inner cavity, which reduces the number of water coordinating to Cl^-.

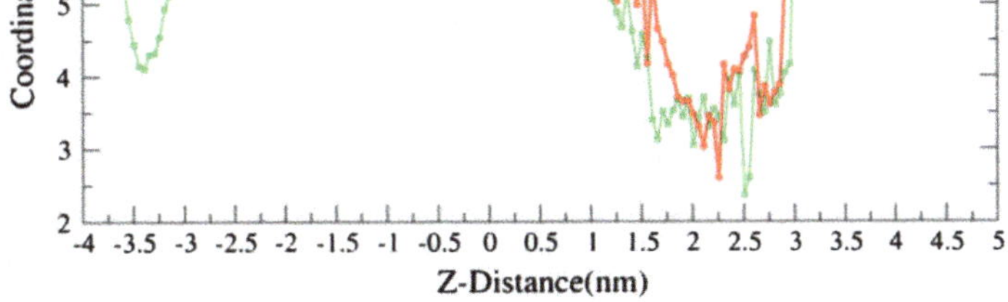

Fig. 12.5 Average water coordination numbers of Cl^- along the permeation pathway in the presence of Ca^{2+} (red) and Na^+ (green)

12.3.2.5 Ca^{2+} Binding Sites Along the Permeation Pathway

After further analyzing the MD trajectories, it can be found that, in addition to bind the Ca^{2+} clasp sites, Ca^{2+} can also tightly bind to the conserved acid residues (E74, E98, E213 and D203) along the anion permeation channel, whereas the stable binding of Na^{+} at those sites are not observed because of the weaker coulomb interactions between the monocation and the carboxyl groups of glutamate and aspartate. A snapshot shown in Fig. 12.6 demonstrates the interactions between Ca^{2+} and these carboxyl groups of the acid residues along the channel. It is worth noting that the interactions between Ca^{2+} and E74 of Best1cryst (Q74 in hBEST1), just locating above the hydrophobic filter, may play an essential role in gating the channel. Because of the narrowness of the region just above the neck, five Ca^{2+} ions may not be accommodated simultaneously at this site to bind the pentamer's five carboxyl groups belonging to E74. Alternatively, the binding of Ca^{2+} at this site adopts a triangular pattern. As shown in Fig. 12.6, only three Ca^{2+} ions would be tightly trapped here; two Ca^{2+} ions were grasped by each of two carboxyl groups and the rest Ca^{2+} was coordinated by the fifth carboxyl group. Such a binding fashion at this narrow region perfectly resolves the collision problem and prevents additional Ca^{2+} ions to bind. Moreover, two carboxyl groups grasping one Ca^{2+} ion, not only tightens the binding, but enhances the local concentration of Cl^{-} above the neck as well. Therefore, the permeation of chloride through the neck is facilitated if Ca^{2+} is present. On the contrary, the stable binding of Na^{+} to the carboxyl group of E74 is not identified, which explains the higher free energy barrier when Cl^{-} passes through the peak1 when Na^{+} is present.

There are three more Ca^{2+} binding sites at the cytosolic region along the permeation channel. Two of them locate in the inner cavity (E98 and E213) and one

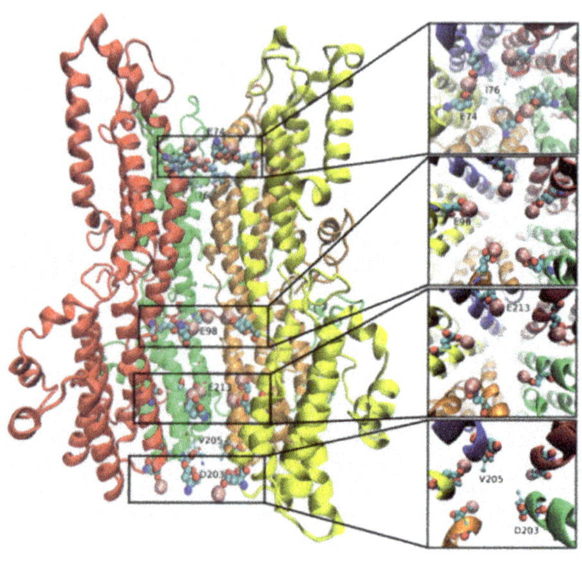

Fig. 12.6 Snapshot of Ca^{2+} binding to the conversed acid residues (E74, E98, E213 and D203) along the channel (left: side view, right: top view). Ca^{2+} ions are shown in wheat sphere, the acid residues are shown in van der Waals spheres and the gate residues (I76 and V205) are depicted in CPK. The figure was rendered using VMD [145]

locates at the bottom of the protein (D203). Each E98 and E213 of the pentamer traps one Ca^{2+} ion, leading to a high Ca^{2+} concentration in the cavity. This is the reason why the inner cavity of BEST might play as a Ca^{2+} reservoir to help accumulate and release Ca^{2+} from ER stores.

In addition, the binding of Ca^{2+} to D203 at the bottom of protein is not so tight. According to the MD trajectories, the Ca^{2+} ion binding to D203 will be frequently exchanged with the free Ca^{2+} in solution, indicating Ca^{2+} could be easily released at this site. Moreover, umbrella sampling simulations show that as Cl^- permeates from the inner cavity through the aperture to the bulk solution, a free Ca^{2+} ion plays the role of carrier. The residues E213 and D203 on the two sides of the aperture reduce the transporting barrier of the ions, explaining the free energy discrepancy for Cl^- passing through the aperture between Ca^{2+} and Na^+. This result is also in line with the hypothesis that bestrophin might conduct chloride as counter ion for Ca^{2+} uptake into cytosolic Ca^{2+} stores.

12.3.2.6 The Binding of Ca^{2+} Altering the Electrostatic Environment Along the Channel

In order to answer the question that why the free energy barrier decreases so dramatically (about 12 kcal/mol) for Cl^- passing through peak2 when Ca^{2+} is present, the Adaptive Poisson-Boltzmann Solver (APBS) package [146] in VMD [145] is employed to perform electrostatics calculations in the presence of Ca^{2+} and Na^+, respectively. In Fig. 12.7, 3D charge densities are compared in the presence of Ca^{2+} and Na^+, respectively. The result clearly shows the presence of Ca^{2+} radically changes the electrostatic properties along the channel. In the presence of Na^+, the extracellular region exhibits a favorable environment for positively charged ions, thus raising the free energy barrier for Cl^- passing through the neck. On the contrary, the binding of Ca^{2+} to E74 flips the electrostatic environment around the outer entryway to favor negatively charged ions and enhance the local anion concentration, and therefore faciliting the anion permeation. In the inner cavity, the presence of Ca^{2+} and Na^+ exhibits the same electrostatic properties. However, the binding of Ca^{2+} to E98 and E213 in the inner cavity dilates the charge densities to favor anions in the cavity (Fig. 12.7b), especially for the region just below the neck. When Na^+ is present, the charge densities are not seen below the neck. This is key evidence to explain why the free energy barrier for Cl^- passing through peak2 is dramatically reduced in the presence of Ca^{2+}.

At the bottom of the protein below the aperture, a small volume of positive charge density could be still identified when Na^+ is present, indicating the repulsion of the anions here. This result is also consistent with the PMF profiles that Cl^- would experience a higher energetic barrier passing through the aperture.

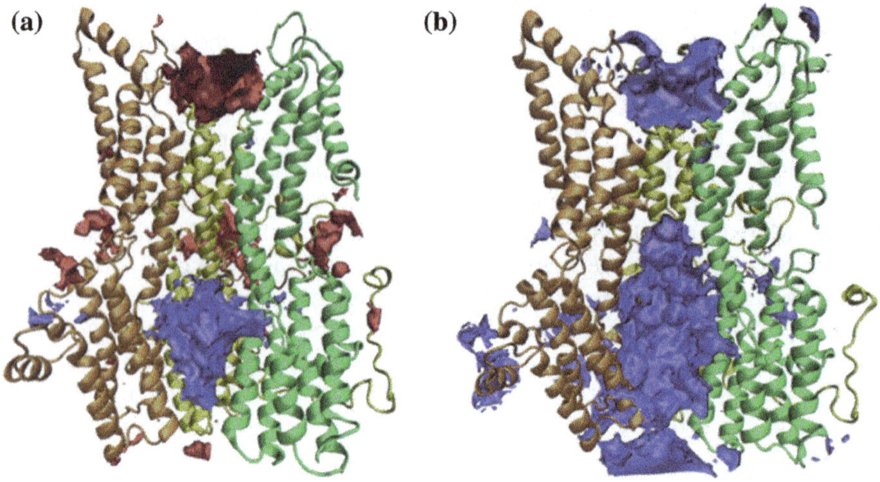

Fig. 12.7 Three dimensional charge densities along the anion permeation channel in the presence of Na^+ (**a**) and Ca^{2+} (**b**). The positive charge densities are depicted in red using charge density isovalue of +0.5 and the negative charge densities are depicted in blue using charge density isovalue of −1.2

12.4 Protocol

In this section, we briefly introduce the procedure of using CHARMM-GUI interface to build membrane protein systems for MD simulations.

CHARMM-GUI (http://www.charmm-gui.org) [102], is a web-based graphical user interface to prepare complex bio-molecular systems for molecular dynamic simulations. During the last decades, a range of capabilities has been consistently extended since its original announced in 2006 and now it contains a number of different modules designed to set up a broad range of simulations [147].

One of the most prominent features of CHARMM-GUI is that the interface would provide input files for a majority MD simulation engines such as CHARMM, NAMD, GROMACS, AMBER, LAMMPS, Desmond and OpenMM, and help users to build a sophisticated membrane/protein system easily and interactively. Here, as shown in Fig. 12.8, we illustrate the utilization of the Membrane Builder model to generate a protein/membrane system in six subsequent steps.

12.4.1 PDB Reading

Reading a PDB file is generally considered the first hurdle to initialize a simulation project. Since many PDB files may miss residues in loop region, especially for the membrane proteins, or introduce mutations to facilitate crystallization, therefore,

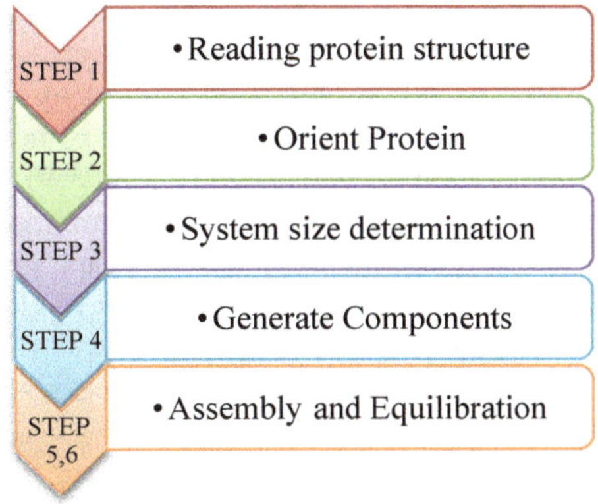

Fig. 12.8 A schematic workflow of the six subsequent steps

before uploading the PDB file in step 1, it is highly recommended for the users to convert the PDB file into sequence format (such as using pdb2fasta) and then blasting [148] the sequence on NCBI to check the sequence completeness of the structure. The missing loop region and mutations can be completed and recovered using homology modeling package like MODELLER [149].

In addition, disulfide bonds, different protonation states of titratable residues, other post-translational modifications (such as phosphorylation, glycosylation, and lipid-tail linkers) may be easily handled in this step using PDB Reader and Manipulator.

12.4.2 Orient Protein

Generally, the PDB file of a membrane protein does not have proper information on relative disposition in a membrane bilayer. In Membrane Builder, users can place the protein appropriately in a lipid bilayer by aligning its principal axis or a vector between two specific C-alpha atoms with respect to the membrane normal. It is assumed that the membrane normal is parallel to the Z-axis and the center is located at $Z = 0$ Å. Users can either upload their own pre-oriented structure handled by external package like orient in VMD [145], or specify PDB entry ID of a database (PDB database [150] or OPM database [151]). Protein structures from OPM database are pre-oriented, therefore, users do not need any modification of the protein orientation.

12.4.3 Determine System Size

As of 2016, there has been 295 lipid types supported by Membrane Builder in the context of CHARMM additive Force Field including phosphoinositides, cardiolipin, sphingolipids, bacterial lipids, sterols, and fatty acids [102].

After alignment in the previous step, the protein cross-sectional along the *Z*-axis is calculated and the protein areas in the top and bottom lipid leaflets are used to determine the system size. Users can specify the type and the number of lipid molecules to build a homogeneous or heterogeneous system. If a user specifies the number of lipid molecules in a bilayer, the system size in *XY* is determined by a ratio of the *XY* dimension. It is recommended to have the same *XY* lengths, unless users have specific reasons. Because a membrane is allowed to have different types and amounts of lipid molecules for the lower and upper leaflets, the resulting lipid bilayer probably has a different system size in *XY* for each leaflet. To avoid such situations, proceeding to the next step is not allowed until the difference in area of each leaflet is less than the smallest surface area among the lipid molecules used for the lipid bilayer. Then, the size along the *Z* axis is determined by specifying the thickness of bulk water from the protein extent along *Z*. In the case of some membrane proteins or peptides that do not span the bilayer, the size along Z is determined by the specified water thickness from $Z = \pm 20$ Å, approximately from the lipid headgroup.

12.4.4 Build Components

In this step, Membrane Builder will generate individual components to fully solvate the protein, including lipid bilayer, bulk water, and counter ions. Any complex (homogeneous or heterogeneous) bilayer system can be generated by the so-called "replacement method" that first packs the lipid-like pseudo atoms, and then replaces them with lipid molecules one at a time by randomly selecting a lipid molecule from a lipid structural library. Using the replacement method, it generates nicely packed lipid molecules around a protein, although Membrane Builder provides an insertion method for limited homogeneous bilayer system building.

If the ion concentration is specified, the numbers of ions are determined by the ion-accessible volume and the total charges of the system are neutralized. The initial configuration of ions is then determined through Monte Carlo simulations using a primitive model, i.e., scaled Coulombic and van der Waals interactions.

12.4.5 Assembly

Each component generated in the previous step will be assembled here and this procedure will take minutes to hours depending on the system size. One of the most significant advantages of using the web environment is that, if a problem is found, users can go back and re-generate the whole system again before quitting the browser. Therefore, the visualization of the initially assembled structure is important to verify if the system is reasonable.

12.4.6 Equilibration

After assembly is accomplished, the equilibration must be performed to relax the uncorrelated initial system before MD production simulations. Membrane Builder provides six consecutive input files for widely used MD simulation engines such as CHARMM, NAMD, GROMACS, AMBER and OpenMM. To assure gradual equilibration of the initially assembled system, various restraints are applied to the protein, water, ions, and lipid molecules during the equilibration: (1) harmonic restraints to ions and heavy atoms of the protein, (2) repulsive planar restraints to prevent water from entering into the membrane hydrophobic region, and (3) planar restraints to hold the position of head groups of membranes along the *Z*-axis. These restraint forces are slowly reduced as the equilibration progresses. To warrant the successful equilibration, i.e., to avoid instability of dynamics integrations during equilibration, the NVT dynamics (constant volume and temperature) is used for the first and second steps with integration time step of 1 fs, and the NPAT (constant pressure, area, and temperature) dynamics for the rest equilibrations.

Acknowledgements This work was supported by the National Natural Science Foundation of China (No. 91430110 and 31370714).

References

1. Gullingsrud J, Schulten K (2004) Lipid bilayer pressure profiles and mechanosensitive channel gating. Biophys J 86:3496–3509
2. Pastor RW (1994) Molecular-dynamics and Monte-Carlo simulations of lipid bilayers. Curr Opin Struct Biol 4:486–492
3. Feller SE (2000) Molecular dynamics simulations of lipid bilayers. Curr Opin Colloid Interface Sci 5:217–223
4. Forrest LR, Sansom MSP (2000) Membrane simulations: bigger and better? Curr Opin Struct Biol 10:174–181
5. Jiang YX, Lee A, Chen JY, Ruta V, Cadene M, Chait BT, MacKinnon R (2003) X-ray structure of a voltage-dependent K+ channel. Nature 423:33–41
6. Riviere S, Challet L, Fluegge D, Spehr M, Rodriguez I (2009) Formyl peptide receptor-like proteins are a novel family of vomeronasal chemosensors. Nature 459:574–577
7. Lin JC, Duell K, Konopka JB (2004) A microdomain formed by the extracellular ends of the transmembrane domains promotes activation of the G protein-coupled alpha-factor receptor. Mol Cell Biol 24:2041–2051
8. Davis JE, Raharnan O, Patel S (2009) Molecular dynamics simulations of a DMPC bilayer using nonadditive interaction models. Biophys J 96:385–402
9. Dickson CJ, Rosso L, Betz RM, Walker RC, Gould IR (2012) GAFFlipid: a General Amber Force Field for the accurate molecular dynamics simulation of phospholipid. Soft Matter 8:9617–9627
10. Dickson CJ, Rosso L, Walker RC, Gould IR (2012) Lipid bilayer simulations: Expanding time and space with the General Amber Force Field. Abstr Pap Am Chem Soc 243
11. Lim JB, Rogaski B, Klauda JB (2012) Update of the cholesterol force field parameters in CHARMM. J Phys Chem B 116:203–210

12. Yin DX, Mackerell AD (1998) Combined ab initio empirical approach for optimization of Lennard-Jones parameters. J Comput Chem 19:334–348
13. Klauda JB, Venable RM, Freites JA, O'Connor JW, Tobias DJ, Mondragon-Ramirez C, Vorobyov I, MacKerell AD Jr, Pastor RW (2010) Update of the CHARMM all-atom additive force field for lipids: validation on six lipid types. J Phys Chem B 114:7830–7843
14. Sun DL, Forsman J, Woodward CE (2015) Evaluating force fields for the computational prediction of ionized arginine and lysine side-chains partitioning into lipid bilayers and octanol. J Chem Theory Comput 11:1775–1791
15. Kukol A (2009) Lipid models for united-atom molecular dynamics simulations of proteins. J Chem Theory Comput 5:615–626
16. Lopes PE, Roux B, Mackerell AD Jr (2009) Molecular modeling and dynamics studies with explicit inclusion of electronic polarizability. Theory and applications. Theoret Chem Acc 124:11–28
17. Chowdhary J, Harder E, Lopes PEM, Huang L, MacKerell AD Jr, Roux B (2013) A polarizable force field of dipalmitoylphosphatidylcholine based on the classical drude model for molecular dynamics simulations of lipids. J Phys Chem B 117:9142–9160
18. Vorobyov IV, Anisimov VM, MacKerell AD (2005) Polarizable empirical force field for alkanes based on the classical drude oscillator model. J Phys Chem B 109:18988–18999
19. Lamoureux G, MacKerell AD, Roux B (2003) A simple polarizable model of water based on classical Drude oscillators. J Chem Phys 119:5185–5197
20. Roux B, Berneche S, Egwolf B, Lev B, Noskov SY, Rowley CN, Yu H (2011) Ion selectivity in channels and transporters. J Gen Physiol 137:415–426
21. Kroese DP, Brereton T, Taimre T, Botev ZI (2014) Why the Monte Carlo method is so important today. Wiley Interdisc Rev: Comput Stat 6:386–392
22. Del Moral P, Doucet A, Jasra A (2006) Sequential Monte Carlo samplers. J Roy Stat Soc B (Stat Methodol) 68:411–436
23. Alder BJ, Wainwright TE (1959) Studies in molecular dynamics. 1. General method. J Chem Phys 31:459–466
24. Rahman A (1964) Correlations in the motion of atoms in liquid argon. Phys Rev 136:A405–A411
25. Andrews DH (1930) The relation between the Raman spectra and the structure of organic molecules. Phys Rev 36:0544–0554
26. Lifson S, Warshel A (1968) Consistent force field for calculations of conformations vibrational spectra and enthalpies of cycloalkane and n-alkane molecules. J Chem Phys 49:5116–5129
27. McDaniel JG, Schmidt JR (2016) Next-generation force fields from symmetry-adapted perturbation theory. Annu Rev Phys Chem 67(67):467–488
28. Karplus S, Lifson S (1971) Consistent force field calculations on 2,5-diketopiperazine and its 3.6-dimethyl derivatives. Biopolymers 10:1973–1982
29. Warshel A (1973) Quantum-mechanical consistent force-field (Qcff/Pi) method—calculations of energies, conformations and vibronic interactions of ground and excited-states of conjugated molecules. Isr J Chem 11:709–717
30. Warshel A, Levitt M, Lifson S (1970) Consistent force field for calculation of vibrational spectra and conformations of some amides and lactam rings. J Mol Spectrosc 33:84–99
31. Warshel A, Lifson S (1970) Consistent force field calculations. 2. Crystal structures, sublimation energies, molecular and lattice vibrations, molecular conformations, and enthalpies of alkanes. J Chem Phys 53:582–594
32. Allinger NL, Yuh YH, Lii JH (1989) Molecular mechanics—the MM3 force-field for hydrocarbons. 1. J Am Chem Soc 111:8551–8566
33. Halgren TA (1992) The Merck molecular-force field—form, scope, parameterization and performance. Abstr Pap Am Chem Soc 204:38-Comp
34. Halgren TA, Bush BL (1996) The Merck molecular force field (MMFF94). Extension and application. Abstr Pap Am Chem Soc 212:2-Comp

35. Halgren TA, Nachbar RB (1996) MMF94: the Merck molecular force field. Bridging the gap —from small organics to proteins. Abstr Pap Am Chem Soc 211:70-Comp
36. Rappe AK, Casewit CJ, Colwell KS, Goddard WA, Skiff WM (1992) Uff, a full periodic-table force-field for molecular mechanics and molecular-dynamics simulations. J Am Chem Soc 114:10024–10035
37. Mayo SL, Olafson BD, Goddard WA (1990) Dreiding—a generic force-field for molecular simulations. J Phys Chem 94:8897–8909
38. Case DA, Cheatham TE III, Darden T, Gohlke H, Luo R, Merz KM Jr, Onufriev A, Simmerling C, Wang B, Woods RJ (2005) The Amber biomolecular simulation programs. J Comput Chem 26:1668–1688
39. Cornell WD, Cieplak P, Bayly CI, Gould IR, Merz KM, Ferguson DM, Spellmeyer DC, Fox T, Caldwell JW, Kollman PA (1996) A second generation force field for the simulation of proteins, nucleic acids, and organic molecules (vol 117, p 5179, 1995). J Am Chem Soc 118:2309
40. Best RB, Mittal J, Feig M, MacKerell AD Jr (2012) Inclusion of many-body effects in the additive CHARMM protein CMAP potential results in enhanced cooperativity of alpha-helix and beta-hairpin formation. Biophys J 103:1045–1051
41. Guvench O, Hatcher ER, Venable RM, Pastor RW, Mackerell AD (2009) CHARMM additive all-atom force field for glycosidic linkages between hexopyranoses. J Chem Theory Comput 5:2353–2370
42. Hart K, Foloppe N, Baker CM, Denning EJ, Nilsson L, Mackerell AD Jr (2012) Optimization of the CHARMM additive force field for DNA: Improved treatment of the BI/BII conformational equilibrium. J Chem Theory Comput 8:348–362
43. MacKerell AD Jr, Banavali N, Foloppe N (2000) Development and current status of the CHARMM force field for nucleic acids. Biopolymers 56:257–265
44. Mallajosyula SS, Guvench O, Hatcher E, Mackerell AD Jr (2012) CHARMM additive all-atom force field for phosphate and sulfate linked to carbohydrates. J Chem Theory Comput 8:759–776
45. Raman EP, Guvench O, MacKerell AD Jr (2010) CHARMM additive all-atom force field for glycosidic linkages in carbohydrates involving furanoses. J Phys Chem B 114:12981–12994
46. Vanommeslaeghe K, Hatcher E, Acharya C, Kundu S, Zhong S, Shim J, Darian E, Guvench O, Lopes P, Vorobyov I, Mackerell AD Jr (2010) CHARMM general force field: a force field for drug-like molecules compatible with the CHARMM all-atom additive biological force fields. J Comput Chem 31:671–690
47. Yu W, He X, Vanommeslaeghe K, MacKerell AD Jr (2012) Extension of the CHARMM general force field to sulfonyl-containing compounds and its utility in biomolecular simulations. J Comput Chem 33:2451–2468
48. Daura X, Oliva B, Querol E, Aviles FX, Tapia O (1996) On the sensitivity of MD trajectories to changes in water-protein interaction parameters: the potato carboxypeptidase inhibitor in water as a test case for the GROMOS force field. Proteins 25:89–103
49. Hansen HS, Hunenberger PH (2011) A reoptimized GROMOS force field for hexopyranose-based carbohydrates accounting for the relative free energies of ring conformers, anomers, epimers, hydroxymethyl rotamers, and glycosidic linkage conformers. J Comput Chem 32:998–1032
50. Horta BAC, Lin ZX, Huang W, Riniker S, van Gunsteren WF, Hunenberger PH (2012) Reoptimized interaction parameters for the peptide-backbone model compound N-methylacetamide in the GROMOS force field: influence on the folding properties of two beta-peptides in methanol. J Comput Chem 33:1907–1917
51. Kouwijzer MLCE, vanEijck BP, Kooijman H, Kroon J (1995) Extension of the GROMOS force field for carbohydrates, resulting in improvement of the crystal structure determination of alpha-D-galactose. AIP Conf Proc 330:393
52. Lins RD, Hunenberger PH (2005) A new GROMOS force field for hexopyranose-based carbohydrates. J Comput Chem 26:1400–1412

53. Oostenbrink C, Soares TA, van der Vegt NFA, van Gunsteren WF (2005) Validation of the 53A6 GROMOS force field. Eur Biophys J 34:273–284
54. Ott KH, Meyer B (1996) Parametrization of GROMOS force field for oligosaccharides and assessment of efficiency of molecular dynamics simulations. J Comput Chem 17:1068–1084
55. Pol-Fachin L, Rusu VH, Verli H, Lins RD (2012) GROMOS 53A6(GLYC), an improved GROMOS force field for hexopyranose-based carbohydrates. J Chem Theory Comput 8:4681–4690
56. Reif MM, Hunenberger PH, Oostenbrink C (2012) New interaction parameters for charged amino acid side chains in the GROMOS force field. J Chem Theory Comput 8:3705–3723
57. Smith MD, Rao JS, Segelken E, Cruz L (2015) Force-field induced bias in the structure of A beta(21-30): a comparison of OPLS, AMBER, CHARMM, and GROMOS force fields. J Chem Inf Model 55:2587–2595
58. Soares TA, Hunenberger PH, Kastenholz MA, Krautler V, Lenz T, Lins RD, Oostenbrink C, Van Gunsteren WF (2005) An improved nucleic acid parameter set for the GROMOS force field. J Comput Chem 26:725–737
59. Suardiaz R, Maestre M, Suarez E, Perez C (2006) Parameterization and validation of Gromos force field to use in conformational analysis of epoxidic systems. J Mol Struc-Theochem 778:21–25
60. Jorgensen WL, Maxwell DS, TiradoRives J (1996) Development and testing of the OPLS all-atom force field on conformational energetics and properties of organic liquids. J Am Chem Soc 118:11225–11236
61. Kaminski GA, Friesner RA, Tirado-Rives J, Jorgensen WL (2001) Evaluation and reparametrization of the OPLS-AA force field for proteins via comparison with accurate quantum chemical calculations on peptides. J Phys Chem B 105:6474–6487
62. Gu R-X, Liu LA, Wei D-Q, Du J-G, Liu L, Liu H (2011) Free energy calculations on the two drug binding sites in the M2 proton channel. J Am Chem Soc 133:10817–10825
63. Lian P, Wei D-Q, Wang J-F, Chou K-C (2011) An allosteric mechanism inferred from molecular dynamics simulations on phospholamban pentamer in lipid membranes. PLoS One 6:e18587
64. Wang J, Zhu W, Li G, Hansmann UHE (2011) Velocity-scaling optimized replica exchange molecular dynamics of proteins in a hybrid explicit/implicit solvent. J Chem Phys 135:08B625
65. Mortier WJ, Ghosh SK, Shankar S (1986) Electronegativity-equalization method for the calculation of atomic charges in molecules. J Am Chem Soc 108:4315–4320
66. Cieplak P, Dupradeau FY, Duan Y, Wang J (2009) Polarization effects in molecular mechanical force fields. J Phys Condens Matter: Inst Phys J 21:333102
67. Mulliken RS (1934) A new electroaffinity scale; together with data on valence states and on valence ionization potentials and electron affinities. J Chem Phys 2:782–793
68. Parr RG, Pearson RG (1983) Absolute hardness: companion parameter to absolute electronegativity. J Am Chem Soc 105:7512–7516
69. Nalewajski RF, Korchowiec J, Zhou Z (1988) Molecular hardness and softness parameters and their use in chemistry. Int J Quantum Chem 34:349–366
70. Martyna GJ, Tuckerman ME, Tobias DJ, Klein ML (1996) Explicit reversible integrators for extended systems dynamics. Mol Phys 87:1117–1157
71. Patel S, Mackerell AD, Brooks CL (2004) CHARMM fluctuating charge force field for proteins: II protein/solvent properties from molecular dynamics simulations using a nonadditive electrostatic model. J Comput Chem 25:1504–1514
72. Chelli R, Procacci P, Righini R, Califano S (1999) Electrical response in chemical potential equalization schemes. J Chem Phys 111:8569–8575
73. Stern HA, Kaminski GA, Banks JL, Zhou R, Berne B, Friesner RA (1999) Fluctuating charge, polarizable dipole, and combined models: parameterization from ab initio quantum chemistry. J Phys Chem B 103:4730–4737

74. Banks JL, Kaminski GA, Zhou R, Mainz DT, Berne BJ, Friesner RA (1999) Parametrizing a polarizable force field from ab initio data. I. The fluctuating point charge model. J Chem Phys 110:741–754
75. Baker CM (2015) Polarizable force fields for molecular dynamics simulations of biomolecules. Wiley Interdisc Rev: Comput Mol Sci 5:241–254
76. Bauer BA, Patel S (2012) Recent applications and developments of charge equilibration force fields for modeling dynamical charges in classical molecular dynamics simulations. Theoret Chem Acc 131:1153
77. Zhao D-X, Liu C, Wang F-F, Yu C-Y, Gong L-D, Liu S-B, Yang Z-Z (2010) Development of a polarizable force field using multiple fluctuating charges per atom. J Chem Theory Comput 6:795–804
78. Yang Z-Z, Wang J-J, Zhao D-X (2014) Valence state parameters of all transition metal atoms in metalloproteins—development of ABEEM sigma pi fluctuating charge force field. J Comput Chem 35:1690–1706
79. Davis JE, Patel S (2009) Charge equilibration force fields for lipid environments: applications to fully hydrated DPPC bilayers and DMPC-embedded gramicidin A. J Phys Chem B 113:9183–9196
80. Patel S, Davis JE, Bauer BA (2009) Exploring ion permeation energetics in gramicidin A using polarizable charge equilibration force fields. J Am Chem Soc 131:13890–+
81. Bauer BA, Lucas TR, Meninger DJ, Patel S (2011) Water permeation through DMPC lipid bilayers using polarizable charge equilibration force fields. Chem Phys Lett 508:289–294
82. Lamoureux G, Bt Roux (2003) Modeling induced polarization with classical Drude oscillators: theory and molecular dynamics simulation algorithm. J Chem Phys 119:3025–3039
83. Rick SW, Stuart SJ (2002) Potentials and algorithms for incorporating polarizability in computer simulations. Rev Comput Chem 18:89–146
84. Allen TW, Andersen OS, Roux B (2006) Ion permeation through a narrow channel: using gramicidin to ascertain all-atom molecular dynamics potential of mean force methodology and biomolecular force fields. Biophys J 90:3447–3468
85. Dorairaj S, Allen TW (2007) On the thermodynamic stability of a charged arginine side chain in a transmembrane helix. Proc Natl Acad Sci USA 104:4943–4948
86. Lopes PE, Huang J, Shim J, Luo Y, Li H, Roux B, MacKerell AD Jr (2013) Polarizable force field for peptides and proteins based on the classical drude oscillator. J Chem Theory Comput 9:5430–5449
87. He X, Lopes PEM, MacKerell AD Jr (2013) Polarizable empirical force field for acyclic polyalcohols based on the classical drude oscillator. Biopolymers 99:724–738
88. Patel DS, He X, MacKerell AD Jr (2015) Polarizable empirical force field for hexopyranose monosaccharides based on the classical drude oscillator. J Phys Chem B 119:637–652
89. Jana M, MacKerell AD Jr (2015) CHARMM drude polarizable force field for aldopentofuranoses and methyl-aldopentofuranosides. J Phys Chem B 119:7846–7859
90. Savelyev A, MacKerell AD Jr (2014) Balancing the interactions of ions, water, and DNA in the Drude polarizable force field. J Phys Chem B 118:6742–6757
91. Savelyev A, MacKerell AD Jr (2014) All-atom polarizable force field for DNA based on the classical drude oscillator model. J Comput Chem 35:1219–1239
92. Lemkul JA, Huang J, Roux B, MacKerell AD (2016) An empirical polarizable force field based on the classical drude oscillator model: development history and recent applications. Chem Rev 116:4983–5013
93. Robinson D (2013) A polarizable force-field for cholesterol and sphingomyelin. J Chem Theory Comput 9:2498–2503
94. Straatsma TP, McCammon JA (1990) Molecular dynamics simulations with interaction potentials including polarization development of a noniterative method and application to water. Mol Simul 5:181–192

95. Cieplak P, Caldwell J, Kollman P (2001) Molecular mechanical models for organic and biological systems going beyond the atom centered two body additive approximation: aqueous solution free energies of methanol and N-methyl acetamide, nucleic acid base, and amide hydrogen bonding and chloroform/water partition coefficients of the nucleic acid bases. J Comput Chem 22:1048–1057
96. Ren P, Ponder JW (2002) Consistent treatment of inter-and intramolecular polarization in molecular mechanics calculations. J Comput Chem 23:1497–1506
97. Thole BT (1981) Molecular polarizabilities calculated with a modified dipole interaction. Chem Phys 59:341–350
98. Wang QT, Rackers JA, He C, Qi R, Narth C, Lagardere L, Gresh N, Ponder JW, Piquemal JP, Ren PY (2015) General model for treating short-range electrostatic penetration in a molecular mechanics force field. J Chem Theory Comput 11:2609–2618
99. Piquemal JP, Gresh N, Giessner-Prettre C (2003) Improved formulas for the calculation of the electrostatic contribution to the intermolecular interaction energy from multipolar expansion of the electronic distribution. J Phys Chem A 107:10353–10359
100. Marrink SJ, Tieleman DP (2013) Perspective on the Martini model. Chem Soc Rev 42:6801–6822
101. Qi YF, Cheng X, Im W (2015) CHARMM-GUI martini maker for coarse-grained simulations. Biophys J 108:161a–161a
102. Im W (2016) CHARMM-GUI 10 years for biomolecular modeling and simulation. Biophys J 110:328a–328a
103. Qi YF, Ingolfsson HI, Cheng X, Lee J, Marrink SJ, Im W (2015) CHARMM-GUI martini maker for coarse-grained simulations with the martini force field. J Chem Theory Comput 11:4486–4494
104. Fredriksson R, Lagerstrom MC, Lundin LG, Schioth HB (2003) The G-protein-coupled receptors in the human genome form five main families. Phylogenetic analysis, paralogon groups, and fingerprints. Mol Pharmacol 63:1256–1272
105. Gudermann T, Nurnberg B, Schultz G (1995) Receptors and G-proteins as primary components of transmembrane signal-transduction. 1. G-protein-coupled receptors—structure and function. J Mol Med 73:51–63
106. Drews J (2000) Drug discovery: a historical perspective. Science 287:1960–1964
107. Cherezov V, Rosenbaum DM, Hanson MA, Rasmussen SGF, Thian FS, Kobilka TS, Choi H-J, Kuhn P, Weis WI, Kobilka BK, Stevens RC (2007) High-resolution crystal structure of an engineered human beta(2)-adrenergic G protein-coupled receptor. Science 318:1258–1265
108. Pucadyil TJ, Chattopadhyay A (2006) Role of cholesterol in the function and organization of G-protein coupled receptors. Prog Lipid Res 45:295–333
109. Taylor MRG (2007) Pharmacogenetics of the human beta-adrenergic receptors. Pharmacogenomics J 7:29–37
110. Milligan G, Svoboda P, Brown CM (1994) Why are there so many adrenoceptor subtypes. Biochem Pharmacol 48:1059–1071
111. Takeda S, Kadowaki S, Haga T, Takaesu H, Mitaku S (2002) Identification of G protein-coupled receptor genes from the human genome sequence (vol 520, p 97, 2002). FEBS Lett 523:257
112. Harrison T, Samuel BU, Akompong T, Hamm H, Mohandas N, Lomasney JW, Haldar K (2003) Erythrocyte G protein-coupled receptor signaling in malarial infection. Science 301:1734–1736
113. Watts A, Volotovski ID, Marsh D (1979) Rhodopsin-lipid associations in bovine rod outer segment membranes—identification of immobilized lipid by spin-labels. Biochemistry 18:5006–5013
114. Fretten P, Morris SJ, Watts A, Marsh D (1980) Lipid-lipid and lipid-protein interactions in chromaffin granule membranes—a spin label electron-spin-resonance study. Biochem Biophys Acta 598:247–259
115. Engelman DM (2005) Membranes are more mosaic than fluid. Nature 438:578–580

116. Zocher M, Zhang C, Rasmussen SGF, Kobilka BK, Mueller DJ (2012) Cholesterol increases kinetic, energetic, and mechanical stability of the human beta(2)-adrenergic receptor. Proc Natl Acad Sci USA 109:E3463–E3472
117. Periole X, Cavalli M, Marrink S-J, Ceruso MA (2009) Combining an elastic network with a coarse-grained molecular force field: structure, dynamics, and intermolecular recognition. J Chem Theory Comput 5:2531–2543
118. Hanson MA, Cherezov V, Griffith MT, Roth CB, Jaakola V-P, Chien EYT, Velasquez J, Kuhn P, Stevens RC (2008) A specific cholesterol binding site is established by the 2.8 angstrom structure of the human beta(2)-adrenergic receptor. Structure 16:897–905
119. Marrink SJ, de Vries AH, Mark AE (2004) Coarse grained model for semiquantitative lipid simulations. J Phys Chem B 108:750–760
120. Wassenaar TA, Ingolfsson HI, Boeckmann RA, Tieleman DP, Marrink SJ (2015) Computational lipidomics with insane: a versatile tool for generating custom membranes for molecular simulations. J Chem Theory Comput 11:2144–2155
121. Kusalik PG, Svishchev IM (1994) The spatial structure in liquid water. Science 265:1219–1221
122. Hartzell C, Putzier I, Arreola J (2005) Calcium-activated chloride channels. Annu Rev Physiol 67:719–758
123. Hartzell HC, Qu Z, Yu K, Xiao Q, Chien LT (2008) Molecular physiology of bestrophins: multifunctional membrane proteins linked to best disease and other retinopathies. Physiol Rev 88:639–672
124. Kunzelmann K (2015) TMEM16, LRRC8A, bestrophin: chloride channels controlled by Ca (2+) and cell volume. Trends Biochem Sci 40:535–543
125. Gomez NM, Tamm ER, Straubeta O (2013) Role of bestrophin-1 in store-operated calcium entry in retinal pigment epithelium. Pflugers Arch Eur J Physiol 465:481–495
126. Bakall B, Marknell T, Ingvast S, Koisti MJ, Sandgren O, Li W, Bergen AA, Andreasson S, Rosenberg T, Petrukhin K, Wadelius C (1999) The mutation spectrum of the bestrophin protein–functional implications. Hum Genet 104:383–389
127. Sun H, Tsunenari T, Yau KW, Nathans J (2002) The vitelliform macular dystrophy protein defines a new family of chloride channels. Proc Natl Acad Sci USA 99:4008–4013
128. Kane Dickson V, Pedi L, Long SB (2014) Structure and insights into the function of a Ca(2+)-activated Cl(-) channel. Nature 516:213–218
129. Yang T, Liu Q, Kloss B, Bruni R, Kalathur RC, Guo Y, Kloppmann E, Rost B, Colecraft HM, Hendrickson WA (2014) Structure and selectivity in bestrophin ion channels. Science 346:355–359
130. Davidson AE, Millar ID, Urquhart JE, Burgess-Mullan R, Shweikh Y, Parry N, O'Sullivan J, Maher GJ, McKibbin M, Downes SM, Lotery AJ, Jacobson SG, Brown PD, Black GC, Manson FD (2009) Missense mutations in a retinal pigment epithelium protein, bestrophin-1, cause retinitis pigmentosa. Am J Hum Genet 85:581–592
131. Gifford JL, Walsh MP, Vogel HJ (2007) Structures and metal-ion-binding properties of the Ca2+-binding helix-loop-helix EF-hand motifs. Biochem J 405:199–221
132. Yuan P, Leonetti MD, Hsiung YC, MacKinnon R (2012) Open structure of the Ca2+ gating ring in the high-conductance Ca2+-activated K+ channel. Nature 481:94–U105
133. Qu Z, Cheng W, Cui Y, Cui Y, Zheng J (2009) Human disease-causing mutations disrupt an N-C-terminal interaction and channel function of bestrophin 1. J Biol Chem 284:16473–16481
134. Jo S, Im W (2011) CHARMM-GUI: brining advanced computational techniques to web interface. Biophys J 100:156
135. Van der Spoel D, Lindahl E, Hess B, Groenhof G, Mark AE, Berendsen HJC (2005) GROMACS: fast, flexible, and free. J Comput Chem 26:1701–1718
136. MacKerell AD, Bashford D, Bellott M, Dunbrack RL, Evanseck JD, Field MJ, Fischer S, Gao J, Guo H, Ha S, Joseph-McCarthy D, Kuchnir L, Kuczera K, Lau FTK, Mattos C, Michnick S, Ngo T, Nguyen DT, Prodhom B, Reiher WE, Roux B, Schlenkrich M, Smith JC, Stote R, Straub J, Watanabe M, Wiorkiewicz-Kuczera J, Yin D, Karplus M (1998)

All-atom empirical potential for molecular modeling and dynamics studies of proteins. J Phys Chem B 102:3586–3616
137. Hockney RW, Goel SP, Eastwood JW (1974) Quiet high-resolution computer models of a plasma. J Comput Phys 14:148–158
138. Darden T, York D, Pedersen L (1993) Particle mesh Ewald—an N.Log(N) method for Ewald sums in large systems. J Chem Phys 98:10089–10092
139. Nose S, Klein ML (1983) Constant pressure molecular-dynamics for molecular-systems. Mol Phys 50:1055–1076
140. Nose S (1984) A molecular-dynamics method for simulations in the canonical ensemble. Mol Phys 52:255–268
141. Bonomi M, Branduardi D, Bussi G, Camilloni C, Provasi D, Raiteri P, Donadio D, Marinelli F, Pietrucci F, Broglia RA, Parrinello M (2009) PLUMED: a portable plugin for free-energy calculations with molecular dynamics. Comput Phys Commun 180:1961–1972
142. Grossfield A. WHAM: the weighted histogram analysis method, version 2.0.6, http://membrane.urmc.rochester.edu/content/wham
143. Timko J, De Castro A, Kuyucak S (2011) Ab initio calculation of the potential of mean force for dissociation of aqueous Ca-Cl. J Chem Phys 134:204510
144. Lamoureux G, Roux B (2006) Absolute hydration free energy scale for alkali and halide ions established from simulations with a polarizable force field. J Phys Chem B 110:3308–3322
145. Humphrey W, Dalke A, Schulten K (1996) VMD: visual molecular dynamics. J Mol Graph Model 14:33–38
146. Baker NA, Sept D, Joseph S, Holst MJ, McCammon JA (2001) Electrostatics of nanosystems: application to microtubules and the ribosome. Proc Natl Acad Sci USA 98:10037–10041
147. Kumar R, Iyer VG, Im W (2007) CHARMM-GUI: a graphical user interface for the CHARMM users. Abstr Pap Am Chem Soc 233:273
148. Altschul SF, Gish W, Miller W, Myers EW, Lipman DJ (1990) Basic local alignment search tool. J Mol Biol 215:403–410
149. Sali A, Potterton L, Yuan F, Vanvlijmen H, Karplus M (1995) Evaluation of comparative protein modeling by modeler. Proteins 23:318–326
150. Deshpande N, Addess KJ, Bluhm WF, Merino-Ott JC, Townsend-Merino W, Zhang Q, Knezevich C, Xie L, Chen L, Feng ZK, Green RK, Flippen-Anderson JL, Westbrook J, Berman HM, Bourne PE (2005) The RCSB Protein Data Bank: a redesigned query system and relational database based on the mmCIF schema. Nucleic Acids Res 33:D233–D237
151. Lomize MA, Lomize AL, Pogozheva ID, Mosberg HI (2006) OPM: orientations of proteins in membranes database. Bioinformatics 22:623–625

Chapter 13
Other Modern Methods for Studying Biomembranes

Matthias Amrein, Tie Xia and Yan Shi

13.1 Introduction

Biological and biomedical research has been mostly relying on biochemical and genetic tools. This practice has generated the bulk of our knowledge in life sciences. One of the central assumptions here is the specific interactions among proteins and nucleic acids. Often, these molecules are presumed to operate in an imaginary isolation with limited interferences or constrains from environmental factors. It is further assumed that all these interactions could be measured in terms of biochemical interactions. The most popular tools, such as Western blotting, flow cytometry, surface plasma resonance are all suited for these two basic premises. With our understanding deepening, we are realizing that this is simply not the case. The biological world at the microscopic levels is highly chaotic. This can be viewed from two different angles. First, very few biological interactions happen on their own. They usually take place in a membrane-bound state or in a highly complex environment, with varying salt levels, electrobiophysic field strengths, and structural hindrances. These factors fundamentally change/modify a "simple" interaction. Second, a biological interaction is far beyond a simple contact/binding. It is often a process of biophysical changes, registered as force, velocity, binding

M. Amrein (✉)
Department of Cell Biology and Anatomy, Snyder Institute of Chronic Diseases,
University of Calgary, Calgary, AB, Canada
e-mail: mamrein@ucalgary.ca

T. Xia · Y. Shi (✉)
School of Medicine, Tsinghua-Peking Center for Life Sciences, Institute for Immunology
and Department of Basic Medical Sciences, Tsinghua University, Beijing, China
e-mail: yanshi@biomed.tsinghua.edu.cn

Y. Shi
Department of Microbiology, Immunology & Infectious Diseases and Snyder Institute,
University of Calgary, Calgary, AB, Canada

H. Wang and G. Li (eds.), *Membrane Biophysics*,
https://doi.org/10.1007/978-981-10-6823-2_13

kinetics, and temporal/spatial changes. The deeper we go into the inner work of biology, the higher biophysical impacts felt. This is particularly true for events occurring at the plasma surface. At the higher order of cellular contact, it is inevitable that multiple surface binding partners participate in the overall interactions, with each pair influences and is influenced by other binding pairs. To resolve these intricate interactions, membrane biology and cell surface biophysics are rapidly moving to the center of biological/medical research. A set of newly developed tools have been designed to help us move to these new fields.

To understand the underlying biology for this conceptual leap of faith, let us dissect a generic biological signaling event. When a receptor is engaged by its ligand, a successful signaling is impacted by the feature of the supporting bilayer membrane. This impact can come from several different parameters, i.e., the fluidity of the lipid, the organizing platforms of lipid domains and the successful aggregation of costimulatory molecule and organization of underlying cytoskeleton. These factors contribute to the singular "specificity" afforded by the receptor/ligand interaction. The "activation" is therefore eventually net outcome of all these forces. On the other hand, these signaling events are defined by a set of physical parameters, such as a force change, spatial displacement, the timing, and the overall half-lives of the interactions. For instance, integrin binding involves multistep conformational changes as directed by antigen receptor signaling strength. These changes are characterized by increasing force engagement over time and better association with cortical cytoskeleton. Most biochemical tools are not designed to reveal these subtleties.

This chapter will focus on the applications of modern biophysical methods in studying biological events at the cell membrane. Some methods not mentioned in previous chapters will be briefly introduced. For convenience, the application examples will be categorized into three main groups per their detection modes: force, fluorescence, or electrical signal (patch clamp). It should be noted that many examples have used more than one approach either simultaneously or sequentially to obtain multidimensional information on the membrane events of interest. The aim is to show that how these advanced biophysical tools gain the knowledge that conventional biochemical approaches could not access to and what could be achieved with them in future biological and biomedical research.

13.2 Force-Based Techniques

13.2.1 Atomic Force Microscopy-Based Single-Cell Force Spectroscopy

AFM-based single-cell/single molecule analyses are a defined set of tools suitable for studying the biophysical properties of living cells. AFM-based single-cell force spectroscopy (AFM-SCFS) is one of the prototypical instruments to reveal interaction strengths and dynamics on a live-cell membrane, including those between cells, receptor/ligands, and many other variations. This technique has in recent years morphed into an accepted method to study immune cell activation, adhesion strength, phagocytosis and has even evolved into a precision readout of interaction forces involving multiple participants. Several features of this technique are highly desirable in cellular biophysics, including its ideal force measurement range and minimal needs of sample preparation. As a microscopy-based technique, it is conceptually straightforward to couple the force measurement to imaging analysis, which is expected to be an exceedingly robust combination in the future. However, the most outstanding power is that the interacting pairs work in the support environment of live-cell membrane with sophisticated physiological and spatial features difficult to reproduce in artificial systems.

The principle of AFM-SCFS is explained in Fig. 13.1. A particle or cell is attached to a (tip-less) cantilever of an AFM. Alternatively, cantilevers with integrated beads or tips of all shapes may be functionalized. The cantilever with

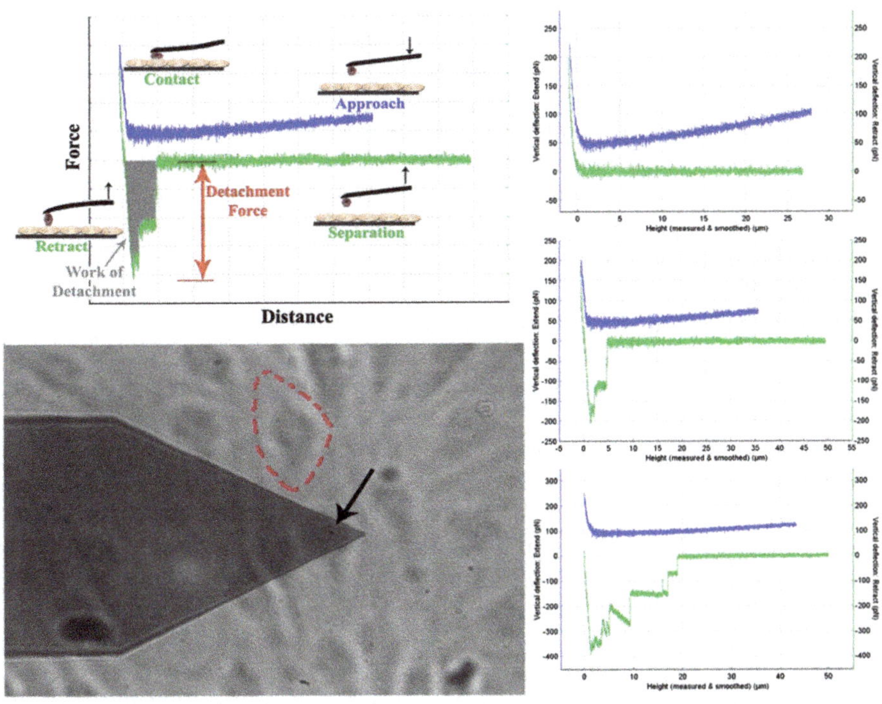

Fig. 13.1 General scheme of AFM-SCFS. Upper left: the illustration of a cell glued to a cantilever approaching and retracting from a contact surface. Blue: approach curve; green, retracting curve; red, the maximal detaching force; gray, over all work used for the separation. Lower left: a live view of cell attached to a cantilever (arrow) ready to make contact with a cell grown on a flat surface (outline). Right: top to bottom: representative force curves with no interaction, relatively short and single interactions and complex, multistep binding partnership. Drawing by Dr. Shaun Davis, used with permission

attached particle or tip approaches toward and then retracts from the cell of interest. Fluorescence imaging can be obtained from below the sample. Force–distance curves are explained in Fig. 13.1. Upon approach, the particle first experiences a range repulsive force before the tip and sample are in close physical contact, such as from penetrating the glycocalyx or electrostatic barrier (electrical double-layer repulsion). Close to the sample, the tip may snap into contact (because of van der Waals and/or an electrostatic force). Once in contact with the cell, the cantilever is deflected steeply by the approaching scanner until a user-defined set point of load is reached. At this point, the particle can be set to remain in contact with the membrane for a given time or immediately be retracted. On the way back, the particle or cell remains adherent to the sample until the pull from the cantilever forces it out of contact. The retraction curve is of interest as they reflect a number of events, occurring at the same time. Firstly, the cell cortex will deform in response to the pull. As a result, the adhesive force increases less sharply as the cantilever is being pulled back. Once a threshold force has been reached, competitive phenomena will start to occur. Either one or a number of receptor–ligand pairs will separate, allowing the tension to relax sharply (i.e., the force to make a step toward baseline). For a solid particle bound to a cell, a patch of lipids adsorbed to the particle may let go, again leading to a sudden relaxation of the tension. Often, receptor–ligand separation (or lipid–particle separation) is preceded by the formation of a membrane tether whereby lipids are pulled out of the plasma membrane into a tube of nanometer diameters and micrometer lengths. In the force curve, the occurrence of a tether is seen as a plateau whereby the force remains constant as the cantilever retracts, before the receptor–ligand pair is separated and the tension stepwise relaxed. Common readings in SCFS are maximum adhesion force or the work of adhesion. In these cases, the entirety of events is collapsed into one value per force curve, representing the strength of the interaction. However, force curves may also be analyzed for individual unbinding events between receptors and ligands or other parameters. Analysis of sequential unbinding events one after another may reveal a progression of the interaction.

Cell-to-cell binding, especially those involving cytoskeletal changes, is in hundreds of pN to a few dozen nN. Cellular attachment to anchoring surfaces can often reach μN. With separation forces accessible from 10 pN (sufficient to resolve rupture of single hydrogen bonds [1] to μN (well beyond a force required to separate cells), this method allows for the broadest force range and permits almost unlimited scenarios of binding partnerships. Because force measurement is a dynamic process rather than single-point readouts commonly obtained in other biological assays under equilibrium, AFM-based SCFS can be used to study energy barriers of binding under stress and total energies required to reach detachment. As cellular functions are vulnerable to excessive manipulations, one additional advance of SCFS is the minimal processing prior to force recording. The high cellular viability allows the interaction to be measured in real life, likely the "closest" to the physiologic situations.

At the very minimum, AFM-SCFS covers measurements related to receptor function, adhesion, phagocytosis/antigen uptake, cytoskeleton, and topological deformation. In its most native applications or as a pure tool to measure binding affinity, AFM-SCFS is most relevant in measuring cellular adhesion strength and shape change of cells. Cell adhesion molecules (CAMs) are the foremost problem studied by SCFS (for a comprehensive recent review, see single-cell force spectroscopy [2]). SCFS provides insight into the detailed molecular mechanisms of cell adhesion and allows determining thermodynamic quantities not otherwise attainable. Here, we use CAM SCFS analysis to illustrate the theoretical insights of this class of force interactions. Specifically, the affinity K (or, inversely, the dissociation constant) is the traditional measure to describe binding partnerships (antibody–antigen, receptor–ligand) and is derived from equilibrium binding assays. K is reflective of the change in Gibbs energy upon binding $\Delta G = -RT\ lnK$ (R = gas constant, T = absolute temperature) and allows calculating concentration of bound versus unbound receptor–ligand pairs (for given receptor and ligand concentrations). However, for CAMs, the dissociation rate $k_{dissociation}$ (or its inverse, the bond survival time τ) is physiologically more meaningful than the affinity. The dissociation rate allows prediction of how long a cell is held in place that has caught on to another cell, for example, and how much force is needed to break the bond immediately. While affinity can be expressed as the ratio of the rates of association and dissociation $K = k_{association}/k_{dissociation}$, neither $k_{association}$ or $k_{dissociation}$ are individually revealed. Moreover, CAMs are designed to perform under external force with some CAM only active under tension (catch bonds). These properties too are not apparent from K and, hence, not revealed from an equilibrium binding assay. In the body, tension on the binding partners maybe the result of the viscous drag experienced by a cell that has caught onto the endothelium in the blood stream. Another scenario is tissue forces in muscle or forces between cells during organogenesis.

On the other hand, the dissociation rate $k_{dissociation}$ (or the bond survival time τ) and how it depends on an external force can be computed from SCFS data, by means of a set of equations that are based on a model of the nature of bonds. Bell in 1978 introduced such model to describe the biophysical function of adhesive bonds, with the bound state and the unbound state being separated by an activation energy barrier (the peak of the activation energy being termed "transition energy" and the state of the bond in this situation "transition state") [3]. Without pulling the binding partners apart, dissociation will occur when thermal oscillations "push" the binding partners over the activation barrier, leading to a rate of dissociation k_0 that depends on the height of the transition energy barrier and the temperature, among other parameters. Receptor and ligand may then remain separated or immediately rebind. Bell predicts that applying tension to the bond will lead to a new dissociation rate $k(F)$ that increases logarithmically with increasing force (i.e., bond survival exponentially decays with increasing force). According to Bell, the work (i.e., energy) $F.\ \Delta x$ that is done by an external force (in case of SCFS by the cantilever spring) as it pulls the binding partners apart along the binding axis x can directly be subtracted from the bond energy landscape (Fig. 13.2) [4]. The consequence is twofold. Firstly, pulling lowers the activation barrier ΔG^{ts} to increase the frequency of bond

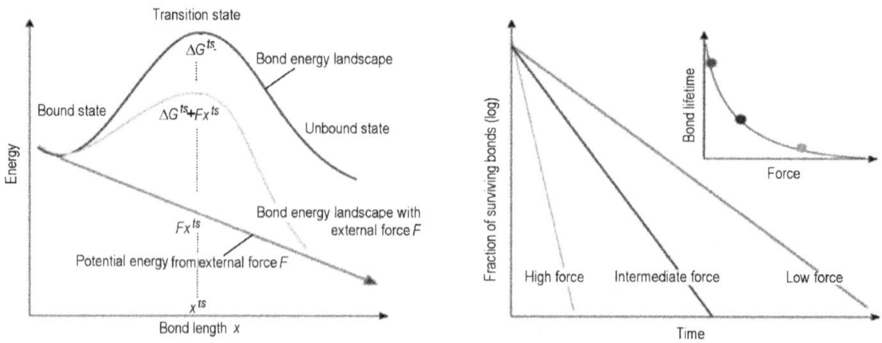

Fig. 13.2 Left: Potential bond energy of the bond in the direction of force. The dotted line indicates the transition state. x^{ts} is the bond length at the transition state. The potential energy of the bond under tension is obtained by adding the potential energy of an external force, acting along the bond separation axis, from the bond energy. Right: The predicted survival of bonds over time at low, medium, and high levels of constant force. The average bond lifetimes are shown in the inset

dissociation, and secondly, once separated, the spring pulls the partners further apart so that they cannot rebind after dissociation.

Evans and Ritchie [5] as well as Szabo et al. [6] have evolved and applied Bell's theory to derive a set of equations suitable to compute the dissociation constant, bond life time but also the bond length at the transition state x^{ts} and the transition energy barrier ΔG^{ts} directly from quantities measured by SCFS (Fig. 13.3). Data obtained in two different manners each depend on a different set of equations. In the "steady ramp" method, the binding partners (after binding) are pulled apart by moving the cantilever back at a constant pulling rate r_F and the force F_{mp} that breaks the bond is measured. To compute the dissociation rate (or the bond survival time) as a function of tension, one needs to measure the breaking force for a number of different pulling rates (i.e., the experiment is repeated at a range of different speeds of retracting the cantilever, Fig. 13.3a). In the "force clamping" method, the cantilever is rapidly stepped back to bring the bond under a given constant tension (clamping force, Fig. 13.3b). Now, the bond survival time $\tau(F)$ is measured for a number of repeats. The experiment is repeated for different clamping forces, all being below the force under which the bond is ruptured immediately. The spectroscopic data, from either method ($Fmp(rF1, 2, 3, ...)$ for the "steady ramp" method and τ ($F1, 2, 3, \ldots$) for the "force clamping method") are then entered into the equations (Fig. 13.3a, b), and the respective quantities are obtained from curve fitting.

The type of bond described in the above section and in Fig. 13.3 is termed slip bond where binding partners are held together in a single energy trough. In summary in this case, the equations allow deriving an expression for $k_{dissociation}$ in dependence of tension from $F_{mp}(r_{F1,\ 2,\ 3,\ \ldots})$, including k_0, the dissociation rate under no tension. Other equations reveal bond survival time in dependence of the

(a) Steady Ramp

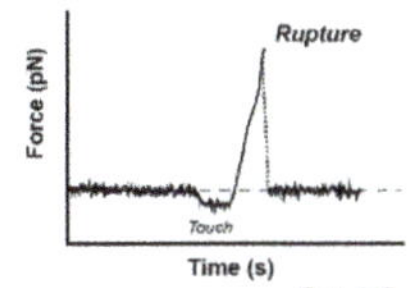

Bell
$$k(F) = k_0 \exp\left[\frac{Fx^{ts}}{k_B T}\right]$$

Evans & Ritchie
$$F_{mp} = \frac{k_B T}{x_\beta} \ln\left(\frac{x_\beta \times r_F}{k_B T \times k_0}\right)$$

Activation free energy ΔG^{ts}:
$$k_0 = \phi \exp\left[-\frac{\Delta G^{ts}}{k_B T}\right]$$

Pulling rate:
$$dF/dt = r_F$$

F_{mp} versus $log(r_F)$: dynamic force spectra to extract x_β and k_0 from the slope and x -intercept

Szabo, Hummer & Dudko:
$$\langle F \rangle \cong \frac{\Delta G^{ts}}{v x^{ts}}\left\{1 - \left[\frac{1}{\Delta G^{ts}}\right] \ln \frac{k_0 e^{\Delta G^{ts}+\gamma}}{x^{ts} KV}\right\}$$

$\langle F \rangle$ versus $log(r_F)$: to extract ΔG^{ts} and k_0 from slope and x -intercept

(b) Force clamp

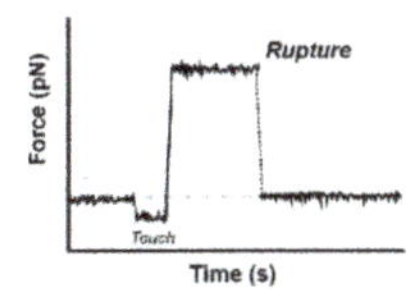

Bond life time (Richie & Evans):
Inverse dissociation rate $1/k(F)$
$$\tau(F) = \tau_0 \exp\left[-\frac{Fx^{ts}}{k_B T}\right]$$

Bond survival probalitity:
$$P(t) = A \exp\left(-\frac{t}{\tau(F)}\right)$$

Bond life time (Szabo, Hummer & Dudko):
$$\tau(F) = \tau_0 \left(1 - \frac{vFx^{ts}}{\Delta G^{ts}}\right)^{1-1/v} \exp\left[-\frac{\Delta G^{ts}}{k_B T}\left(1 - (1 - vFx^{ts}/\Delta G^{ts})^{1/v}\right)\right]$$

v depends on the potential and is 2/3 for most intermolecular potentials and 1 for Bell's model. The parameters τ_0, x^{ts}, and ΔG^{ts} are obtained from nonlinear least squares fits of this equation to the lifetimes obtained at different (constant) forces.

Fig. 13.3 **a** In the "steady ramp" method, the binding partners are pulled apart with a constant pulling rate until the bond breaks. Bell, Evans and Ritchie as well as Szabo et al. provide a mathematical framework to compute the dissociation constant, bond lifetime but also the bond length at the transition state x^{ts} and the transition energy barrier ΔG^{ts} directly from quantities measured by SCFS. **b** shows the approach for the "force clamp" approach, whereby binding partners are rapidly subjected to predetermined load and the time measured from the onset of the load until the bond breaks. For both A and B, the text in red explains how the quantities of interest are obtained

clamping force, including $\tau(0)$ at no tension. The equations also allow computing the height of the activation energy barrier and the physical width of the binding potential trough.

Aside from receptor–ligand interactions, AFM-SCFS has revealed a number of phenomena on how a cell responds to the force of separation of a cell surface receptor and a ligand. Pulling on the adhesion spot locally deforms the cell cortical cytoskeleton in proportion to the force. When the membrane becomes stretched locally, tension starts to build, as the lipid molecules are forced apart. Moreover, the cytoskeleton to which the membrane is anchored will come under tension too. Aside of these elastic phenomena, the cytoskeleton may remodel, allowing for large cellular deformations over time. For a force spectrum to only reflect the membrane binding events of interest, it is important to prevent large-scale viscous deformation of the cell that otherwise compounds the data. Upon fast pull, the elastic contributions to cell deformation will be dominant and act like a second spring in parallel with the cantilever. Thus, for the "steady rate" method, one only needs to compensate for this additional spring by adjusting the nominal pulling rate. The force clamp method takes care of the additional spring on its own.

The first reported success in using AFM to measure cell–cell interaction force was in 2000. Benoit et al. used WGA to bind a Dictyostelium discoideum for interaction with an identical cell. The experiment was designed as a prototypical SCFS platform and used adhesion molecule Contact site A (CsA) as the target of investigation. With a constant contact at 150 pN for 20 s, they found that the rupture force extended above 1 nN. To establish discreet and presumably single molecular interaction, the contact force was reduced (30–40 pN) and time extremely shortened (0.2 s). Interestingly, these time force events characterized by a minimal of 7 pN rupture force showed a Poisson distribution centered around 23 ± 8 (s.d.) pN. This work was very significant as it established some of the basic premises of cell–cell contact study with SCFS [7].

In 2008, Shi et al. reported that binding forces between T cells and antigen-presenting cells (APCs) were determined by the presence of specific epitope peptides [8]. It was for the first time to demonstrate the binding intensity between a T cell and an APC, revealing a biophysical aspect of antigen presentation. To avoid potential interference of lectin-based adhesion, Cell Tak was used to glue the T cells to a flat tip. Without any environmental control, the rupture forces were in 100–200 pN. Two technical details were noted in the work. 1. APCs grown on the glass disks must be very sparse, ideally with distance between individual cells more than several cell diameters. Otherwise, high levels of noise and unstable force measurements would result. 2. The most reproducible force traces were generated if the T cells approached and retracted from the APC in a cyclic pattern, with a defined set point (0.5–1 nN) and intermediate contact time (3–10 s). At that time, the analysis was to provide a qualitative comparison that suggested phagocytic forces were substantially stronger than interactions between T cells and dendritic cells. A more systematic analysis was carried out in an environmental controlled setting with CO_2 and temperature supportive of long term cell culture. There they found that the forces were from 0.5 to several nN. Non-specific interactions (without epitope peptide) were found to be below 200 pN [9].

Shi's group recently made a significant progress in dissecting complex force interactions in T cell activation/suppression involving multiple cells. Regulatory

T cells (Treg) have been found to interact with DCs in vivo and in vitro. In one model, this contact dampens DC's ability to activate other conventional T cells (Tconv). However, a biophysical demonstration of such a model at work is still not available, which would ideally show that Tconv interactions with DCs would be blocked by the latter's simultaneous contact with Treg. They developed a system to study this possibility. Basically a Treg was placed in one side of DC with another Tconv approaching the same DC from the opposite side. In several modes tested, this was the only viable way to reduce the noise among this three-cell setup. Here, the preloading of Treg significantly reduced the force interactions between the Tconv and the DC. With biochemical and genetic tools, they further revealed that it was caused by DC cytoskeleton polarization under Treg binding, which diminished the structural resources in the DC to optimally engage other T cells. Amazingly, in this setup if Treg was flushed away, the Tconv/DC regained the ability to interact after several minutes. This result suggests occupancy/probability-based Treg suppression, revealing a new regulatory dimension [10]. Importantly, this in vitro observation was correlated by a similar Treg occupancy-based blockage of DC interaction with other T cells in vivo [11]. Therefore, this special design of SCFS was instrumental in revealing an important inhibitory mechanism of Treg cells.

Hosseini et al. in 2009 first used AFM force recording to correlate the immune synapse (IS) formation. They found that in the presence of specific epitope peptide, the force went over 10 nN. This is correlated by an IS formation, which was completely abolished by LFA-1 inhibition [12]. In a later report, they confirmed the LFA-1/ICAM-1-dependent IS formation with a different antigen presentation system [13]. Lim et al. used setups similar to those of Hosseini and found that specific epitope peptides must be present on DCs for proper force induction. The resulting strong/stable binding was a prerequisite for Ca^{2+} signaling in T cells. The binding intensity was not only dependent on TCR (T cell receptor) and LFA-1, but also required a complete system of cortical actin cytoskeleton, microtubules as well as undisturbed membrane lipid domains [14]. They also found that the binding strengths between T cell and antigen-loaded DCs were stronger than those with B cells. This was not solely as the result of CD80/CD86 expression, as the blocking antibodies, while lowering the binding intensity, did not negate the difference, suggesting that additional features, possibly adhesion molecules and cytoskeleton changes unique in DCs are significant considerations in T cell activation [15]. Pusch et al. revealed something slightly different from the discussion above. They discovered that the CD8 served as a "hook" to stabilize TCR/DC interaction regardless of epitope peptides, and this allowed a prolonged window/increases probability of TCR to sample the antigen [16].

On a single molecule level, Xu's group recently reported using single molecule force spectroscopy (SMFS) to study the CD3 chain conformational change with respect to the plasma membrane. They found that CD3 chain is buried in the lipid bilayer via more than one motif. Exerting different pulling forces on an AFM tip functionalized with this protein yields several sequential relaxation steps. By studying the relative distances among those steps, they suggested that there are at least three motifs at work for its association with a bilayer membrane. Studies of this nature may reveal how a TCR is progressively activated in response to the external binding forces and different conformational states dictated by the lipid binding in a TCR signaling event [17]. All these reports clearly demonstrate the prowess of AFM-SCFS as a novel and exceptionally revealing technique in immunology research.

13.2.2 Biomembrane Force Probe-Based Single-Cell Force Spectroscopy

The biomembrane force probe (BFP) technique uses a red blood cell as the force transducer held by a glass micropipette. The red blood cell works as a spring with a known constant *k*, which can be tuned by adjusting the aspiration pressure applied though the holding micropipette. The surface of the red blood cell is often biotinylated to facilitate the attachment of a streptavidin-coated microsized glass bead. The glass bead can be further functionalized with ligands of interest, resulting in a fully functional and specific force probe. When the glass bead is brought into contact with a target such as a living cell held by another micropipette, cognate receptors present on the cell surface may form specific bonding with the ligands. Similar to AFM-SCFS, an adhesion event(s) takes place when the probe is retracted. The adhesion force can be calculated from the deformation of the spring (the red blood cell) which can be precisely measured by video camera. To ensure that the force is generated from single molecular bonds, the density of ligands on the bead surface is controlled at a low level so that roughly a single bonding event is registered in 10 contacts. The BFP technique allows a force range from 1 to 20 pN. Because the experiments are performed on an inverted light microscope, adding fluorescence detection module into the system is straight forwards. The combination of force and fluorescence probing is ideal for single receptor–ligand binding kinetics under force control while simultaneously monitoring specific binding-triggered intracellular events (fluorescent analysis) on a single living cell. This methodology termed fBFP has been used to study T cell receptor signaling [18].

The T cell receptor (TCR) is a multimeric transmembrane complex consisting of an antigen-binding heterodimer ($\alpha\beta$ or $\gamma\delta$) in association with the signal-transducing CD3 subunits. These CD3 chains have immunoreceptor tyrosine-based activation motifs (ITAMs) in their cytoplasmic domains. Phosphorylation of the ITAMs is believed to be the initiation event of TCR-mediated signaling cascade. However, previous reports suggest that these domains are buried in the inner leaflet of the plasma membrane through electrostatic interactions between the basic amino acid residues and the acidic phospholipids [19]. The central question is therefore how the TCR ligation by specific peptide-major histocompatibility complex (pMHC) can trigger the dissociation of these domains and render them accessible for kinase activities. It has been accepted that mechanical force is involved in T cell antigen recognition. Yet the role of mechanical force in this recognition process is not known. Zhu and his colleagues studied the kinetics of TCR-pMHC interactions by using the BFP technique (Fig. 13.4) [18]. To monitor the force-induced intracellular response in real time, they also integrated fluorescence imaging into their setup to detect the change of Ca^{2+} signal in the cytosol. They showed that with agonist pMHCs, a pulling force exerted on the TCR actually created a long half-life bonding, while antagonist peptides created slip bonds with short and unpredictable half-lives. The catch bond, which presumably designates a TCR activation event, is accompanied by simultaneous Ca^{2+} signaling. The study indicates that TCR works as a mechanosensor and force acts as an additional checkpoint for TCR to further discriminate the specific antigen from irrelevant ones.

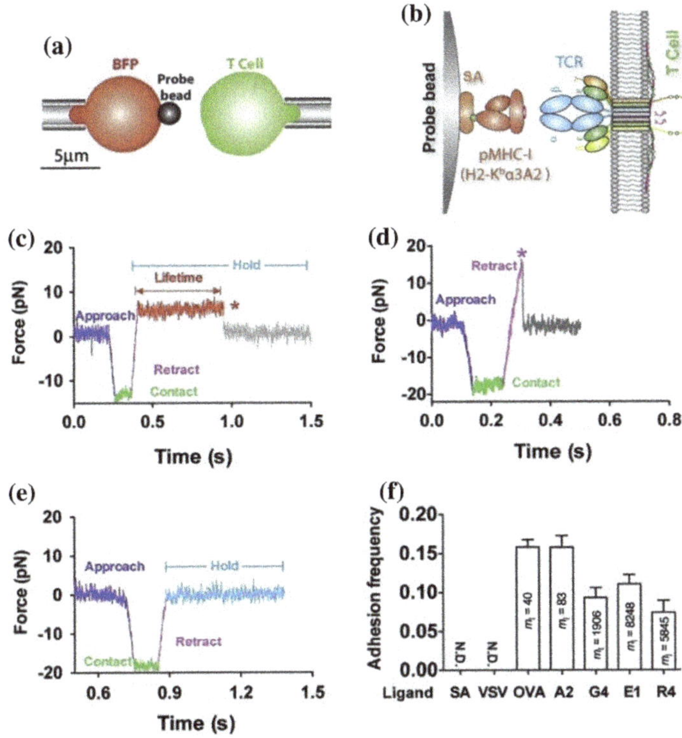

Fig. 13.4 **a** BFP schematic. A micropipette-aspirated RBC with a probe bead attached to the apex (left) was aligned against a T cell held by an apposing pipette (right). **b** BFP functionalization. The probe bead was covalently linked to SA to capture pMHC (left) to interact with TCR (right). **c**–**e** Representative force traces of measurement cycles showing adhesion that survived ramping and sustained a preset level of force until dissociation (marked by a red star), enabling bond lifetime measurement (**c**), adhesion ruptured by a ramp force (marked by a magenta star) before reaching the set force or in force–ramp assay (**d**), or no adhesion (**e**). **f** Binding specificity. Mean ± SEM of adhesion frequencies of >10 T cell-bead pairs with 50 contacts for each. Densities of pMHCs (ml) are indicated inside of each bar. N.D., not detected. Reprinted from Ref. [18], Copyright 2014, with permission from Elsevier

13.2.3 Optical Tweezers-Based Single-Cell Force Spectroscopy

Optical tweezers, or optical trap, use a highly focused laser beam to trap and manipulate microsized dielectric particles. The attractive or repulsive force produced by the trap is typically on the order of piconewtons depending on the intensity of laser power, the shape of laser focus (objective-dependent), the size and shape of the trapped particles, and the refractive index mismatch between the trapped particles and the surrounding medium. The trapped particles experience two types of forces, namely the scattering force produced by the photons striking the particle along the light propagation direction and the gradient force produced by a gradient of electric field intensity of the laser focus (Fig. 13.5). These two forces are defined by the wavelength (λ) of the laser beam and the particle size (r). The trapping scenarios can be divided into three regimes: Mie regime ($r \gg \lambda$); Rayleigh regime ($r \ll \lambda$); and regime in between them ($r \sim \lambda$). Since most of the biological applications deal with cells and microspheres whose sizes are much greater than the wavelength of light, the Mie regime is discussed here.

In the Mie regime, the trapping mechanism can be explained using ray optics or the conservation of momentum. As shown in the Fig. 13.5, individual ray of light is refracted when passing through a dielectric particle. Thus, the ray will exit in a different direction from which it originated. Because light has a momentum associated with it, this change in direction means a change in momentum. According to Newton's third law, there should be an equal and opposite momentum change on the particle. Most optical tweezers operated with a Gaussian beam profile intensity. Under this condition, if the particle is displaced from the center of the beam, the particle has a net force returning it to the center of the trap because more intense beams impart a larger momentum change toward the center than less intense beam, which impart a smaller momentum change away from the center. The net force restores the particle to the trap center. If the particle is right at the center of the beam, the net lateral force is zero since all the refracted rays are symmetrical and momentum changes cancel each other out. The net force in this case is along the axial direction, which cancels out the scattering force of the laser light. If the particle is displaced away from the focal center along the axial direction in either way, the net force will bring the particle back to the laser focus.

The trapping capability allows for physical traction of microscopic objects such as cell or functionalized microspheres in a liquid medium without contact. This unique feature is highly desirable for cell manipulation and optical sorting in sterile conditions. In addition, if the trap stiffness (k) is known, the restoring force (F) from the optical tweezers can be calculated by $F = kx$ where x is the displacement measured from the trap center. The typical force range (1–200 pN) offered by optical tweezers is well suited for probing the weak interactions on the cell surface at the single molecule level. Weisel and his colleagues studied the binding intensity and activation state of single fibrinogen–integrin pairs on live-cell surfaces [20]. The platelet integrin, αIIbβ3, is an adhesion receptor involved in the formation of

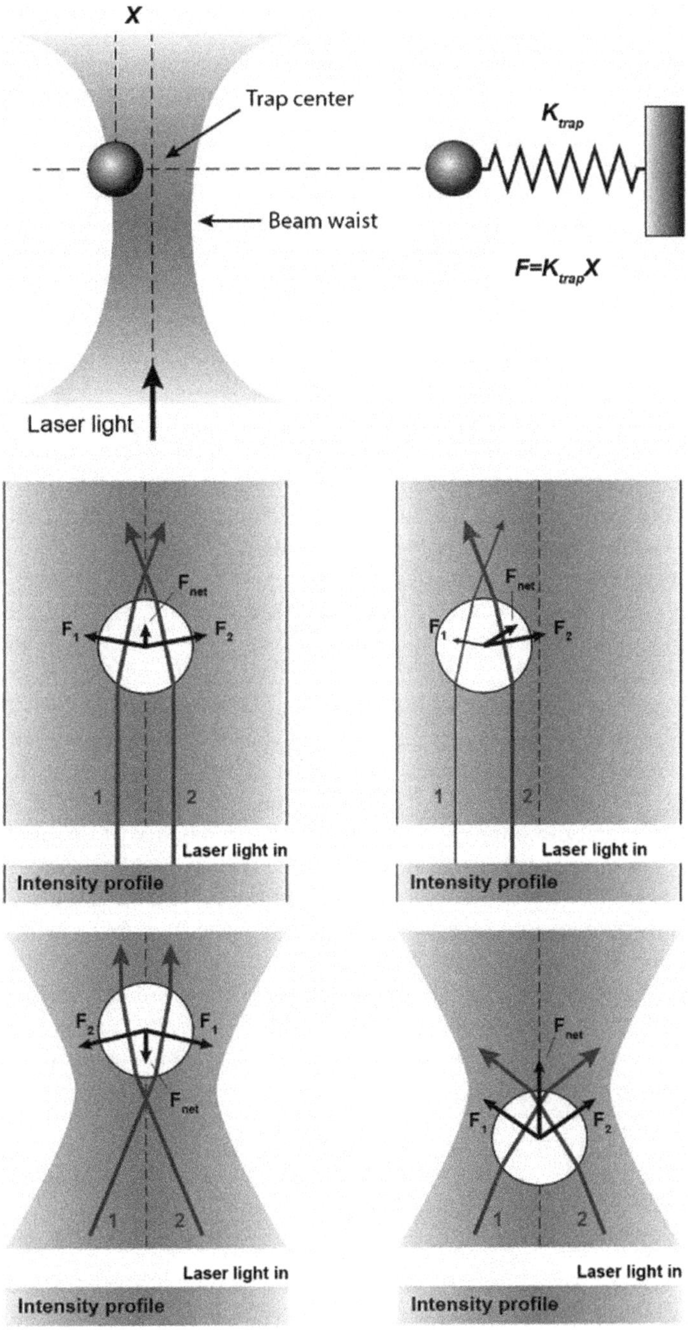
X
Trap center
K_{trap}
Beam waist
$F=K_{trap}X$
Laser light
F_{net}
F_1
F_2
1
2
Laser light in
Intensity profile
F_{net}
F_1
F_2
1
2
Laser light in
Intensity profile
F_2
F_1
F_{net}
1
2
Laser light in
Intensity profile
F_{net}
F_1
F_2
1
2
Laser light in
Intensity profile

◂**Fig. 13.5** (upper) A dielectric particle is attracted to the center of the beam, slightly above the beam waist. The force applied on the particle depends linearly on its displacement from the trap center just as with a simple spring system. (middle) Ray optics explanation for unfocused laser. When the particle is displaced from the beam center (right image), the larger momentum change of the more intense rays causes a net force to be applied back toward the center of the laser. When the particle is laterally centered on the beam (left image), the resulting lateral force is zero. But an unfocused laser still causes a force pointing away from the laser. (lower) Ray optics explanation for focused laser. In addition to keeping the bead in the center of the laser, a focused laser also keeps the bead in a fixed axial position: The momentum change of the focused rays causes a force toward the laser focus, both when the particle is in front (left image) and behind (right image) the laser focus. So, the particle will stay slightly behind the focus, where this force compensates the scattering force

thrombi. It is inactive on resting platelets and can be activated by agonists such as ADP and thrombin. The activity of αIIbβ3 needs to be tightly regulated to prevent the spontaneous thrombi formation. The authors employed optical tweezers to quantify the rupture force between individual fibrinogen molecules and either purified αIIbβ3 or αIIbβ3 on the surface of living platelets. They found that the rupture forces ranged from 60 to 150 pN with a peak value at 80–100 pN. Interestingly, platelet stimulation by either ADP or the thrombin receptor-activating peptide only enhanced the accessibility but not the binding strength of single αIIbβ3 molecules, suggesting that there may exist two conformational states of αIIbβ3 activation.

The forced-based techniques afford two main advantages into live-cell studies. First, because the mechanical stimuli are applied to single cells, specific binding events can be manually controlled at the single-cell level unlike most biochemical ensemble assays where binding interactions take place randomly and synchronization is inherently impossible. Second, the precise control of the binding events allows the subsequent cell responses to be probed hierarchically, particularly in combination with fluorescence imaging. In addition, spatial manipulation of individual cells using optical tweezers greatly facilitates the study of more sophisticated scenarios involving sequential/multiple cell interactions such as DC-Treg-Tconv interactions discussed before. Thus, the force-based techniques will continue to expand and develop as indispensable tools to not only probe the forces involved in membrane events, but also deliver both mechanical and chemical stimuli in a spatiotemporally controlled manner for cell activation studies.

13.3 Light-Based Techniques

Among optical signals, fluorescence is the most favorable readout because of its high sensitivity and selectivity. The nearly noninvasive nature of fluorescence is particularly appreciated in cell biology. The advance on live-cell single molecule fluorescence detection together with super-resolution imaging has pushed the optical probing limit to the molecular scales (see Chaps. 5 and 6). Here, we focus on light-based sensing and activating methods for cell signaling studies at the plasma membrane.

13.3.1 FRET-Based Tension Sensor

Focal adhesions (FAs) are large and complex macromolecular structures on the cell membrane through which mechanical forces are transmitted between intracellular actin bundles and the extracellular substrate. The structure of FAs is highly dynamic and undergoes assembly and disassembly in response to intensity of force experienced by the cell. Vinculin is an intracellular FA protein bearing a head domain (Vh) and a tail domain (Vt) with a flexible linker placed in between. Vh binds to talin, whereas Vt binds to F-actin and paxillin. The recruitment of vinculin to FAs is regulated by mechanical forces produced either externally or internally. However, where and when forces effects on vinculin take place at the subcellular level is not known. Ha and Schwartz groups designed a protein-based tension sensor module for vinculin [21]. The sensor has a 40-aminoacid-long elastic domain flanked by two fluorescent proteins (mTFP1 and venus (A206 K)) that undergo efficient fluorescence resonance energy transfer (FRET) (Fig. 13.6). The elastic linker domain is derived from the spider silk protein flagel liformand acts as a nanospring with piconewton force sensitivity. This sensor sequence was then inserted between the head and tail domains of vinculin, resulting in a complete vinculin tension sensing construct. If the construct experiences any tension, the elastic linker changes its length accordingly, which will lead to the change of FRET efficiency. Certainly, to link the FRET efficiency to the exact force, the construct needs to be

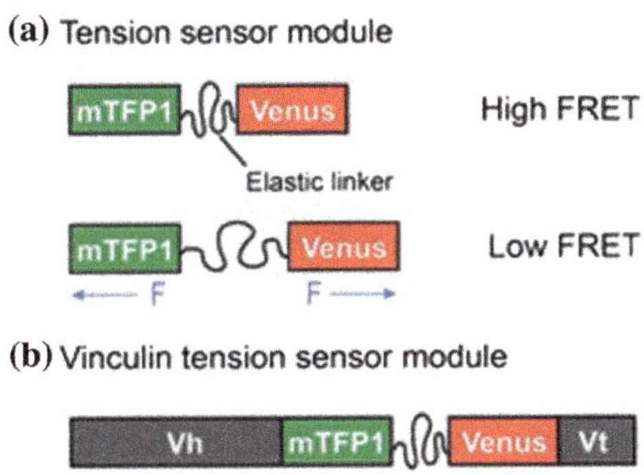

Fig. 13.6 Vinculin tension sensor constructs. **a** The tension sensor module contains two fluorophores separated by an elastic linker. When a stretching force acts on the tension sensor module, the elastic linker will extend, resulting in a low FRET efficiency (F, force). **b** The tension sensor module is then inserted between the Vh domain and Vt domain of vinculin, forming a functional vinculin tension sensor module

calibrated in vitro first by optical tweezers-based single molecule fluorescence force spectroscopy. With this new fluorescence-based tension sensor, the authors were able to determine the tension at about 2.5 pN across vinculin in stable FAs. The highest tension is associated with FA assembly and enlargement, whereas low tension is linked to FA disassembly at the trailing edge of migrating cells. Surprisingly, vinculin recruitment and force transmission are controlled independently. This work reveals an unexpected regulatory mechanism in which the fate of FAs under tension is dependent on whether vinculin bears force or not. Noninvasive nature of fluorescence together with the modular design of tension sensor makes this method very appealing for intracellular force probing at the molecular level.

Shortly after, Salaita et al. reported a molecular tension sensor that can be used to map forces exerted by cell surface receptors [22]. The sensor consists of a flexible linker covalently connected with a ligand at one end and anchored onto a surface through the other end (Fig. 13.7). The linker is a short piece of polyethylene glycol (PEG) polymer whose force–extension behavior has been well characterized, thus acting as a spring with a known constant. The ligand and the anchor of the linker are functionalized with fluorophore and quencher molecules, respectively. Cellular forces exerted on the ligand will stretch the linker and bring the fluorophore away from the quencher, thus leading to increased fluorescence intensity due to low FRET or quenching efficiency. This provides a signal to map mechanical tension transduced through specific receptors. In addition, precise quantification of the force magnitude can be performed since the conversion of the distance from FRET measurements to force is possible due to the well-known elastic property of PEG. With this approach, the authors were able to show that approximately 4 pN force was exerted through EGF receptor upon binding to its cognate ligand, corresponding to the lower-bound ensemble average force applied by the EGF receptor.

The above two methods share similar physics, but they are different in two ways. The first sensor is protein-based and designed to probe tensions inside the cell. The second sensor is polymer-based and suitable for probing tensions on the cell

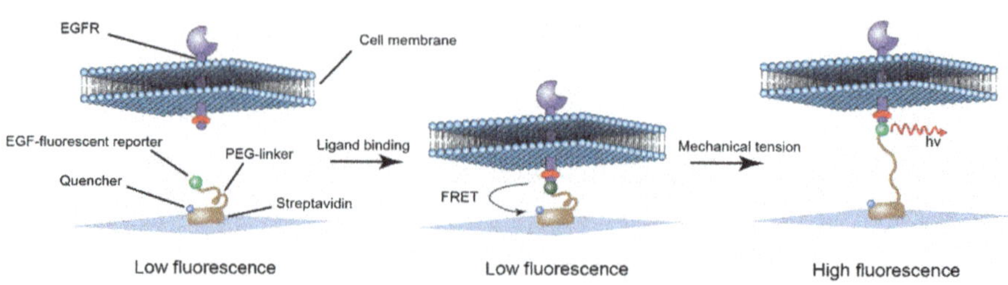

Fig. 13.7 Schematic of the EGF-PEG tension sensor and the sensing mechanism. The sensor comprised of a PEG polymer flanked by fluorescently labeled EGF ligand and a biotin moiety for surface immobilization via streptavidin capture. The unbounded sensor has a low fluorescence emission due to the close proximity between the quencher and reporter (high FRET efficiency). Upon ligand binding, the pulling force exerted by EGFR will stretch the PEG linker. The displacement of the ligand leads to an increase in the measured fluorescence intensity, thus reporting the transmission of mechanical tension through the EGF-EGFR complex. hν, emission of a photon

surface. The first sensor requires fluorescence lifetime imaging because low emission rate of fluorescent proteins hinders precise FRET efficiency quantification from intensity domain. The second sensor employs much brighter organic dyes, and FRET value can be readily probed by intensity measurements. Overall, both sensors are able to offer piconewton force sensitivity once properly calibrated and have high extendibility. Lastly, they are functionally complementary in molecular tension probing.

13.3.2 DNA-Based Tension Gauge Tether

Single molecule force spectroscopy has been very useful to quantify binding intensity at the single molecule level. However, it fails to reveal the single molecule forces required for physiological functions. Ha et al. proposed DNA-based tension gauge tether (TGT) to determine the single molecule forces required for mechanical signaling in cells (Fig. 13.8) [23]. Double-stranded DNA is chosen as a rupturable

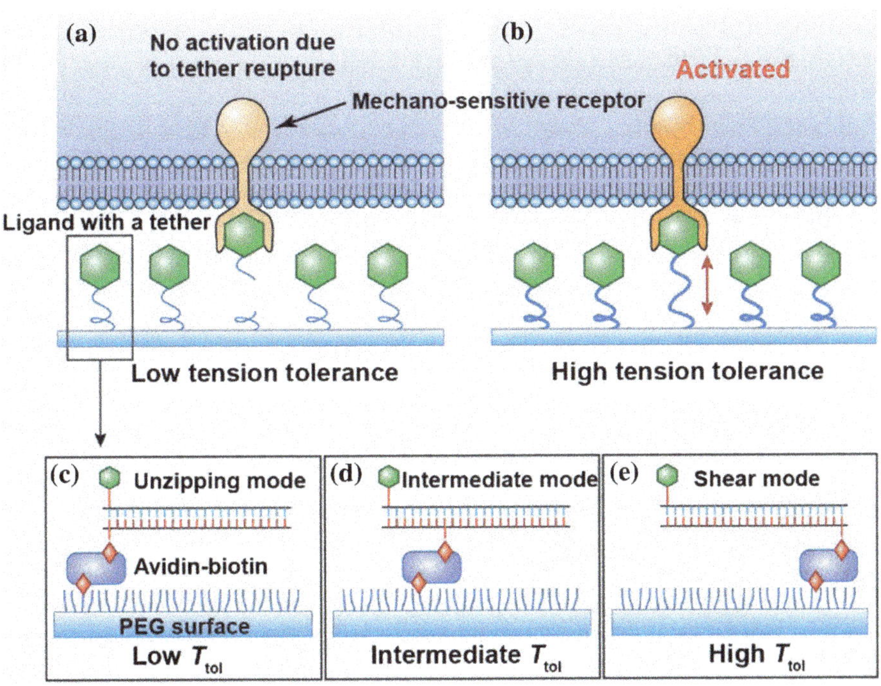

Fig. 13.8 Working principle of tension gauge tether (TGT). A ligand is anchored on the surface via a tether that ruptures if the tension applied by the cell through the receptor is larger than its tension tolerance (T_{tol}). If the tension required for activation exceeds T_{tol} and ruptures the tether, signaling through the receptor is not activated (**a**) and is activated if T_{tol} is larger than the required tension (**b**). **c–e** DNA duplex helix as a tether with tunable T_{tol} values

tether because the measured rupture force is dependent on force application geometry. For example, the rupture force of a 21-base-pair DNA is about 12 pN in the unzipping geometry and is about 56 pN in the shear geometry, determined by the nature of double-stranded separation. Intermediate rupture forces can be found by applying forces in an internal position on the DNA duplex. If a ligand is attached to different tethers with a range of critical rupture forces or "tension tolerances," the specific ligand–receptor force required to activate signaling events can be tested at the single molecule level. The authors were able to show that cells apply a universal peak tension of $\sim$40 pN to single integrin–ligand bonds during initial adhesion. In addition, they found that the force required to activate Notch receptors is less than 12 pN.

Strictly speaking, TGT is not a fluorescence-based technique, but very often the activation readout or cell response is probed with fluorescence, and thus, it is discussed in this section. Recent work by Tolar et al. indicates that B cell receptors are also influenced by physical cues from the antigen [24]. However, it is not understood how the mechanical force is involved in B cell activation? Liu et al. employed similar TGT approach to investigate how the different B cell receptors (BCRs) are activated mechanically [25]. They conjugated a BCR-specific antigen, 4-hydroxy-3-nitrophenylacetyl (NP), to the ligand chain of the TGT system and prepared 8 TGT sensors with predefined forces of 12, 16, 23, 33, 43, 50, 54, and 56 pN, respectively. The activation of BCRs in response to these sensors was probed by real-time monitoring the accumulation of fluorescently labeled BCRs with high-speed TIRF imaging. They showed that the activation of IgM-BCRs depends on the magnitude of mechanical force applied. Low levels of force only weakly activated the receptors, whereas higher levels of force led to more robust activation. Strikingly, both IgG-BCR and IgE-BCR on memory B cells can be triggered by small amounts of mechanical force (less than 12 pN) or no tension at all. This may explain the low threshold that memory B cells used to be quickly activated upon the same antigenic challenge.

13.3.3 FRET-Based Charge Sensor

Reversible association of proteins with the plasma membrane has been a key regulatory feature in many cellular events such as endocytosis, exocytosis, and signaling [26–28]. Protein–membrane associations are governed by both hydrophobic and electrostatic interactions. The former involves the insertion of aromatic amino acid into the hydrophobic core of the lipid–bilayer and lipid modification of protein such as palmitoylation, myristoylation, and farnesylation [29]. The latter refers to the fact that negatively charged phospholipid head group at the inner surface of the membrane can attract proteins with polybasic motifs. Prior to the work by Xu's group discussed in AFM-SCFS section, they used FRET to probe the distance between CD3 chain and the plasma membrane. They showed that Ca^{2+} influx triggered by TCR activation induces dissociation of CD3 tail from the membrane, thus exposing the buried tyrosine residues for subsequent phosphorylation [30]. The effect of Ca^{2+} on facilitating CD3 phosphorylation is mainly due to the charge of this ion that binds to the negatively charged phospholipids competitively. Similar mechanism was also found in IgG-B cell receptor activation where membrane-sequestered mIgG tail via ionic protein–lipid interaction can be released by antigen engagement or Ca^{2+} mobilization in the initiation of B cell activation [31].

Anionic phospholipids at the inner leaflet of the plasma membrane generate a negative potential of ~30 mV. Although there are fluorescence reporters that can be used to probe changes in membrane potential, quantification of the effective electrostatic potential of cellular membrane has been challenging. Gaus et al. developed a FRET-based fluorescence membrane charge sensor (MCS) that can report on the change of membrane charges in live cells [32]. MCS is based on the earlier sensor R-pre designed by Grinstein et al. The sensor contains two membrane attachment units (MA1 and MA2) on either side of the two fluorescent proteins (FP1 and FP2) that are connected with a short linker (Fig. 13.9). MA1 has a permanent, hydrophobic membrane anchor that stably associates the sensor with the plasma membrane and comprises of either a myristoylation and two palmitoylation groups or a myristoylation group and a stretch of positively charged residues. MA2 is an electrostatic anchor, consisting of the original or modified R-pre. If the membrane potential is highly negative, the dipoles of two fluorescent proteins are aligned, leading to high FRET efficiency. When the membrane charge is reduced, or less negative, MA2 detaches from the membrane, while MA1 remains. The detachment of MA2 will cause a change in the dipole alignment of the two emitters and thus a reduced FRET value. It should be noted that the change of FRET efficiency here is mainly due to the change of relative orientation of the two fluorophores instead of the distance between them. MCS allows for quantification

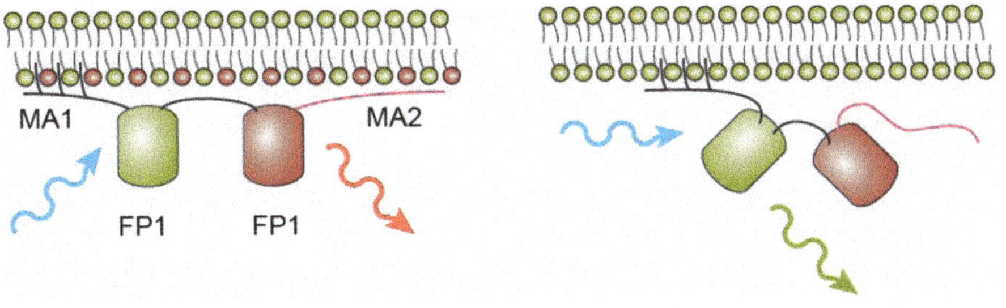

Fig. 13.9 FRET-based membrane charge sensor. The charge sensor comprises two membrane attachment (MA) units and two fluorescent proteins (FP) separated by a short linker. MA1 is a hydrophobic anchor containing a palmitoylation and myristoylation group that permanently associates with the membrane. MA2 (pink) is an electrostatic anchor that associates with the inner surface of the plasma membrane in a charge-dependent manner. When MA2 is attached to the membrane, the FRET efficiency is high; when MA2 is detached from the membrane, the FRET efficiency is low

of the effective electrostatic potential modified by Ca^{2+} influx, hypo and hypertonic treatments, redistribution of PS, and changes in PI phosphorylation, and even can be used to distinguish charged membrane domains in the immunological synapse.

13.3.4 Photoactivatable Triggering System

In conventional cell activation assays, the engagement of receptors to agonist ligands is a diffusion-limited process. This will at times make synchronized cell activation difficult to predict real-time optical observation of cell responses challenging. Therefore, a triggering system that can be controlled spatiotemporally is highly desired for precise cell activation and response studies. Although the force-based techniques meet this requirement perfectly, these experiments normally involve specialized and sometime expensive setups, thus not very practical in reality. To overcome this limitation, a controllable activation mechanism based on light-induced chemical reaction was proposed. The basic idea is that a reactive chemical group or molecule is first chemically modified by adding a light-sensitive protection structure, forming an inactive state or a caged state. The protected reactive group can regain its activity, becoming uncaged, upon UV illumination which induces specific chemistry to remove the "cage." Such photoactivatable systems have been used to study the kinetics of second messenger activity via caged calcium [33] and caged cAMP [34]. The relevant applications were also found in immunoreceptor activations. Davis et al. investigated TCR signaling events with major histocompatibility complexes bearing a caged antigenic peptide that can be photoactivated by UV light [35]. UV irradiation of these complexes in contact with cognate T cells allows downstream signaling events to be analyzed with high temporal resolution. The authors showed that phosphorylation of the LAT adaptor protein occurred in 4 s and diacylglycerol production and calcium flux was observed in 6–7 s. TCR activation also led to cytoskeletal polarization within 2 min. CD4 antibody blocking reduced LAT phosphorylation level and reduced the speed of calcium flux. This work establishes the speed and localization of early signaling steps and has important implications regarding the overall hierarchy of TCR signaling pathway.

Liu et al. recently investigated the dynamic responses during the B cell activation with caged model antigen NP (caged-NP) (Fig. 13.10) [36]. This photoactivatable antigen system in combination with live-cell single molecule fluorescence imaging enabled them to identify previously uncharacterized B cell probing termination behavior and reveal the precise BCR sorting mechanisms during B cell activation. The resting B cells exhibit probing behavior with constant extension of membrane pseudopods in random directions, which is strictly associated with F-actin remodeling. Such probing behavior was terminated within 4 s after

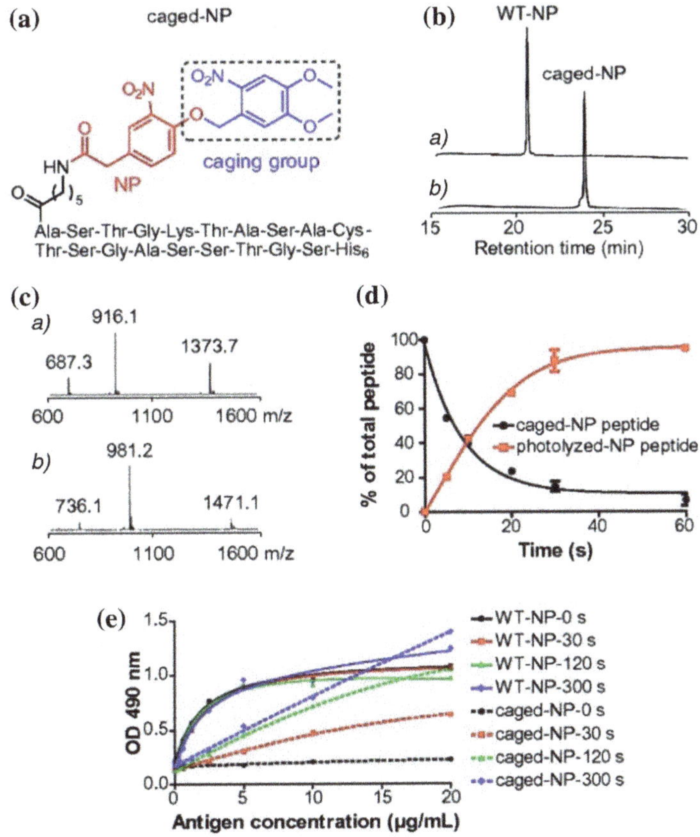

Fig. 13.10 Development of a caged-NP-based photoactivatable antigen system. **a** The conjugation of a UV-sensitive moiety, DMNB (caging group), to the -OH group of the NP hapten antigen to generate DMNB-NP. **b** The RP-HPLC analysis of WT-NP or caged-NP. **c** The ESI-MS analysis of WT-NP or caged-NP. **d** The photolysis kinetics of the caged-NP peptide on photoactivation with UV for different times as examined by RP-HPLC to quantify the percentage of photolyzed peptides in the total peptide population. **e** The ELISA evaluation of the binding capacity of anti-NP antibodies to WT-NP- or caged-NP-peptide before or after photoactivation for different time. Reprinted from Ref. [36] with permission by Wang et al.

photoactivation, indicating that this response was sensitive and specific to BCR activation. The termination of B cell probing was accompanied with the synaptic accumulation of the BCRs due to the decreased lateral mobility of antigen-engaged BCRs. Thus, the antigen-binding-induced trapping of BCRs into microclusters is a fundamental mechanism for B cells to acquire antigens.

Photoactivatable triggering systems not only offers a spatiotemporal control on cell activations, but also allows for probing the very same cells before and after activation, which cannot be achieved with conventional approaches.

13.4 Electrical Signal-Based Techniques

13.4.1 Patch-Clamp Techniques

Many ions have a concentration gradient across the cell membrane. For instance, potassium (K^+) is at high concentration inside the cell and a low concentration outside the cell membrane. Sodium (Na^+) and chloride (Cl^-) ions on the other hand have a reversed gradient. These concentration gradients on opposite sides of the cell membrane produce a voltage across the membrane, known as the membrane potential. The concentration differentials are established by actively transporting ions through ion channels on the membrane. Depending on what triggers the opening and closing of channels, ion channels can be classified into two groups, i.e., voltage-gated ion channels, whose gating is controlled by the voltage across the membrane (membrane potential), and ligand-gated ion channels that open or close depending on binding of ligands to the channel. Eukaryotic cells have a negative voltage in the cell interior as compared to the cell exterior ranging from −40 to −80 mV. The membrane potential creates an electrical field across the membrane and transmembrane domains of any proteins bearing charged residues will sense the field and thus adopt certain conformation to minimize the free energy. Thus, the change of membrane potential will induce the conformational change of the charged domain. In all, membrane potential functions as a power source to drive "molecular devices" embedded in the membrane such as voltage-gated ion channels. In addition, in electrically excitable cells such as neurons and muscle cells, membrane potential is used for transmitting signals between different parts of a cell.

The flow of ions can be thought as electrocurrent flow (*I*), whereas the membrane potential driving the current flow is voltage (*V*). The ion channel itself or the path for current flow is resistance (*R*). The relationship among them follows Ohm's law: $V = IR$. Ohm's law predicts that if the voltage across a membrane is held constant (voltage clamp), changes in current through the membrane will reflect changes in the resistance of the membrane, i.e., the opening and closing of ion channels. This provides the basis for the voltage clamp technique.

The first attempt to characterize the biophysical property of ion channels was carried out on the squid giant axon by Hodgkin and Huxley in 1952 [37]. Later Neher and Sakmann made a breakthrough by developing a new technique with the use of micropipettes to form high-resistance seals on tiny patches of cell membrane, thus reducing the surface area of the membrane being probing [38]. This allowed, for the first time, high-fidelity recordings of single ion channels. This new technique was termed patch clamp, a refinement of the voltage clamp. Patch-clamp recording uses a glass micropipette (patch pipette) as a recording electrode and another electrode in the bath around the cell as a reference ground electrode. Depending on questions of interest, patching can be performed in several configurations (Fig. 13.11). The inside-out and outside-out configurations are called "excised patch" because the patch is excised or removed from the cell body. Cell attached and both excised patch are used to probe the behavior of single ion channels in the

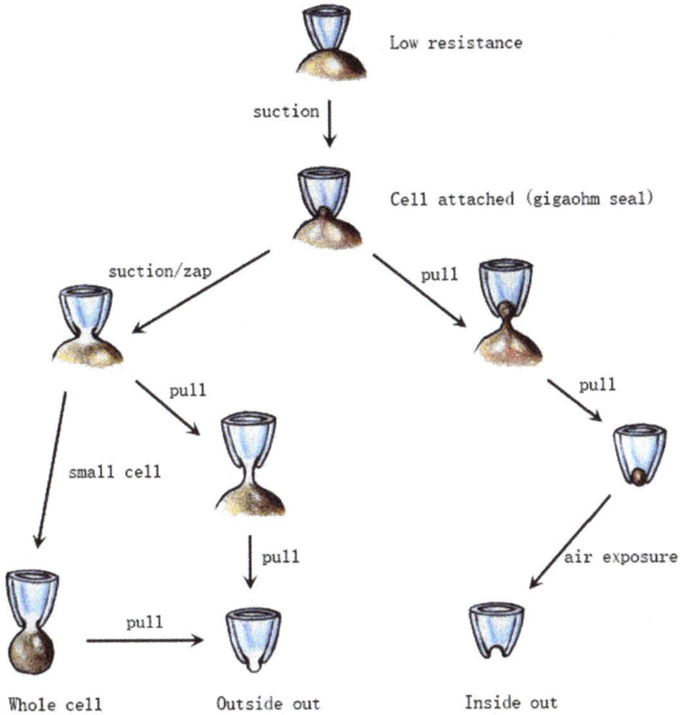

Fig. 13.11 Diagram showing various configurations of the patch-clamp technique and how they are obtained. Reproduced from Ref. [39], Copyright 2007, with permission from Elsevier

membrane attached to the electrode. Whole cell patch allows the study of the ensemble response of all ion channels within a cell's membrane. All these patching methods involve the initial formation of a high-resistance gigaseal between the tip of the recording pipette and the cell membrane. This high-resistance seal prevents the leakage of ion between the pipette and the membrane, so that any detected current signal can only originate from the ion movement through the ion channels.

As a particular study topic, ion channels that can be activated by mechanical force have been implicated in numerous biological processes such as sensing touch/pain (somatosensation), sound (hearing), and shear stress (cardiovascular physiology). However, the identities of these ion channels have remained elusive. In 2010, Patapoutian et al. identified two candidate proteins, mouse Piezo1 (MmPiezo1) and MmPiezo2, required for mechanically activated (MA) current through expression profiling and RNA interference knockdown [40]. These two proteins are very large proteins with 30 putative transmembrane domains. They do not resemble any known ion channels or other protein classes. Two years later, the same group also demonstrated pore-forming properties including sensitivity to the pore blocker ruthenium red and single-channel conductance on *Drosophila melanogaster* Piezo (DmPiezo) [41]. Purified MmPiezo1 reconstituted into asymmetric lipid bilayers and liposomes forms ruthenium-red-sensitive ion channels.

The gating of ion channels involves conformational changes of the pore-forming proteins. These conformational changes can be probed with fluorescence-based techniques. When combined with patch-clamp methods, both the function and structure of ion channels can be monitored simultaneously. This combinational approach, or patch-clamp fluorometry (PCF), has advanced our knowledge of ion channel biophysics [42]. The classical PCF is based on the inside-out configuration of the patch-clamp technique. The conformational probing is achieved by FRET sensors either with two fluorophores or with only one fluorescent reporter plus ionic quenchers. The first application of this kind was demonstrated by Zheng and Zagotta in 2000 [43]. They studied the conformational rearrangements during cyclic nucleotide-gated (CNG) channel activation. Alexa Fluor 488 was labeled via cysteine residue near the residue C481 in the C-linker region of CNGA1 subunits (Fig. 13.12) after inside-out patching. Quenching was assessed by two ionic quenches, iodide and thallium, respectively. They showed that the anionic quencher iodide had a higher quenching efficiency in the channel's closed state, whereas thallium ion, a cationic quencher, had a higher quenching efficiency in the open state. The results indicate that positively charged or dipolar residues in the C-linker near C481 in the closed state reorganize upon channel activation. Later on, a more detailed structural characterization of CNG channels was carried out with conventional two-fluorophore-based FRET by this same group in 2007 [44]. This time FRET was measured between either GFP or Alexa Fluor 488 and Alexa Fluor 568 inserted at different sites in the C-terminus (donor) and a small, negatively charged nonfluorescent membrane probe, dipicrylamine (acceptor). Again, structural dynamics via FRET probing and channel function through current recording were

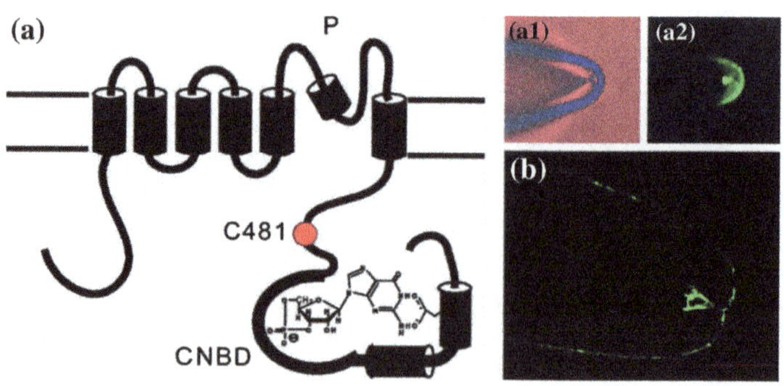

Fig. 13.12 Membrane topology of a cyclic nucleotide-gated channel subunit illustrating the position of cysteine C481 (left) and imaging CNG channels in inside-out patches (right). **a** Pipette with patch membrane from an oocyte expressing C481 channels after labeling with 2 mM Alexa 488 for 3 min in 1 m McGMP solution. **a1** bright field; **a2** fluorescence. **b** Confocal image of a patch pipette after Alexa 488 labeling. Scale bar is 50 μm. Reprinted from Ref. [43], Copyright 2012, with permission from Elsevier

simultaneously performed. They were able to show that the domains within the cytoplasmic region of the channel did not move perpendicular to the membrane during gating. There was only a subtle lateral movement within the C-linker upon channels opening. So far, PCF has only been performed on a small ensemble level. The next challenge would be applications of PCF at the single-molecule level even in the cell-attached patching configuration.

13.4.2 Scanning Ion Conductance Microscopy

Scanning ion conductance microscopy or SICM is a non-contact scanning probe microscopy technique which uses a glass nanopipette as a sensitive probe that detects the proximity of a surface via a decrease in the ion current flowing through the pipette without any physical contact with the surface. SICM allows visualization of the three-dimensional surface topography of living cells with a resolution typically 50–100 nm (depending on the inner diameter of the nanopipette). SICM can be operated in four imaging modes: constant-z mode, direct current (constant distance) mode, alternating current mode, and hopping mode. In constant-z mode, the probing pipette is held at a constant-z height, while it is scanned laterally and the resistance is monitored. Variations of the resistance can be used to reconstruct the topography of the sample. This mode is fast but is barely used since it only works on very flat sample, not suitable for complex biological samples such as cells. In direct current (DC) mode, the distance between the pipette and the sample during scanning is kept constant by maintaining a predefined resistance. Thus, the Z-position of the pipette determines the topography of the sample. This mode does not work on samples with steep slopes and is prone to electrode drift. In alternating current (AC) mode, the pipette oscillates vertically in addition to its usual movement. Thus, the resistance also oscillates and the amplitude of this oscillation serves as feedback to control the height of the pipette. AC mode responses much faster than DC mode and allows for the recording of more complex samples. Hopping mode works similar to DC mode, expect that in each probing cycle, the pipette, once reaching a given resistance, is dragged back before moving to the next position laterally. Although the scanning speed is very low in hopping mode, it is able to image complex topography without distorting the sample surface.

Apart from imaging capability, SICM can also be used as patch clamp to probe electrical properties in various subcellular regions with distinct structure since these two techniques share the same physics [45]. When combined with fluorescence imaging, SICM can help to correlate subcellular signaling response with the membrane topography. For example, Gorelik et al. determined both position and function of endogenous β-adrenergic receptor (AR) subtypes by combining SICM with FRET-based measurements of cAMP production after local receptor stimulation [46]. They found that β_2ARs were selectively localized in the T-tubules of cardiomyocytes from healthy adult rats and mice, whereas β_1ARs were evenly distributed all over the cell membrane. In cardiomyocytes from a rat model of chronic heart failure, β_2ARs were relocated from the T-tubules to the cell crest, resulting in diffuse receptor-mediated cAMP signaling. Therefore, the redistribution of β_2ARs in heart failure alters compartmentation of cAMP, which might play an important role in the development of heart disease.

13.5 Perspectives

The increasing use of biophysical probes in living cell studies comes at a time when biophysical aspects of cellular functions are moving to the center of cell biology research, no longer as a "supplementary" investigation piggybacked to other mainstream technologies. These technologies are each suitable for a specific set of membrane biology. Using SCFS as an example, different from single molecule force spectroscopy or other techniques such as plasma resonance, the studies with SCFS are in theory more "holistic" as they take into the consideration a living cell environment. Combined use of fluorescent labeling or FRET and a cohort of force recording devices is most ideal in analyzing the sequential and spatial arrangements of signaling cascades. With our deepening understanding of membrane biology, a receptor–ligand interaction on a cell surface is known to be affected by underlying lipid motion, composition, surface charge, and interactions with other surface structures. With technological advancements, these biophysical studies are becoming more accessible to most researchers and are in some ways analogous to early development of flow cytometry and fluorescence imaging. However, in preparation of this new dawn, several limitations should be systematically addressed. For instance, there needs to be a collective effort to standardize the protocols and the analytical methods. The field also needs to work on more streamlined coupling force/current measurements to biochemical signaling recording (i.e., simultaneous fluorescence imaging). These efforts will greatly speed up the extended use of these new tools in laboratory-based biology research, paving the way for its future developments in clinical diagnosis and pharmacological/therapeutic research.

Acknowledgements The authors thank Dr. Shaun Davis, University of Calgary, for producing Fig. 13.1 and Ms. Libing Mu, Tsinghua University, for producing Figs. 13.5 and 13.7–13.9.

References

1. Boland T, Ratner BD (1995) Direct measurement of hydrogen bonding in DNA nucleotide bases by atomic force microscopy. Proc Natl Acad Sci U S A 92(12):5297–5301
2. Helenius J, Heisenberg CP, Gaub HE, Muller DJ (2008) Single-cell force spectroscopy. J Cell Sci 121(11):1785–1791. doi:10.1242/jcs.030999
3. Bell GI (1978) Models for the specific adhesion of cells to cells. Science 200(4342):618–627
4. Thomas WE, Vogel V, Sokurenko E (2008) Biophysics of catch bonds. Annu Rev Biophys 37:399–416. doi:10.1146/annurev.biophys.37.032807.125804

5. Evans E, Ritchie K (1997) Dynamic strength of molecular adhesion bonds. Biophys J 72 (4):1541–1555. doi:10.1016/S0006-3495(97)78802-7
6. Dudko OK, Hummer G, Szabo A (2008) Theory, analysis, and interpretation of single-molecule force spectroscopy experiments. Proc Natl Acad Sci U S A 105(41): 15755–15760. doi:10.1073/pnas.0806085105
7. Benoit M, Gabriel D, Gerisch G, Gaub HE (2000) Discrete interactions in cell adhesion measured by single-molecule force spectroscopy. Nat Cell Biol 2(6):313–317. doi:10.1038/35014000
8. Ng G, Sharma K, Ward SM, Desrosiers MD, Stephens LA, Schoel WM, Li T, Lowell CA, Ling CC, Amrein MW, Shi Y (2008) Receptor-independent, direct membrane binding leads to cell-surface lipid sorting and Syk kinase activation in dendritic cells. Immunity 29(5):807–818. doi:10.1016/j.immuni.2008.09.013
9. Flach TL, Ng G, Hari A, Desrosiers MD, Zhang P, Ward SM, Seamone ME, Vilaysane A, Mucsi AD, Fong Y, Prenner E, Ling CC, Tschopp J, Muruve DA, Amrein MW, Shi Y (2011) Alum interaction with dendritic cell membrane lipids is essential for its adjuvanticity. Nat Med 17(4):479–487. doi:10.1038/nm.2306
10. Chen J, Ganguly A, Mucsi AD, Meng J, Yan J, Detampel P, Munro F, Zhang Z, Wu M, Hari A, Stenner MD, Zheng W, Kubes P, Xia T, Amrein MW, Qi H, Shi Y (2017) Strong adhesion by regulatory T cells induces dendritic cell cytoskeletal polarization and contact-dependent lethargy. J Exp Med 214(2):327–338. doi:10.1084/jem.20160620
11. Yan J, Liu B, Shi Y, Qi H (2017) Class II MHC-independent suppressive adhesion of dendritic cells by regulatory T cells in vivo. J Exp Med 214(2):319–326. doi:10.1084/jem.20160629
12. Hosseini BH, Louban I, Djandji D, Wabnitz GH, Deeg J, Bulbuc N, Samstag Y, Gunzer M, Spatz JP, Hammerling GJ (2009) Immune synapse formation determines interaction forces between T cells and antigen-presenting cells measured by atomic force microscopy. Proc Natl Acad Sci U S A 106(42):17852–17857. doi:10.1073/pnas.0905384106
13. Hoffmann S, Hosseini BH, Hecker M, Louban I, Bulbuc N, Garbi N, Wabnitz GH, Samstag Y, Spatz JP, Hammerling GJ (2011) Single cell force spectroscopy of T cells recognizing a myelin-derived peptide on antigen presenting cells. Immunol Lett 136(1): 13–20. doi:10.1016/j.imlet.2010.11.005
14. Lim TS, Mortellaro A, Lim CT, Hammerling GJ, Ricciardi-Castagnoli P (2011) Mechanical interactions between dendritic cells and T cells correlate with T cell responsiveness. J Immunol 187(1):258–265. doi:10.4049/jimmunol.1100267
15. Lim TS, Goh JK, Mortellaro A, Lim CT, Hammerling GJ, Ricciardi-Castagnoli P (2012) CD80 and CD86 differentially regulate mechanical interactions of T-cells with antigen-presenting dendritic cells and B-cells. PLoS One 7(9):e45185. doi:10.1371/journal.pone.0045185
16. Puech PH, Nevoltris D, Robert P, Limozin L, Boyer C, Bongrand P (2011) Force measurements of TCR/pMHC recognition at T cell surface. PLoS One 6(7):e22344. doi:10.1371/journal.pone.0022344
17. Guo X, Yan C, Li H, Huang W, Shi X, Huang M, Wang Y, Pan W, Cai M, Li L, Wu W, Bai Y, Zhang C, Liu Z, Wang X, Zhang XF, Tang C, Wang H, Liu W, Ouyang B, Wong CC, Cao Y, Xu C (2017) Lipid-dependent conformational dynamics underlie the functional versatility of T-cell receptor. Cell Res 27(4):505–525. doi:10.1038/cr.2017.42
18. Liu B, Chen W, Evavold BD, Zhu C (2014) Accumulation of dynamic catch bonds between TCR and agonist peptide-MHC triggers T cell signaling. Cell 157(2):357–368. doi:10.1016/j.cell.2014.02.053
19. Xu CQ, Gagnon E, Call ME, Schnell JR, Schwieters CD, Carman CV, Chou JJ, Wucherpfennig KW (2008) Regulation of T Cell receptor activation by dynamic membrane binding of the CD3 epsilon cytoplasmic tyrosine-based motif. Cell 135(4):702–713. doi:10.1016/j.cell.2008.09.044

20. Litvinov RI, Shuman H, Bennett JS, Weisel JW (2002) Binding strength and activation state of single fibrinogen-integrin pairs on living cells. Proc Natl Acad Sci U S A 99(11):7426–7431. doi:10.1073/pnas.112194999
21. Grashoff C, Hoffman BD, Brenner MD, Zhou R, Parsons M, Yang MT, McLean MA, Sligar SG, Chen CS, Ha T, Schwartz MA (2010) Measuring mechanical tension across vinculin reveals regulation of focal adhesion dynamics. Nature 466(7303):263–266. doi:10.1038/nature09198
22. Stabley DR, Jurchenko C, Marshall SS, Salaita KS (2011) Visualizing mechanical tension across membrane receptors with a fluorescent sensor. Nat Methods 9(1):64–67. doi:10.1038/nmeth.1747
23. Wang X, Ha T (2013) Defining single molecular forces required to activate integrin and notch signaling. Science 340(6135):991–994. doi:10.1126/science.1231041
24. Natkanski E, Lee W-Y, Mistry B, Casal A, Molloy JE, Tolar P (2013) B cells use mechanical energy to discriminate antigen affinities. Science 340(6140):1587–1590. doi:10.1126/science.1237572
25. Wan Z, Chen X, Chen H, Ji Q, Chen Y, Wang J, Cao Y, Wang F, Lou J, Tang Z, Liu W (2015) The activation of IgM- or isotype-switched IgG- and IgE-BCR exhibits distinct mechanical force sensitivity and threshold. Elife 4. doi:10.7554/eLife.06925
26. McLaughlin S, Smith SO, Hayman MJ, Murray D (2005) An electrostatic engine model for autoinhibition and activation of the epidermal growth factor receptor (EGFR/ErbB) family. J Gen Physiol 126(1):41–53. doi:10.1085/jgp.200509274
27. Yeung T, Terebiznik M, Yu L, Silvius J, Abidi WM, Philips M, Levine T, Kapus A, Grinstein S (2006) Receptor activation alters inner surface potential during phagocytosis. Science 313(5785):347–351. doi:10.1126/science.1129551
28. Yeung T, Grinstein S (2007) Lipid signaling and the modulation of surface charge during phagocytosis. Immunol Rev 219:17–36. doi:10.1111/j.1600-065X.2007.00546.x
29. McLaughlin S, Aderem A (1995) The myristoyl-electrostatic switch: a modulator of reversible protein-membrane interactions. Trends Biochem Sci 20(7):272–276
30. Shi X, Bi Y, Yang W, Guo X, Jiang Y, Wan C, Li L, Bai Y, Guo J, Wang Y, Chen X, Wu B, Sun H, Liu W, Wang J, Xu C (2013) Ca2+ regulates T-cell receptor activation by modulating the charge property of lipids. Nature 493(7430):111–115. doi:http://www.nature.com/nature/journal/v493/n7430/abs/nature11699.html#supplementary-information
31. Chen X, Pan W, Sui Y, Li H, Shi X, Guo X, Qi H, Xu C, Liu W (2015) Acidic phospholipids govern the enhanced activation of IgG-B cell receptor. Nat Commun 6:8552. doi:10.1038/ncomms9552
32. Ma Y, Yamamoto Y, Nicovich PR, Goyette J, Rossy J, Gooding JJ, Gaus K (2017) A FRET sensor enables quantitative measurements of membrane charges in live cells. Nat Biotechnol 35(4):363–370. doi:10.1038/nbt.3828
33. Ellis-Davies GC, Barsotti RJ (2006) Tuning caged calcium: photolabile analogues of EGTA with improved optical and chelation properties. Cell Calcium 39(1):75–83. doi:10.1016/j.ceca.2005.10.003
34. Nerbonne JM, Richard S, Nargeot J, Lester HA (1984) New photoactivatable cyclic nucleotides produce intracellular jumps in cyclic AMP and cyclic GMP concentrations. Nature 310(5972):74–76
35. Huse M, Klein LO, Girvin AT, Faraj JM, Li QJ, Kuhns MS, Davis MM (2007) Spatial and temporal dynamics of T cell receptor signaling with a photoactivatable agonist. Immunity 27(1):76–88. doi:10.1016/j.immuni.2007.05.017
36. Wang J, Tang S, Wan Z, Gao Y, Cao Y, Yi J, Si Y, Zhang H, Liu L, Liu W (2016) Utilization of a photoactivatable antigen system to examine B-cell probing termination and the B-cell receptor sorting mechanisms during B-cell activation. Proc Natl Acad Sci U S A 113(5):E558–E567. doi:10.1073/pnas.1517612113
37. Hodgkin AL, Huxley AF (1952) A quantitative description of membrane current and its application to conduction and excitation in nerve. J Physiol 117(4):500–544

38. Neher E, Sakmann B (1976) Single-channel currents recorded from membrane of denervated frog muscle fibres. Nature 260(5554):799–802
39. Kornreich BG (2007) The patch clamp technique: principles and technical considerations. J Vet Cardiol 9(1):25–37. doi:10.1016/j.jvc.2007.02.001
40. Coste B, Mathur J, Schmidt M, Earley TJ, Ranade S, Petrus MJ, Dubin AE, Patapoutian A (2010) Piezo1 and Piezo2 are essential components of distinct mechanically activated cation channels. Science 330(6000):55–60. doi:10.1126/science.1193270
41. Coste B, Xiao B, Santos JS, Syeda R, Grandl J, Spencer KS, Kim SE, Schmidt M, Mathur J, Dubin AE, Montal M, Patapoutian A (2012) Piezo proteins are pore-forming subunits of mechanically activated channels. Nature 483(7388):176–181. doi:10.1038/nature10812
42. Kusch J, Zifarelli G (2014) Patch-clamp fluorometry: electrophysiology meets fluorescence. Biophys J 106(6):1250–1257. doi:10.1016/j.bpj.2014.02.006
43. Zheng J, Zagotta WN (2000) Gating rearrangements in cyclic nucleotide-gated channels revealed by patch-clamp fluorometry. Neuron 28(2):369–374
44. Taraska JW, Zagotta WN (2007) Structural dynamics in the gating ring of cyclic nucleotide-gated ion channels. Nat Struct Mol Biol 14(9):854–860. doi:10.1038/nsmb1281
45. Miragoli M, Moshkov A, Novak P, Shevchuk A, Nikolaev VO, El-Hamamsy I, Potter CM, Wright P, Kadir SH, Lyon AR, Mitchell JA, Chester AH, Klenerman D, Lab MJ, Korchev YE, Harding SE, Gorelik J (2011) Scanning ion conductance microscopy: a convergent high-resolution technology for multi-parametric analysis of living cardiovascular cells. J R Soc Interface 8(60):913–925. doi:10.1098/rsif.2010.0597
46. Nikolaev VO, Moshkov A, Lyon AR, Miragoli M, Novak P, Paur H, Lohse MJ, Korchev YE, Harding SE, Gorelik J (2010) Beta2-adrenergic receptor redistribution in heart failure changes cAMP compartmentation. Science 327(5973):1653–1657. doi:10.1126/science.1185988

CPSIA information can be obtained
at www.ICGtesting.com
Printed in the USA
LVHW06*0452140718
583664LV00001B/18/P